Fungicide Resistance in Plant Pathogens

Hideo Ishii • Derek William Hollomon
Editors

Fungicide Resistance in Plant Pathogens

Principles and a Guide to Practical Management

Springer

Editors
Hideo Ishii
Kibi International University
Minami-Awaji, Hyogo, Japan

National Institute for
Agro-Environmental Sciences
Tsukuba, Ibaraki, Japan

Derek William Hollomon
Pest Management Science
Bristol, UK

ISBN 978-4-431-55641-1 ISBN 978-4-431-55642-8 (eBook)
DOI 10.1007/978-4-431-55642-8

Library of Congress Control Number: 2015949140

Springer Tokyo Heidelberg New York Dordrecht London

Printed on acid-free paper

Springer Japan KK is part of Springer Science+Business Media (www.springer.com)

Preface

Modern fungicides have greatly contributed to the control of crop diseases. However, the development of resistant fungal strains has caused a failure of disease control by many fungicides. By the early 1970s, resistance became a practical problem in many countries, and extensive research has been done since then. Dekker and Georgopoulos (1982), the two pioneer scientists in fungicide resistance, published the first textbook on this subject, but it is now timely to update our understanding of the principles underpinning sound resistance management.

Nearly 50 years have passed since we first recognized the importance of fungicide resistance. Since then, the Fungicide Resistance Action Committee (FRAC) established by manufacturers, public organizations including the Fungicide Resistance Action Group in the UK (FRAG-UK), and similar groups in many other countries and learned societies such as the Phytopathological Society of Japan all exist to alert users to resistance problems and advise on anti-resistance management strategies. Nevertheless, resistance remains a serious problem and can emerge rapidly.

In recent years, the development of new fungicides has become more difficult as increasing amounts of environmental and toxicological data are needed to satisfy regulatory authorities. In addition, many existing fungicides may be banned in the near future due to suspected toxicological reasons. Concern over the loss of key modes of action was expressed in the Declaration of Ljubljana, which states that "In order to safeguard the production of food at affordable prices, it is essential to provide farmers with access to sufficient diversity of crop protection solutions. This is essential to prevent or delay the development of resistant pests, and to maintain the efficacy of remaining crop protection products" (Bielza et al. 2008).

Against this background, the editors proposed publishing an updated text on resistance for students and researchers because we believe that regulation based on the precautionary principle involving hazard rather than scientific-based risk assessment of fungicide use will reduce the diversity of modes of action and increase resistance in the near future. To manage fungicide resistance successfully will require the promotion of integrated disease management, involving not just chemical fungicides but also host plant resistance, agronomic factors, and reliable biological control agents where these are available.

This book comprises four parts: Development of Fungicide Resistance, Mechanisms of Resistance, Monitoring Resistance, and Resistance Management in Major Crops. In total, 29 chapters have been written by representative scientists in this field worldwide. The chapters cover the most important fungicide groups that have caused resistance on various crops. This book includes descriptions of the basics of fungicide resistance including the history, genetics, evolution, and also up-to-date information on mechanisms and management of resistance.

It is a great pleasure for the editors to draw on their experience to create a book that we believe will help readers understand more about fungicide resistance and its management. We must also take this opportunity to thank Ms. Fumiko Yamaguchi and Dr. Mei Hann Lee, from Springer Japan, both for their assistance in editorial matters and for overseeing the production of the book.

Minami-Awaji, Japan Hideo Ishii
Bristol, UK Derek William Hollomon

References

Bielza P, Denholm I, Sterk G, Leadbeater A, Leonard P, Jørgensen LN (2008) Declaration of Ljubljana – the impact of a declining European pesticide portfolio on resistance management. Outlooks Pest Manag 19:246–248

Dekker J, Georgopoulos SG (1982) Fungicide resistance in crop protection. In: Dekker J, Georgopoulos SG (eds) PUDOC, Wageningen, pp 1–265

Contents

Part I Development of Fungicide Resistance

1 **Fungicide Resistance: 40 Years on and Still a Major Problem** 3
Derek William Hollomon

2 **Genetics of Fungicide Resistance** 13
Rita Milvia De Miccolis Angelini, Stefania Pollastro, and Franco Faretra

3 **Stability of Resistance** 35
Hideo Ishii

4 **The Use of Mathematical Models to Guide Fungicide Resistance Management Decisions** 49
Frank van den Bosch, Bart Fraaije, Richard Oliver, Femke van den Berg, and Neil Paveley

5 **Evidence-Based Resistance Management: A Review of Existing Evidence** 63
Frank van den Bosch, Neil Paveley, Bart Fraaije, Femke van den Berg, and Richard Oliver

6 **Fitness Cost of Resistance: Impact on Management** 77
Alexey Mikaberidze and Bruce A. McDonald

Part II Mechanisms of Resistance

7 **Anti-tubulin Agents** 93
David H. Young

8 **Respiration Inhibitors: Complex II** 105
Gerd Stammler, Antje Wolf, Alice Glaettli, and Kristin Klappach

9 **Respiration Inhibitors: Complex III** 119
Helge Sierotzki

10 **Oomycete Fungicides: Phenylamides, Quinone Outside Inhibitors, and Carboxylic Acid Amides** 145
Ulrich Gisi and Helge Sierotzki

11 **Melanin Biosynthesis Inhibitors** 175
Makiichi Takagaki

12 **Histidine Kinase Inhibitors** 181
Makoto Fujimura, Shinpei Banno, Akihiko Ichiishi, and Fumiyasu Fukumori

13 **Sterol Biosynthesis Inhibitors: C14 Demethylation (DMIs)** 199
Basil Nicholas Ziogas and Anastasios Andreas Malandrakis

14 **Sterol Biosynthesis Inhibitors: C-4 Demethylation** 217
Danièle Debieu and Pierre Leroux

15 **Multidrug Efflux Transporters** 233
Matthias Hahn and Michaela Leroch

Part III Monitoring Resistance

16 **Monitoring Resistance in Obligate Pathogens by Bioassays Relating to Field Use: Grapevine Powdery and Downy Mildews** 251
Marie-France Corio-Costet

17 **Monitoring Resistance by Bioassay: Relating Results to Field Use Using Culturing Methods** 281
Guido Schnabel, Meng-jun Hu, and Dolores Fernández-Ortuño

18 **Monitoring Resistance Using Molecular Methods** 295
Derek William Hollomon and Hideo Ishii

Part IV Resistance Management in Major Crops

19 **Wheat Pathogens in China** 313
Ming-guo Zhou and Xiao-jing Jia

20 **Origin of Fungicide-Resistant Barley Powdery Mildew in Western Australia: Lessons to Be Learned** 329
Madeline Ann Tucker, Francisco Lopez-Ruiz, Kithsiri Jayasena, and Richard Peter Oliver

21 **Rice Pathogens in Japan** 341
Hideo Ishii

22 Potato Pathogens in Northern and Western Europe 355
Huub T.A.M. Schepers and Louise R. Cooke

23 Sugar Beet Diseases: *Cercospora* Leaf Spot 379
Mohamed F.R. Khan

24 Oilseed Rape Pathogens in France 389
Annette Penaud and Anne-Sophie Walker

25 Cucurbit Powdery Mildew in the USA 401
Margaret Tuttle McGrath

26 Grapevine Gray Mold in France 419
Anne-Sophie Walker and Pierre Leroux

27 Fungicide Resistance in *Venturia inaequalis*, the Causal Agent of Apple Scab, in the United States 433
Kerik D. Cox

28 Resistance in Postharvest Pathogens of Citrus in the United States 449
James E. Adaskaveg and Helga Förster

29 Banana Diseases 467
Andreas Mehl and Friederike Manger-Jacob

Concluding Remarks 481

Index 483

Part I
Development of Fungicide Resistance

Chapter 1
Fungicide Resistance: 40 Years on and Still a Major Problem

Derek William Hollomon

Abstract Fungicide resistance emerged as a practical disease control problem in the 1970s, but it was the outcome of two workshops in Wageningen in 1980 and 1981 that set the framework for research to tackle the problem. Some but not all fungicides quickly select already-existing resistant mutants from within target pathogen populations. Several mechanisms contribute to resistance, but where target-site changes predominate, cross resistance does not extend to other modes of action. Field efficacy and bioassay are key to confirming resistance, but molecular techniques are increasingly used to detect resistance and to augment biochemistry to determine mechanisms. Resistance is not inevitable but depends on the impact of both pathogen and fungicide properties on pathogen populations. Some factors can be manipulated to minimise resistance risk, and a cornerstone of anti-resistance strategies combines treatments with more than one mode of action, either in mixtures or in alternation. Controlled release formulations may also help reduce selection. Resistance has a financial cost to users and manufacturers and seriously reduces available modes of action. Consequently to combat resistance, fungicides should be embedded in integrated disease management systems.

Keywords Cereal eyespot • Controlled release formulations • Evolution of resistance • Fungicides • Griseofulvin • Monitoring

1.1 Introduction

My interest in fungicide resistance began whilst at Rothamsted in the 1970s, when resistance was emerging as a serious disease control problem. Workshops at Wageningen in 1980 and 1981 strengthened the resistance agenda, resulting in the establishment of the industry-based Fungicide Resistance Action Committee (FRAC) and in the publication of a book based on the workshop's proceedings, edited by Johan Dekker and Spiros Georgopoulos (1982). This book quickly became the standard text for those involved in fungicide resistance. Inevitably things have

D.W. Hollomon (✉)
Pest Management Science, Bristol BS40 8UB, UK
e-mail: orchardhse@ukgateway.net

H. Ishii, D.W. Hollomon (eds.), *Fungicide Resistance in Plant Pathogens*,
DOI 10.1007/978-4-431-55642-8_1

moved on since the early 1980s. A glance through the book's pages finds no mention of molecular techniques which now impact greatly on resistance research, especially in the areas of mode of action, resistance mechanisms and monitoring. Epidemiologically based modelling of the evolution of resistance and anti-resistance management strategies have also expanded greatly since the 1980s. To illustrate these changes I have included some key references in this chapter that have shaped the direction of resistance research over the past 40 years. This helps set the scene for the chapters that follow and which provide an up-to-date account of where we are today in dealing with the resistance problem.

1.2 History

Chemistry has a long history of involvement in disease control. The use of lime sulphur sprays for powdery mildew control became popular from 1800 onwards (Forsyth 1802), but the discovery of Bordeaux mixture by Millardet (1885) and use of copper compounds for control of grape downy mildew (*Plasmopara viticola*) and potato late blight (*Phytophthora infestans*) marked the beginning of fungicide research within the agrochemical industry. It prompted research that eventually led to the introduction of organic compounds and especially dithiocarbamates (thiram, mancozeb), phthalimides (captafol) and chlorothalonil, which are now very widely used as sprays or seed treatments throughout the world. Despite extensive use over many years, resistance has not been a problem with these largely nonsystemic protectant fungicides because of their multisite modes of action.

A serious limitation of protectant fungicides is that they hardly penetrate into plants, and so established infections are not controlled, and losses through weathering coupled with the need to protect new foliage require frequent applications, often at weekly intervals. A landmark step in overcoming this problem was the demonstration that the antifungal antibiotic (griseofulvin) isolated from *Penicillium griseofulvum* was translocated within plants and not only controlled established infections but protected new growth (Brian et al. 1951). Unfortunately its high cost prevented use in crop production, although it found use as a medical fungicide. Nevertheless, it illustrated that systemic protection was possible and stimulated the search for novel systemic compounds, many of which are now key components in the successful management of otherwise damaging crop diseases. Because systemic fungicides necessarily have a close association with the biochemistry and physiology of plants, their modes of action are specific and usually involve just one biochemical target site.

To date 42 different modes of action (FRAC 2014) are identified for fungicides.

1.3 Evolution of Resistance

Darwin viewed all organisms as survivors which, through natural selection, competed with others in their particular environment (Darwin 1859). Fungicides disrupt metabolism and threaten survival, so it is no surprise that pathogens can initiate mechanisms to resist lethal effects. Fungal genomes are very plastic and may contain many thousands of polymorphisms (Cuomo et al. 2007). Large pathogen populations will inevitably contain rare distinct genetic, and stable, individuals able to counter to varying degrees any unfavourable metabolic impact and increase in response to selection at the expense of sensitive components of the population. What form resistance will take depends on the resistance mechanism selected. Where the fungicide target is a specific biochemical step, a single point mutation causing one amino acid change can rapidly and effectively block fungicide binding within the target site (single-site inhibitors) and generally causes high levels of resistance. Fungicides that target many biochemical steps (multisite inhibitors) require a combination of many mutations and so resistance evolves slowly, if at all.

An effective resistance mechanism does not ensure that a practical disease control problem will evolve. In treated crops resistant individuals will be more fit than the wild-type sensitive population. But initially, in the absence of the fungicide, resistant individuals must have a lower relative fitness; otherwise, any new fungicide would not offer control benefits. Although any fitness penalty may be linked to decreased enzyme efficiency inherent in the target-site change (Nicholas et al. 2004), many environmental factors also influence fitness, and mutations can alter the general genetic background so that resistant individuals are no longer at a fitness disadvantage.

A core feature of fungicide resistance is that products with the same mode of action, and hence the same specific resistance mechanism, show cross resistance, but not resistance to other modes of action. Active efflux mechanisms may also contribute to resistance. This phenomenon known as multidrug resistance (MDR) may generate resistance between products with different modes of action, especially in laboratory assays. However, resistance levels are low, perhaps up to 20-fold in some pathogens (Kretschmer 2012), so in practice MDR only augments target-site resistance.

But even products in the same mode of action group (FRAC 2014) may interact somewhat differently to a particular change in the target site, resulting in differences in resistance levels which in turn impact on the evolution of resistance within pathogen populations. Prothioconazole, for instance, interacts with the haem component of the target-site sterol 14α-demethylase (CYP51) differently from other azoles (Parker et al. 2011), showing lower resistance and still effective control of some cereal diseases. Extensive analysis of azole resistance in *Mycosphaerella graminicola*, the cause of wheat leaf blotch, has shown that different target-site mutations alone, or in combination, generate different cross resistance patterns (Cools et al. 2011) and indeed improved the performance of prochloraz (Leroux and Walker 2011). A structural analysis of the impact of these CYP51 changes on azole sensitivity provides a potential insight to manage resistance through new chemistries (Kelly and Kelly 2013).

1.4 Detecting Resistance

Resistance above all is a field-based problem recognised by a decline in fungicide performance, to which growers may often respond by increasing dose rate and/or treatment frequency. Poor performance can be caused by a host of factors, including poor application and timing, wrong dose rate or very exceptional disease pressure. So anecdotal evidence from growers must be backed up by a programme of field work supported by glasshouse and laboratory assays.

Development of a new fungicide involves many efficacy trials involving different dose rates and carried out under a range of environmental conditions. If resistance is a problem, repeating these field trials should show a decline in performance. But this approach to confirming resistance requires more than just a single season's work.

A more common approach to confirming resistance involves comparing the sensitivity of isolates obtained from sites where performance has eroded with the sensitivity of isolates never exposed to the at-risk fungicide. Underpinning this approach is the need to have developed suitable bioassays in which there is a clear relationship between dose rate and response which, depending on the pathogen and the fungicide mode of action, may involve measuring germination, germ-tube or mycelial growth rate or, especially for obligate pathogens, infection levels. Ideally the existence of a sensitivity distribution of the target fungal population established prior to widespread use of a new fungicide will allow a meaningful confirmation of resistance. The key role of a "baseline" sensitivity distribution in various aspects of resistance management was discussed in detail by Russell (2003), and its importance is recognised in many countries where a baseline sensitivity distribution is a requirement for registration of a new fungicide. The ability to confirm resistance through comparison with a baseline sensitivity will depend on the sample size from the suspected resistant population and inclusion of at least one reference isolate to check for variation between assay tests. In practice where baseline sensitivity data do not exist, comparisons can be made between isolates obtained from at-risk sites with those collected from untreated areas. Often researchers obtain baseline data using "historic" isolates which have been maintained in culture collections, sometimes for many years, and which were isolated before the at-risk fungicide was used.

The first attempts to diagnose fungicide resistance using molecular techniques were reported in the early 1990s and involved monitoring benomyl resistance (Koenraadt and Jones 1992; Koenraadt et al. 1992). Since then tremendous advances have been made in polymerase chain reaction (PCR) and sequencing technologies, and which now allow rapid detection of single-nucleotide polymorphism (SNP) mutations causing resistance. Indeed the literature is full of different molecular techniques used to monitor resistance, and certainly the most well documented is perhaps detection of the mutation generating the G143A amino acid change in the target b-type cytochrome of complex III of respiration, causing resistance to QoI fungicides (Di Rago et al. 1989).

However, molecular techniques are only useful after resistance has been confirmed using bioassays, the resistance mechanism determined and the DNA change causing resistance identified. But ample evidence suggests that a target-site change causing resistance in one pathogen will occur in other pathogens, so molecular techniques are being used to monitor for resistance in pathogen populations that have not yet evolved resistance in the field to a particular mode of action. Molecular technologies present a different concept of "baseline" from that understood from bioassay data. Resistance can be defined in terms of the frequency of the resistance-causing mutation compared with wild-type frequency, but it is not necessarily clear what frequency of the resistance mutation will cause disease control problems in the field. Indeed, molecular techniques can be extremely sensitive, detecting perhaps 1 in 10,000 mutations in target populations that are clearly sensitive (Windass et al. 2000). Equally relevant is that other point mutations may cause resistance (eg. F129L and G137R Leadbeater 2012) in the case of QoI resistance, and which requires a battery of molecular assays where one bioassay would suffice. Furthermore, in diploid or polyploid *Oomycetes* "fungi", mutations may be recessive (Gisi et al. 2007) and, therefore, simply detecting a mutation may not be sufficient to confirm resistance.

A final step towards confirming resistance requires rigorously conducted *in planta* assays involving a range of dose rates, preferably using single spore isolates, and comparing a wild-type isolate with at least one suspected of being resistant. Generating a dose–response relationship will provide not only measure of sensitivity of each isolate (commonly the effective dose needed to reduce infection 50 %, i.e. EC_{50}) but also a resistance factor (RF).

A detailed characterisation in this way of one or more resistant isolates obtained from the field provides a platform to determine the biochemical and molecular mechanism of resistance, which may not be the same as mechanisms identified in resistant mutants generated in the laboratory during the development programme. It also provides standard resistant isolates available for use in monitoring surveys and for other research programmes.

1.5 Likelihood of Resistance

The likelihood of resistance is the outcome of the impact of fungicide treatment on the target population and depends on both biological and chemical factors. Many of the pathogen, or intrinsic, properties (Table 1.1) contribute to the "pathogen risk" and are mostly outside the control of the grower. But many of the treatment measures (Table 1.2) provide opportunities for growers to adjust the risk of resistance for a particular pathogen/fungicide combination.

Historically, resistance emerged quickly where growers were cultivating disease-susceptible crops (especially cereals, cucurbits and vines) and relying extensively and repeatedly on fungicides with a single mode of action to control disease. Choosing less-susceptible cultivars where possible and operating an integrated

Table 1.1 Pathogen properties influencing evolution and spread of resistance

Biochemical
Dependence on disruptible biochemical steps
Availability of resistance mechanisms
Epidemiological
Dispersal method, e.g. wind, rain splash, soilborne
Abundance of sporulation
Pathogen life cycle: short or long generation time
Ability to infect all crop stages, requiring repeated treatment
Isolation of pathogen populations preventing re-entry of more competitive sensitive genotypes
Genetic
Relative abundance of genotypes with different sensitivities
Fitness properties of different genotypes
Sexual or asexual reproduction: influence on inheritance of resistance
Mutation rate
If relevant dominance of resistance alleles

Table 1.2 Fungicide properties influencing evolution and spread of resistance

Biochemical
Interaction with target metabolism and its susceptibility
Physicochemical/toxicological
Stability, solubility, volatility, polarity
Partition and transport properties
Application
Initial dose and distribution
Formulation
Exclusive and repeated use of at-risk mode of action
Extent of area treated
Integration with other disease management tools, including biofungicides, resistant varieties, crop rotation and crop hygiene

disease management (IDM) programme employing different modes of action, either in mixtures or in alternation, has become the cornerstone of anti-resistance management systems. Kable and Jeffrey (1980) were perhaps the first to employ a modelling approach to management of fungicide resistance. More recent publications discuss in detail the assessment and management of resistance risk (Kuck 2005; Brent and Hollomon 2007a, b).

One question that occurs repeatedly in discussions with growers about resistance risk relates to the impact of the dose rate (van den Bosch et al. 2011). A factor which receives little attention in this debate, but which could impact significantly on the likelihood of resistance, involves how pathogens are actually exposed to fungicides. Ideally, fungicide doses should be just sufficient to kill enough of the wild-type population to provide acceptable control levels. In practice, because of the exponential kinetics of decay due to evaporation, metabolic degradation in both the host plant and the pathogen and poor rain fastness, initial dose rates are much higher than needed to inhibit pathogen growth. High initial doses, especially in an eradicant mode, certainly imply a high selection pressure and, depending on the range of rare resistant individuals in the population, could increase the likelihood of resistance. One way to reduce initial dose rates would be to use controlled release formulations which maintained a lower but steady dose rate, over a longer period of time. It suggests that more attention needs to be given to formulation in the management of resistance and how it could be used to achieve a gradual release of an active ingredient held in the leaf surface. A somewhat old (but very relevant) discussion of formulation in relation to efficacy was given by Graham-Bryce (1987), and Shephard (1985) presented evidence that release of different azoles from reservoirs in the cuticular leaf layers depends on formulation,

1.6 What Does Resistance Cost?

Resistance does not come without a cost to both growers and manufacturers. However, useful economic data are seldom available, especially from naturally infected field or glasshouse trials where losses can be accurately quantified. Although costs will probably be similar for other cereal diseases, my own experience is from a series of trials in the 1980s, following failure of MBC (carbendazim, benomyl) fungicides to control cereal eyespot (*Oculimacula yallundae*) because of resistance. Without treatment eyespot causes wheat losses of 10 %, whilst MBCs reduced losses to 3 % (Pavely et al. 2011). At current wheat prices (£120/ton), this failure to control eyespot equates to a loss of £170 m per annum for UK growers, compared with a £49 m loss before carbendazim resistance. Fortunately an alternative, but more expensive, eyespot fungicide (prochloraz) was available in the 1980s; otherwise, without effective eyespot control, wheat would have been an unprofitable crop for some growers. Similar economic losses no doubt occurred as a result of phenylamide resistance in *Phytophthora infestans*, the cause of late blight of potato.

Although manufacturers' losses are equally serious, no detailed costs are available. Registration authorities have responded to resistance by requiring additional data before giving approval for new products (Kuck 2005). Not only are baseline sensitivity distributions needed for target pathogens but also information on mode of action, cross resistance, assessment of resistance risk and proposed anti-resistance management strategies. To generate all this information requires a substantial commitment of resources, not only within the development programme for a new product but also in support of an existing product when its use increases as a replacement for fungicides no longer effective because of resistance. Add to this loss of sales revenue, coupled with redirecting resources to monitoring resistance, stewarding product use, and adapting chemical plant for other uses, leaves less available for research and development of products with new modes of action. This last point emphasises a major cost of resistance, not just for growers and manufacturers but also for consumers, in so far as resistance reduces the modes of action available to combat the problem. To respond to these many challenges, resistance management teams have been expanded in all the major manufacturing companies.

1.7 Future Directions

The following chapters amply illustrate that, whilst significant progress has been made in understanding and managing fungicide resistance, the problem remains. Despite the fact that effective anti-resistance management requires access to different modes of action, governments, and especially the European Union, continue to enact legislation, without much scientific evidence, that reduces the number of modes of action, including possibly azoles! Against this background future anti-resistance strategies will be embedded in integrated disease management (IDM) systems, which combine conventional chemical fungicides with biofungicides and host plant resistance, generated either by conventional plant breeding or by GM technologies. To what extent biofungicides (usually bacteria and fungi) offer new modes of action is unclear, although the ease with which it is now possible to follow changes in expression levels of genes involved in the activation of host resistance suggests that the mode of action of some may involve systemic acquired resistance (SAR). Evidence from the long-term and effective use of the rice blast fungicide probenazole which activates SAR suggests this mode of action is not easily overcome by the development of resistance. But whatever the modes of action of biofungicides, it is quite possible that pathogens will eventually evolve resistance to them. Finally, complex IDM strategies will challenge pathologists to define treatment thresholds and monitor changes in pathogen populations and will challenge growers to maintain production and profitability.

Acknowledgements I am greatly indebted to both Dr Keith Brent and Professor Phil Russell for their many helpful suggestions during the preparation of this chapter.

References

Brent KJ, Hollomon DW (2007a) Fungicide resistance in crop pathogens: how can it be managed? FRAC monograph 1, 2nd edn. CropLife International, Brussels, p 55

Brent KJ, Hollomon DW (2007b) Fungicide resistance: the assessment of risk. FRAC Monograph 2, 2nd edn. CropLife International, Brussels, p 52

Brian PW, Wright JM, Stubb J, Way AM (1951) Uptake of antibiotic metabolites of soil micro-organisms by plants. Nature 167:347–348

Cools HJ, Mullins JGL, Fraaije BA, Parker JE, Lucas JA et al (2011) Impact of recently emerged sterol 14α-demethylase (CYP51) variants of *Mycosphaerella graminicola* on azole fungicide sensitivity. Appl Environ Microbiol 77:3830–3837

Cuomo CA, Güldener U, Xu JR, Trail F et al (2007) The *Fusarium graminearum* genome reveals a link between localised polymorphism and pathogen speciation. Science 317:1400–1402

Darwin C (1859) On the origin of species. Murray, London

Dekker J, Georgopoulos SG (1982) Fungicide resistance in crop protection. Pudoc, Wageningen, p 273

Di Rago JP, Coppée JY, Colson AM (1989) Molecular basis for resistance to myxothiazol, mucidin (strobilurin A) and stigmatellin. Cytochrome b inhibitors acting at the centre of the mitochondrial ubiquinol-cytochrome c reductase in *Saccharomyces cerevisiae*. J Biol Chem 264:14543–14548

Forsyth W (1802) A treatise on the culture and management of fruit trees. Nichols, London

FRAC (Fungicide Resistance Action Committee) (2014) FRAC list of plant pathogenic organisms resistant to disease control agents. www.frac.info

Gisi U, Waldner M, Kraus N, Dubuis PH, Sierotzki H (2007) Inheritance of resistance to carboxylic acid amide (CAA) fungicides in *Plasmopara viticola*. Plant Pathol 56:199–208

Graham-Bryce IJ (1987) Resistance to pesticides and antibiotics: how far is it comprehensible and manageable? In: Ford MG, Hollomon DW, Khambay BPS, Sawicki RM (eds) Combating resistance to xenobiotics: biological and chemical approaches. Ellis Horwood, Chichester, pp 11–25

Kabel PF, Jeffery H (1980) Selection for tolerance in organisms exposed to sprays of biocide mixtures: a theoretical model. Phytopathology 70:8–12

Kelly SL, Kelly DE (2013) Microbial cytochromes P450: biodiversity and biotechnology. Where do cytochromes P450 come from and what can they do for us. Philos Trans R Soc Series B 368. doi: 10.1098/rstb.2012.0476

Koenraadt H, Jones AL (1992) The use of allele-specific oligonucleotide probes to characterise resistance to benomyl in field strains of *Venturia inaequalis*. Phytopathology 82:1354–1358

Koenraadt H, Somerville SC, Jones AL (1992) Characterisation of mutations in the beta-tubulin gene of benomyl-resistant field strains of *Venturia inaequalis* and other plant pathogenic fungi. Phytopathology 82:1348–1354

Kretschmer M (2012) Emergence of multi-drug resistance in fungal pathogens. A potential threat to fungicide performance in agriculture. In: Thind TS (ed) Fungicide resistance in crop protection. CABI, Wallingford, pp 251–267

Kuck KH (2005) Fungicide resistance management in a new regulatory environment. In: Dehne HW, Gisi U, Kuck KH, Russell PE, Lyr H (eds) Modern fungicides and anti-fungal compounds IV. BCPC, Alton, UK, pp 35–43

Leadbeater A (2012) Resistance risk to QoI fungicides and anti-resistance strategies. In: Thind TS (ed) Fungicide resistance in crop protection. CABI, Wallingford, pp 141–154

Leroux P, Walker AS (2011) Multiple mechanisms account for resistance to sterol 14α-demethylation inhibitors in field isolates of *Mycosphaerella graminicola*. Pest Manag Sci 67:44–59

Millardet PMA (1885) Traitement du mildou par le melange de sulphate de cuivre et de chaux. J d'Agric Practique 49:513–516 (In French)

Nicholas F, Brown AC, Sexton G, Cook A, Windass J, Meunier B (2004) Modelling the Qo site of crop pathogens in *Saccharomyces cerevisiae* cytochrome b. Eur J Biochem 271:2264–2271

Parker JE, Warrilow AGS, Cools HJ, Martel CM, Nes WD, Fraaije BA, Lucas JA, Kelly DE, Kelly SL (2011) Prothioconazole binds to *Mycosphaerella graminicola* CYP51 by a different mechanism compared to other azole antifungals. Appl Environ Microbiol 77:1460–1465

Pavely N, Blake J, Gladders P, Cockerall V (2011) HGCA wheat management guide. Agriculture and Horticulture Development Board, Stonleigh Park, p 19

Russell PE (2003) Sensitivity baselines in fungicide resistance research and management. FRAC monograph 3. Crop Life International, Brussels, p 56

Shephard MC (1985) Fungicide behaviour in the plant systemicity. In: Smith IM (ed) Fungicides (BCPC Monograph 31), Alton, UK, pp 99–106

van den Bosch F, Pavely N, Shaw M, Hobbelen P, Oliver R (2011) The dose rate debate: does the risk of fungicide resistance increase or decrease with dose? Plant Pathol 60:597–606

Windass JD, Heaney SP, Renwick A, Whitcombe DM, Little S, Gibson NJ, Theaker J, Stanger CP (2000) Methods for detecting low frequencies of mutations in mitochondrially encoded genes. International Patent WO00/66773

Chapter 2
Genetics of Fungicide Resistance

Rita Milvia De Miccolis Angelini, Stefania Pollastro, and Franco Faretra

Abstract Acquired resistance to fungicides in fungal plant pathogens is a challenge in modern crop protection. Fungi are indeed very able to adapt to changing environmental conditions, such as the introduction of a new fungicide in the agricultural practice. Several genetic mechanisms may underlay fungicide resistance and influence the chance and time of its appearance and spreading in fungal populations. Resistance may be caused by mutations in major genes (monogenic or oligogenic resistance) or in minor genes (polygenic resistance) which may occur in nuclear genes as well as in cytoplasmic genes. They are immediately expressed in haploid fungi, while they may be dominant or recessive in diploid fungi. Allelic variants may cause different levels of resistance and/or different negative pleiotropic effects on the fitness of resistant mutants. The sexual process, where occurring, plays an important role in releasing new recombinant genotypes in fungal populations. Heterokaryosis provides multinucleate fungi with a further mechanism of adaptation. Resistant mutants can be obtained from samples representative of field population of a pathogen or under laboratory conditions through selection of spontaneous mutations or following chemical or physical mutagenesis. Nowadays, molecular tools, such as gene cloning, sequencing, site-directed mutagenesis and gene replacement, make genetic studies on fungicide resistance amenable even in asexual fungi for which classical genetic analysis of meiotic progeny is not feasible.

Keywords Mutations • Major genes • Minor genes • Cytoplasmic genes • Ploidy • Heterokaryosis • Population genetics

---

R.M. De Miccolis Angelini • S. Pollastro (✉) • F. Faretra
Department of Soil, Plant and Food Sciences, University of Bari,
via Amendola 165/A, 70126 Bari, Italy
e-mail: stefania.pollastro@uniba.it

H. Ishii, D.W. Hollomon (eds.), *Fungicide Resistance in Plant Pathogens*,
DOI 10.1007/978-4-431-55642-8_2

2.1 Introduction

Resistance to chemicals in microorganisms is a very common phenomenon occurring whenever antimicrobial compounds are used against pathogens of plants, animals or humans. Natural or innate resistance refers to intrinsic features (e.g. the lack of a specific molecular target and/or a metabolic pathway) protecting the organism from the effects of antimicrobials. For example, strobilurin-producing organisms, including wood-degrading *Basidiomycetes*, such as *Strobilurus tenacellus*, have innate resistance to their own strobilurins that show, instead, activity against a very broad spectrum of fungi and *Oomycetes*. Acquired resistance refers to organisms that in their wild-type form are sensitive and may develop resistance after their exposure to an antimicrobial compound. Acquired resistance is due to genetic modifications transmissible to the progeny so that a chemical that was once effective against the organism is no longer effective.

Resistance to fungicides used in agriculture as well as in animal or human health care is a more recent phenomenon than resistance to antibiotics (Coplin 1989; Cookson 2005) and insecticides (Brown 1977). Until the late 1960s, fungicides used in crop protection (e.g. sulphur, copper derivatives, dithiocarbamates) were indeed essentially multisite inhibitors, affecting multiple target sites and hence interfering with many metabolic processes of the pathogen. Despite their protracted and widespread use, acquired resistance to multisite fungicides is still a rare event. This is because there is a low probability that a number of mutations at different loci, needed for the onset of the resistance, simultaneously occur in fungal cells and, if this happens, the mutated isolates remain viable. Afterwards, with the introduction of single-site fungicides and as a consequence of their frequent and repeated use, fungicide resistance has become a major concern in modern crop protection seriously threatening effectiveness of several fungicides (Brent and Hollomon 2007a, b).

Fungicide resistance is hence a result of adaptation of a fungus to a fungicide due to a stable and inheritable genetic change, leading to the appearance and spread of mutants with reduced fungicide sensitivity (Delp and Dekker 1985).

2.2 Genetic Bases of Fungicide Resistance

Genetics of fungicide resistance have been previously reviewed by Grindle (1987), Grindle and Faretra (1993), Steffens et al. (1996) and Ma and Michailides (2005), and deeper information is available on the website of the Fungicide Resistance Action Committee (www.frac.info).

Fungal genetic backgrounds and genetic bases of resistance are key factors in the intrinsic risk of resistance and influence its evolution in the pathogen populations. For example, the occurrence of genetic recombination through the sexual process, where it regularly occurs in nature, or parasexuality, in essentially asexual fungi, may greatly influence the dynamics of resistant subpopulations producing new combinations of resistance and fitness traits originally occurring in separate individuals.

Most genetic studies on fungicide resistance have been carried out on 'model' saprophytic *Ascomycetes*, such as *Aspergillus nidulans*, *Neurospora crassa* and *Saccharomyces cerevisiae*. Nevertheless, the genetics of fungicide resistance has been investigated in several pathogenic fungi (Table 2.1).

Key factors in the genetic bases of fungicide resistance are (1) the number of loci involved, (2) the number of allelic variants at each locus, (3) the existence and relevance of dominant or recessive relationship between resistant and wild-type alleles (Borck and Braymer 1974) and (4) the additive or synergistic interactions between resistance genes.

Genes responsible for fungicide resistance may be located on chromosomes inside the nucleus or on extrachromosomal genetic determinants. Nuclear and cytoplasmic genes can be distinguished by their inheritance patterns. Nuclear genes typically show classical biparental (disomic) inheritance in sexual crosses, i.e. the zygote receives one allele of each gene from each of its parents. In contrast, genetic material in the cytoplasm has a non-Mendelian inheritance and is characterized by uniparental (usually maternal) transmission (Griffiths 1996). In addition, cytoplasmic genes differ from nuclear genes in showing vegetative segregation and intracellular selection potentially affecting resistance stability (Birky 2001; Ziogas et al. 2002).

Most fungicide-resistance genes are located on nuclear chromosomes. In most cases, there is only one copy of resistance gene in the genome and mutations are usually located in gene sequences encoding enzymatic or structural proteins. However, multidrug resistance (MDR) in *B. cinerea* and other fungi is caused by overexpression of membrane efflux transporter genes resulting in an increased efflux of toxicants that reduces fungal sensitivity to several unrelated fungicides as well as plant defence chemicals (reviewed by Kretschmer 2012). In MDR1 strains of *B. cinerea*, resistance is conferred by mutations in the regulator *mrr1* gene encoding a transcription factor controlling the ABC transporter *AtrB* gene, whereas in MDR2 strains resistance is caused by an insertion of a retrotransposon-derived sequence in the promoter region of the facilitator superfamily (MFS) transporter gene *mfsM2* (Kretschmer et al. 2009).

Fungicide resistance may result from mutations in single major genes (Georgopoulos 1988) or from additive (Kalamarakis et al. 1991; Lasseron-de Farandre et al. 1991) or synergistic interactions (Molnar et al. 1985) between several mutant genes.

Monogenic and oligogenic resistance are caused, respectively, by one or few major genes. Major genes have an appreciable influence on the phenotype, and resistance mutations cause a qualitative change in the response to a fungicide with the appearance in the field of new fungicide-resistant subpopulation(s) well distinguishable from the wild-type sensitive one (Fig. 2.1). Most cases of fungicide resistance are due to mutations in major genes (Table 2.1). Mutations in major genes conferring resistance to fungicides having different modes of action may also occur in a same isolate, causing multiple resistance. In oligogenic resistance, several different major genes are involved, any one of which can mutate to cause an increase

Table 2.1 Most relevant examples of the genetic bases of resistance to fungicides in fungal plant pathogens

Fungicides (chemical groups)	Mode of action and target site	Genetic bases of the resistance	Target site mutations	References
Benzimidazoles and thiophanates [MBC (methyl benzimidazole carbamates)]	Mitosis and cell division: ß-tubulin assembly in mitosis	Monogenic resistance due to mutations in single major genes. Multiallelic resistance. Negative cross resistance between benzimidazoles and *N*-phenylcarbamates observed in *Botrytis cinerea* or between MBCs and the fungicides diethofencarb and zoxamide in *Pyrenopeziza brassicae*	Several SNPs, mostly E198A/G/K, F200Y in β-tubulin gene (e.g. *Mbc1* gene in *Botrytis cinerea*)	Faretra and Pollastro (1991), Yarden and Katan (1993), Nakazawa and Yamada (1997), and Ishii (2012b)
N-Phenylcarbamates				
Dicarboximides	Signal transduction: MAP/ osmosensing class III histidine kinase	Monogenic resistance due to mutations in single major genes. Multiallelic resistance	Several mutations in *os-1* (*Daf1*) gene, mostly I365S or different substitutions in conserved amino acid domains	Faretra and Pollastro (1991), Cui et al. (2002), Oshima et al. (2002), Yoshimi et al. (2003), Dry et al. (2004), and Fillinger et al. (2012)
		Hypersensitivity to high osmolarity frequently associated to the resistance		
PP fungicides (phenylpyrroles)	Signal transduction (mechanism speculative): MAP/osmosensing class III histidine kinase	Monogenic resistance due to mutations in single major genes also conferring resistance to dicarboximides and increased osmotic sensitivity. Additional mechanisms (i.e. overexpression of efflux transporters in MDR mutants of *B. cinerea*)	Mutations in osmosensing class III histidine kinase genes (*os-1*; *os-2*, *HOG1*)	Faretra and Pollastro (1993b), Ochiai et al. (2001), Zhang et al. (2002), Avenot et al. (2005), and Fillinger et al. (2012)

PA fungicides (phenylamides)	Nucleic acid synthesis: RNA polymerase I	Monogenic resistance based on a single incompletely dominant gene or both a major and several minor genes	Unknown	Shattock (1988), Crute and Harrison (1988), and Hermann and Gisi (2012)
SDHI (succinate dehydrogenase inhibitors)	Respiration: complex II (succinate dehydrogenase)	Monogenic resistance due to mutations in single major genes. Multiallelic resistance. Partial cross resistance between SDHIs	SNPs in SDH genes (*SdhB*, *SdhC* and *SdhD*), e.g. H/Y (or H/L) at 257, 267, 272 or P225L/T/F in *SdhB* gene, dependent on fungal species	Skinner et al. (1998), De Miccolis Angelini et al. (2010a), and Sierotzki and Scalliet (2013)
QoI fungicides (quinone outside inhibitors)	Respiration: complex III: [cytochrome bc1 (ubiquinol oxidase) at Qo site (*cyt*b gene)]	Point mutations in a mitochondrial gene or additional mechanisms (i.e. alternative respiration and efflux transporters)	SNPs in *cyt*b gene, mostly G143A (in several fungal species), F129L or G137R	Gisi et al. (2002), Kim et al. (2003), and Sierotzki et al. (2007)
AP fungicides (anilinopyrimidines)	Amino acids and protein synthesis: methionine biosynthesis (proposed) (*cgs* gene)	Monogenic resistance due to mutations in single major genes. Alternative mechanisms (i.e. overexpression of efflux transporters in MDR mutants of *B. cinerea*)	No mutations associated with AP resistance in *cbl*, *cgs* or other key genes involved in the biosynthesis and metabolism of methionine or sulphate assimilation	Hilber and Hilber-Bodmer (1998), Chapeland et al. (1999), De Miccolis Angelini et al. (2010b), and Liu et al. (2014)

(continued)

Table 2.1 (continued)

Fungicides (chemical groups)	Mode of action and target site	Genetic bases of the resistance	Target site mutations	References
DMI fungicides (demethylation inhibitors) SBI: class I	Sterol biosynthesis in membranes: C14-demethylase in sterol biosynthesis (erg11/cyp51)	Mostly polygenic resistance (mutations in unlinked genes which show an additive effect). Resistance also conferred by single major genes (monogenic, e.g. in *Erysiphe necator*) or by few resistance genes (oligogenic, e.g. in *Blumeria graminis* f.sp. *hordei*). Additional mechanisms (e.g. ABC transporters)	Mutations in *cyp51* (*erg11*) gene, e.g. V136A, Y137F, A379G, I381V, or in *cyp51* promotor	Kalamatrakis et al. (1991), Peever and Milgroom (1992), Délye et al. (1997), Brown et al. (1992), Blatter et al. (1998), and Cools and Fraaije (2013)
Amines ('morpholines') SBI: class II	Sterol biosynthesis in membranes: Δ14-reductase and Δ8 → Δ7-isomerase in sterol biosynthesis (erg24, erg2)	Polygenic resistance due to mutations in unlinked genes which show an additive effect. No cross resistance to other SBI classes	Unknown	Lasseron-de Falandre et al. (1991) and Markoglou and Ziogas (1999, 2000, 2001)
Hydroxyanilides SBI: class III	Sterol biosynthesis in membranes: 3-keto reductase, C4-demethylation (erg27)	Monogenic resistance due to mutations in single major genes. Additional mechanisms (e.g. P450-mediated detoxification of fenhexamid) have been proposed	Several mutations in *Erg27* gene, mostly F412S/I/V in *B. cinerea* HydR3^{+} strains	Leroux et al. (2002), De Guido et al. (2007), and Fillinger et al. (2008)
CAA fungicides (carboxylic acid amides)	Cell wall biosynthesis: cellulose synthase	Monogenic resistance due to point mutations in a recessive nuclear gene	SNPs in *CesA3* gene leading to amino acid changes in cellulose synthase 3, i.e. G1105A/S/V/W, V1109L/M or Q1077K	Gisi et al. (2007), Blum et al. (2010, 2012), and Pang et al. (2013)

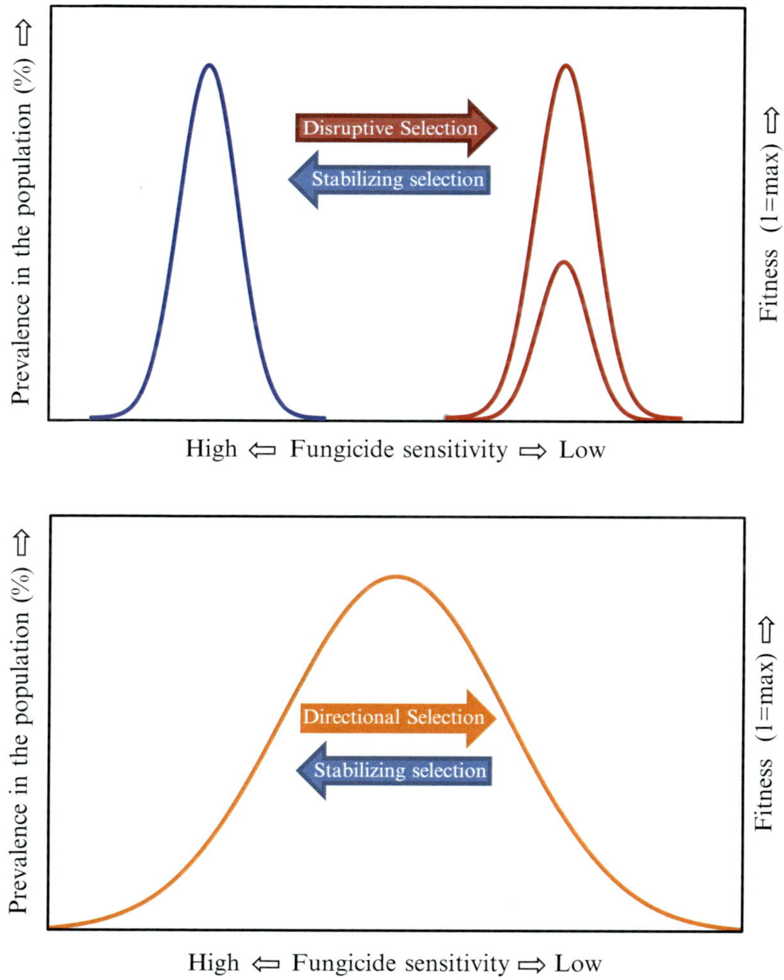

Fig. 2.1 Population dynamics of fungicide resistance in monogenic resistance (*upper*) and polygenic resistance (*bottom*). Disruptive or directional selection is caused by the usage of fungicides having the same mode of action at risk of resistance. Stabilizing selection is due to possible reduction of fitness of fungicide-resistant mutants

in resistance to a same fungicide. For instance, kasugamycin resistance in *Pyricularia oryzae* as well as resistance to the two fungicides ethirimol and triadimenol in *Blumeria graminis* f.sp. *hordei* may be controlled by three different loci where a resistance allele at any one locus confers resistance (Taga et al. 1979; Brown et al. 1992). Furthermore, differently from what is usually observed in most fungi where a single multiallelic gene is responsible for resistance to benzimidazole fungicides, the resistance of *Fusarium oxysporum* to benzimidazoles is caused by mutations in two major genes which interact synergistically conferring high degrees of fungicide resistance (Molnar et al. 1985).

Different mutations in a same gene may cause different levels of resistance to a particular fungicide; this is known as multiallelic resistance. In the past, multiallelic resistance could be assessed only on the ground of phenotypic differences in the level of resistance and/or pleiotropic effects of mutations. With the availability of molecular and sequencing tools, nowadays it is clear that multiallelic resistance is quite common (Table 2.1).

Each mutant allele can be partially/completely dominant or partially/completely recessive to its wild-type allele. That is, when mutant and wild-type alleles of the same gene are combined in the same fungal cells or hyphae, the phenotype may be fungicide resistant (mutant) or fungicide sensitive (wild type).

Combinations of major genes may interact when they are present in the same fungal cells, so that the phenotype of a double mutant may be different from either single-gene mutants (Molnar et al. 1985). Usually, however, one mutant gene is epistatic to another mutant gene, which means that the double mutant has the same level of resistance of the single-gene mutants (Kappas and Georgopoulos 1970; Van Tuyl 1977). The presence of modifier genes affecting phenotypic response of resistant mutants has been suggested to influence the expression of response to phenylamides in *Oomycete* pathogens (Crute and Harrison 1988) or to mediate fitness of resistant mutants as found in mutants of *N. crassa* resistant to dicarboximides (Grindle and Dolderson 1986) and *A. nidulans* resistant to imazalil (van Tuyl 1977). The consequent increase in fitness will result in better survival and possible selection of resistant subpopulations in the field.

Polygenic resistance is due to mutations in minor genes. Those have individually a little effect on the phenotype and cause hence a negligible reduction in the sensitivity to a fungicide. However, numerous mutated minor genes may contribute, with an additive effect, to produce an appreciable increase of the level of resistance. In the field, the result is a quantitative decrease of the sensitivity to a fungicide with a slow, continuous and gradual shift of the fungal population towards increasing resistance levels (Fig. 2.1). Polygenic resistance is much more difficult to be detected and ascertained in the field. Polygenic resistance was demonstrated in *B. graminis* f.sp. *hordei* to ethirimol (Hollomon 1981) and triadimenol (Hollomon et al. 1984). Resistance to dodine is polygenic in *Nectria haematococca* var. *cucurbitae* (Kappas and Georgopoulos 1970). Ultraviolet-induced mutants of *N. haematococca* var. *cucurbitae* also show polygenic inheritance for resistance to fenarimol (Kalamarakis et al. 1991), fenpropimorph and terbinafine (Lasseron-deFalandre et al. 1991).

Cytoplasmic genes are present in mitochondria, plasmids and viruses. Mitochondrial genome, which contains mitochondrial rRNA genes and some of the proteins of the respiratory chain, is the most relevant among fungal extrachromosomal genetic elements affecting resistance to chemicals. However, antibiotic-resistance genes have been located on fungal episomes, plasmids or viruses (Guerineau et al. 1974).

Natural or induced resistance to QoI fungicides, inhibitors of mitochondrial respiration at the Qo site of the cytochrome *bc1* complex (complex III), is usually conferred by point mutations in the mitochondrial *cytb* gene causing amino acid substitutions in the target protein. In particular, at least three possible codon changes have been associated to a moderate (F129L or G137R) or, more frequently, high (G143A) level of resistance to QoIs in several fungal species (Grasso et al. 2006; Fernández-Ortuño et al. 2008). The presence of a G143-associated group I-like intron in the *cytb* gene in some fungal species (i.e. *Puccinia* spp., *Uromyces appendiculatus*, *Alternaria solani*) or isolates (i.e. *B. cinerea*) prevents the occurrence of the G143A mutation and QoI resistance, since it would be lethal because it would be affecting the correct intron splicing process (Grasso et al. 2006).

Analysis of meiotic progenies of appropriate crosses between sensitive and resistant strains confirmed cytoplasmic (maternal) inheritance of QoI resistance in *B. graminis* (Robinson et al. 2002), *Venturia inaequalis* (Steinfeld et al. 2002) and *B. cinerea* (De Miccolis Angelini et al. 2012a). The segregation pattern in randomly collected progenies is expected to be in a phenotypic 1:0 ratio in most fungal species showing a uniparental, anisogamous inheritance of mitochondrial genome or 1:1 ratio in species, such as *A. nidulans* and *B. graminis* f.sp. *tritici*, showing an hermaphroditic, isogamous mitochondrial inheritance (Robinson et al. 2002).

Wild-type and mutated mitochondrial DNA carrying the G143A mutation in the *cytb* gene may coexist in heteroplasmic state within a single isolate, as demonstrated in several species, including *V. inaequalis* (Zheng et al. 2000), *B. cinerea* (Ishii et al. 2009) and other fungal pathogens (Ishii et al. 2007). Equilibrium between resistant and sensitive mitochondria depends on the strength of selective pressure (Ishii 2010). In *Podosphaera leucotricha*, the relative proportion of mutated and wild-type mitochondria is associated with differences in QoI sensitivity levels of the isolates (Lesemann et al. 2006). An instability of QoI resistance in heteroplasmic isolates grown in absence of selective pressure has been frequently reported (Ishii 2012a) suggesting a fitness cost associated to the resistance (Markoglou et al. 2006).

2.3 Ploidy Level

Differences in ploidy level, affecting the number of alleles at each locus, constitute a major genomic trait influencing the onset and subsequent evolution of fungicide resistance. Firstly, frequency of mutations that may arise in single individuals is directly related to the ploidy level as a result of the different numbers of mutational targets (Otto and Gerstein 2008).

Most phytopathogenic fungi are in haploid state for the major part of their life cycle. In contrast, *Oomycetes* typically show a diploid life cycle and the haploid phase is restricted to the gametes (Fincham et al. 1979). Furthermore, polyploids have been frequently identified among *Oomycetes*, such as *Plasmopara viticola* and *Phytophthora* spp. (Rumbou and Gessler 2006; Bertier et al. 2013).

In haploid fungi, mutations conferring resistance are immediately expressed and then directly exposed to selection, while in diploids or polyploids, mutations first appear in heterozygotic state and their phenotypic effects can be masked by dominant wild-type alleles on the homologous chromosome. For this reason, resistance mutations spread more rapidly in haploid than in diploid or polyploid populations. Fixation time may be reduced and selection against deleterious pleiotropic effects of mutations is more effective in haploids than in diploids (Anderson et al. 2004; Otto and Gerstein 2008).

CAA (carboxylic acid amide) fungicides, inhibitors of cellulose biosynthesis in *Oomycete* phytopathogens, are considered at low to medium resistance risk depending on the fungal species. Resistance to CAAs in *P. viticola* is controlled by one or two recessive nuclear genes, as demonstrated through sexual crosses between CAA-sensitive and CAA-resistant isolates and analysis of segregation patterns of sensitive and resistant phenotypes in F1 and F2 progenies (Gisi et al. 2007; Blum and Gisi 2008) and by sequence analysis of putative resistance genes (Blum et al. 2010). Classic genetic analysis also showed that resistance to all CAA fungicides co-segregates and has thus the same genetic basis (Young et al. 2005; Gisi et al. 2007). However, no cross resistance exists between CAA and other fungicides currently available against *Oomycetes*, such as phenylamides and QoI fungicides, where the intrinsic risk of resistance is estimated to be significantly higher than CAA due to their genetic differences. Resistance to phenylamides is indeed a monogenic trait, conferred by a semidominant chromosomal gene (Gisi and Cohen 1996; Knapova et al. 2002), while QoI resistance is due to mutations in the mitochondrial *cytb* gene (Gisi and Sierotzki 2008).

Similar to CAA, resistance to the new benzamide zoxamide in isolates of *Phytophthora capsici* is recessive and is conferred by two nontarget nuclear genes (Bi et al. 2014). This implies that resistance phenotype is expressed only in homozygous mutants, thus limiting resistance spreading and risk.

Nevertheless, the risk of resistance is significantly increased by the occurrence of gene recombination, even if several cycles of sexual process may be required for making resistance fixed and fully expressed in phenotypically aggressive and well-adapted isolates of the pathogen. Sexual recombination naturally occurring under field conditions has been proposed, for instance, as a possible explanation of the higher risk of CAA resistance assessed in field populations of *Pseudoperonospora cubensis* as compared to in vitro estimations (Zhu et al. 2007). Moreover, CAA resistance has been experienced in *P. viticola* field populations since shortly after their introduction, while no reduced sensitivity to CAA has been detected in other *Oomycetes*, such as the late blight pathogen, *Phytophthora infestans*, despite their intensive usage against these pathogens and extensive monitoring. It has been suggested that the lower risk of CAA resistance in *P. infestans* may be due to the lower frequency of sexual recombination under field conditions, as well as to polyploidy, heterokaryosis (Catal et al. 2010) and chromosomal aberrancies (Gisi 2012).

2.4 Heterokaryosis and Nuclear Number

The presence of two or more genetically different haploid nuclei, coexisting in a common hyphal compartment, occurs frequently in some fungal *taxa* and is a potential source of genetic variation. In multinucleate *Ascomycetes*, this condition, known as heterokaryosis, often permits changes in the proportions of different nuclei in response to selection and is a prerequisite to parasexual recombination (Davis 1966). In heterothallic *Basidiomycetes*, two distinct parental haploid nuclei coexist without fusion in each cell establishing a stable dikaryotic state. The dikaryon is genetically equivalent to a diploid, as two haploid genomes of different origins exist in each cell even if they remain separated in different nuclei.

Heterokaryons and dikaryons, harbouring several nuclei, offer the opportunity of genes to complement each other (genetic complementation). Heterokaryons harbouring both fungicide-resistant and fungicide-sensitive nuclei may be able to grow in the presence or absence of fungicides (Grindle 1987). They can adapt to fluctuations in fungicide exposure as a result of changes in the proportions and distribution of resistant-sensitive nuclei within cells (Meyer and Parmeter 1968; Ogden and Grindle 1983). Nucleotypic competition and selection after exposure to fungicides and the ability of heterokaryons to adapt to modified environmental conditions have been demonstrated, for instance, in *B. cinerea* strains resistant to dicarboximides (Summers et al. 1984) or to anilinopyrimidines (Santomauro et al. 2000).

Dominance or recessivity can be tested by inducing hyphal anastomosis between one strain carrying the wild-type allele and the other the mutant allele of a resistance gene to form heterokaryotic mycelium. Incompatibility impeding heterokaryon establishment can be overcome by fusion of protoplasts. The mutant allele is completely (or partially) dominant if the heterokaryon is phenotypically identical (or similar) to the mutant parent; it is completely (partially) recessive if the heterokaryon is phenotypically similar to the wild-type parent (Grindle 1987; Grindle and Faretra 1993).

2.5 Level of Resistance and Pleiotropic Effects of Resistance Mutations

Levels of resistance are usually quantified by determining from dose–response curves the concentration of fungicide needed to reduce ‘life’ parameters, such as colony or mycelium growth or spore germination by 50 % (effective concentration 50; EC_{50}) and the minimal inhibitory concentration (MIC). A mutant can be designated resistant to a fungicide if its EC_{50} value is at least twice the EC_{50} value of sensitive wild-type isolates (Delp and Dekker 1985). However, small differences in EC_{50} values among resistant mutants and sensitive isolates may not be detected unless environmental variables are rigorously controlled and/or data from dose–response experiments are subjected to statistical analysis. Moreover, with small differences it may be difficult to establish whether resistance is due to a single major gene or to polygenes (Grindle and Faretra 1993).

A mutant gene conferring resistance to a particular fungicide often confers positive cross resistance to other fungicides having the same or related mode of action. On the contrary, mutant genes causing resistance to one fungicide may increase sensitivity to other chemicals (negative cross resistance) (Brent and Hollomon 2007a, b).

Resistance mutations may have deleterious pleiotropic effects on unrelated phenotypic characters, such as competitiveness, virulence, survival and reproductive success. Physiological mechanisms underlying resistance to fungicides may also be associated with a metabolic cost. Hence, resistant isolates may have lower fitness than wild-type sensitive isolates. Differences in fitness can be experimentally measured as reduction in mycelial growth rate, sporulation and conidial germination, pathogenicity, survival under stressing conditions, etc., in a fungicide-free environment. For instance, fitness penalty was observed in (1) DMI resistance in powdery mildews (Gisi et al. 2002); (2) resistance to dicarboximides and phenylpyrroles in several fungi, such as *B. cinerea* (Pollastro et al. 1996; Ochiai et al. 2001), *N. crassa* (Hollomon et al. 1997) and *Monilinia laxa* (Katan and Shabi 1982); (3) resistance to QoIs (G143A replacement) in field populations of *P. viticola* (Fernández-Ortuño et al. 2008) and *Pyricularia grisea* (Avila-Adame and Köller 2003) and in laboratory mutants of *B. cinerea* (Markoglou et al. 2006), *C. beticola* (Malandrakis et al. 2006) and *Ustilago maydis* (Ziogas et al. 2002), but not in *B. graminis* f.sp. *tritici* (Heaney et al. 2000; Chin et al. 2001), *Mycosphaerella graminicola* (Miguez et al. 2004) and *Magnaporthe grisea* (Avila-Adame and Köller 2003); (4) most of the mutations in the *SdhB* gene conferring resistance to SDHIs in *B. cinerea* except for $SdhB^{H272Y}$ (Lalève et al. 2014a; Veloukas et al. 2014); and (5) *Penicillium expansum* resistant to tebuconazole, fludioxonil and iprodione, but not to cyprodinil, coupled with reduction in patulin production (Karaoglanidis et al. 2011).

Mutations responsible for fungicide resistance may influence mycotoxin production. For instance, the production of 3-acetyl deoxynivalenol (3-ADON) was altered in isolates of *Fusarium culmorum* resistant to the DMI fungicide difenoconazole (D'Mello et al. 1997). A higher production of T-2 toxin, 4,15-diacetoxyscirpenol and neosolaniol was found in a carbendazim-resistant strain of *Fusarium sporotrichioides* (D'Mello et al. 1998, 2000). More recently, Zhang et al. (2009) found that benzimidazole resistance increased trichothecene production in *F. graminearum*. Laboratory mutants of *Aspergillus parasiticus* resistant to phenylpyrroles and dicarboximides produced more aflatoxins than the parental wild-type strain (Markoglou et al. 2008a). Similarly, laboratory mutant strains of *A. parasiticus*, *A. ochraceus* and *F. verticillioides* resistant to triazoles (epoxiconazole and flusilazole) and mutant strains of *A. carbonarius* and *P. expansum* resistant to fludioxonil produced significantly higher levels of mycotoxins (ochratoxins, patulin and fumonisins) compared to the parental sensitive strains (Doukas et al. 2008; Markoglou et al. 2008b, 2009).

It is generally assumed that fitness costs of resistance are invariable. However, Chin et al. (2001) showed that the cost of resistance to QoI fungicides in *B. graminis* varies with environmental conditions, such as temperature, being more costly under suboptimal conditions for the fungus.

2.6 Population Genetics

To develop effective resistance management strategies, it is crucial to know all the factors influencing relationship between sensitive and resistant strains.

The prevailing model explaining the selection of fungicide-resistant fungal populations considers random and rare mutations as the cause for pre-existing but infrequent resistant phenotypes prior to the introduction of a new fungicide (Torriani et al. 2009; Camps et al. 2012). Nevertheless, the evolutionary question on how populations adapt to novel environments, such as new antimicrobials, through de novo mutations or through selection from standing genetic variation, which affect the probability and speed of emergence of resistant alleles, is still debated (Hermisson and Pennings 2005; Hawkins et al. 2014).

Anyway, rare resistant mutants gain in competitiveness under the selection force of fungicide sprays and are selected to frequencies at which disease control becomes unsatisfactory (Milgroom et al. 1989; Skylakakis 1987; Wolfe 1982; Hobbelen et al. 2014). The shift towards resistance occurs at different rates depending on the number of genes conferring resistance. In monogenic resistance, a rapid shift towards resistance may occur, leading to discrete resistant subpopulation(s), while in polygenic resistance, the shift towards resistance progresses slowly, leading to a reduced sensitivity of the entire population. Resistant and wild-type subpopulations are in a dynamic equilibrium due to two selective pressures: i) the disruptive selection (directional selection in polygenic resistance), favouring resistant subpopulation(s), is due to repeated sprays with fungicides having the same mode of action at risk of resistance, and ii) stabilizing selection, favouring the wild-type sensitive populations, is caused by possible negative pleiotropic effect of resistance mutations leading to a reduced fitness (Fig. 2.1). Unfit mutants compete well only under the selection pressure of fungicide sprays, and, hence, resistance is at least partially reversible when the selection pressure is removed or minimized by applying resistance management strategies.

2.7 Obtainment of Resistant Mutants

Field isolates collected from diseased plants, plant debris, soil or air may include fungicide-resistant mutants, particularly if crops have been sprayed intensively with single-site fungicides. Resistant field isolates may be selected on appropriate agar media amended with a fungicide at a concentration inhibiting germination of conidia and/or mycelium growth of wild-type sensitive isolates. In choosing agar medium the mode of action of the fungicide must be complained. In the case of obligate biotrophic pathogens, plants or parts of plants must replace agar media. A number of monitoring methods are available (www.frac.info). Field isolates may display a broad variation making their genetic analysis more difficult than laboratory mutants (Grindle and Faretra 1993). It is advisable to obtain 'monoconidial' or 'single hyphal tip' isolates rather than 'mass-conidial' or 'mass-hyphal' isolates since they are likely to be genetically more homogeneous and stable. These traits are improved by repeated subculturing monoconidial isolates selecting the 'most typical' progeny.

Experiments under laboratory conditions are useful because it is possible to replicate them, to control the strength of selection and to use defined reference strains (Cowen et al. 2002). Resistant laboratory mutants can be generated in vitro from wild-type strains of known phenotype, and all mutants deriving from a same strain are near isogenic since their genomes are virtually identical, except for mutant gene(s) conferring resistance.

Selection of spontaneous mutations may be achieved by growing fungal colonies on media added with sublethal fungicide concentration; resistant hyphae grow better than the sensitive ones and produce vigorous sectors from slow-growing colonies. Alternatively, a high number of conidia can be plated on fungicide-amended media, and growing colonies can be singly transferred to fresh media. For instance, in *B. cinerea* it is relatively easy to get spontaneous mutants resistant to dicarboximides, phenylpyrroles, anilinopyrimidines and QoIs (Faretra and Pollastro 1991, 1993a, b; De Miccolis Angelini et al. 2002, 2012a).

Mutations can be induced by exposing conidia or hyphae to chemical (e.g. N-methyl-N-nitro-*n*-nitrosoguanidine) or physical mutagens (e.g. UV light) causing established proportions of lethality, before incubation on selective media. Mutagenesis greatly increases the yield of mutants but may cause unwished mutations in the genome which may interfere with identification and analysis of gene(s) causing fungicide resistance. Physical or chemical mutagenesis have been used successfully to produce resistant mutants in numerous fungi, including *B. cinerea* (De Miccolis Angelini et al. 2002, 2010a, b, 2012a, b), *F. graminearum* NRRL 13383 (Becher et al. 2010), *P. capsici* and *P. infestans* (Young et al. 2001), *Ustilago maydis* (Orth et al. 1994) and *V. inaequalis* (Zheng et al. 2000). In fungi with multinucleate conidia, laboratory mutants are frequently heterokaryons containing both mutated and wild-type nuclei so that they are often phenotypically unstable and produce both resistant and wild-type progeny during subculturing. Hence, at least initially, the selective pressure exerted by fungicide is crucial for the stability of the resistance trait.

The availability of molecular techniques has made it possible to investigate the genomes of pathogenic fungi which are not amenable to classical Mendelian analysis of meiotic progeny. For instance, genetic differences between isolates can be detected by RFLP (restriction fragment length polymorphisms) or various PCR-based techniques suitable for evidencing SNPs (single nucleotide polymorphisms) and allelic variants (AS-PCR, allele-specific PCR). Individual genes can be dissected out of the genome, then cloned and sequenced or altered genetically and put back into the genome. Cells containing cloned genes can be used to obtain large amounts of protein for amino acid sequencing. In the last fifteen years, site-direct mutagenesis has been used for studies of fungicide resistance. For instance, *N. crassa* mutants in the osmosensing histidine kinase *os-1* gene exhibit resistance to dicarboximides, aromatic hydrocarbons and phenylpyrroles. The *os-1* mutants can be classified into two groups: type I are null mutants highly resistant to iprodione and fludioxonil and moderately sensitive to osmotic stress, and type II carry single amino acid changes and are moderately resistant to both fungicides and highly sensitive to osmotic stress. This suggests that Os1p is essential for the antifungal activity of these fungicides and that amino acid repeats have an important function in osmoregulation (Ochiai et al. 2001). Site-directed mutagenesis followed by gene replacement was used to introduce mutations in different codons of the β-tubulin gene of a carbendazim-sensitive field strain of *Gibberella zeae*. All the mutants were resistant to carbendazim, but the level of resistance was depending on the mutations (Qiu et al. 2011). Site-directed mutagenesis of the *SdhB* gene was applied to confirm that each of the mutations identified in field strains conferred resistance to boscalid in *B. cinerea* and partial cross resistance to other SDHIs (fluopyram, carboxin) (Lalève et al. 2014b).

References

Anderson JB, Sirjusingh C, Ricker N (2004) Haploidy, diploidy and evolution of antifungal drug resistance in *Saccharomyces cerevisiae*. Genetics 168:1915–1923

Avenot H, Simoneau P, Iacomi-Vasilescu B, Bataillé-Simoneau N (2005) Characterization of mutations in the two-component histidine kinase gene *AbNIK1* from *Alternaria brassicicola* that confer high dicarboximide and phenylpyrrole resistance. Curr Genet 47:234–243

Avila-Adame C, Köller W (2003) Characterization of spontaneous mutants of *Magnaporthe grisea* expressing stable resistance to the Qo-inhibiting fungicide azoxystrobin. Curr Genet 42:332–338

Becher R, Hettwer U, Karlovsky P, Deising HB, Wirsel SGR (2010) Adaptation of *Fusarium graminearum* to tebuconazole yielded descendants diverging for levels of fitness, fungicide resistance, virulence, and mycotoxin production. Phytopathology 100:444–453

Bertier L, Leus L, D'hondt L, de Cock AWAM, Höfte M (2013) Host adaptation and speciation through hybridization and polyploidy in Phytophthora. PLoS ONE 8, e85385

Bi Y, Chen L, Cai M, Zhu S, Pang Z, Liu X (2014) Two non-target recessive genes confer resistance to the anti-oomycete microtubule inhibitor zoxamide in *Phytophthora capsici*. PLoS ONE 9, e89336

Birky CW Jr (2001) The inheritance of genes in mitochondria and chloroplasts: laws, mechanisms, and models. Annu Rev Genet 35:125–148

Blatter RHE, Brown JKM, Wolfe MS (1998) Genetic control of the resistance of *Erysiphe graminis* f.sp. *hordei* to five triazole fungicides. Plant Pathol 47:570–579

Blum M, Gisi U (2008) Inheritance of resistance in *Plasmopara viticola*. In: Dehne HW, Gisi U, Kuck KH, Russell PE, Lyr H (eds) Modern fungicides and antifungal compounds V, BCPC, DPG, Braunschweig, Germany, pp 101–104

Blum M, Waldner M, Gisi U (2010) A single point mutation in the novel *PvCesA3* gene confers resistance to the carboxylic acid amide fungicide mandipropamid in *Plasmopara viticola*. Fungal Genet Biol 47:499–510

Blum M, Gamper HA, Waldner M, Sierotzki H, Gisi U (2012) The cellulose synthase 3 (*CesA3*) gene of oomycetes: structure, phylogeny and influence on sensitivity to carboxylic acid amide (CAA) fungicides. Fungal Biol 116:529–542

Borck K, Braymer HD (1974) The genetic analysis of resistance to benomyl in *Neurospora crassa*. J Gen Microbiol 85:51–56

Brent KJ, Hollomon DW (2007a) Fungicide resistance in crop pathogens: how can it be managed? FRAC monograph no. 1. Global Crop Protection Federation, Brussels

Brent KJ, Hollomon DW (2007b) Fungicide resistance: the assessment of risk. FRAC monograph no. 2. Global Crop Protection Federation, Brussels

Brown AWA (1977) Epilogue: resistance as a factor in pesticide management. In: Mattson WJ (ed) The role or arthropods in forest ecosystems. Proceedings, 15th international congress of entomology, Washington, DC, 19–27 August 1976. Springer, New York

Brown JKM, Jessop AC, Thomas S, Rezanoor HN (1992) Genetic control of the response of *Erysiphe graminis* f.sp. *hordei* to ethirimol and triadimenol. Plant Pathol 41:126–135

Camps SM, Rijs AJ, Klaassen CH, Meis JF, O'Gorman CM, Dyer PS, Melchers WJ, Verweij PE (2012) Molecular epidemiology of *Aspergillus fumigatus* isolates harboring the TR34/L98H azole resistance mechanism. J Clin Microbiol 50:2674–2680

Catal M, King L, Tumbalam P, Wiriyajitsomboon P, Kirk WW, Adams GC (2010) Heterokaryotic nuclear conditions and a heterogeneous nuclear population are observed by flow cytometry in *Phytophthora infestans*. Cytometry A 77:769–775

Chapeland F, Fritz R, Lanen C, Gredt M, Leroux P (1999) Inheritance and mechanisms of resistance to anilinopyrimidine fungicides in *Botrytis cinerea* (*Botryotinia fuckeliana*). Pestic Biochem Physiol 64:85–100

Chin KM, Chavaillaz D, Kaesbohrer M, Staub T, Felsenstein FG (2001) Characterizing resistance risk of *Erysiphe graminis* f.sp. *tritici* to strobilurins. Crop Prot 20:87–96

Cookson B (2005) Clinical significance of emergence of bacterial antimicrobial resistance in the hospital environment. J Appl Microbiol 99:989–996

Cools HJ, Fraaije BA (2013) Update on mechanisms of azole resistance in *Mycosphaerella graminicola* and implications for future control. Pest Manag Sci 69:150–155

Coplin D (1989) Plasmids and their role in evolution of plant pathogenic bacteria. Annu Rev Phytopathol 27:187–212

Cowen LE, Anderson JB, Kohn LM (2002) Evolution of drug resistance in *Candida albicans*. Annu Rev Microbiol 56:139–165

Crute IR, Harrison JM (1988) Studies on the inheritance of resistance to metalaxyl in *Bremia lactucae* and on the stability and fitness of field isolates. Plant Pathol 37:231–250

Cui W, Beever RE, Parkes SL, Weeds PL, Templeton MD (2002) An osmosensing histidine kinase mediates dicarboximide fungicide resistance in *Botryotinia fuckeliana* (*Botrytis cinerea*). Fungal Genet Biol 36:187–198

Davis RH (1966) Mechanisms of inheritance. 2. Heterokaryosis. In: Ainsworth GC, Sussman AS (eds) The fungi, an advanced treatise, vol 2, The fungal organism. Academic Press, New York, pp 567–588

De Miccolis Angelini RM, Santomauro A, De Guido MA, Pollastro S, Faretra F (2002) Genetics of anilinopyrimidine-resistance in *Botryotinia fuckeliana* (*Botrytis cinerea*). In: Abstract Book of the 6th European Conference on Fungal Genetics, Pisa, Italy, 6–9 April 2002

De Guido MA, De Miccolis Angelini RM, Pollastro S, Santomauro A, Faretra F (2007) Selection and genetic analysis of laboratory mutants of *Botryotinia fuckeliana* resistant to fenhexamid. J Plant Pathol 89:203–210

De Miccolis Angelini RM, Habib W, Rotolo C, Pollastro S, Faretra F (2010a) Selection, characterization and genetic analysis of laboratory mutants of *Botryotinia fuckeliana* (*Botrytis cinerea*) resistant to the fungicide boscalid. Eur J Plant Pathol 128:185–199

De Miccolis Angelini RM, Rotolo C, Pollastro S, Faretra F (2010b) Phenotypic and molecular characterization of fungicide-resistant field isolates of *Botryotinia fuckeliana* (*Botrytis cinerea*). In: Abstracts of the XV international botrytis symposium, Càdiz, Spain, 30 May–4 June 2010

De Miccolis Angelini RM, Rotolo C, Masiello M, Pollastro S, Ishii H, Faretra F (2012a) Genetic analysis and molecular characterisation of laboratory and field mutants of *Botryotinia fuckeliana* (*Botrytis cinerea*) resistant to QoI fungicides. Pest Manag Sci 68:1231–1240

De Miccolis Angelini RM, Pollastro S, Faretra F (2012b) Genetics of fungicide resistance in *Botryotinia fuckeliana* (*Botrytis cinerea*). In: Thind TS (ed) Fungicide resistance in crop protection: risk and management. CAB International, Wallingford, pp 237–250

Delp CJ, Dekker J (1985) Fungicide resistance: definitions and use of terms. Bull OEPP 15:333–335

Délye C, Laigret F, Corio-Costet MF (1997) A mutation in the 14 alpha-demethylase gene of *Uncinula necator* that correlates with resistance to a sterol biosynthesis inhibitor. Appl Environ Microbiol 63:2966–2970

D'Mello JP, Macdonald AM, Postel D, Hunter EA (1997) 3-Acetyl deoxynivalenol production in a strain of *Fusarium culmorum* insensitive to the fungicide difenoconazole. Mycotoxin Res 13:73–80

D'Mello JPF, MacDonald AMC, Postel D, Dijksma WTP, Dujardin A, Plactina CM (1998) Pesticide use and mycotoxin production in *Fusarium* and *Aspergillus* phytopathogens. Eur J Plant Pathol 104:741–751

D'Mello JP, Macdonald AM, Briere L (2000) Mycotoxin production in a carbendazim-resistant strain of *Fusarium sporotrichioides*. Mycotoxin Res 16:101–111

Doukas EG, Markoglou AN, Ziogas BN (2008) Biochemical and molecular study of triazole-resistance and its effect on ochratoxin production by *Aspergillus ochraceus* Wilh. In: Abstracts of the 9th international congress of plant pathology. ICPP, Torino, Italy, 24–29 August 2008. J Plant Pathol 90:S2.319

Dry IB, Yuan KH, Hutton DG (2004) Dicarboximide resistance in field isolates of *Alternaria alternata* is mediated by a mutation in a two-component histidine kinase gene. Fungal Genet Biol 41:102–108

Faretra F, Pollastro S (1991) Genetic basis of resistance to benzimidazole and dicarboximide fungicides in *Botryotinia fuckeliana* (*Botrytis cinerea*). Mycol Res 95:943–951

Faretra F, Pollastro S (1993a) Genetics of sexual compatibility and resistance to benzimidazole and dicarboximide fungicides in isolates of *Botryotinia fuckeliana* (*Botrytis cinerea*) from nine countries. Plant Pathol 42:48–57

Faretra F, Pollastro S (1993b) Isolation, characterization and genetic analysis of laboratory mutants of *Botryotinia fuckeliana* (*Botrytis cinerea*) resistant to the phenylpyrrole fungicide CGA 173506. Mycol Res 97:620–624

Fernández-Ortuño D, Torés JA, de Vicente A, Pérez-García A (2008) Mechanisms of resistance to QoI fungicides in phytopathogenic fungi. Int Microbiol 11:1–9

Fillinger S, Ajouz S, Nicot PC, Leroux P, Bardin M (2012) Functional and structural comparison of pyrrolnitrin- and iprodione-induced modifications in the class III histidine-kinase Bos1 of *Botrytis cinerea*. PLoS ONE 7, e42520

Fillinger S, Leroux P, Auclair C, Barreau C, Al Hajj C, Debieu D (2008) Genetic analysis of fenhexamid-resistant field isolates of the phytopathogenic fungus *Botrytis cinerea*. Antimicrob Agents Chemother 52:3933–3940

Fincham JRS, Day PR, Radford A (1979) Fungal genetics, 4th edn. University of California Press, Berkeley, p 636

Georgopoulos SG (1988) Genetics and population dynamics. In: Delp CJ (ed) Fungicide resistance in North America. APS Press, St. Paul, pp 12–13

Gisi U (2012) Resistance to carboxylic acid amide (CAA) fungicides and anti-resistance strategies. In: Thind TS (ed) Fungicide resistance in crop protection: risk and management. CAB International, Wallingford, pp 96–103

Gisi U, Cohen Y (1996) Resistance to phenylamide fungicides: a case study with *Phytophthora infestans* involving mating type and race structure. Annu Rev Phytopathol 43:549–72

Gisi U, Sierotzki H (2008) Fungicide modes of action and resistance in downy mildews. Eur J Plant Pathol 122:157–167

Gisi U, Sierotzki H, Cook A, McCaffery A (2002) Mechanisms influencing the evolution of resistance to Qo inhibitor fungicides. Pest Manag Sci 58:859–867

Gisi U, Waldner M, Kraus N, Dubuis PH, Sierotzki H (2007) Inheritance of resistance to carboxylic acid amide (CAA) fungicides in *Plasmopara viticola*. Plant Pathol 56:199–208

Grasso V, Palermo S, Sierotzki H, Garibaldi A, Gisi U (2006) Cytochrome b gene structure and consequences for resistance to Qo inhibitor fungicides in plant pathogens. Pest Manag Sci 62:465–472

Griffiths AJF (1996) Mitochondrial inheritance in filamentous fungi. J Genet 75:403–414

Grindle M (1987) Genetics of fungicide resistance. In: Ford MG, Hollomon DH, Khambay BPS, Sawiki RM (eds) Combating resistance to xenobiotics – biological and chemical approaches. Ellis Horwood, Chichester, pp 75–93

Grindle M, Dolderson GH (1986) Effects of a modifier gene on the phenotype of a dicarboximide-resistant mutant of *Neurospora crassa*. Trans Br Mycol Soc 87:457–460

Grindle M, Faretra F (1993) Genetic aspects of fungicide resistance. In: Lyr H, Polter C (eds) Modern fungicides and antifungal compounds. Proceedings of the 10th international symposium, Reinhardsbrunn, Germany, May 1992. Ulmer, Stuttgart, Germany, pp 33–43

Guerineau M, Slonimski PP, Avner PR (1974) Yeast episome: oligomycin resistance associated with a small covalently closed non-mitochondrial circular DNA. Biochem Biophys Res Commun 61:462–469

Hawkins NJ, Cools HJ, Sierotzki H, Shaw MW, Knogge W, Kelly SL, Kelly DE, Fraaije BA (2014) Paralog re-emergence: a novel, historically contingent mechanism in the evolution of antimicrobial resistance. Mol Biol Evol 31:1793–1802

Heaney SP, Hall AA, Davies SA, Olaya G (2000). Resistance to fungicides in the QoI-STAR cross-resistance group: current perspectives. In: Proceedings of the British crop protection conference: pests and diseases, 13–16 November 2002, vol 2. BCPC, Farnham, Surrey, UK, pp 755–762

Hermann D, Gisi U (2012) Fungicide resistance in oomycetes with special reference to *Phytophthora infestans* and phenylamides. In: Thind TS (ed) Fungicide resistance in crop protection: risk and management. CAB International, Wallingford, pp 133–140

Hermisson J, Pennings PS (2005) Soft sweeps: molecular population genetics of adaptation from standing genetic variation. Genetics 169:2335–2352

Hilber UW, Hilber-Bodmer M (1998) Genetic basis and monitoring of resistance of *Botryotinia fuckeliana* to anilinopyrimidines. Plant Dis 82:496–500

Hobbelen PH, Paveley ND, van den Bosch F (2014) The emergence of resistance to fungicides. PLoS One 9, e91910

Hollomon DW (1981) Genetic control of ethirimol resistance in a natural population of *Erysiphe graminis* f.sp. *hordei*. Phytopathology 71:536–540

Hollomon DW, Butters J, Clark J (1984) Genetic control of triadimenol resistance in barley powdery mildew. In: Proceedings of the British crop protection conference: pests and diseases, 19–22 November 1984, vol 2. BCPC, Croydon, UK, pp 477–482

Hollomon DW, Butters JA, Kendall SJ (1997) Mechanism of resistance to fungicides. In: Sjut V (ed) Molecular mechanisms of resistance to agrochemicals, vol 13, Chemistry of plant protection. Springer, Heidelberg, pp 1–20

Ishii H (2010) QoI fungicide resistance: current status and the problems associated with DNA-based monitoring. In: Gisi U, Chet I, Gullino ML (eds) Recent developments in management of plant diseases, plant pathology in the 21st century, vol 1. Springer, Dordrecht, pp 37–45

Ishii H (2012a) Resistance to QoI and SDHI fungicides in Japan. In: Thind TS (ed) Fungicide resistance in crop protection: risk and management. CAB International, Wallingford, pp 237–250

Ishii H (2012b) Resistance in *Venturia nashicola* to benzimidazoles and sterol demethylation inhibitors. In: Thind TS (ed) Fungicide resistance in crop protection: risk and management. CAB International, Wallingford, pp 21–31

Ishii H, Yano K, Date H, Furuta A, Sagehashi Y, Yamaguchi T, Sugiyama T, Nishimura K, Hasama W (2007) Molecular characterization and diagnosis of QoI resistance in cucumber and eggplant fungal pathogens. Phytopathology 97:1458–1466

Ishii H, Fountaine J, Chung WH, Kansako M, Nishimura K, Takahashi K, Oshima M (2009) Characterization of QoI resistant field isolates of *Botrytis cinerea* from citrus and strawberry. Pest Manag Sci 65:916–922

Kalamarakis AE, de Waard MA, Ziogas BN, Georgopoulos SG (1991) Resistance to fenarimol in *Nectria haematococca* var. *cucurbitae*. Pestic Biochem Physiol 40:212–220

Kappas A, Georgopoulos SG (1970) Genetic analysis of dodine resistance in *Nectria haematococca* (Syn. *Hypomyces solani*). Genetics 66:617–622

Karaoglanidis GS, Markoglou AN, Bardas GA, Doukas EG, Konstantinou S, Kalampokis JF (2011) Sensitivity of *Penicillium expansum* field isolates to tebuconazole, iprodione, fludioxonil and cyprodinil and characterization of fitness parameters and patulin production. Int J Food Microbiol 145:195–204

Katan T, Shabi E (1982) Characterization of a dicarboximide-fungicide-resistant laboratory isolate of *Monilinia laxa*. Phytoparasitica 10:241–245

Kim YS, Dixon EW, Vincelli P, Farman ML (2003) Field resistance to strobilurin (QoI) fungicides in *Pyricularia grisea* caused by mutations in the mitochondrial cytochrome b gene. Phytopathology 93:891–900

Knapova G, Schlenzig A, Gisi U (2002) Crosses between isolates of *Phytophthora infestans* from potato and tomato and characterization of F1 and F2 progeny for phenotypic and molecular markers. Plant Pathol 51:698–709

Kretschmer M (2012) Emergence of multi-drug resistance in fungal pathogens: a potential threat to fungicide performance in agriculture. In: Thind TS (ed) Fungicide resistance in crop protection: risk and management. CAB International, Wallingford, pp 251–267

Kretschmer M, Leroch M, Mosbach A, Walker AS, Fillinger S, Mernke D, Schoonbeek HJ, Pradier JM, Leroux P, De Waard MA, Hahn M (2009) Fungicide-driven evolution and molecular basis of multidrug resistance in field populations of the grey mould fungus *Botrytis cinerea*. PLoS Pathog 5, e1000696

Lalève A, Fillinger S, Walker AS (2014a) Fitness measurement reveals contrasting costs in homologous recombinant mutants of *Botrytis cinerea* resistant to succinate dehydrogenase inhibitors. Fungal Genet Biol 67:24–36

Lalève A, Gamet S, Walker AS, Debieu D, Toquin V, Fillinger S (2014b) Site-directed mutagenesis of the P225, N230 and H272 residues of succinate dehydrogenase subunit B from *Botrytis cinerea* highlights different roles in enzyme activity and inhibitor binding. Environ Microbiol 16:2253–2266

Lasseron-De Falandre A, Daboussi MJ, Leroux P (1991) Inheritance of resistance to fenpropimorph and terbinafine, two sterol biosynthesis inhibitors, in *Nectria haematococca*. Phytopathology 81:1432–1438

Leroux P, Fritz R, Debieu D, Albertini C, Lanen C, Bach J, Gredt M, Chapeland F (2002) Mechanisms of resistance to fungicides in field strains of *Botrytis cinerea*. Pest Manag Sci 58:876–888

Lesemann SS, Schimpke S, Dunemann F, Deising HB (2006) Mitochondrial heteroplasmy for the cytochrome b gene controls the level of strobilurin resistance in the apple powdery mildew fungus *Podosphaera leucotricha* (Ell. & Ev.) E.S. Salmon. J Plant Dis Prot 113:259–266

Liu Y, Liu Z, Hamada MS, Yin YN, Ma ZH (2014) Characterization of laboratory pyrimethanil-resistant mutants of *Aspergillus flavus* from groundnut in China. Crop Prot 60:5–8

Ma Z, Michailides TJ (2005) Advances in understanding molecular mechanisms of fungicide resistance and molecular detection of resistant genotypes in phytopathogenic fungi. Crop Protect 24:853–863

Malandrakis AA, Markoglou AN, Nikou DC, Vontas JG, Ziogas BN (2006) Biological and molecular characterization of laboratory mutants of *Cercospora beticola* resistant to Qo inhibitors. Eur J Plant Pathol 116:155–166
Markoglou AN, Ziogas BN (1999) Genetic control of resistance to fenpropimorph in *Ustilago maydis*. Plant Pathol 48:521–530
Markoglou AN, Ziogas BN (2000) Genetic control of resistance to tridemorph in *Ustilago maydis*. Phytoparasitica 28:349–360
Markoglou AN, Ziogas BN (2001) Genetic control of resistance to the piperidine fungicide fenpropidin in *Ustilago maydis*. J Phytopathol 149:551–559
Markoglou AN, Malandrakis AA, Vitoratos AG, Ziogas BN (2006) Characterization of laboratory mutants of *Botrytis cinerea* resistant to QoI fungicides. Eur J Plant Pathol 115:149–162
Markoglou AN, Doukas EG, Ziogas BN (2008a) Phenylpyrrole-resistance and aflatoxin production in *Aspergillus parasiticus* Speare. Int J Food Microbiol 127:268–275
Markoglou AN, Vattis K, Dimitriadis K, Doukas EG, Ziogas BN (2008b) Effect of phenylpyrrole resistance mutations on mycotoxin production by *Aspergillus carbonarius* and *Penicillium expansum*. In: Abstracts of the 9th international congress of plant pathology. ICPP, Torino, Italy, pp 24–29 August 2008. J Plant Pathol 90:S2.322
Markoglou AN, Vitoratos AG, Doukas EG, Ziogas BN (2009) Phytopathogenic and mycotoxigenic characterization of laboratory mutant strains of *Fusarium verticillioides* resistant to demethylation inhibiting fungicides. In: Proceedings of the DPG-BCPC 3rd international symposium-plant protection and plant health in Europe, Berlin, Germany, 14–16 May 2009
Meyer RW, Parmeter JR (1968) Changes in chemical tolerance associated with heterokaryosis in *Thanatephorus cucumeris*. Phytopathology 58:472–475
Miguez M, Reeve C, Wood PM, Hollomon DW (2004) Alternative oxidase reduces the sensitivity of *Mycosphaerella graminicola* to QoI fungicides. Pest Manag Sci 60:3–7
Milgroom MG, Levln SA, Fry WE (1989) Population genetics theory and fungicide resistance. In: Leonard KJ, Fry WE (eds) Plant disease epidemiology: genetics, resistance, and management. McGraw-Hill Publishing Company, New York, pp 340–367
Molnar A, Hornok L, Pesti M (1985) The high level of benomyl tolerance in *Fusarium oxysporum* is determined by the synergistic interaction of two genes. Exp Mycol 9:326–333
Nakazawa Y, Yamada M (1997) Chemical control of grey mould in Japan. A history of combating resistance. Agrochem Japan 71:2–6
Ochiai N, Fujimura M, Motoyama T, Ichiishi A, Usami R, Horikoshi K, Yamaguchi I (2001) Characterization of mutations in the two-component histidine kinase gene that confer fludioxonil resistance and osmotic sensitivity in the *os-1* mutants of *Neurospora crassa*. Pest Manag Sci 57:437–442
Ogden JE, Grindle M (1983) Changes in genetic constitution and sterol composition during growth of nystatin-resistant heterokaryons of *Neurospora crassa*. Genet Res 42:91–103
Orth AB, Sfarra A, Pell EJ, Tien M (1994) Characterization and genetic analysis of laboratory mutants of *Ustilago maydis* resistant to dicarboximide and aromatic hydrocarbon fungicides. Phytopathology 84:1210–1214
Oshima M, Fujimura M, Banno S, Hashimoto C, Motoyama T, Ichiishi A, Yamaguchi I (2002) A point mutation in the two-component histidine kinase *BcOS-1* gene confers dicarboximide resistance in field isolates of *Botrytis cinerea*. Phytopathology 92:75–80
Otto SP, Gerstein AC (2008) The evolution of haploidy and diploidy. Curr Biol 18:R1121–4
Pang Z, Shao J, Chen L, Lu X, Hu J, Qin Z, Liu X (2013) Resistance to the novel fungicide pyrimorph in *Phytophthora capsici*: risk assessment and detection of point mutations in CesA3 that confer resistance. PLoS One 8, e56513
Peever TL, Milgroom MG (1992) Inheritance of triadimenol resistance in *Pyrenophora teres*. Phytopathology 82:821–828

Pollastro S, Faretra F, Santomauro A, Miazzi M, Natale P (1996) Studies on pleiotropic effects of mating type, benzimidazole-resistance and dicarboximide-resistance genes in near-isogenic strains of *Botryotinia fuckeliana* (*Botrytis cinerea*). Phytopathol Medit 35:48–57

Qiu J, Xu J, Yu J, Bi C, Chen C, Zhou M (2011) Localisation of the benzimidazole fungicide binding site of *Gibberella zeae* β2-tubulin studied by site-directed mutagenesis. Pest Manag Sci 67:191–198

Robinson HL, Riduct CJ, Sierotzki H, Gisi U, Brown JKM (2002) Isogamous, hermaphroditic inheritance of mitochondrion-encoded resistance to Qo inhibitor fungicides in *Blumeria graminis* f.sp. *tritici*. Fungal Genet Biol 36:98–106

Rumbou A, Gessler C (2006) Particular structure of *Plasmopara viticola* populations evolved under Greek island conditions. Phytopathology 96:501–509

Santomauro A, Pollastro S, De Guido MA, De Miccolis Angelini RM, Natale P, Faretra F (2000) A long-term trial on the effectiveness of new fungicides against grey mould on grapevine and on their influence on the pathogen's population. In: Abstract Book of the XII international botrytis symposium. Reims, France, p P75

Shattock RC (1988) Studies on the inheritance of resistance to metalaxyl in *Phytophthora infestans*. Plant Pathol 37:4–11

Sierotzki H, Scalliet G (2013) A review of current knowledge of resistance aspects for the next-generation succinate dehydrogenase inhibitor fungicides. Phytopathology 103:880–887

Sierotzki H, Frey R, Wullschleger J, Palermo S, Karlin S, Godwin J, Gisi U (2007) Cytochrome *b* gene sequence and structure of *Pyrenophora teres* and *P. tritici-repentis* and implications for QoI resistance. Pest Manag Sci 63:225–33

Skinner W, Bailey A, Renwick A, Keon J, Gurr S, Hargreaves J (1998) A single amino-acid substitution in the iron-sulphur protein subunit of succinate dehydrogenase determines resistance to carboxin in *Mycosphaerella graminicola*. Curr Genet 34:393–398

Skylakakis G (1987) Changes in the composition of pathogen populations caused by resistance to fungicides. In: Wolfe MS, Caten CE (eds) Populations of plant pathogens: their dynamics and genetics. Blackwell, Oxford, pp 227–237

Steffens JJ, Pell EJ, Tien M (1996) Mechanisms of fungicide resistance in phytopathogenic fungi. Curr Opin Biotechnol 7:348–355

Steinfeld U, Sierotzki H, Parisi S, Gisi U (2002) Comparison of resistance mechanisms to strobilurin fungicides in *Venturia inaequalis*. In: Lyr H, Russell PE, Dehne HW, Gisi U, Kuck KH (eds) Modern fungicides and antifungal compounds III. Proceedings of the 13th international reinhardsbrunn symposium, Reinhardsbrunn, Germany, May 2001. AgroConcept GmbH, Bonn, Germany, pp 167–176

Summers RW, Heaney SP, Grindle M (1984) Studies of a dicarboximide resistant heterokaryon of *Botrytis cinerea*. In: Proceedings, Brighton crop protection conference, Brighton, UK, pp 453–458

Taga M, Nakagawa H, Tsuda M, Ueyama A (1979) Identification of three different loci controlling kasugamycin resistance in *Pyricularia oryzae*. Phytopathology 69:463–466

Torriani SFF, Brunner PC, McDonald BA, Sierotzki H (2009) QoI resistance emerged independently at least 4 times in European populations of *Mycosphaerella graminicola*. Pest Manag Sci 65:155–162

van Tuyl JM (1977) Genetic aspects of resistance to imazalil in *Aspergillus nidulans*. Neth J Plant Path 83:169–176

Veloukas T, Kalogeropoulou P, Markoglou AN, Karaoglanidis GS (2014) Fitness and competitive ability of *Botrytis cinerea* field isolates with dual resistance to SDHI and QoI fungicides, associated with several *sdh*B and the *cyt*b G143A mutations. Phytopathology 104:347–356

Wolfe MS (1982) Dynamics of the pathogen population in relation to fungicide resistance. In: Dekker J, Georgopoulos SG (eds) Fungicide resistance in crop protection. Centre for Agricultural Publishing and Documentation, Wageningen, pp 139–148

Yarden O, Katan T (1993) Mutations leading to substitutions at amino acids 198 and 200 of beta-tubulin that correlate with benomyl-resistance phenotypes of field strains of *Botrytis cinerea*. Phytopathology 83:1478–1483

Yoshimi A, Imanshi J, Gafur A, Tanaka C, Tsuda M (2003) Characterisation and genetic analysis of laboratory mutants of *Cochliobolus heterostrophus* resistant to dicarboximide and phenylpyrrole fungicides. J Gen Plant Pathol 69:101–108

Young DH, Spiewak SL, Slawecki RA (2001) Laboratory studies to assess the risk of development of resistance to zoxamide. Pest Manag Sci 57:1081–1087

Young DH, Kemmitt GM, Owen J (2005) A comparative study of XR-539 and other oomycete fungicides: similarity to dimethomorph and amino acid amides in its mechanism of action. In: Dehne HW, Gisi U, Kuck KH, Russell PE, Lyr H (eds) Modern fungicides and antifungal compounds IV. BCPC, Alton, pp 145–52

Zhang Y, Lamm R, Pillonel C, Lam S, Xu JR (2002) Osmoregulation and fungicide resistance: the *Neurospora crassa os-2* gene encodes a *HOG1* mitogen-activated protein kinase homologue. Appl Environ Microbiol 68:532–538

Zhang YJ, Yu JJ, Zhang YN, Zhang X, Cheng CJ, Wang JX, Hollomon DW, Fan PS, Zhou MG (2009) Effect of carbendazim resistance on trichothecene production and aggressiveness of *Fusarium graminearum*. Mol Plant Microbe Interact 22:1143–1150

Zheng D, Olaya G, Köller W (2000) Characterization of laboratory mutants of *Venturia inaequalis* resistant to the strobilurin-related fungicide kresoxim-methyl. Curr Genet 38:148–155

Zhu SS, Liu XL, Wang Y, Wu XH, Liu PF, Li JQ, Yuan SK, Si NG (2007) Resistance of *Pseudoperonospora cubensis* to flumorph on cucumber in plastic houses. Plant Pathol 56:967–975

Ziogas BN, Markoglou AN, Tzima A (2002) A non-Mendelian inheritance of resistance to strobilurin fungicides in *Ustilago maydis*. Pest Manag Sci 58:908–916

Chapter 3
Stability of Resistance

Hideo Ishii

Abstract The fitness of filamentous fungi is defined as the survival and reproductive success of an allelic, individual, or group (Pringle and Taylor 2002). Fungal strains resistant to fungicides are considered to appear first in a very low frequency by spontaneous mutations in nature irrespective of fungicide use. Resistant strains are then selected by applications of at risk fungicides resulting in the increase of their populations, and efficacy is decreased or eventually lost. It is therefore important to manage resistance before fungicide efficacy is influenced by resistance. However, in many cases, effective countermeasures are taken only when growers or related authorities noticed the decrease of field performance of fungicides. Fitness of resistant isolates is often examined in laboratory and/or greenhouse experiments by measuring, e.g., mycelial growth, spore productivity and germination ability, pathogenicity, and competitiveness. Despite that, wide ranges of genetic diversity of pathogen strains make it difficult to assess the fitness of resistant strains precisely even if field isolates are employed. Stability of resistance is particularly crucial in order to judge whether growers can expect to reuse fungicides in question after their withdrawal for some years. It is thus essential to monitor fluctuation of resistance in the field when discussing on stability of resistance in a practical sense.

Keywords Benzimidazoles • Cytochrome *b* gene • Fitness • Heteroplasmy • QoIs • SDHIs • Stability • Sterol demethylation inhibitors

3.1 Introduction

Stability of fungicide resistance is defined as the ability of the pathogen to retain the same level of fungicide insensitivity after successive generations of either exposure or no exposure to the target fungicide (Vega and Dewdney 2014). In addition, fitness

H. Ishii (✉)
Kibi International University, Minami-Awaji, Hyogo 656-0484, Japan

National Institute for Agro-Environmental Sciences,
Kannondai 3-1-3, Tsukuba, Ibaraki 305-8604, Japan
e-mail: hi481204@yahoo.co.jp

H. Ishii, D.W. Hollomon (eds.), *Fungicide Resistance in Plant Pathogens*,
DOI 10.1007/978-4-431-55642-8_3

is defined as the survival and reproductive success of an allele, individual, or group (Pringle and Taylor 2002). There is no doubt that fungicide resistance can cause a problem as resistant fungal strains have a level of fitness needed for their survival in the environment. Otherwise, they could not appear and increase their populations under the selection pressure by fungicide use. However, differential fitness between resistant and sensitive populations, if any, seems to be closely related with the stability of resistance in the field. If resistant populations have a serious fitness penalty in comparison with sensitive ones and/or other surrounding microbe antagonists, they would decline sooner or later in the absence of fungicide selection pressure, and then we might be able to reuse the problem fungicide. Here, I review the stability of fungicide resistance mainly focusing on situations in the field.

Stability of resistance in *Phytophthora infestans*, the pathogen of potato late blight, to the oomycete fungicides phenylamides, metalaxyl in particular, is well known in the field and is discussed in Chaps. 10 (by Gisi and Sierotzki) and 22 (by Schepers and Cooke) of this book. Furthermore, instability of dicarboximide resistance together with a fitness penalty of resistant strains has been well characterized in the gray mold fungus *Botrytis cinerea* (Waker et al. 2013). See Chap. 26 (by Walker and Leroux) of this book.

3.2 Benzimidazole Fungicide Resistance

When thinking about the stability of resistance, the most well-known cases of persistence will be the resistance to benzimidazole fungicides. In the 1970s, benzimidazole resistance occurred in plenty of pathogens in the field worldwide. In Japanese pear scab, caused by the ascomycete *Venturia nashicola*, when the application of benzimidazoles was stopped and alternative fungicides alone were applied in orchards where highly resistant isolates predominated, the proportion of highly resistant isolates gradually decreased and that of intermediately resistant, weakly resistant, and sensitive isolates increased (Fig. 3.1, Ishii et al. 1985; Ishii 2011). This phenomenon was thought to be an example of genetic homeostasis within microbial populations; however, resistant isolates comprised 80 % of the total 5 years after the last application of benzimidazoles, demonstrating clearly that benzimidazole resistance tends to persist for a long time. The mechanism of resistance has also been elucidated, and single-point mutations in β-tubulin gene encoding the fungicide target protein govern differential levels of resistance, i.e., high and intermediate resistance, respectively, as reviewed recently (Ishii 2011).

Ishizaki et al. (1983) also reported persistence of thiophanate-methyl resistance in *V. nashicola*. But in their monitoring, the majority of isolates, 125 out of 132 isolates tested, maintained very high level of resistance, MIC values higher than >1,000 ppm, even 5 years after withdrawal of benzimidazole fungicide use. The level of fungicide selection pressure given to resistant fungal populations prior to its withdrawal might have differed depending on the field, and such difference may

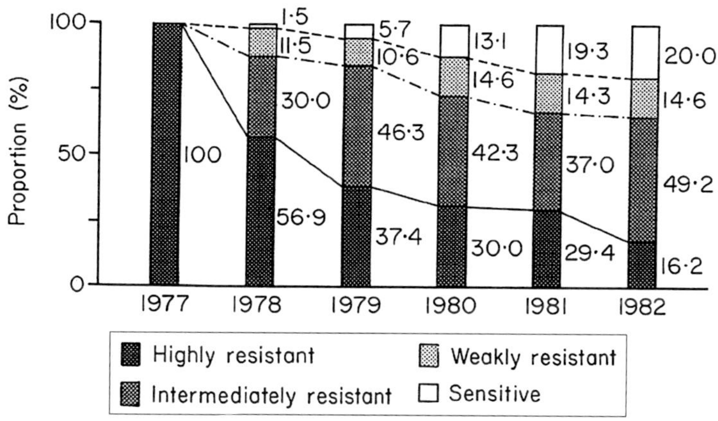

Fig. 3.1 Fluctuation of fungicide resistance in an orchard where benzimidazole fungicides are replaced by other fungicides in 1977 (Reprinted from Ishii et al. (1985))

involve in the speed of a shift back to sensitive populations. Alternatively, less fit resistant populations might have acquired additional fitness while they were predominant in the field.

In the apple scab fungus *V. inaequalis,* carbendazim-resistant populations sometimes showed a fairly substantial decrease, but in other cases no decrease occurred after the application of benzimidazole fungicides was stopped (Schwabe 1979). Although the reason for these differences is uncertain, it is reasonable to suspect that the fitness of fungal populations in the field greatly varies with environmental factors. Concerning this, Cox et al. (2007) stated that "vegetative and reproduction fitness costs associated with fungicide resistance in ascomycetes are highly specific to the fungal species, fitness parameters, and the fungicide in question, as studies indicate no consistent trends with regard to fungicide resistance and fitness." Recent use of benzimidazole fungicides for summer diseases control on apple in the USA and its relationship with the persistence of resistance in scab fungus is described by Cox in Chap. 27 of this book.

Benzimidazole-resistant isolates of *Gloeosporium theae-sinensis* were first found in tea-growing areas in 1978, 3 years after the introduction of thiophanate-methyl (Nonaka 1984). Although resistant strains declined in the fungal populations in the absence of fungicide selection pressure, more than 5 years were needed until they disappeared. When resistant populations were reduced to about 50 %, thiophanate-methyl was applied again only once a year, but the decline of resistance stopped. In contrast, the detection ratio of resistant strains continued to decrease when resistant strains were in a low frequency even if benzimidazole fungicides were subsequently used once a year. Stability of resistance is thus influenced by the initial proportion of resistant strains.

In 1992, it was confirmed that benzimidazole-resistant strains of *Cercospora kikuchii* (the pathogen of purple stain of soybean) were widely distributed. The use of benzimidazole fungicides was stopped and alternative fungicides were used since then. However, resistant strains have been detected in a high frequency in monitoring conducted both in 1997 and 2001 making it difficult to reuse benzimidazole fungicides (Hasegawa 2003). On the contrary, in other cases, frequency of resistant isolates decreased to 10–20 % after stopping the use, although the frequency remained in the same level subsequently. Importantly, the frequency rapidly increased when they used benzimidazole fungicides again. Reuse of benzimidazole fungicides was judged to be difficult also in this case (Mukobata 2004).

Resistance monitoring for benzimidazole fungicides has been carried out in French vineyards for *B. cinerea* (Waker et al. 2013). As the BenR1 phenotype which displayed high resistance to benzimidazoles but remained sensitive to the *N*-phenylcarbamate diethofencarb (negative cross-resistance) was predominant, carbendazim plus diethofencarb mixture was widely used to control BenR1 strains. But this led to the rapid selection of BenR2 strains which were highly resistant to both compounds. Since the 1990s, these fungicides have been replaced by fungicides with other modes of action. However, the frequency of BenR1 has remained high in populations indicating that the fitness penalty of these strains is small (Fig. 3.2, Waker et al. 2013). The frequency of BenR2 strains fell to low levels when the use of the mixture ceased suggesting high fitness penalty of these strains.

Stability of benzimidazole resistance is also described by Walker and Leroux in Chap. 26 and Cox in Chap. 27, respectively, of this book.

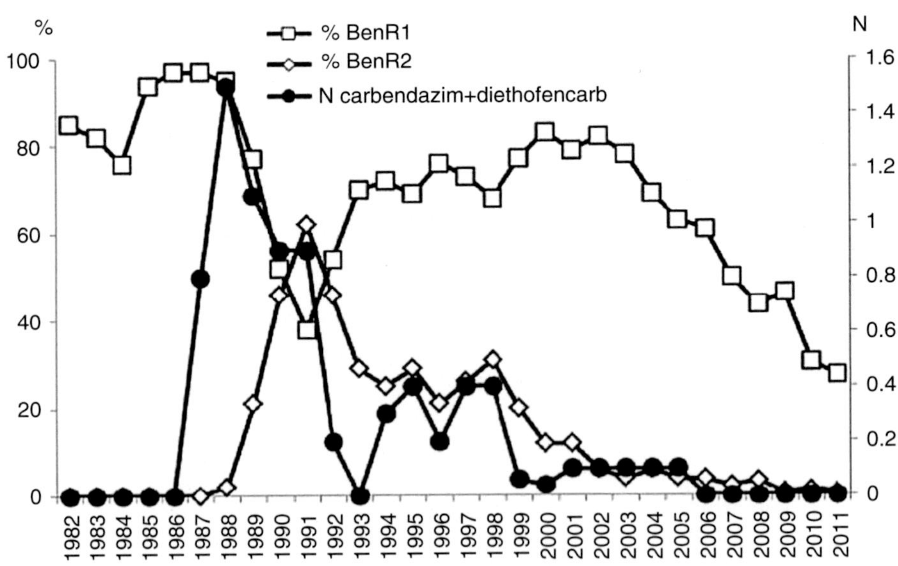

Fig. 3.2 Evolution and decline of benzimidazole resistance (mean frequency of BenR1 and BenR2 strains in the whole samples) in gray mold fungus in Champagne vineyards. *N*, mean number of sprays of carbendazim + diethofencarb per season (Reprinted from Walker et al. (2013))

3.3 DMI Fungicide Resistance

In contrast to the durability mostly found in resistance to benzimidazole and phenylamide fungicides, the stability of DMI resistance is more variable. In *Blumeria graminis* f. sp. *tritici* which causes powdery mildew on wheat, the sensitivity pattern to DMIs has remained stable for several years (FRAG-UK 2013). *Cercospora beticola* strains isolated from leaf-spot diseased sugar beet in the fields showed resistance to epoxiconazole and flutriafol (Nikou et al. 2009). No fitness penalties were associated with resistance mutations in the highly resistant phenotype, and most isolates retained their resistance levels even after four generations on fungicide-free medium. The C14α-demethylase gene (*cyp51*) was isolated and a study of transcriptional levels of the *cyp51* gene showed that overexpression was strongly associated with the highly DMI-resistant phenotype. Subsequently, Karaoglanidis and Thanassoulopoulos (2002) tested phenotypic stability of resistance to flutriafol with strains of *C. beticola*. After successive transfers of isolates either in vitro or in vivo in the absence of flutriafol, sensitivity was unchanged. Conversely, when fungal strains were exposed to cold conditions, either as mycelial isolates or as conidia, an increase in sensitivity was observed. This increase was most pronounced for strains with moderate and high levels of resistance to DMIs.

In *Monilinia fructicola*, the cause of brown rot disease of peach, sensitivity to propiconazole increased when resistant isolates transferred repeatedly onto fungicide-free PDA medium (Cox et al. 2007). EC_{50} values for resistant isolates returned to almost baseline levels at the end of the consecutive transfer and storage experiments (Zhu et al. 2012). The instability of propiconazole resistance may have important implications for disease management in that a reversion to sensitivity could potentially occur in the absence of DMI fungicide pressure in the field.

A variety of models have been reported as mechanisms for resistance to DMIs (see Chap. 13 of this book), but the following three are the major mechanisms: (a) target-site modifications in the *CYP51* gene resulting in decreased affinity of fungicides, (b) overexpression of the target gene *CYP51* during sterol formation, and (c) decreased intracellular fungicide accumulation mediated by an increased energy-dependent fungicide efflux.

In *M. fructicola*, the expression of 14α-demethylase gene (*MFCYP51*) is a major molecular mechanism of DMI resistance, and the presence of the transposable element Mona, a 65 bp insertion sequence located immediately upstream of *MFCYP51*, was strongly associated with the overexpression of this gene, as the Mona element contains a putative promoter sequence and has been found in resistant isolates (Luo and Schnabel 2008).

DMI fungicides are most important for scab control on apple and pear in Japan. Immediately after registration in mid 1980s, application of these fungicides was strictly limited to two to three times per year to avoid the practical problem of resistance development (Ishii 1997). Many reports have been available on DMI field resistance in apple scab worldwide, but in Japan, no loss of DMI efficacy has been experienced on apple so far.

In contrast, DMI resistance has caused a control failure recently in *V. nashicola*, the scab fungus of Asian pears (Ishii and Kikuhara 2007; Ishii 2011). In those orchards, DMI fungicides have been used slightly more frequently than on apple resulting in the selection of less-sensitive strains. However, recent studies indicated that methods for testing sensitivity, such as mycelial growth tests on DMI-amended PDA medium, are not appropriate to determine resistance in this fungus. Reduced fenarimol sensitivity of single-spore isolates often returned to sensitivity when stored and/or subcultured on fungicide-free PDA medium. Instability of reduced DMI sensitivity was reported earlier in *V. inaequalis* (Köller et al. 1991). Due to this, it is doubtful how precisely mycelial growth tests on culture reflect the real situation of DMI sensitivity of the fungal populations in the field. This phenomenon, recovery of DMI sensitivity, is probably related with the mechanism of resistance in these slow-growing fungi.

To assess field relevance of the instability of resistance, it will be essential to monitor yearly and/or seasonal changes of resistance in field populations. In an experimental orchard located in Nova Scotia, Canada, a complete loss of apple scab control was experienced in 1987 when 67.1 and 99.4 % of the leaves and fruit treated with bitertanol were infected. It occurred 10 years after DMI fungicides were first used and 5 years after the first sign of resistance (Braun 1994). In this orchard, DMI-sensitive and DMI-resistant populations were present in similar proportions 3 years after the selection pressure was removed suggesting that the resistant populations are fit. However, the mean ED_{50} values of myclobutanil for the resistant populations decreased over the 3 years, also suggesting that the resistant populations are slightly less fit than the wild-type populations. Overwintering activity of this fungus might be related with the stability of resistance as ascospore formation, and its discharge is important for primary infection in early spring.

The seasonal distribution of DMI resistance has been monitored in apple scab (Marine et al. 2011). Percent growth suppression (PGS) – the difference in colony growth on 0- and 1-ppm myclobutanil at 28 days – was used to assess fungicide resistance in culture. In the May and June sampling intervals, the average PGS hovered around 60 % regardless of cultivar or tree treatment. In contrast, the average PGS in the July sampling interval was around 30 % (i.e., more resistant to myclobutanil). The average PGS in the August sampling interval (around 50 %) was more similar to that seen early in the summer. In other studies, surprisingly, the pathogen populations shifted toward significantly reduced DMI sensitivities over the course of the summer in the absence of DMI exposure (Köller et al. 1995). The authors assumed that the observed population response might indicate a higher virulence of isolates with reduced DMI sensitivities on leaf tissue present in summer.

It is generally considered that resistance to DMI fungicides proceeds slowly and gradually in the field. In *V. inaequalis*, inheritance of DMI resistance was demonstrated by the cross of a DMI-resistant with a DMI-sensitive isolates followed by progeny tests conducted in culture. Sensitivity distribution of single ascospore isolates was clearly divided into two groups, resistant and sensitive, and fit into 1:1 ratio. However, when the level of sensitivity of progenies were compared with their parental isolates, mean EC_{50} values for fenarimol in resistant progeny populations were lower, but the range was broader than in their resistant parents. Similarly in sensitive progeny populations, the mean EC_{50} was higher and the distribution broader than in their sensitive parents. It was thus suspected that DMI resistance in apple scab fungus is governed by a single major gene in addition to some minor genes.

3.4 QoI Fungicide Resistance

It is well known that QoI fungicides generally carry a high risk of pathogen resistance development. Actually, resistant strains have been detected in over 60 plant diseases in the world so far and field performance of QoI fungicides has decreased in many cases (Ishii 2012). Although instability of QoI resistance has already been reviewed by Ishii (2010), it is summarized here and further information is added.

QoI-resistant populations of *Plasmopara viticola*, the pathogen of grapevine downy mildew, gradually reverted to full sensitivity following consecutive transfers to untreated plants, but when transferred on QoI-treated plants, sensitivity decreased resulting in almost full resistance (Genet et al. 2006). Similarly, but under practical greenhouse conditions, QoI-resistant populations of cucumber powdery mildew fungus *Podosphaera xanthii* declined gradually. However, resistant strains were rapidly recovered when azoxystrobin was applied again, but only once (Ishii et al. 2007). These findings indicated that resistant populations seemed to be somewhat less fit in the absence of QoI fungicide selection pressure, but it will be difficult to expect constant activity from QoI fungicides after resistance has developed. Hence, we can easily understand why and how precautionary measures are important in resistance management.

Thus, instability of QoI resistance has been noticed in individuals as well as pathogen populations. Heteroplasmy (a natural mixture of A143-mutated sequence and G143 wild-type sequence) of mitochondrial cytochrome *b* gene, conferring high QoI resistance, is also related with resistance instability. Heteroplasmy has been often detected in fungi such as *Colletotrichum gloeosporioides* (anthracnose fungus), *B. cinerea*, and others (Ishii 2010). In *C. gloeosporioides*, for example, RFLP (restriction fragment length polymorphism) analysis of PCR products was conducted using many single-spore isolates, and with the restriction enzyme *Ita*I (or *Fnu*4HI) which recognizes the resistance point mutation (5′-GC↓NGC-3′) at the position 143 in cytochrome *b* gene. However, both digested and undigested PCR products were seen on an agarose gel by electrophoresis indicating the phenomenon of heteroplasmy in resistant isolates (Uzuhashi et al. unpublished).

In highly QoI-resistant isolates, the G143A mutation in cytochrome *b* gene was confirmed by PCR-RFLP and/or sequencing. However, in some cases, we could not identify this mutation when isolates were subcultured without fungicide selection pressure. Moreover, reversion of resistance to sensitive was experienced in isolates of the cucumber powdery mildew fungus by subculture on fungicide untreated detached leaves for 8 years (Ishii 2012). Decline of mutated type seems to have occurred in an original heteroplasmic cytochrome *b* gene complement within mitochondrial genomes.

QoI-resistant isolates of *Corynespora cassiicola* (cucumber *Corynespora* leaf spot), *Mycovellosiella nattrassii* (eggplant leaf mold), and *C. gloeosporioides* were subcultured on PDA plates in the presence or absence of QoI fungicides. It was found that the proportion of mutated sequences in cytochrome *b* gene decreased over time in the absence of fungicide selection pressure in particular (Ishii et al. 2007). From this, the following hypothesis for heteroplasmic stages was proposed: (1) in the absence of fungicidal selection pressure, the mutated homoplasmic sequences of the cytochrome *b* gene in QoI-resistant strains may gradually revert to wild-type sequences resulting in transient heteroplasmic stages, and (2) the proportion of mutated sequences in heteroplasmic stages rapidly increases by higher copy numbers when receiving strong selection pressure by QoI fungicides. Resistant isolates of *B. cinerea*, in which G143A mutation was not detected by direct DNA sequencing of PCR products, have been maintained on a QoI-amended culture medium. Surprisingly, G143A mutation has been recovered subsequently in the same isolates (De Miccolis Angelini et al. 2012) strengthening the hypothesis mentioned above.

In *Magnaporthe grisea*, azoxystrobin-resistant spontaneous mutants were generated in the laboratory (Avila-Adame and Köller 2003). The two mutants characterized showed high resistance to azoxystrobin, with resistance factors exceeding 1,000. From these isolates, two different mutations of the cytochrome *b* gene were identified leading to a G143A or a G143S amino acid exchange. Resistance of both target-site mutants remained stable during four consecutive disease cycles on inoculated barley leaves in the absence of azoxystrobin. Several parameters tested to measure fitness penalties revealed that the G143A mutant was not compromised. In contrast, the conidial production of the G143S mutant was significantly lower than parental sensitive isolates.

As described in Chap. 21 of this book, QoI resistance has caused a serious problem in the control of blast disease not only on rice in Japan but also on wheat in Brazil. Using two isolates of *M. oryzae* with reduced-sensitivity to orysastrobin, coincidentally isolated from rice, fitness parameters and stability of resistance were examined (Nakamura et al. 2011). These isolates were inferior to sensitive ones in regard to both mycelial growth and appressorrial formation on culture medium. Moreover, they changed to sensitive to QoI fungicides when maintained on fungicide-free medium, which might suggest that QoI-resistant isolates are less fit in the environment. Further studies are required to relate these results to the situation in the field. Stability of azoxystrobin resistance in rice blast fungus has been tested on fungicide-amended and fungicide-unamended PDA plates, where two resistant and two sensitive monoconidial isolates were subcultured every 2 weeks. However, no change has occurred in the response to azoxystrobin after 25 transfers (Ishii unpublished).

Development of azoxystrobin resistance has also been reported in *M. oryzae* from perennial ryegrass in the USA (Ma and Uddin 2009). A fitness comparison experiment using detached perennial ryegrass blades indicated that the disease severity of the wild-type strain was significantly higher than that of the G143A-resistant mutant. When inoculated with three mixed populations of resistant and wild-type strains at different ratios, the production of conidia by the wild-type strain increased and that of the mutant decreased after infection in all three populations tested.

In *Alternaria alternata* tangerine pathotype, stability of resistance to azoxystrobin and pyraclostrobin was determined after consecutive transfers on fungicide-free PDA medium, but QoI sensitivity did not change in both QoI-sensitive and QoI-resistant isolates after 10 transfers with few exceptions, showing that resistance was generally stable in the absence of QoI selection pressure. It was also suggested that resistance development did not affect the fitness of resistant isolates (Vega and Dewdney 2014). In their paper, however, great variability among isolates within the same sensitivity group was found in mycelial growth, conidial production, and conidial germination. Slight variability was also detected in virulence among isolates within the same sensitivity group. Considering the large population size of fungal pathogens and their great variability, a limitation will exist to evaluate and predict fitness of fungicide-resistant strains in field populations.

Most recently, fitness studies have been conducted with *Erysiphe necator* (*Uncinula necator*), the obligate pathogen of grapevine powdery mildew (Rallos et al. 2014). To determine persistence of QoI resistance when QoI fungicides are withdrawn, competition assays were performed on unsprayed grape plants by cycling mixtures of resistant and sensitive isolates characterized as genetically diverse based on microsatellite analyses. Under laboratory conditions, %G143A, determined by quantitative polymerase chain reaction (qPCR), increased significantly, indicating the competitiveness of the resistant fraction. The persistence of resistant populations was documented in a vineyard that withdrew QoI fungicides from its spray program for four consecutive years. QoI-resistant populations with >5 % G143A also harbored Y136F in the *cyp51* gene that also confers resistance to DMI fungicides. Hence, double resistance could have been partly responsible for persistence of QoI resistance (Rallos et al. 2014).

3.5 SDHI Fungicides

New generation of SDHI (succinate dehydrogenase inhibitor) fungicides has been intensively developed in recent years, but field resistance to this class of single-site fungicides has also occurred in several pathogens. Field monitoring for the stability of SDHI resistance is still very limited, but the fluctuation of SDHI resistance has been examined in greenhouse populations of *C. cassiicola* (Ishii 2014). *Corynespora* leaf spot, caused by this fungus, is now one of the most serious diseases on cucumber in Japan. Strains multiple resistant to benzimidazole, dicarboximide, QoI, and SDHI fungicides, such as boscalid, have developed and are widely distributed (Miyamoto et al. 2010). The molecular mechanism of boscalid resistance has been characterized, and very high resistance (VHR) and high resistance (HR) are conferred by a single-point mutation of H278Y and H278R in *sdhB* gene, respectively (Miyamoto et al. 2010). Additionally, other mutations of S73P in *sdhC*, S89P, and G109V in *sdhD* have been detected in not all but some of the moderately boscalid-resistant isolates.

Subsequently, Ishii et al. (2011) found in *C. cassiicola* that isolates of very high resistance (SDHB-H278Y) or high resistance (SDHB-H278R) to boscalid exhibit high sensitivity to a newly developed SDHI fungicide fluopyram. The lack of cross-resistance to fluopyram in VHR and HR isolates is unique and confirmed not only on YBA agar medium [yeast extract 10, peptone 10, sodium acetate 20, and agar 15 g L^{-1} in distilled water (DW)], only slightly modified after Stammler and Speakman (2006), but also on fungicide-sprayed potted cucumber plants. Then monitoring tests have been conducted using commercial cucumber greenhouses to determine the stability of boscalid resistance as well as fluopyram sensitivity in practical conditions (Ishii 2014).

In the spring of 2010, spray applications of SDHI fungicides were stopped and conventional fungicides including iminoctadine albesilate and chlorothalonil have been used alternatively since then. Where a grower switched cucumber cultivars to ones less susceptible (tolerant) to *Corynespora* leaf spot, disease occurrence was not very severe, but infected cucumber leaves were nevertheless sampled every year between 2010 and 2013. Single conidium isolates were obtained and their mycelial growth tested on fungicide-amended YBA agar medium to determine the situation after withdrawal of SDHI fungicide applications. All isolates collected in 2013, 3 years after the withdrawal, still showed very high resistance to boscalid indicating that this type of resistant strains were stable in a greenhouse in the absence of selection pressure by SDHI fungicides (Table 3.1). In contrast, these isolates all exhibited sensitivity to fluopyram which further confirmed the lack of cross-resistance in very highly boscalid-resistant isolates to fluopyram (Table 3.2). Moderately boscalid-resistant isolates cross-resistant to fluopyram have not increased in the greenhouses tested.

Stability, fitness, and competitive ability of pyraclostrobin and boscalid-resistant isolates of *B. cinerea* from apple were investigated (Kim and Xiao 2011). Stability of resistance was determined after consecutive transfers on PDA or being cycled on apple fruit. Resistance to the two fungicides remained at levels similar to that of the

Table 3.1 Monitoring for boscalid resistance of *Corynespora cassiicola* in a cucumber greenhouse after withdrawal of boscalid

	Number of isolates determined[a]				
Time of fungus isolation	VHR	HR	MR	S	Total
May 2010	6	0	2	2	10
June 2011	19	0	0	1	20
July 2012	12	0	0	2	14
June 2013	21	0	0	0	21

[a]*VHR* very highly resistant, *HR* highly resistant, *MR* moderately resistant, *S* sensitive to boscalid

Table 3.2 Monitoring for fluopyram sensitivity of *Corynespora cassiicola* in a cucumber greenhouse after withdrawal of boscalid

	Number of isolates determined[a]				
Time of fungus isolation	VHR	HR	MR	S	Total
May 2010	0	0	2	8	10
June 2011	0	0	0	20	20
July 2012	0	0	0	14	14
June 2013	0	0	0	21	21

[a]*VHR* very highly resistant, *HR* highly resistant, *MR* moderately resistant, *S* sensitive to fluopyram

initial generation after 20 transfers on PDA and five disease cycles on apple fruit. Great variability in individual fitness components tested was observed among isolates within the same phenotype groups either sensitive or resistant to the fungicides. Resistant isolates were as pathogenic and virulent on apple fruit as sensitive isolates. The results suggested that resistance to pyraclostrobin and boscalid was stable in the absence of the fungicides and that resistance did not significantly impair individual fitness components tested. However, both pyraclostrobin- and boscalid-resistant isolates exhibited competitive disadvantage over the dual-sensitive isolate on apple fruit.

Most recently, using *sdhB* mutants constructed by homologous recombination, fitness penalty of SDHI-resistant isolates has been evaluated in *B. cinerea* (Lalève et al. 2014). This approach was based on site-directed mutagenesis and demonstrated that the mutations detected in *sdhB* gene were really responsible for SDHI resistance in this fungus (Lalève et al. 2013). The advantage of using these transformants is that they all have the same genetic background. They differ only in terms of their *sdhB* alleles, making it possible to evaluate the precise correlation between resistance cost and each *sdhB* mutation in the recipient strain. In this research, the *sdhB*^{P225L} mutants were most severely affected for most of the fitness parameters measured but displaying the highest resistance factor for SDHIs. It was further suggested that different *sdhB* mutations have different effects on fitness but an inhibition of SDH activity is not always consistent with fitness defect and mutant frequency in the field. Compensatory system might operate *in natura*.

The *sdhB*^{H272Y} mutant is the most frequent allele in the field, conferring resistance to boscalid, and also is detected in French and German vineyards with the *sdhB*^{H272R} allele (Leroux et al. 2010). As discussed above, Lalève et al. (2014) compared the *sdhB*^{H272Y}, *sdhB*^{H272R}, *sdhB*^{P225L}, and other recombinant mutants. In vitro, they mimicked evolution in the field, by mixing conidia of wild-type (sensitive to boscalid) and mutant transformants (resistant to boscalid) over 7–10 cycles of coculture and estimated the proportions of wild-type and mutated conidia at the end of each cycle (Fig. 3.3). When started with 75 % of boscalid-resistant conidia (*sdhB*^{H272Y}) in the mixture, the ratio of resistant versus sensitive conidia did not significantly change from its initial proportion. The second mixture was composed of 40 % of boscalid-resistant conidia at the starting point (cycle 0), and this proportion reached 70 % after seven cycles. Importantly, the ratio of *sdhB*^{P225L} mutants,

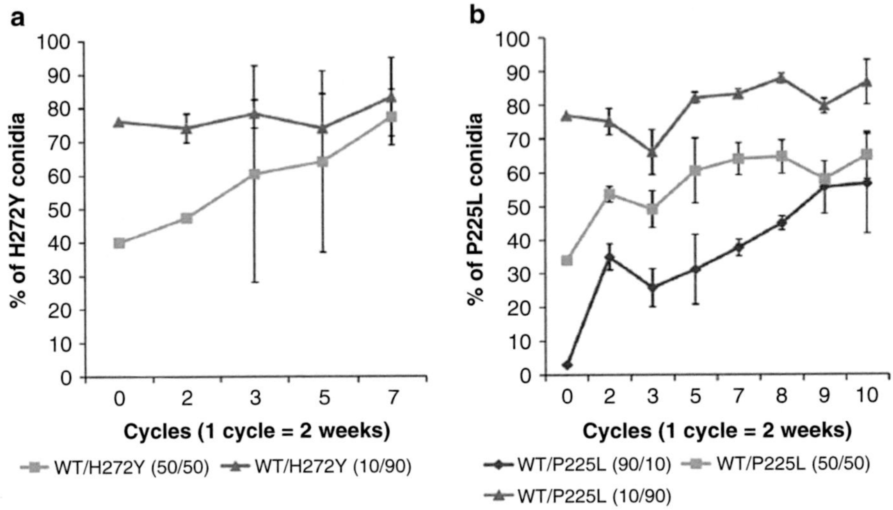

Fig. 3.3 Changes over time in the proportion of resistant (*sdhB*^{H272Y} or *sdhB*^{P225L}) conidia in coculture with wild-type (WT) conidia (From Lalève et al. (2014))

displaying the highest resistance factor for boscalid and fluopyram (Lalève et al. 2013) increased over time. The distribution of *sdhB*^{P225L} mutants is still limited in field populations. However, the concurrent use of boscalid and fluopyram may also lead to the selection of other types of mutants, cross-resistant to these two fungicides and harboring changes at position P225 in particular, with no major effect on fitness, like *sdhB*^{P225F} and *sdhB*^{P225T} (Lalève et al. 2014).

References

Avila-Adame C, Köller W (2003) Characterization of spontaneous mutants of *Magnaporthe grisea* expressing stable resistance to the Qo-inhibiting fungicide azoxystrobin. Curr Genet 42:332–338

Braun PG (1994) Development and decline of a population of *Venturia inaequalis* resistant to sterol-inhibiting fungicides. Norweg J Agric Sci (Suppl 17):173–184

Cox KD, Bryson PK, Schnabel G (2007) Instability of propiconazole resistance and fitness in *Monilinia fructicola*. Phytopathology 97:448–453

De Miccolis Angelini RM, Rotolo C, Masiello M, Pollastro S, Ishii H, Faretra F (2012) Genetic analysis and molecular characterisation of laboratory and field mutants of *Botryotinia fuckeliana* (*Botrytis cinerea*) resistant to QoI fungicides. Pest Manag Sci 68:1231–1240

FRAG-UK (2013) Fungicide resistance management in cereals. http://www.pesticides.gov.uk/Resources/CRD/Migrated-Resources/Documents/F/FRAGCerealsResistanceGuidelinesv6April2013.pdf

Genet JL, Jaworska G, Deparis F (2006) Effect of dose rate and mixtures of fungicides on selection for QoI resistance in populations of *Plasmopara viticola*. Pest Manag Sci 62:88–194

Hasegawa M (2003) Abstr 13th Symp Res Com Fungic Resist. Phytopathol Soc Jpn:9–16 (in Japanese with English abstr)

Ishii H (1997) Resistance management in diseases of top fruit in Japan. Pestic Sci 51:383–386
Ishii H (2010) QoI fungicide resistance: current status and the problems associated with DNA-based monitoring. In: Gisi U, Chet I, Gullino ML (eds) Recent developments in management of plant diseases, plant pathology in the 21st century 1. Springer, Dordrecht, pp 37–45
Ishii H (2011) Resistance of *Venturia nashicola* to benzimidazoles and sterol demethylation inhibitors. In: Thind TS (ed) Fungicide resistance in crop protection. CAB International, Wallingford, pp 21–31
Ishii H (2012) Current situation of QoI and SDHI resistance and guideline for fungicide use. Plant Protect 66:481–487 (in Japanese)
Ishii H (2014) Stability of boscalid resistance and fluopyram sensitivity of *Corynespora cassiicola* strains in commercial cucumber greenhouses. Phytopathology 104(3):S3.53 (Abstr)
Ishii H, Kikuhara K (2007) Occurrence of DMI resistance in *Venturia nashicola*, the scab fungus of Asian pears. Abstr 17th Symp Res Commit Fungic Resist. Phytopathol Soc Jpn:49–60 (in Japanese with English abstr)
Ishii H, Udagawa H, Yanase H, Yamaguchi A (1985) Resistance of *Venturia nashicola* to thiophanate-methyl and benomyl: build-up and decline of resistance in the field. Plant Pathol 34:363–368
Ishii H, Yano K, Date H, Furuta A, Sagehashi Y, Yamaguchi T, Sugiyama T, Nishimura K, Hasama W (2007) Molecular characterization and diagnosis of QoI resistance in cucumber and eggplant fungal pathogens. Phytopathology 97:1458–1466
Ishii H, Miyamoto T, Ushio S, Kakishima M (2011) Lack of cross-resistance to a novel succinate dehydrogenase inhibitor, fluopyram, in highly boscalid-resistant isolates of *Corynespora cassiicola* and *Podosphaera xanthii*. Pest Manag Sci 67:474–482
Ishizaki H, Kohno M, Tsuchida M, Umino M, Hazumi Y, Katoh S, Kunoh H (1983) Occurrence of thiophanate-methyl-resistant strains of the pear scab fungus in pear orchards at Karasu Town, Mie Prefecture. Ann Phytopathol Soc Jpn 49:347–351 (in Japanese with English abstr)
Karaoglanidis G, Thanassoulopoulos C (2002) Phenotypic instability of *Cercospora beticola* Sacc. strains expressing resistance to the sterol demethylation-inhibiting (DMI) fungicide flutriafol after cold exposure. J Phytopathol 150:692–696
Kim YK, Xiao CL (2011) Stability and fitness of pyraclostrobin and boscalid-resistant phenotypes in field isolates of *Botrytis cinerea* from apple. Phytopathology 101:1385–1391
Köller K, Smith FD, Reynolds KL (1991) Phenotypic instability of flusilazole sensitivity in *Venturia inaequalis*. Plant Pathol 40:608–611
Köller W, Smith FD, Reynolds KL, Wilcox WF, Burr JA (1995) Seasonal changes of sensitivities to sterol demethylation inhibitors in *Venturia inaequalis* populations. Mycol Res 99:689–692
Lalève A, Gamet S, Walker AS, Debieu D, Toquin V, Fillinger S (2013) Site-directed mutagenesis of the P225, N230 and H272 residues of succinate dehydrogenase subunit B from *Botrytis cinerea* highlights different roles in enzyme activity and inhibitor binding. Environ Microbiol 16:2253–2266
Lalève A, Fillinger S, Walker AS (2014) Fitness measurement reveals contrasting costs in homologous recombinant mutants of *Botrytis cinerea* resistant to succinate dehydrogenase inhibitors. Fungal Genet Biol 67:24–36
Leroux P, Gredt M, Leroch M, Walker AS (2010) Exploring mechanisms of resistance to respiratory inhibitors in field strains of *Botrytis cinerea*, the causal agent of gray mold. Appl Environ Microbiol 76:6615–6630
Luo CX, Schnabel G (2008) The cytochrome P450 lanosterol 14-alpha demethylase gene is a demethylation inhibitor fungicide resistance determinant in *Monilinia fructicola* field isolates from Georgia. Appl Environ Microbiol 74:359–366
Ma B, Uddin W (2009) Fitness and competitive ability of an azoxystrobin-resistant G143A mutant of *Magnaporthe oryzae* from perennial ryegrass. Plant Dis 93:1044–1049
Marine SC, Schmale DG, Yoder KS (2011) Seasonal distribution of SI fungicide resistance in apple scab in Virginia. Phytopathology 101:S274 (Abstr)

Miyamoto T, Ishii H, Stammler G, Koch A, Ogawara T, Tomita Y, Fountaine JM, Ushio S, Seko T, Kobori S (2010) Distribution and molecular characterization of *Corynespora cassiicola* isolates resistant to boscalid. Plant Pathol 59:873–881

Mukobata H (2004) Occurrence and chemical control of thiophanate-methyl-resistant strains of soybean purple seed stain fungus, *Cercospora kikuchii*. Plant Protect 58:97–101 (in Japanese)

Nakamura N, Miyagawa N, Uchida K, Fuji M, Takeda T (2011) QoI sensitivity of rice blast fungus isolated from Japan pad field. Jpn J Phytopathol 77:163 (Abstr in Japanese)

Nikou D, Malandrakis A, Konstantakaki M, Vontas J, Markoglou A, Ziogas B (2009) Molecular characterization and detection of overexpressed C-14 alpha-demethylase-based DMI resistance in *Cercospora beticola* field isolates. Pestic Biochem Physiol 95:18–27

Nonaka T (1984) Tolerance of tea anthracnose fungus, *Gloeosporium theae-sinensis*, to thiophanate-methyl and benomyl fungicides. (4) Fluctuations in the occurrence of resistant fungi under restricted use of fungicides. Proc Assoc Pl Prot Kyushu 30:73–76 (in Japanese)

Pringle A, Taylor JW (2002) The fitness of filamentous fungi. Trends Microbiol 10:474–481

Rallos LEE, Johnson NG, Schmale G III, Prussin AJ II, Baudoin AB (2014) Fitness of *Erysiphe necator* with G143A-based resistance to quinone outside inhibitors. Plant Dis 98:1494–1502

Schwabe WFS (1979) Resistance of the apple scab fungus (*Venturia inaequalis*) to benzimidazole fungicides. Deciduous Fruit Grower 29:418–422

Stammler G, Speakman J (2006) Microtiter method to test the sensitivity of *Botrytis cinerea* to boscalid. J Phytopathol 154:508–510

Vega B, Dewdney MM (2014) QoI-resistance stability in relation to pathogenic and saprophytic fitness components of *Alternaria alternata* from citrus. Phytopathology 98:1371–1378

Waker AS, Micoud A, Rémuson F, Grosman J, Gredt M, Leroux P (2013) French vineyards provide information that opens ways for effective resistance management of *Botrytis cinerea* (grey mould). Pest Manag Sci 69:667–678

Zhu F, Bryson PK, Schnabel G (2012) Influence of storage approaches on instability of propiconazole resistance in *Monilinia fructicola*. Pest Manag Sci 68:1003–1009

Chapter 4
The Use of Mathematical Models to Guide Fungicide Resistance Management Decisions

Frank van den Bosch, Bart Fraaije, Richard Oliver, Femke van den Berg, and Neil Paveley

Abstract Historically, models have had little influence on decision-making in fungicide resistance management. The reasons are found in the level of abstraction of these models making it difficult for stakeholders to interpret them and inadequate connection between modelling and experimentation. Recently, however, the authors of this chapter have developed models in close collaboration with stakeholders and experimenters. These models range from simple functions representing governing principles of resistance evolution to complex models for quantitative studies. In this chapter we discuss the development and testing of these models. A governing principle is discussed predicting whether a change in a fungicide application programme will increase or decrease the rate of selection for fungicide resistance. Complex models are discussed that reflect sufficient biological detail to study specific plant-pathogen-fungicide combinations. Ultimately we describe the combined experimental and modelling work that is currently undertaken to informing resistance management methods.

Keywords Resistance management • Governing principle • Strategy • Tactic • Selection coefficient • Exposure time • Dose • Mixture • Alternation • Spray timing

---

F. van den Bosch (✉) • B. Fraaije • F. van den Berg
Rothamsted Research, Harpenden AL5 2JQ, UK
e-mail: frank.vandenbosch@rothamsted.ac.uk

R. Oliver
Environment & Agriculture, Centre for Crop and Disease Management (CCDM), Curtin University, Bentley, WA 6102, Australia

N. Paveley
ADAS High Mowthorpe, Duggleby YO17 8BP, UK

H. Ishii, D.W. Hollomon (eds.), *Fungicide Resistance in Plant Pathogens*, DOI 10.1007/978-4-431-55642-8_4

4.1 Introduction

A range of models has been published, both in the plant pathology and the biomathematics literature, evaluating fungicide resistance management methods (reviewed in: van den Bosch and Gilligan 2008). These models have had little influence on decisions taken on practical issues around fungicide resistance management. The reasons for this are (in our view) fourfold:

1. Published models were used to compare different fungicide treatments (e.g. using fungicides as mixtures compared to solo use) which differed substantially in their disease control efficacy or which would not provide acceptable efficacy. Hence, the comparisons were of limited practical relevance.

   Related to this, most models considered a measure of success of a resistance management method that is of limited practical value. Most model studies consider the frequency of the resistance in the pathogen population. The measure of success then is defined as the time it takes to build up a frequency of resistance to a preset level. The longer it takes for the resistance to increase to this level, the more effective the resistance management method is. This criterion of success however ignores that fungicides are used to control disease and that any resistance management method that does not provide effective disease control is of no practical value.
2. Published models are often quite abstract. The models had strongly simplified biology, so that model parameters could not be interpreted easily by plant pathologists. Also the level of abstraction made it difficult to test model predictions against experimental data. Although there is nothing against abstract models, and many of the model predictions do still hold up against recent evidence, it is plant pathologists and others working in the area of resistance management policy that have to judge the relevance of a study. If it is difficult or impossible to see that the relevant biology of the interactions between crop pathogen and fungicides is correctly represented in a model, the mosel results will not be convincing to those taking practical decisions on resistance management.
3. Field data for testing models were scarce or not available until recently, due to the time and costs of developing such data sets. Models were therefore published largely untested. Only recently has a fully detailed test of a fungicide resistance model been published (Hobbelen et al. 2011a).
4. The development of the models studied was often done in isolation, not consulting adequately with the groups of people that would ultimately use the information and guidance resulting from the modelling exercise.

The poor connection between modelling, experimentation and practice means that the potential value of modelling to help guide resistance management decisions has not been exploited. This is a waste, because experiments can only study a limited number of the great range of possible management tactics and (generally) only measure selection within one season. Models enable comparisons of many tactics over many seasons. They can also be used to assess management strategies when selection is occurring concurrently against more than one mode of action (for which there is very little experimental data).

The authors of this chapter have, over the past 10 years, worked on models that are well rooted in the biology of a range of patho-systems, have been developed in close interaction with those people that will ultimately use the outcomes and are validated with field experiments (Hobbelen et al. 2011a, b, 2013, 2014; van den Bosch et al. 2011, 2014; van den Berg et al. 2013). Findings from the work are now informing and influencing practical decision-making. This chapter will discuss these recent modelling approaches and their integration with field experimental work on fungicide resistance management.

Mathematical models of biological systems range from very complex, incorporating large ranges of detailed knowledge of the system, to very simplified generic models. There is a lively debate among modellers which type of model is the most useful. We will here largely ignore this debate but simply conclude that it depends on the question to be answered whether a detailed complex model or a simple generic model is needed. Or as Einstein phrased it: 'Everything should be made as simple as possible, but not simpler'. We will start this chapter with simple and end the chapter with complex models. These approaches are tailored to the questions asked. The chapter will finish with a discussion of a current project aiming at developing advise for the management of possible resistances that may develop against the succinate dehydrogenase inhibitor (SDHI) fungicides and the use of models in that research.

List of Terms Used in the Chapter

Fungicide resistance management methods	We use method rather than strategy so that we can use the words strategy and tactic in their exact and specific meaning in the chapter.
Principle	A general rule that has a range of applications across a wide field. A principle of fungicide resistance management is a rule that applies to a wide range of specific resistance management methods.
Strategy	The "what" aspect of resistance management. What do we want to achieve with resistance management?
Tactics	The "how" aspects of resistance management. How do we achieve our strategy of resistance management?
Emergence	When a new fungicide mode of action is introduced and the population is uniformly sensitive, resistance has to arise through mutations in sensitive strain(s). Emergence occurs when the resistant population increases so that it is large enough not to die out because of demographic stochasticity.

Selection	The process that determines the relative contribution of a genotype to the next generation. When a fungicide is used and a resistant genotype is present in the population, its proportional contribution to the next pathogen generation is larger relative to that of the fungicide-sensitive strain(s) in the population.
Effective disease control	The disease control of a sufficient level to keep yield losses below a required level.
Fungicide effective life	The time from introduction of the fungicide to the point at which effective control can no longer be obtained.

4.2 Governing Principles

It would be useful in the development of anti-resistance methods if there was a governing principle to guide whether a proposed change in a fungicide application programme increases or decreases selection for fungicide resistance. Based on the work by Milgroom and Fry (1988) and Milgroom et al. (1989), Van den Bosch et al. (2014) recently published such a governing principle with surprisingly strong predictive power.

The Governing Principle When a fungicide is used, it reduces the growth rate of the population of fungicide-sensitive pathogen strains. That is basically the reason to use a fungicide. Any strains present in the population that are resistant/insensitive to the fungicide will be unaffected or less affected by the fungicide and therefore have a higher growth rate than the sensitive strains. The result of this is that the population size of the fungicide resistant strain increases faster than that of the sensitive strains, resulting in the resistant strain increasing in frequency in the population. That is exactly what selection is about, a phenotype increasing in frequency. We can thus measure the selection for fungicide resistance by the difference in population growth rate of the sensitive and the resistant pathogen strains. We formalise the selection coefficient, s, as $s = (r_R - r_S)$, where r_R and r_S are the population growth rate of the resistant and the sensitive strain, respectively (Crow and Kimura 1970).

Staub and Sozzi (1983) and Dekker (1986) recognised that the amount of time a fungal population is exposed, T, to a fungicide also determines the extent to which the frequency of resistance increases. When a fungicide is applied, the dose of the fungicide on or in the leaf increases sharply. After the application the fungicide dose decreases gradually due to plant metabolism and effects of weather such as UV exposure, rain, etc. until the dose becomes too low to be biologically active. The amount of time, T, a fungicide poses a selection pressure thus is limited. Combining this with the selection coefficient, we find that a measure of selection for resistance of a fungicide application programme is

$$sT = (r_R - r_S)T \tag{4.1}$$

We refer the reader to van den Bosch et al. (2014) for a more formal mathematical derivation of the selection coefficient and exposure time principle.

The Fungicide Resistance Management Strategies From this governing principle we can derive three basic strategies that reduce the selection for fungicide resistance in an application programme. All anti-resistance measures fit into one or more of these strategies.

Strategy 1: Selection for fungicide resistance is reduced when the per capita rate of increase of both the sensitive and resistant strains (r_R and r_S, respectively) are reduced.

Strategy 2: Selection for fungicide resistance is reduced when the per capita rate of increase of the resistant strain (r_R) is reduced relative to that of the sensitive strain (r_S).

Strategy 3: Selection for fungicide resistance is reduced when the time span over which selection takes place (exposure time) is reduced.

These three simple strategies can help guide the development of fungicide application programmes that minimise the selection for fungicide resistance.

Putting the Strategies to Work: The Fungicide Resistance Management Tactics The governing principle can be used to predict the effect of a change in an application programme on the selection for fungicide resistance. The predicted effects have been compared to published data (van den Bosch et al. 2014). We will discuss some specific examples here, but most of the results will be included in the Chap. 5 'Evidence-Based Resistance Management'.

Consider a situation where resistance is developing against a fungicide A which is used in an application programme where the solo product is used twice a year. Now change this application programme by adding a mixing partner B with a different mode of action to fungicide A and we do not change the dose of fungicide A. Does this change increase or decrease the selection for fungicide resistance? Pathogen strains resistant to fungicide A are expected to be sensitive to fungicide B. Hence, the mixing partner affects both the growth rate of the strains sensitive to A and resistant to A in the same way. This is a clear case of Strategy 1, and we expect this change in application programme to decrease selection for resistance.

We have found 34 papers where selection for resistance of a treatment programme with a solo product was compared with a mixture programme where a mixing component was added without reducing the dose of the fungicide of which resistance was measured. In these papers a total of 51 pathogen-fungicide-mixing partner combinations were tested. In 44 of these combinations, the mixture treatment programme showed a lower rate of selection for fungicide resistance. Only in two cases a larger selection was measured. These two cases could subsequently be related to cross resistance to the fungicides (van den Bosch et al. 2014 for further detail).

As a second example consider a spray programme with four applications with a solo product A each growing season. Consider a change in the treatment programme where the second and the fourth application of A are replaced by a fungicide B with a different mode of action. The change in application programme causes the pathogen population to be exposed to the fungicide A for half as long as in the original treatment programme. This is thus an example of strategy 3, and we expect this change in treatment programme to reduce the selection for resistance.

We have found 11 publications where these spray programmes are compared, with a total of 15 pathogen-fungicide combinations. In 12 cases the change in spray programme reduced the selection for fungicide resistance, as predicted by the governing principle. In the other three cases no difference between the two treatment programmes was measured.

Van den Bosch et al. (2014) describe another seven changes in treatment programme, use the governing principle to predict the direction of change of the selection, and compare this prediction with published evidence. In 84 % of the cases, the data agree with the prediction. In only 5 % of the cases do the data contradict the predictions. This simple governing principle thus shows to be a robust predictor and can thus help select spray programmes that put a minimal selection pressure on a pathogen to develop resistance.

The governing principle is simple in form and simple to use. It is however a qualitative instrument only. It gives an accurate prediction of whether a change in an application programme will increase or decrease selection. It does not tell us whether the change in selection is large enough to be of relevance in practice. It also does not tell us whether the change in application programme is feasible in practice, i.e. whether the changed application programme actually provides effective disease control, which is the reason to apply the fungicide in the first place. To study whether a change in treatment programme reduces selection to an extent that is of relevance in practice, as well as continuing to provide effective disease control, we will need to use more detailed models. This will be the topic in the next section.

4.3 From Qualitative to Quantitative Modelling

This section will deal with developing and validating models that are sufficiently realistic to provide quantitative guidance on the development of resistance management tactics. As discussed in the introduction, such models need to be developed in close interaction with field studies as well as in close interactions with stakeholders, and we will start with discussing this.

Developing the Research, Stakeholder Engagement and Advice Cycle 'Stakeholder engagement' is often seen as a tick-box exercise in research projects. Our experience is that involving policymakers and industry from the start means that the research is more likely to develop insight relevant to policymakers and industry, gives policymakers and industry the opportunity to engage with the work and significantly increases the uptake of the messages even if the messages are not always popular with the stakeholders (see Chap. 5).

Figure 4.1 shows the cycle of field experiments helping to develop modelling approaches and modelling approaches developing ideas to be tested in the field. From this research cycle, both fundamental insight and practical advice are derived. The outcomes of the work can subsequently be used by industry and regulatory stakeholders to develop fungicide resistance management policies.

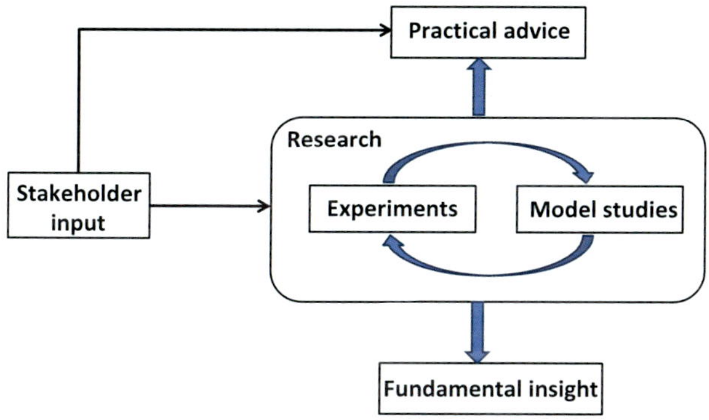

Fig. 4.1 The research cycle of experimentation informing models and models suggesting resistance management strategies to be tested in the field. The research is developed in close discussions with stakeholders. The researchers and stakeholders then use the outcomes to formulate practical advice on fungicide resistance management methods

How to Measure the Success of a Resistance Management Method The way the success of a resistance management method is measured will affect the assessment of the usefulness of a resistance management method. For example, we could measure success as the time it takes for the frequency of the resistance in the pathogen population to increase to a preset level, say 50 %. The spray programme with the longest time till 50 % resistance is then considered to be the best resistance management method. However, this can be misleading because a key aspect of the use of fungicides, to provide effective control of a pathogen population, is not included in this measure of success. For example, applying this measure of success to the dose of a fungicide, it is immediately clear that dose zero, not using the fungicide, is the optimal fungicide resistance management method as no resistance will develop.

In general, fungicide programmes that are less effective in their control of the pathogen usually cause less selection. So if the only metric of success is selection, models are prone to identifying ineffective treatments as being optimal for resistance management. We thus need to define a measure of success that incorporates the fact that fungicides are used to control disease and that therefore any resistance management methods that are being compared must provide effective control. We have used the 'effective life', which is defined as the time from introduction of the fungicide to the time where effective disease control is lost due to the development of resistance, as a measure of success (van den Bosch and Gilligan 2008; Hobbelen et al. 2011a, b).

Model Development The model structure is shown in Fig. 4.2a. The crop canopy is represented by growth and senescence of leaf material through the season. Spores of the pathogen can land on a leaf and cause an infection. The infection causes a lesion to develop and progresses through the latent stage to become infectious, producing spores that can cause new infections. The end of the infectious period is marked by the cessation of spore production. For each pathogen strain in the population, the model keeps track of spore production and the density of latent and infectious lesions.

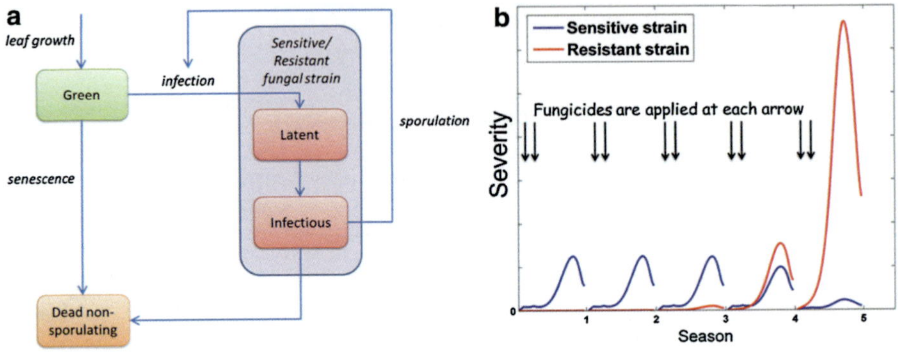

Fig. 4.2 *Panel a*: the structure of the model, which describes the selection for insensitive pathogen strains in a sensitive population in response to applications of fungicide. For each pathogen strain in the population, there is one pathogen box with latent and infectious lesions. *Panel b*: an example output is shown, where the frequency of the resistant strain is very small initially. In a series of seasons, the resistant strain builds up a large population and finally outcompetes the subpopulation of sensitive strains

Application of a fungicide affects life-cycle parameters of the pathogen by: (1) reducing the proportion of pathogen spores which infect the plant (infection efficiency), (2) increasing the time from infection to spore production (the latent period) and (3) reducing the sporulation rate or duration (the infectious period). The extent to which these pathogen life-cycle parameters are affected by the fungicide depends both on the type of fungicide and its application dosage. A derivation of the model equations can be found in Hobbelen et al. (2011a, b) and van den Berg et al. (2013).

The model is parameterised on the basis of published data. We have mainly studied the case of *Zymoseptoria tritici* (*Mycosphaerella graminicola*) on wheat. Parameter values for crop growth, spore production, latent and infectious period and infection efficiency are available in the literature or can be derived from published data.

Figure 4.2b shows a typical output of the model, in this case parameterised for the use of a full dose of a QoI fungicide in two sprays (T1 and T2) during a growing season. The figure shows a gradual build-up of the density of the resistant strain in the first 3 years, due to its selective advantage over the sensitive strain. In the following years the resistant strain outcompetes the sensitive strain and takes over the entire population. Because the resistant strain is not influenced by the fungicide, the population size develops into much larger values than when the sensitive strain dominated the population. Clearly the effective control of the pathogen is lost, in this specific example, in year 5. The effective life of this fungicide thus is 4 years. (A more formal definition of effective life can be found in van den Bosch and Gilligan (2008) and Hobbelen et al. (2011a, b).)

Model Validation The model was validated using data of a field study of the development of resistance to the QoI fungicide azoxystrobin of the barley powdery mildew pathogen *Blumeria graminis* f.sp. *hordei* (Fraaije et al. 2006; Hobbelen et al 2011a). Resistance was determined by measuring the frequency of the mutation resulting in the replacement of glycine by alanine at codon 143 (G143A) of the mitochondrially encoded cytochrome *b* QoI target protein in DNA samples of infected leaves using allele-specific real-time PCR (Fraaije et al. 2006). The model was parameterised using a combination of published data and data sets available from previous research by the authors. The experiment consisted of treatment programmes with one spray, with two sprays and with three sprays during a growing season. The total dose of the spray programme was 1, 2 and 3 full dosages. Using real-time PCR the frequency of the resistant strain was estimated at several time points during the experiment. From the data the selection ratio, SR, was calculated as

$$SR = \frac{\text{Frequency of the resistant strain at the end of the experiment}}{\text{Frequency of the resistant strain at the start of the experiment}}$$

Model predictions were compared to field results. An example of one site year is shown in Fig. 4.3. Generally, model predictions closely agreed with field measurements. Although further validation studies are needed, we can conclude that the model seems to be an accurate instrument for predicting the development of fungicide resistance. Clearly a model test using experimental measurement of effective life estimates is not feasible due to the difficulties of experimentation over a long sequence of years.

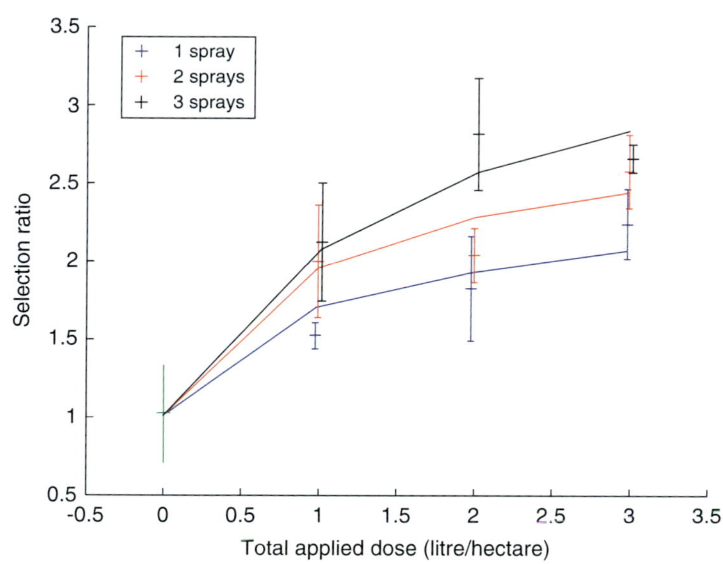

Fig. 4.3 Example of results of the model validation. Observed and predicted response of the selection ratio to variation in total applied dose and number of sprays. *Vertical bars* indicate 95 % confidence intervals of observed selection ratios. *Lines* indicate model predictions for the one-, two- and three-spray programmes. Data are obtained from mildew-infected barley leaf samples collected at the ADAS experimental site in Terrington (Norfolk, UK) in 2003 (see Fraaije et al. 2006)

Table 4.1 The selection coefficient for resistance and the effective life of azoxystrobin (the at-risk fungicide) in a range of mixtures with chlorothalonil

	Selection coefficient			Effective life		
Dose of mixing partner	Dose of at risk, 25 %	Dose of at risk, 50 %	Dose of at risk, 100 %	Dose of at risk, 25 %	Dose of at risk, 50 %	Dose of at risk, 100 %
0	0.70	0.84	0.96	6	5	5
20	0.50	0.61	0.72	9	8	7
40	0.45	0.55	0.65	11	9	8
60	0.42	0.52	0.62	12	10	9
80	0.41	0.50	0.61	13	11	9
100	0.41	0.49	0.59	13	11	9

The calculations were done with the model of van den Berg et al. (2013) parameterised for *Z. tritici* on wheat. Exponential curves were fitted to the resistance frequency at harvest, and the parameter then is the selection coefficient in units of per year. The effective life has units of years.

Some Key Findings of the Model Study The model was used to study a range of anti-resistance methods including application dose, alternations, mixtures, application timing and application number. We refer the reader for more detail to the publications (Hobbelen et al. 2011a, b, 2013, 2014; van den Bosch et al. 2011, 2014; van den Berg et al. 2013) and Chap. 5 on 'Evidence-Based Resistance Management'. We will restrict the examples considered here to the case of mixtures, as also discussed for the governing principle.

The model was used to study the selection coefficient and the effective life of a QoI fungicide in mixtures with chlorothalonil. As shown in the discussion on the governing principle, such a mixture will reduce the selection for fungicide resistance. The key question then is whether this reduction in selection is large enough to increase the effective life of the fungicide. Table 4.1 summarises a subset of the results. The table shows that mixing can significantly increase the effective life of an at-risk fungicide, even without reducing the dose of the at-risk fungicide.

Furthermore, the table and further results not shown here show that the larger the doses of the mixing partner, the longer the effective life of the at-risk fungicide will last. We note here that this uses a fungicide (chlorothalonil is a multisite inhibitor) as a mixing partner to which no resistance is developing. For the evidence on mixtures of two at-risk fungicides, see Chap. 5 on 'Evidence-Based Resistance Management'.

4.4 An Example of the Use of the Model and Field Experiments to Develop Practical Advice

We have come full circle around the research cycle of modelling and experimentation and are now using the model to generate hypotheses and promising resistance management tactics that can be field tested.

In the project "Improved tools to rationalise and support stewardship programmes for SDHI fungicides to control cereal diseases in the UK" (funded by DEFRA,

BASF, Bayer, DuPont, Syngenta and HGCA), we have tested hypotheses developed from the modelling studies. Being introduced onto the market for use in cereals, the SDHI will need to be supported by a resistance management method to maximise the effective life of this group of chemicals.

At the time of writing, although a small number of insensitive mutants have been found in Europe, no field resistance to the SDHIs has developed in the pathogen *Z. tritici*, the most important pathogen in UK wheat. We therefore, in our experiments, used SDHIs as mixing partners to study the development of azole resistance. The azole fungicides are a key chemical for the control of *Z. tritici* by interfering with the sterol biosynthesis pathway. But their effectiveness in the UK is eroding due to multiple alterations in the transcriptional and/or coding regions of the sterol 14α-demethylase encoding *CYP51* gene (Cools and Fraaije 2008; Cools and Fraaije 2013). In a series of field experiments at three sites over three seasons, we have tested the following hypotheses derived from the modelling work:

Hypothesis 1: Using SDHIs as a mixing partner will reduce the selection for reduced sensitivity to the azole fungicides. The larger the SDHI dose in the mixture, the smaller the rate of selection for reduced azole sensitivity.

Hypothesis 2: There will be an additional effect on selection for azole resistance, from reducing the dose of the azole component in the mixture. The size and direction of the effect depend on the direction and strength of the effect of dose on selection.

The experiments consisted of sprays with mixtures of prochloraz (PZ) and the SDHI isopyrazam (IZM) using a range of dose combinations (Fig. 4.4). The frequency of a mutation resulting in the CYP51 amino acid substitution valine to alanine (V136A), reducing the sensitivity of *Z. tritici* to prochloraz, was monitored using a SNP detection pyrosequencing assay as described by Fraaije et al. (2011). The prochloraz effect on V136A was selected as the experimental system because: (1) the frequency of the mutation in the natural *Z. tritici* population is at a level which allows treatment effects to be quantified readily and (2) the active substance is moderately effective, thus generating good treatment effects but still allowing sufficient disease expression to enable posttreatment sampling. Data from the 2012 experiment at the ADAS site Rosemaund (Herefordshire, UK) are shown in Fig. 4.4 (the full data set will be published (Fraaije et al. in preparation)). From the data presented in Fig. 4.4, we can conclude that model predictions were good. Specifically:

- Increasing the dose of isopyrazam in the mixture and keeping the dose of prochloraz constant (1 full dose), selection for the V136A mutation decreased with increasing dose. Reducing the dose of prochloraz while increasing the IZM dose (so that a constant level of disease control is maintained through the range of treatments), a reduction of the prochloraz dose helps reduce selection.
- Comparing the PZ1-IZM1 with the PZ0.25-IZM1 treatment, the additional benefit of reducing the prochloraz dose for alleviating selection for the V136A mutation is quite small. This follows from the generally weak effect of prochloraz dose on selection seen experimentally.

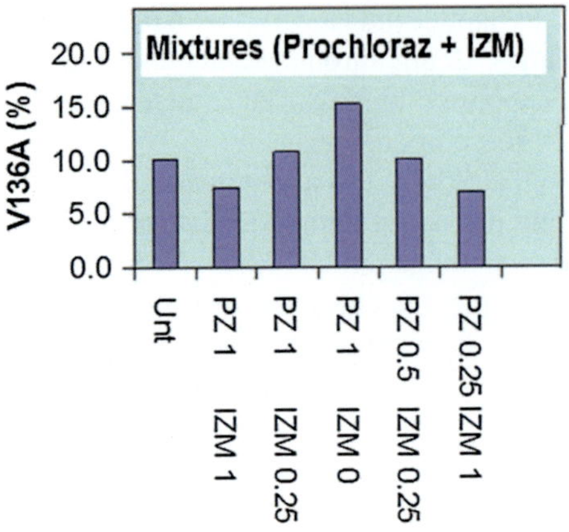

Fig. 4.4 The frequency of strains carrying CYP51 V136A in sampled *Z. tritici* leaf populations after various treatments with mixtures of prochloraz (Poraz 450; 450 g L^{-1} prochloraz (BASF)) and isopyrazam (straight product 125 g L^{-1} supplied by Syngenta). Unt, untreated plot; PZ1 IZM1, mixture contains one full dose of prochloraz (0.9 L ha^{-1} Poraz) and one full dose of isopyrazam (1.0 L ha^{-1}); PZ1 IZM 0.25, mixture contains a full dose of prochloraz and a quarter dose of isopyrazam; PZ1 IZM0, full dose of solo prochloraz; PZ0.5 IZM0.25, mixture contains a half dose of prochloraz and a quarter dose of isopyrazam; PZ0.25 IZM1, mixture contains a quarter dose of prochloraz and a full dose of isopyrazam. Data are obtained from *Septoria*-infected wheat leaf samples collected at the ADAS experimental site in Rosemaund (Herefordshire, UK) in 2012

In the experiments we have also tracked the selection of three more key CYP51 mutations by a range of azole active substances. Moreover, the development of the sensitivity phenotype of the population has been tracked. All site years show patterns of treatment effects which are in agreement with qualitative predictions from modelling.

Thinking this through for the development of an SDHI resistance management programme, we can conclude that if our findings also hold true for the most commonly used azoles, epoxiconazole and prothioconazole, then mixtures of SDHIs with robust doses of azoles will help increase the effective life of the SDHI fungicides with little deleterious effect on the effective life of the azole component in the mixture. We have thus used hypotheses generated by the modelling work, to select resistance management tactics to be tested out in field experiments. This is now leading to the development of advice on SDHI and azole resistance management.

4.5 Discussion

In this chapter we have discussed recent developments of models for fungicide resistance. The models have led to simple governing principles that predict whether a change in a spray programme will increase or decrease selection for resistance. Such simple, rule-of-thumb models are useful to separate tactics for resistance management that are worth field-testing from those that are likely not to be effective.

We also have discussed recent development of more complex models aimed at studying resistance management tactics and their effect on the effective life of the fungicides in the application programme. This work has led to the application to the specific example of resistance management methods to protect the recently introduced new generation of broad-spectrum SDHI fungicides. The preliminary key conclusion from the combined modelling and fieldwork is that SDHI mixtures with robust doses of azole fungicides present a suitable anti-resistance strategy for the SDHIs without putting a large selection pressure for azole resistance onto *Z. tritici.*

We have developed approaches to calculate the effectiveness of fungicide anti-resistance strategies (e.g. use of alternation, mixtures, choice of dose, etc.). This modelling work is strongly interlinked with field experimental studies. We have used data sets from past funded fungicide resistance research for model testing. Currently, hypotheses from the modelling are being tested in an SDHI resistance management project. We have thus developed the cycle where experimental data inform modelling, and models suggest the resistance management strategies that are good candidates to be tested and implemented in the field.

The more complex models were all built around cereal diseases, where there are usually between one and four treatments in a season. For many diseases many more fungicide treatments are the usual practice (e.g. potato blight, banana sigatoka, apple scab and powdery and downy mildew on cucurbits). We are currently extending the models to such crops, for which the possible options of mixtures and alternation are more numerous, and the opportunities for using forecasting to target treatments are well established. The experience gained in cereals suggests that resistance modelling should play a valuable role in identifying good anti-resistance strategies for multi-spray crops.

Acknowledgements Rothamsted Research receives support from the Biological and Biotechnological Sciences Research Council (BBSRC) of the UK. Parts of this work were funded by the Chemicals Regulation Directorate (CRD) and Department of Environment, Food and Rural Affairs (Defra) of the UK and the Grains Research and Development Corporation (GRDC) of Australia. Part of this work was supported by Defra through the Sustainable Arable LINK programme, project LK09133, in collaboration with HGCA, BASF, Bayer CropScience, DuPont, Syngenta, Velcourt, SRUC and ADAS.

References

Cools HJ, Fraaije BA (2008) Are azole fungicides losing ground against Septoria wheat disease? Resistance mechanisms in *Mycosphaerella graminicola*. Pest Manag Sci 64:681–684

Cools HJ, Fraaije BA (2013) Update on mechanisms of azole resistance in *Mycosphaerella graminicola* and implications for future control. Pest Manag Sci 69:150–155

Crow JF, Kimura M (1970) An introduction to population genetic theory. Harper & Row, New York

Dekker J (1986) Preventing and managing fungicide resistance. In: Committee on Strategies Management of Pesticide Resistant Pest Population, National Research Council (eds) Pesticide resistance: strategies and tactics for management. National Academic Press, Washington, DC, pp 347–354

Fraaije BA, Burnett FJ, Clark WS, Lucas JA (2006) Development and field testing of fungicide anti-resistance strategies, with particular reference to Strobilurin QoI Group of fungicides. Home-Grown Cereals Auth, London, 112 pp

Fraaije BA, Burnett FJ, Cools HJ, Clark WS, Paveley N, Norman K, Lucas JA (2011) Understanding evolution and selection of azole resistance mechanisms in UK populations of *Mycosphaerella graminicola*. HGCA Project Report 475, 111 pages

Hobbelen PHF, Fraaije B, Lucas JA, Paveley ND, van den Bosch F (2011a) Derivation and testing of a model to predict selection for fungicide resistance. Plant Pathol 60:304–313

Hobbelen PHF, Paveley ND, van den Bosch F (2011b) Delaying selection for fungicide insensitivity by mixing fungicides at a low and high risk of resistance development: a modelling analysis. Phytopathology 101:1224–1233

Hobbelen PHF, Paveley ND, van den Bosch F (2013) The value of alternation or mixtures of fungicides for delaying the selection of resistance against two modes of action in populations of *Mycosphaerella graminicola* on winter wheat. Phytopathology 101:690–707

Hobbelen PHF, Paveley ND, van den Bosch F (2014) The emergence of resistance to fungicides. PLoS ONE 9, e91910. doi:10.1371/journal.pone.0091910

Milgroom MG, Fry WE (1988) A simulation analysis of the epidemiological principles for fungicide resistance management in pathogen populations. Phytopathology 78:565–570

Milgroom MG, Levin SA, Fry WE (1989) Population genetics theory and fungicide resistance. In: Leonard KJ, Fry WE (eds) Plant disease epidemiology 2. Genetics, resistance and management. McGraw Hill, New York, pp 340–367

Staub T, Sozzi D (1983) Recent practical experiences with fungicide resistance. In: Proceedings of the international congress on plant protection, 10th. BCPC Publication, Alton, UK, pp 591–598

van den Berg F, van den Bosch F, Paveley ND (2013) Optimal fungicide application timings for disease control are also an effective anti-resistance strategy: a case study for *Zymoseptoria tritici* (*Mycosphaerella graminicola*) on wheat. Phytopathology 103:1209–19

van den Bosch F, Gilligan CA (2008) Models of fungicide resistance. Annu Rev Phytopathol 46:123–47

van den Bosch F, Paveley ND, Shaw MW, Hobbelen P, Oliver R (2011) The dose rate debate: does the risk of fungicide resistance increase or decrease with dose? Plant Pathol 60:597606

van den Bosch F, Oliver R, van den Berg F, Paveley N (2014) Governing principles can guide fungicide resistance management tactics. Annu Rev Phytopathol, in press

Chapter 5
Evidence-Based Resistance Management: A Review of Existing Evidence

Frank van den Bosch, Neil Paveley, Bart Fraaije, Femke van den Berg, and Richard Oliver

Abstract The control of fungal plant pathogens has been characterised by repeated cycles of introduction of new fungicides and for many of them a subsequent loss of efficacy due to the emergence and selection of resistant pathogen strains. Several strategies have been proposed to prevent, or at least delay, resistance problems. Such resistance management strategies should be based on evidence interpreted within a sound experimental and theoretical framework. Industry and regulatory decisions about fungicide resistance management often cannot wait for the accumulation of new evidence, so decisions should be taken by weighing the existing evidence. In discussions on resistance management, it is often not explicit what the evidence is. In this chapter, we review experimental and modelling evidence on (1) the choice of application dose, (2) the number of applications, (3) the use of fungicide mixtures, (4) the use of fungicide alternation and (5) protectant versus curative fungicide application. At several places in the text, we stress that resistance management should not compromise effective disease control.

Keywords Evidence • Resistance management • Fungicides • Mixture • Alternation • Dose • Spray timing • Protective • Curative

---

F. van den Bosch (✉) • B. Fraaije • F. van den Berg
Rothamsted Research, Harpenden AL5 2JQ, UK
e-mail: frank.vandenbosch@rothamsted.ac.uk

N. Paveley
ADAS High Mowthorpe, Duggleby YO17 8BP, UK

R. Oliver
Environment & Agriculture, Centre for Crop and Disease Management (CCDM), Curtin University, Bentley, WA 6102, Australia

H. Ishii, D.W. Hollomon (eds.), *Fungicide Resistance in Plant Pathogens*,
DOI 10.1007/978-4-431-55642-8_5

5.1 Introduction

The use of fungicides as crop protection agents is characterised by a cycle of introduction of a fungicide mode of action followed, in many cases, by the build-up of pathogen strains that are resistant or less sensitive to the fungicide and growers eventually abandoning use of the fungicide because it no longer provides effective disease control. Resistance management methods have been suggested ranging from (1) the management of application dose, (2) the management of the number of sprays, (3) the use of fungicide mixtures, (4) the use of fungicide alternation and (5) the avoidance of curative use of fungicides.

The choice of fungicide resistance management methods should be based on evidence. Ideally, the evidence should be used to develop a framework within which each resistance management method can be assessed for its ability to reduce the selection for resistance posed by the fungicide. The authors of this chapter have developed such a framework and have, in a series of reviews, shown that the framework makes predictions that are consistently in close agreement with published evidence (van den Bosch and Gilligan 2008; van den Bosch et al. 2011, 2014a, b).

In this chapter, we summarise the findings in the reviews. We will structure the chapter around the key resistance management methods listed above. For definitions of any unfamiliar words used in this chapter, we refer to the Chap. 4. We will use the "governing principle" discussed in that chapter, so we suggest reading that section first. Instead of repeating the extensive reference lists from the reviews, we refer the reader to these reviews for details on the references in several places.

Important Note Resistance management is only one of the aspects to be considered when developing a fungicide application programme. Fungicides are used to control plant disease, so any resistance management strategy of practical relevance needs to provide effective disease control. It is easy to find ineffective fungicide programmes that minimise selection, but such programmes are of no practical use.

5.2 Practical Use of the Existing Evidence

In advocating evidence-based resistance management, the authors recognise that commercial and regulatory decisions about product development and registration cannot wait for the accumulation of all the evidence that might be desirable. Decisions need to be made on basis of the existing evidence. The review of existing evidence presented in this chapter does, however, lead to a set of conclusions (see Sect. 5.9) that is well underpinned by the existing evidence. These conclusions are underpinned by experimental as well as modelling research on a wide range of very contrasting modes of action and patho-systems. The upshot of this is that these conclusions can thus be used to underpin the development of resistance management plans.

This issue is most acute where a new mode of action fungicide is being introduced, and the manufacturer, regulatory authorities and advisory bodies have to decide on appropriate resistance management strategies. At that point, there is no history of the resistance behaviour, usually no resistant isolates available from the field to help judge how resistance might evolve and no evidence of the effectiveness of different strategies for that specific mode of action. The only relevant guidance at this point thus is the set of conclusions derived from a review of existing evidence and the knowledge gained by the company during the development of the fungicide. For example, we find overwhelming evidence that adding a mixing partner reduces selection for fungicide resistance. Given a new MOA about which nothing is known, the only sensible assumption thus is that for this MOA, mixing will be a useful resistance management strategy. The same holds for the other conclusions on resistance management.

We thus advocate that the set of conclusions reached in this chapter provides the strongest available basis for decisions on resistance management of new modes of action. The strategies can then be refined by specific information on the new mode of action, as it becomes available – for example, from laboratory mutation studies, from specific modelling of resistance evolution based on the known efficacy of the new compounds and their potential mixture partners and then from experience of resistance development on the first patho-systems on which the new mode of action is used.

5.3 Managing the Application Dose

In discussions on fungicide resistance, claims are sometimes made that it is important to use the maximum dose permitted on the product label (labelled dose), in order to prevent, or at least slow down, the development of resistance. The evidence about the effect of dose on selection for fungicide resistance, however, tells a different story.

Van den Bosch et al. (2011, 2014a, b) summarise the available evidence on the effect of dose. Of the experiments on 19 pathogen-fungicide combinations, published in 15 papers, 16 show that an increased dose increases selection for fungicide resistance, and only two show a decreased selection for resistance. In eight of the eight published modelling studies, it was found that an increasing dose increases selection for resistance. How can this be?

Before answering this question, we need to stress that reducing the dose of a fungicide may compromise effective disease control. We are thus not advocating here that dosages should be reduced without careful consideration of the circumstances. But where dose can be reduced appropriately (Paveley et al. 2001), it should be considered as a resistance management option.

The idea that using the maximum permitted dose is a good anti-resistance measure may have originated from insecticide resistance. Consider a diploid sexually reproducing insect and the effect of an insecticide on mortality of the homozygous SS, heterozygous SR and homozygous RR genotypes (S is the sensitive allele and R the resistant allele, Fig. 5.1), where we discuss here a case where the heterozy-

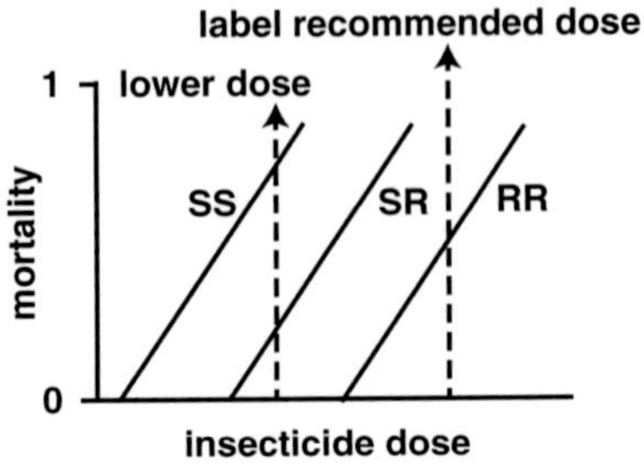

Fig. 5.1 An illustration of the mortality rate of an insect species as a function of the dose of the insecticide applied. The line marked SS is the dose response curve of homozygous sensitive individuals, the line marked SR is the curve for heterozygotes and the line marked RR is the curve for homozygous resistant individual

gous is intermediately sensitive to the fungicide. Applying the maximum permitted dose will cause high mortality of the heterozygous individuals.

When the resistance frequency is small, almost all resistance alleles will be present in heterozygous individuals, because virtually all individuals carrying one or more resistance alleles (SR and RR) will mate with a homozygous sensitive individual, SS. This results in heterozygous individuals and homozygous sensitive individuals. The maximum permitted dose may then delay (or prevent) the development and increase of homozygous resistant, RR, individuals by a high kill rate of the heterozygotes.

In contrast to insects, most of the economically important fungal plant pathogens are haploid when fungicide sprays are applied; many functionally diploid fungi reproduce in a purely clonal manner. The use of the maximum permitted dose tactic can then not be expected to work as it does in diploid sexual species, and the published evidence clearly shows that to be the case.

The available evidence suggesting that, in most cases, a lower dose decreases the selection for resistance may not however tell the whole story. We have determined four hypothetical mechanisms that may lead to the contrary case of a lower dose increasing the rate of resistance selection (see van den Bosch et al. 2011, 2014a, b for more details). These are however hypothetical possibilities, and none of these have yet been proven to operate in reality.

The most likely cases where a reduced dose increases selection for resistance, or where the effect of dose on selection might be broadly neutral, involve partial resistance and/or the involvement of several genes, or mutations within a gene, each with a small effect on the level of resistance. Space restrictions prevent us from discussing this subject in more detail here, and we refer to the papers by van den Bosch et al. (2011, 2014a, b) for more details.

Note: The maximum permitted dose is proposed by the manufacturer and approved by the regulator, taking into account the disease control efficacy, exposure and toxicology to nontarget organisms. The reasoning above about the effect of dose on resistance selection is however not dependent on the permitted dose that is set, because the logic concerns relative, rather than absolute dosages.

5.4 Managing the Number of Applications

There are two cases to consider when discussing the number of applications: firstly, simply increasing or reducing the number of applications, whilst keeping the dose constant and, secondly, splitting dose, for example, by halving the dose of the spray but applying the fungicide twice as often.

Reducing the number of applications is expected to reduce the selection for fungicide resistance because the pathogen is exposed to the selective pressure over a shorter period. The existing evidence, both experimental and modelling, supports this expectation although the number of studies is small. Six publications study the effect of changing the number of applications on selection, and all find that increasing number of applications increases selection. Two modelling studies come to the same conclusion. See van den Bosch et al. (2014a) for references.

Van den Bosch et al. (2014a) show, from the selection coefficient and exposure time governing principle (see the Chap. 4), that splitting dose is expected to increase selection for resistance, because the increase in exposure time outweighs the effect of decreasing dose. There are 11 pathogen-fungicide combinations studied and published (see van den Bosch et al. 2014a for references). Of these, ten cases show increased selection for split dose applications. Only one case shows a reduced selection for resistance with split dose applications. The two existing modelling studies also find that splitting application dose increases the rate of selection for fungicide resistance.

5.5 The Use of Fungicide Mixtures

There is a long standing debate about the usefulness of fungicide mixtures as a resistance management strategy. The evidence accumulated over the last two decades has made it possible to answer some of the unanswered questions about mixtures (van den Bosch et al. 2014a, b).

5.5.1 *Adding a Mixing Partner to an At-Risk Fungicide (Not Reducing the Dose of the At-Risk Fungicide)*

One of the key discussions on the use of mixtures has been whether adding a mixing partner to an at-risk fungicide and not reducing the dose of the at-risk fungicide has any effect on the rate of selection for resistance. Some authors suggest that 'mixing only reduces the build-up of pesticide resistance by reducing the required dose of the pesticides that are mixed' (Birch and Shaw 1997). Others suggested that adding a mixing component without lowering the dose of the at-risk fungicide is a valid resistance management method (FRAC 2010).

When resistance is developing against a fungicide A and we add a mixing partner B with a different mode of action, the mixing partner affects both the pathogen strains which are sensitive and the strains which are resistant to the fungicide A. Hence, the population growth rates of the sensitive and the resistant strains are reduced to the same extent by the mixing partner. The governing principle (see the Chap. 4) then tells us that we can expect the selection for resistance to A to decrease. This prediction has been tested against published evidence.

We have found a total of 51 pathogen-fungicide mixing partner combinations tested. In 44 of these combinations, the mixture treatment programme showed a lower rate of selection for fungicide resistance. This holds for multi-site and single-site mixing partners. Only in two cases was greater selection measured. These cases could subsequently be related to cross-resistance to the fungicides (van den Bosch et al. 2014b for further detail). Of the nine modelling studies, seven show that mixing without reducing the dose of the at-risk fungicide decreases selection. In the two remaining studies, other parameters, such as spray coverage and fitness costs, play a role.

A detailed modelling study by Hobbelen et al. (2011a) shows that the effective life of an at-risk fungicide (defined as the time between introduction of the fungicide and the moment the fungicide can no longer provide effective disease control due to the build-up of resistance) is increased by adding a fungicide with a different mode of action to an at-risk fungicide. The model was parameterised to represent the use of QoI fungicides against infections of wheat with *Zemoseptoria tritici* (formerly known as *Mycosphaerella graminicola*) using a multi-site acting, chlorothalonil-type mixing partner. With a full dose of the mixing partner, the effective life of the QoI fungicide was doubled (see Table 1 in the Chap. 4).

5.5.2 *Adding a Mixing Partner as well as Reducing the Dose of the At-Risk Fungicide*

All current evidence suggests that adding a mixing partner to an at-risk fungicide and lowering the dose of the at-risk fungicide is a valid resistance management tactic. Of the 13 papers with, in total, 20 pathogen-fungicide combinations tested, 15 cases show a reduced selection for resistance of the at-risk fungicide when used as a mixture. The modelling study by Hobbelen et al. (2011) shows that the effective life of an at-risk fungicide is increased by adding a mixing partner and reducing the dose of the at-risk fungicide.

5.5.3 *What to Mix with?*

Companies need to select an appropriate partner fungicide for the mixture. Analysing the existing evidence from both experimental and modelling studies (van den Bosch et al. 2014b), we have found two pieces of clear-cut evidence:

1. Both the existing experimental evidence (Bolton and Smith 1988; English and van Halsema 1954; Genet et al. 2006; Lalancette et al. 1987) and the modelling evidence (Hobbelen et al. 2011b) show that the larger the dose of the mixing partner, the smaller the selection for fungicide resistance and the longer the effective life of at-risk fungicide (see Table 1 in the Chap. 4).
2. A mixing partner fungicide with a lower efficacy will give a smaller reduction in selection rate. Considering two possible mixing partners, to which no resistance has (yet) developed, the best choice would thus be to use the fungicide with the highest efficacy. The existing evidence also implies that a fungicide to which resistance has developed to high levels is not an appropriate mixing partner.

In practice, fungicide manufacturers may be constrained to use their own active substances in the choice of mixing partners, rather than optimal combinations, due to commercial considerations. But agreements to access active substances from other manufacturers may enable a stronger choice in some cases.

5.5.4 *Mixing Two At-Risk Fungicides*

The number of multi-site fungicides is limited, and environmental policy may reduce their number further. This leads to the development of fungicides mixtures where two at-risk fungicides are mixed, to both of which resistance can develop. Such mixtures present two opposing forces:

1. By mixing with fungicide B, as we have seen before, the selection for resistance to fungicide A is reduced and vice versa.
2. But adding an at-risk mixing partner creates a selection pressure for resistance against that mixing partner.

The question thus is whether the gain, in terms of slowing down selection, from adding an at-risk fungicide to the spray programme outweighs the loss of putting an additional fungicide at risk of resistance.

There is very little evidence available to come to a general conclusion on this matter. Only three papers have been published comparing the selection for resistance to two fungicides of a mixture (Brent et al. 1989; Lorenz et al. 1992; Thygesen et al. 2009). All these show that the rate of development of resistance is slowed down for both fungicides in the mixture. Hobbelen et al. (2013) developed a model parameterised for *Zemoseptoria tritici* and two fungicides to which high levels of resistance develop. Key conclusions from this modelling study were that mixing two at-risk fungicides always gave an equal or higher effective fungicide life compared to either concurrent (i.e. applying the two modes of action solo to different fields within the same seasons) or sequential use (i.e. using one mode of action for several seasons until it became ineffective and then using the other mode of action) of the solo products.

Milgroom and Fry (1988) were the first to show that the initial frequency of resistance has a major effect on the success of resistance management strategies. Van den Bosch et al. (2014a) cite various modelling studies corroborating that strategies (including the use of mixtures) are substantially more effective if implemented at the moment a new mode of action is introduced. Waiting until resistance has emerged and is detected in field populations and then putting a resistance management programme in place is unlikely to be effective, unless the resistance is of a slow-shifting type. We are not the first to make this important point, yet discussions about resistance management often start at the moment resistance is found in field.

5.6 The Use of Fungicide Alternation

Alternation is frequently considered as a resistance management tactic. However, there are two different types of alternation that are not always clearly separated in discussions on resistance management:

1. Given a spray programme with fungicide A, we *add* sprays with fungicide B with a different mode of action, between the sprays with A (Fig. 5.2).
2. Given a spray programme with applications of fungicide A, we *replace* part of the applications with sprays with fungicide B (Fig. 5.2).

In the case that an application with fungicide B is added in between applications of A, it is important to recognise that selection for resistance to A will only take place when fungicide A is present. In the figure, we made the decay rate of fungicide A fast enough so that the effect of A has vanished when B is applied. In such cases, fungicide B has no effect on the selection for resistance to fungicide A simply because fungicide A is not operating when B is present. This leads us to predict that

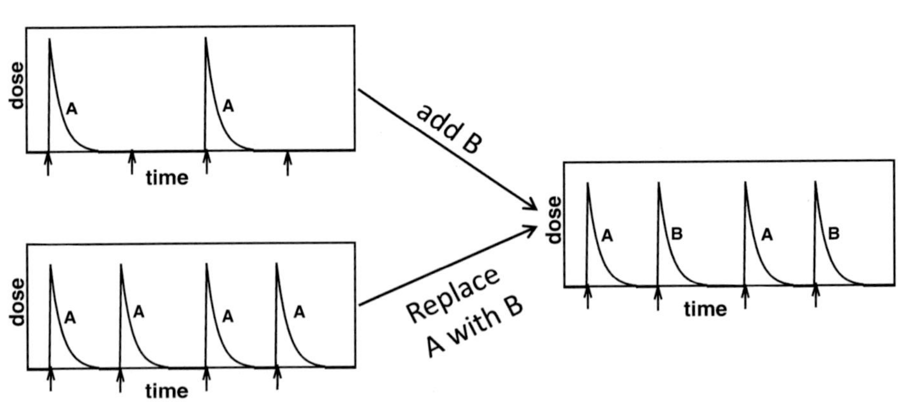

Fig. 5.2 Two different ways to alternate: (1) *add* an application with fungicide *B* in between two applications of fungicide *A* and (2) *replace* applications with fungicide *A* by applications with fungicide *B*. For each panel, time is on the *x*-axis. The time points where fungicide treatments may be applied are marked with arrows pointing to the *x*-axis in all the panels. Dose is plotted on the *y*-axis. For ease of representation, the fungicide dose decays rapidly, so that the time windows in which the fungicide is active do not overlap. There are two fungicides of differing modes of action that can be used in the treatment programme: fungicide *A* and fungicide *B*

adding B has no effect on selection. This prediction is supported by the existing evidence, although the evidence is limited. Of the five pathogen-fungicide combinations that have been tested experimentally (see van den Bosch et al. 2014a for the references), four show no change in selection when B is added to the application programme. In one case, a larger selection for resistance is measured when B is applied, which can be explained by the cross-resistance to the specific fungicides involved (van den Bosch et al. 2014a).

In the case where applications with fungicide A are replaced by applications with B, Fig. 5.2, we expect the selection for resistance to be reduced because the fungal population is exposed to selection pressure for resistance to A only half the time. There are 15 pathogen-fungicide combination cases published (see van den Bosch et al. 2014a for the references) that study this situation. In 12 cases, a reduced selection for resistance was measured when using the alternation; in three cases, no difference was measured.

5.7 Mixture Versus Alternation

Are mixtures better resistance management tactics than alternations? Several authors have studied this question resulting in different answers. To understand the key issues, let's consider a mixture and an alternation using equal amounts of the two fungicides A, an at-risk fungicide to which resistance is developing, and fungicide B with a different mode of action to which no resistance is developing. The alternation consists of an application with fungicide A at dose D_A (which can be a full dose or anything lower than a full dose) followed by an application with fungicide B at dose D_B (again at or below a full dose). To transform this alternation into a mixture, we split the dose of both fungicide A and fungicide B and mix the half doses of A and B and apply this mixture at each spray event. See Fig. 5.3 for an illustration. Which of these two treatment plans selects the least for resistance to fungicide A?

There are two opposing forces:

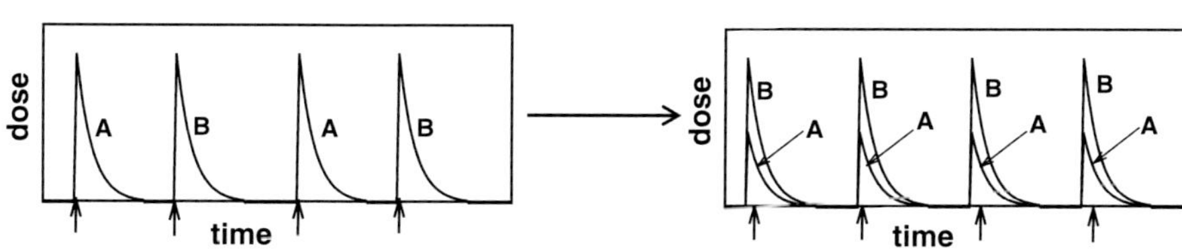

Fig. 5.3 Alternations versus mixtures. For each panel, time is on the x-axis. The time points where fungicide treatments may be applied are marked with *arrows* pointing to the x-axis in all the panels. Dose is plotted on the y-axis. For ease of representation, the fungicide dose decays rapidly, so that the time windows in which the fungicide is active do not overlap. There are two fungicides of differing modes of action that can be used in the treatment programme: fungicide A and fungicide B

1. The dose of at-risk fungicide A is split into two half doses. As discussed, splitting a dose increases the selection for resistance.
2. The half dose of A is mixed with fungicide B. Mixing decreases selection, as discussed previously.

It is thus a question of the balance between the increased selection due to dose splitting and the decreased selection due to mixing that determines whether alternation is a more effective resistance management strategy than mixing or vice versa. This reasoning results in the prediction that whether mixtures or alternations are most effective in reducing selection depends on the particular characteristics of the pathogen-fungicide combination being considered. This finding is reflected in the published evidence. There are eight publications (see van den Bosch et al. 2014a for the references) studying 12 pathogen-fungicide mixing partner combinations. Of these, six show a smaller selection for resistance in the mixture treatment than in the alternation treatment. In two cases, alternation has a smaller selection for resistance than mixtures. In the remaining four cases, no difference was measured between the two tactics.

The modelling evidence (see van den Bosch et al. 2014a, b for the references) leans towards mixtures being the tactic that reduces selection the most. Four papers show this, at least when spray coverage is incomplete, which is always the case in practice. The remaining one study gives a mixed picture with particular circumstances affecting which one is the best anti-resistance tactic.

From a practical perspective, the limited number of spray applications to cereal crops constrains the extent to which alternation can be implemented. In contrast, the multiple-spray programmes applied to control potato late blight (*Phytophthora infestans*), and the wide range of modes of action available, offer many possible options for alternation, mixtures or combinations of the two approaches. These options have not been adequately explored.

5.8 Protective Versus Curative Use

The strict definition of protective use of a fungicide refers to an application timed prior to infection, so that infections are prevented. Curative use refers to the fungicide affecting the pathogen, by increasing the duration of the latent period, constraining lesion size or reducing spore production by an application after infection has occurred. Most resistance management guidance, such as from the Fungicide Resistance Action Committee, advises to use fungicides as protective and avoid curative use because curative use would promote selection for resistant/insensitive strains.

Surprisingly, there is no evidence on the effects of protectant and curative use of fungicides on selection for fungicide resistance. Milgroom and Fry (1988) commented on the absence of studies on protective and curative use. Twenty years later,

Brent and Hollomon (2007) reiterated this and stated that 'to the authors knowledge there is no experimental evidence comparing the resistance risks of prophylactic versus threshold-based schedules, and research on this would be useful'.

We note here again that the primary aim of using fungicides is effective disease control, and the choice of protective and curative sprays or a combination thereof can have major influence on disease control efficacy. To maintain effective disease control, it may, depending on the patho-system and fungicide, be necessary to prevent the epidemic from developing early in the season, making protective fungicide use important. For example, for yellow rust (*Puccinia striiformis*), protective treatment is essential to obtain effective control, and in such cases, curative use may compromise disease control. So, we are not advocating here to use fungicides curatively, but we are saying that using fungicides protectively may be necessary for effective disease control, not for resistance management. There is no evidence that a protective spray is good for fungicide resistance management.

In practice, 'protective' and 'curative' use of fungicides are often loosely translated into use earlier or later in the growing season, respectively. If we compare earlier and later use of fungicides, there are, in total, six pathogen-fungicide resistance cases which have been studied and reported in the literature. In two cases, earlier sprays were found to increase selection for resistance. In three cases, the selection for resistance for earlier sprays was smaller than for later sprays. In one case, no difference was measured (see van den Bosch et al. 2014a for the references). There are two published modelling studies comparing early and later spray applications. One reports an increased selection from earlier sprays and another one reports an increased selection from later sprays.

A factor that may further complicate the issue is that infections in a crop are not synchronised, and the pathogen population is usually a mixture of the various life cycle stages of the pathogen. This makes it difficult to determine whether a spray is (mostly) protective or (mostly) curative.

In conclusion, current evidence does not support early or protective treatment being a resistance management tactic, as it is not possible to conclude whether protective or curative use has the lowest rate or selection for fungicide resistance. Depending on the patho-system-fungicide combination, protective treatment may be essential for effective disease control, however.

Where a mixture is being applied which comprises systemic and nonsystemic components, there is an argument for using the mixture under circumstances where the nonsystemic components will be an effective part of the mixture. In the case of potato late blight, for example, this would imply use early in the season, when initial inoculum is arriving in the crop. However, more generally, there is a resistance downside to early treatment, because prophylactic treatment decisions are more prone to uncertainty about whether treatment is necessary (the disease may fail to develop to damaging levels in the absence of treatment). It is clear that any unnecessary treatments add to selection, even though disease severity is low. Hence, we should avoid anti-resistance guidance which might result in fungicides being applied earlier than the optimum timing for efficacy and economic response.

5.9 Discussion

We have presented the evidence on the effectiveness of a range of resistance management tactics. This chapter combines the information gathered in three recent reviews by van den Bosch and Gilligan (2008) and van den Bosch et al. (2011, 2014a, b). From the evidence discussed, a set of clear conclusions can be derived. These are:

Managing the Application Dose

- The majority of the evidence suggests that an increased dose of fungicide increases selection for fungicide resistance (but note here that the primary aim of effective disease control may make it impossible to reduce fungicide dosages).
- A number of possible mechanisms by which an increased dose may reduce selection have, however, not been studied. Partial resistance and multi-gene/multi-mutation cases are the key examples of this.

Managing the Number of Sprays

- All current evidence suggests that increasing the number of fungicide applications increases selection for fungicide resistance.
- Most evidence suggests that splitting fungicide dosage between two or more applications increases selection.

The Use of Fungicide Mixtures

- The vast majority of the evidence shows that adding a mixing partner to a high-resistance-risk fungicide reduces selection for fungicide resistance, even when the dose of the high-risk fungicide stays the same in the mixture.
- Adding a mixing component to a high-risk fungicide and reducing the dose of the high-risk fungicide further reduces selection for fungicide resistance.
- There is too little evidence on the use of mixtures of two at-risk fungicides, and work in this area is needed. The evidence that does exist suggests that mixing two at-risk fungicides is a valid anti-resistance strategy.

The Use of Fungicide Alternations

- Limited evidence suggests that alternating with a fungicide that has a different mode of action does not alter selection for the high-risk fungicide, if the number of applications of the high-risk fungicide remains constant with and without alternation.
- The evidence suggests that replacing part of the fungicide programme with a fungicide with a different MOA reduces selection.

Alternations Versus Mixtures

- It depends on the balance between increased selection due to dose splitting and decreased selection due to mixing whether mixing reduces selection to a greater or lesser extent than alternation. The experimental and modelling evidence shows that in many cases, mixing is the better strategy, but for any single case, this needs to be established before conclusions can be reached.

Protective Versus Curative Use

- There is no evidence that protective or curative use consistently results in a lower rate of selection for fungicide resistance (but note that protective fungicide applications may be needed for effective disease control).
- The existing evidence suggests that the specific circumstances will determine whether a shift in spray timing will increase or decrease selection for fungicide resistance.

Resistance management should be based on evidence interpreted within a sound experimental, theoretical and practical framework. In discussions on resistance management, it is often not explicit what the evidence is. With this chapter, we hope to contribute to evidence-based resistance management.

References

Birch CPD, Shaw MW (1997) When can reduced doses and pesticide mixtures delay the build-up of pesticide resistance? A mathematical model. J Appl Ecol 34:1032–1042

Bolton NJE, Smith JM (1988) Strategies to combat fungicide resistance in barley powdery mildew. Br Crop Prot Conf Pests Dis: 367–372-1988, Vol 1–3,

Brent KJ, Hollomon DW (2007) Fungicide resistance in crop pathogens: how can it be managed? Fungicide Resistance Action Committee, Brussels

Brent KJ, Carter GA, Hollomon DW, Hunter T, Locke T, Proven M (1989) Factors affecting build-up of fungicide resistance in powdery mildew in spring barley. Neth J Plant Pathol 95(S1):31–41

English AR, van Halsema G (1954) A note on the delay in the emergence of resistant Xanthomonas and Erwinia strains by the use of streptomycin plus terramycin combinations. Plant Dis Report 38:429–431

FRAC (2010) FRAC recommendations for fungicide mixtures designed to delay resistance evolution. http://www.frac.info/

Genet J-L, Jaworska G, Deparis F (2006) Effect of dose rate and mixtures of fungicides on selection for QoI resistance in populations of *Plasmopara viticola*. Pest Manag Sci 62:188–194

Hobbelen PHF, Fraaije B, Lucas JA, Paveley ND, van den Bosch F (2011a) Derivation and testing of a model to predict selection for fungicide resistance. Plant Pathol 60:304–313

Hobbelen PHF, Paveley ND, van den Bosch F (2011b) Delaying selection for fungicide insensitivity by mixing fungicides at a low and high risk of resistance development: a modelling analysis. Phytopathology 101:1224–1233

Hobbelen PHF, Paveley ND, van den Bosch F (2013) The value of alternation or mixtures of fungicides for delaying the selection of resistance against two modes of action in populations of *Mycosphaerella graminicola* on winter wheat. Phytopathology 101:690–707

Lalancette N, Hickey KD, Cole H Jr (1987) Effects of mixtures of benomyl and mancozeb on build-up of benomyl-resistant *Venturia inaequalis*. Phytopathology 77:86–91

Lorenz G, Saur R, Schelberger K, Forster B, Kung R, Zobrist P (1992) Long term monitoring results of wheat powdery mildew sensitivity towards fenpropimorph and strategies to avoid the development of resistance. Brighton Crop Prot Conf Pests Dis 1:171–176

Milgroom MG, Fry WE (1988) A simulation analysis of the epidemiological principles for fungicide resistance management in pathogen populations. Phytopathology 78:565–570

Paveley ND, Sylvester-Bradley R, Scott RK, Craigon J, Day W (2001) Steps in predicting the relationship of yield on fungicide dose. Phytopathology 91:708–716

Thygesen K, Jorgensen LN, Jensen KS, Munk L (2009) Spatial and temporal impact of fungicide spray strategies on fungicide sensitivity of *Mycosphaerella graminicola* in winter wheat. Eur J Plant Pathol 123:435–447

van den Bosch F, Gilligan CA (2008) Models of fungicide resistance. Annu Rev Phytopathol 46:123–147

van den Bosch F, Paveley ND, Shaw MW, Hobbelen P, Oliver R (2011) The dose rate debate: does the risk of fungicide resistance increase or decrease with dose? Plant Pathol 60:597606

van den Bosch F, Oliver R, van den Berg F, Paveley N (2014a) Governing principles can guide fungicide resistance management tactics. Annu Rev Phytopathol 52:175–195

van den Bosch F, Paveley N, van den Berg F, Hobbelen P, Oliver R (2014b) Mixtures as a fungicide resistance management tactic. Phytopathology 104(12):1264–1273

Chapter 6
Fitness Cost of Resistance: Impact on Management

Alexey Mikaberidze and Bruce A. McDonald

Abstract Fungicides are important tools for the management of fungal diseases in many crops. But eventually, most fungicides fail because the treated pathogen population evolves resistance to the fungicide. This chapter focuses on how our knowledge of fitness costs associated with resistance informs strategies of fungicide deployment that help to avoid or delay development of resistance. Many different fungicide deployment strategies should be considered that take into account fungal population genetics as well as the specific agroecosystem. Mono-applications will be replaced by strategies that use several fungicides with different modes of action. Modeling approaches will be needed to inform us regarding the optimum strategies to use under different circumstances. It is clear that fitness costs connected to mutations that encode fungicide resistance will need to be better measured and taken into account in order to design optimum fungicide deployment strategies.

We discuss the importance of fitness costs in assessing the usefulness of fungicide mixtures that contain a high-risk fungicide together with a low-risk fungicide and the role of population dynamical mathematical models of plant–pathogen interaction. According to models, the fitness cost of resistance determines the outcome of competition between the sensitive and resistant pathogen strains. If fitness costs are absent, then the use of the high-risk fungicide in a mixture selects for resistance, and the fungicide eventually becomes nonfunctional. If there is a cost of resistance, then an optimal ratio of fungicides in the mixture can be found, at which selection for resistance is expected to vanish and the level of disease control can be optimized.

Keywords Fitness • Epidemiology • Host–pathogen interaction • Plant disease • Dynamic diversity

---

A. Mikaberidze (✉)
Theoretical Biology, Institute of Integrative Biology, ETH Zurich, Zurich, Switzerland
e-mail: alexey.mikaberidze@env.ethz.ch

B.A. McDonald
Plant Pathology, Institute of Integrative Biology, ETH Zurich, Zurich, Switzerland
e-mail: bruce.mcdonald@usys.ethz.ch

H. Ishii, D.W. Hollomon (eds.), *Fungicide Resistance in Plant Pathogens*,
DOI 10.1007/978-4-431-55642-8_6

6.1 Introduction

Fungicides are important tools for the management of fungal diseases in many crops. Fungicide applications at the farm and landscape scales keep many important fungal diseases in check, sometimes for several years. But eventually, most fungicides fail because the treated pathogen population evolves resistance to the fungicide. In this chapter, fungicide resistance is treated as a population phenomenon, exhibited mainly at the scale of fields, farms, and landscapes, rather than at the level of fungal individuals or cells, even though resistance is usually measured based on growth rates of individual strains. But it should be noted that the emergence of fungicide resistance at the population or landscape level generally begins with mutations that encode resistance at the level of a cell or individual.

Fungicide resistance emerges because pathogens evolve. Evolution requires genetic diversity. Genetic diversity is affected by mutation rate, population size, recombination, gene flow, and selection, the same factors that affect an organism's population genetics. Thus, the key to developing strategies to avoid the emergence of fungicide resistance lies in understanding the population genetics of fungal pathogens.

The root of the problem of evolution of fungal pathogens is the lack of diversity in agricultural ecosystems (agroecosystems). Since the invention of agriculture 12,000 years ago, crop genetic diversity has declined steadily in agroecosystems globally to facilitate gradual improvements in agricultural production systems, including innovations such as tillage, fertilization, controlled irrigation, and mechanization. The decline in crop genetic diversity at the field scale accelerated rapidly during the last 100 years to increase the efficiency of food production and feed the burgeoning human population. The large-scale mechanization of planting and harvesting operations and the green revolutions that swept through agriculture led to the replacement of locally adapted but genetically diverse land races by genetically uniform but broadly adapted, high-yielding, dwarf cultivars, further depleting genetic diversity and increasing the environmental homogeneity present in agroecosystems worldwide. The highly mechanized modern industrial agroecosystems now found around the world are extremely productive food factories but also are highly effective incubators of pathogen evolution (Stukenbrock and McDonald 2008). As a result of the increased planting density and genetic uniformity of host populations in agroecosystems, pathogen population sizes increased, which led to more genetic diversity for the selection to act upon by increasing the total number of mutations available at the field scale while simultaneously lowering the effects of genetic drift. Due to these changes in agroecosystems over time, pathogen evolutionary potential likely increased as agricultural pathogens (including fungi, bacteria, and viruses) became domesticated and adapted to the agroecosystem environment.

The increasing genetic uniformity of the major crops facilitated the large-scale mechanization of agriculture and simplified all aspects of food production and food processing. The basic trade-off in all agroecosystems is that the increased environmental and genetic uniformity that enables more efficient food production also facilitates pathogen evolution and increases the risk of significant losses due to disease. Thus, we are confronted with a dilemma: How can we maximize the many benefits of modern industrial agriculture while minimizing the risks of disease epidemics in these agroecosystems? Is it possible to achieve long-lasting and stable (i.e., durable) disease control, especially if it is based largely on growing resistant cultivars and/or applying fungicides?

Durable disease control will not be achieved unless we increase the overall diversity present in agroecosystems at both the farm and landscape spatial scales. The diversity will need to be dynamic, changing regularly over both time and space, to significantly slow the rate of pathogen adaptation. Fortunately, a wide array of low-technology, medium-technology, and high-technology strategies can be used to increase both spatial and temporal diversity in agroecosystems (McDonald 2014). Many of these strategies are well known and oriented around crop husbandry, requiring no additional breeding effort. These crop husbandry strategies include improved crop rotations, species intercropping, and planting smaller fields to increase crop heterogeneity at the farm and landscape scales. Other possibilities include the introduction of new crop species into existing agroecosystems and increasing the overall genetic diversity present in existing crop species by planting mixtures of host cultivars (Mikaberidze et al. 2014a; Mundt 2002) or introducing new genes from their wild relatives.

In this chapter we focus on another aspect of this diversity, namely, the chemical diversity represented by the applied fungicides. Static, long-term use of any fungicide alone will impose strong directional selection on the corresponding fungal population that will favor evolved pathogen strains that are resistant to the fungicide. The development of resistance to systemic fungicides that target a single molecule used alone has been widely documented for many decades in all major crops (Brent and Hollomon 2007). Fungicides that inhibit more basic biological processes (e.g., based on copper or thiol groups) used alone are also likely to select for less-sensitive populations over time as a result of pathogen evolution, though the "erosion" of the activity of these fungicides may be more difficult to document. Strategies of disease control that use more than one fungicide, such as mixtures and alternations of fungicides, have been proposed to delay the emergence of fungicide resistance. But even these mixed strategies are likely to fail after long-term use. We argue that control strategies based on dynamic turnover of fungicide mixtures, i.e., mixtures with components changing over time and space, would be advantageous for delaying the emergence of fungicide resistance in the longer run. In order to predict the effectiveness and durability of fungicide control strategies and optimize the parameters associated with these strategies, i.e., composition, overall dose, proportions of components, and the rate of turnover, one needs to use biologically relevant and carefully parameterized mathematical models accompanied by extensive field experimentation. One of the key issues here is to characterize pathogen fitness in these different chemical environments. In this respect the concept of fitness costs becomes useful.

6.2 What Are Fitness Costs and How to Measure Them?

In some cases fungicide resistance can be conferred by a single-point mutation (e.g., QoI resistance (Torriani et al. 2009)) that gives mutants full protection from the fungicide. In other cases, several mutations need to be accumulated in the same gene or in different genes in order to gain considerable resistance (Cools 2008; Zhan et al. 2006).

Any mutation that confers fungicide resistance may also disrupt or lower the efficiency of important physiological and biochemical processes (Anderson 2005). Consequently, a resistance mutation may result in lower pathogen fitness. We define the fitness cost of a mutation conferring fungicide resistance, ρ_r, as the difference in fitness between the resistant pathogen strain carrying the resistance mutation and the sensitive pathogen strain in the absence of the fungicide. Thus, in order to determine fitness costs, one needs to measure fitness of both sensitive and resistant pathogen strains in the same, nonselective environment.

The overall fitness of fungal plant pathogens can be comprised of several components that correspond to different stages of the pathogen life cycle. These include, but are not limited to, spore production, spore dispersal, infection efficiency, mycelial growth, and the ability to survive between seasons. The ultimate measure of fitness is based on the ability to compete with other strains in a field environment characterized by fluctuating environmental conditions and interactions with a wide variety of host genotypes and competing microbiota. The strains with the highest reproductive fitness will contribute the most genes to future generations. A mark–release–recapture experimental design has been shown to work well to measure competitive fitness for three different cereal pathogens (Zhan and Mcdonald 2013) though these field experiments were not testing fitness costs associated with fungicide resistance. Similar experimental evolution designs should provide useful insight into fitness costs associated with resistance mutations, though as with all fieldwork, these experiments remain quite costly in terms of resources and labor. Experimental evolution in chemostats, flasks, and Petri dishes can also be used to infer fitness costs (Anderson 2005), is much less resource-intensive, and can be combined with next-generation sequence analyses of entire genomes to identify both primary mutations and compensatory mutations associated with resistance. More traditional methods cited below involve comparing growth rates and spore production of resistant and susceptible strains on Petri plates or in planta to infer fitness costs.

When perennial ryegrass (*Lolium perenne* L.) was inoculated with mixtures of azoxystrobin-resistant and azoxystrobin-sensitive strains of *Magnaporthe oryzae*, sensitive strains produced more conidia and increased in frequency over time in the absence of the fungicide, consistent with a fitness penalty for azoxystrobin resistance (Ma and Uddin 2009). Similarly, the frequency of *Cercospora beticola* isolates resistant to DMI fungicides slightly but significantly decreased during an epidemic in competition with DMI-sensitive strains (Karaoglanidis et al. 2001). In contrast, resistant strains were as fit as sensitive strains of *Phytophthora erythroseptica* and even displayed a competitive advantage when mefenoxam was absent, suggesting no fitness penalty (Chapara et al. 2011). Similarly, no fitness costs were found in *Alternaria alternata* resistant to QoI fungicides (Karaoglanidis et al. 2011) or for *Phytophthora nicotianae* strains resistant to mefenoxam (Hu et al. 2008).

Some studies inferred substantial fitness costs from field monitoring that enabled measurements of changes in frequency of resistant strains over time (e.g., Suzuki et al. 2010 and references in Peever and Milgroom 1995). But these findings could result from other factors, including immigration of sensitive isolates, selection for other traits linked to resistance mutations, or genetic drift (Billard et al. 2012). Though relatively few carefully controlled experiments have been conducted, the majority indicates that fitness costs associated with fungicide resistance are either low (e.g., Kim and Xiao 2011; Billard et al. 2012) or absent (e.g., Corio-Costet et al. 2010; Peever and Milgroom 1994). But in some cases, fitness costs were found to be substantial (e.g., Webber 1988; Kadish and Cohen 1992; Holmes and Eckert 1995; Karaoglanidis et al. 2001; Iacomi-Vasilescu et al. 2008) both in laboratory measurements and in field experiments. Although measurements of fitness costs of resistant mutants performed under laboratory conditions can be informative (e.g., Billard et al. 2012), they do not necessarily reflect the costs connected with resistant mutants selected in the field. This is because field mutants are more likely to possess compensatory mutations improving pathogen fitness (Peever and Milgroom 1995) in the field environment. Moreover, a laboratory setting rarely reflects the balance of environmental and host conditions found throughout the pathogen life cycle, since the field environment is much more complex and fluctuates constantly.

A highly relevant measure of pathogen fitness in the context of anti-resistance management strategies is the growth rate of the pathogen population, r, at the very start of an epidemic. Figure 6.1 illustrates the effect of fitness cost on the selection for fungicide resistance. The growth rate, r, of the sensitive strain decreases as a function of the fungicide dose, C (solid curves).

When resistance is full (left panel), the growth rate of the resistant strain is not affected by the fungicide. Hence, r remains constant versus C with the magnitude that depends on the fitness cost θ (dashed lines). In the absence of a fitness cost (upper dashed line in Fig. 6.1a), resistant mutants have a selective advantage over the sensitive strain as soon as the fungicide is added (i.e., at any $C>0$). When there is a fitness cost, the sensitive strain is favored by selection at small doses (cf. the range of doses, where the solid curve is above the lower dashed curves in Fig. 6.1a). But when the dose exceeds a certain threshold value C_c, the resistant strain becomes fitter than the sensitive strain. The threshold dose C_c depends on the magnitude of the fitness cost and the dose-response parameters of the sensitive strain that determine how fast its fitness declines with the dose.

In the case of partial resistance (Fig. 6.1b), the fitness of resistant mutants does not remain constant but declines with the dose C. However, it declines slower than the fitness of the sensitive strain. As a result, the range of doses over which the sensitive strain has a selective advantage is expanded as compared to the case of full

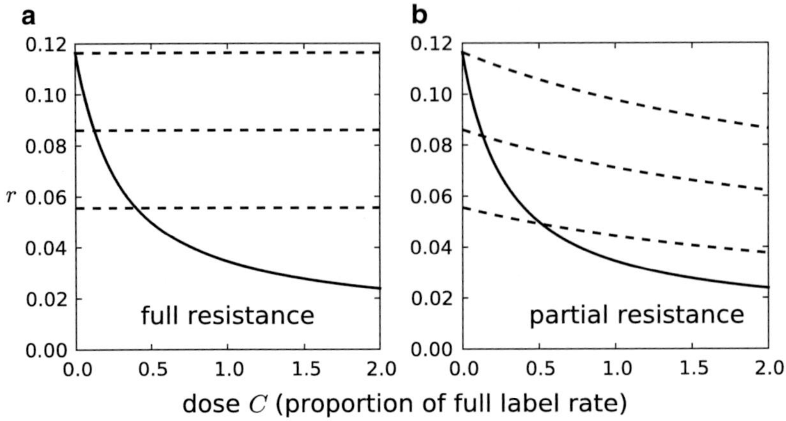

Fig. 6.1 The effect of fitness cost on the selection for fungicide resistance. Apparent growth rate r is plotted as a function of the fungicide dose C for the sensitive pathogen strain [solid curve] and for the resistant pathogen strain [*dashed curves*] with the fitness cost $\rho_r = 0$ (*upper curve*), $\rho_r = 0.2$ (*middle curve*), and $\rho_r = 0.4$ (*lower curve*). Other parameter values are sensitive pathogen's life history traits, $r(C = 0) = 0.1173$, $R_0 = 4.26$, infectious period $1/\mu = 28$ days; dose-response parameters of the sensitive strain, $k_k = 0.7$, $D_{50} = 0.3$; fungicide sensitivities of the resistant strain $\alpha_1 = 0.1$, $\alpha_2 = 0.7$. Parameter values are in the relevant range for *Zymoseptoria tritici*, an important pathogen of wheat

resistance. Here, the threshold dose C_c depends not only on the magnitude of the fitness cost and the dose-response parameters of the sensitive strain but also on the dose-response parameters of the resistant strain.

How does an understanding of fitness costs affect anti-resistance management? If there is no fitness cost associated with a resistance mutation, then the pathogen strain carrying this mutation will persist in fungal populations even in the absence of a fungicide. As a result, the corresponding fungicide may permanently lose its effectiveness, and there will be no benefit associated with using this fungicide in a mixture or alternation. But when fitness costs are present, there are two main conclusions relevant for anti-resistance management. First, selection for resistance can be reversed during the time when the fungicide is absent. However, the rate of reversion is expected to be much slower than the rate of selection for resistance. Also, compensatory mutations that may accompany resistance mutations could make reversion difficult or impossible.

The second conclusion is that low fungicide doses can be applied without selecting for fungicide resistance as long as fitness costs are present. This has a promising application for developing fungicide mixtures that would avoid selection for resistance but at the same time achieve a desired degree of disease control. We elaborate this idea in the next section with the help of a population dynamical mathematical model.

6.3 The Role of Fitness Costs in Selection for Resistance: Insights from Mathematical Modeling

The ranges of fungicide dose and cost of resistance at which the sensitive (white) or resistant (gray) pathogen strain is favored by selection are shown in Fig. 6.2. In all scenarios competitive exclusion is observed: one of the strains takes over the whole pathogen population and the other one is eliminated. If a low-risk fungicide is applied alone, the sensitive strain has a selective advantage across the whole parameter range. When only a high-risk fungicide is applied (Fig. 6.2a), the resistant strain dominates if the fitness cost is lower than the maximum effect of the fungicide $\rho_r < k_k$ and at a fungicide dose higher than a threshold value which increases with the fitness cost (solid curve in Fig. 6.2a). If the fitness cost exceeds k_k (dotted line in Fig. 6.2a), then the sensitive strain dominates at any fungicide dose. Figure 6.2b shows the outcome when the two fungicides are mixed at equal doses. Here the fitness cost at which the sensitive strain dominates is reduced (vertical dotted line is shifted to the left). As expected, without a fitness cost $(\rho_r = 0)$, the resistant strain becomes favored by selection and will eventually dominate the population

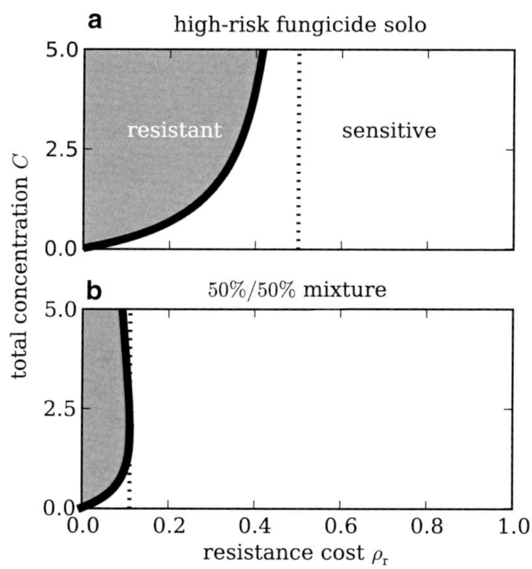

Fig. 6.2 Outcomes of the competition between the sensitive and resistant pathogen strains depending on the fitness cost of resistance ρ_r and the fungicide concentration C when treated with a single fungicide A ($C_A = C$, *panel (a)*) and the combination of fungicides A and B ($C_A = C_B = C/2$, *panel (b)*). The range of the total fungicide concentration C and the fitness cost of resistance ρ_r, in which resistant strain is favored, is shown in gray. The range where selection favors the sensitive strain is shown in white. Fungicides are assumed to have zero interaction $(u = 0)$, and the resistant strain is assumed to be fully protected from fungicide A $(\alpha = 0)$; the fungicide dose-response parameters are $k_k = 0.6$, $C_{50} = 1$. (See Sect. 6.3 for more details)

whenever the high-risk fungicide is applied, alone or in combination with the low-risk fungicide (Fig. 6.2a, b).

It is highly desirable to keep existing fungicides effective for as long as possible. From this point of view, an optimal mixture contains the largest proportion of the high-risk fungicide, at which (1) the resistant pathogen strain is not selected and (2) an adequate level of disease control is achieved. In order to fulfill both of these objectives, the fitness cost of resistance needs to be larger than a threshold value. The threshold ρ_{rb} is shown by the dotted vertical line in Fig. 6.2b.

The threshold ρ_{rb} depends on the proportion of fungicides in the mixture. Adding more of the low-risk fungicide, while keeping the same total dose C, reduces the threshold. This diminishes the range of the values for fitness cost over which the resistant strain dominates. On the other hand, adding less of the low-risk fungicide, while again keeping C the same, increases the threshold, which increases the parameter range over which the resistant strain is favored.

Therefore, at a given fitness cost ρ_r, one can adjust the fungicide ratio r_B such that $\rho_r > \rho_{rb}$. This is shown in Fig. 6.3: the curve shows the critical proportion of the low-risk fungicide r_{Bc}, above which no selection for resistance occurs at any total fungicide dose C. One can see from Fig. 6.3 that if the resistance cost is absent $(\rho_r = 0)$, then the high-risk fungicide should not be added at all if one wants to prevent selection for resistance. At larger fitness costs, the value of r_{Bc} decreases, giving the possibility to use a larger proportion of the high-risk

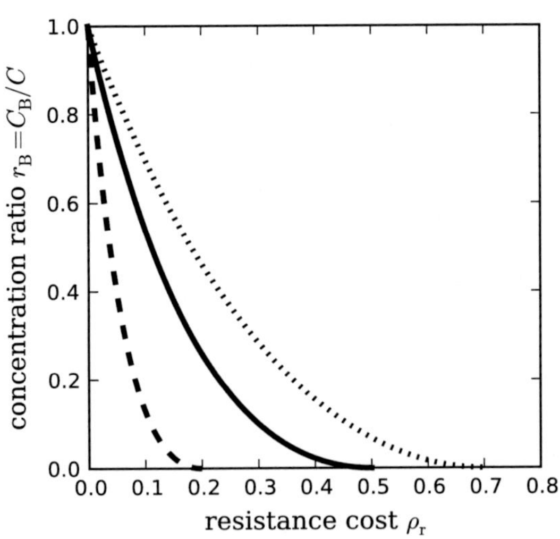

Fig. 6.3 The critical proportion r_{Bc} of fungicide B (low-risk fungicide) in the mixture, above which there is no selection for the resistant strain at any total fungicide dose C, plotted (*black curve*) and a function of the resistance cost ρ_r, assuming no pharmacological interaction $(u = 0)$, full resistance $(\alpha = 0)$, and the maximum fungicide effect $k_k = 0.2$ (*dashed*), $k_k = 0.5$ (*solid*), and $k_k = 0.7$ (*dotted*). *Vertical dashed* lines indicate the estimates for the fitness costs $\rho_r \approx 0.2$ and $\rho_r \approx 0.4$ obtained from the studies by Billard et al. (2012) and Karaoglanidis et al. (2001), correspondingly (see Sect. 6.4.1 for more details)

fungicide without selecting for resistance. Finding an optimum proportion of fungicides requires knowledge of both the fitness cost ρ_r and the maximum effect of the fungicide k_k.

6.4 How Can Knowledge of Fitness Costs Inform Resistance Management Strategies?

In the following sections we discuss how the information about fitness costs combined with the modeling insights from the previous section can help us in determining effective strategies for managing fungicide resistance.

6.4.1 When Are Fungicide Mixtures Effective as a Resistance Management Strategy?

In cases when fungicide resistance confers no fitness cost (e.g., Karaoglanidis et al. 2011), the application of a mixture of high-risk and low-risk fungicides will select for resistance. Consequently, the resistant strain will eventually dominate the pathogen population and the sensitive strain will be eliminated. Because of this, the high-risk fungicide will not affect the amount of disease, and only the low-risk fungicide component of the mixture will be acting against the disease. Hence, the high-risk fungicide becomes nonfunctional in the mixture, and using the low-risk fungicide alone would have the same effect at a lower financial and environmental cost.

The work of Billard et al. (2012) provides a good example of a moderate fitness cost in the grapevine pathogen *B. cinerea*. Three isogenic lines were created carrying different mutations in the gene encoding the 3-ketoreductase target site of fenhexamid. All three mutant lines produced ~20 % fewer spores than the wild-type sensitive strain, indicating a fitness cost of ~20 %. This is because the rate of spore production is a multiplicative factor in the compound transmission rate (Hobbelen et al. 2011), which is a reasonable measure of fitness in this context. If we apply this finding to Fig. 6.3, by drawing a vertical line up from the 0.2 cost on the *X*-axis (see dashed vertical line in Fig. 6.3), we can show that if the fungicides reduces the transmission rate by 70 % (i.e., has 70 % maximum efficacy, $k_k = 0.7$), a mixture can contain up to ~54 % of the high-risk fungicide without emergence of resistance. (This value is obtained from the intersection of the left dashed vertical line with the dotted curve in Fig. 6.3.) If the fungicide has 50 % efficacy, then a mixture can contain up to ~74 % of the high-risk fungicide (similarly, the value is obtained from the intersection of the left dashed vertical line with the solid curve in Fig. 6.3). Finally, if the fungicide has 20 % efficacy, then the mixture can contain up to 100 % of the high-risk fungicide.

Resistance to the DMI flutriafol in the sugarbeet pathogen *C. beticola* presents an example of a high fitness cost (Karaoglanidis et al. 2001). In this case, average spore production was ~40 % lower in nine strains resistant to flutriafol compared to 12 sensitive strains, indicating an average fitness cost of ~40 %. We apply this finding to Fig. 6.3 by drawing a vertical line up from the 0.4 cost on the *X*-axis and find that a fungicide with 70 % efficacy can contain up to 85 % of the high-risk fungicide. With an efficacy of 50 %, the mixture can contain up to ~98 % of the high-risk fungicide, and with an efficacy of 20 %, the mixture can have up to 100 % high-risk fungicide without emergence of resistance.

For each of these examples, we assumed full resistance to the fungicide. If resistance is only partial, then more of the high-risk fungicide can be used without selecting for the emergence of resistance. We also assumed that the fitness cost was a single, constant value for all strains. If the fitness cost exhibits a distribution ranging across higher or lower values than the average shown in our example, then the calculation should be based upon the lowest measured fitness cost in order to avoid selecting for resistant strains.

In the examples discussed above, we used measurements of fitness cost to predict the optimal proportion of the two fungicides in the mixture theoretically. This prediction needs to be tested using field experiments, in which the amount of disease and the frequency of resistance are measured over time at different proportions of the high- and low-risk fungicides in the mixture. From these measurements, the optimal proportions of the fungicides can be obtained empirically. It is this empirically determined optimal proportion of fungicides that is likely to be most useful for growers and fungicide producers to guide decisions about mixture strategies for managing fungicide resistance. Moreover, by comparing the optimal proportions obtained theoretically and empirically, one can evaluate the performance of a mathematical model and identify aspects of the model that need improvement.

6.4.2 Fungicide Mixture Versus Alternation

As we discussed in Sect. 6.2 (also discussed in Shaw MW 2006), fungicide alternation can be effective in the presence of a fitness cost because selection for resistance may be reversed and resistant strains may be eliminated from the population. When using an alternation strategy, the period of selection during which the resistant strain is favored in the presence of the high-risk fungicide is followed by a period during which selection favors the sensitive strain in the absence of this fungicide. The latter period is typically much longer because the selection pressure induced by the high-risk fungicide is much larger than that induced by the fitness cost of resistance. Hence, one needs to wait for quite a long time before the resistant strain disappears and the high-risk fungicide can be used again. Moreover, there are times during which the frequency of the resistant strain becomes high (at the end of the period of the application of the high-risk fungicide), which increases the risk that resistance will spread to other regions. Both of these disadvantages are avoided by using a mixture where the proportion of the low-risk fungicide is above a critical value determined here (Fig. 6.3). In this case there is no need to delay the application of the high-risk fungicide, and the frequency of the resistant strain does not rise above the mutation- or migration-selection equilibrium because the mixture does not induce selection for resistance.

6.4.3 *Dynamic Turnover of Fungicide Mixtures*

In the presence of fitness costs, key advantages of both mixtures and alternations discussed above can be combined using the concept of dynamic turnover. Dynamic turnover means that the fungicides applied will need to change at regular intervals (e.g., within a growing season or between growing seasons) over time and space. It is likely that both mixtures and alternations of fungicides will also fail after prolonged use of the same mixture or alternation, but dynamic turnover of the components in the mixtures will be more likely to provide longer-term solutions. Hence, this approach will be an important aspect of any strategy that leads to durable disease control. The goal of dynamic turnover is to create a diverse chemical environment that imposes disruptive selection on the pathogen population, avoiding the consequences of constant directional selection that favors a particular set of resistance mutations. Dynamic turnover of fungicides presents the pathogen with the evolutionary dilemma of needing to put together a combination of mutations that provides resistance to a diverse set of chemical environments (Mikaberidze et al. 2014b).

6.5 Conclusions

Knowledge of fitness costs is crucial for predicting effectiveness and optimizing disease control strategies based on fungicide combinations, including mixtures, alternations, and dynamic turnover of mixtures. We suggest that dynamic turnover of fungicide mixtures would be a superior strategy in terms of delaying the emergence of fungicide resistance. However, in order to determine optimal parameters for these strategies, more field experimentation as well as additional modeling studies will be necessary. In particular, for polycyclic pathogens, fitness costs of resistant populations should be measured by taking the apparent growth rate, r, as a measure of fitness, because it is most relevant from the point of view of disease control. To the best of our knowledge, the fitness costs of fungicide-resistant strains were not yet measured with respect to r. In the studies cited in Sect. 6.2, different components of fitness were measured that may or may not be related to r. Therefore, we identified a major gap in our knowledge of fitness costs. We hope this chapter will stimulate further experimental investigations to better characterize fitness costs.

References

Anderson JB (2005) Evolution of antifungal-drug resistance: mechanisms and pathogen fitness. Nat Rev Microbiol 3:547–556

Billard A, Fillinger S, Leroux P, Lachaise H, Beffa R, Debieu D (2012) Strong resistance to the fungicide fenhexamid entails a fitness cost in *Botrytis cinerea*, as shown by comparisons of isogenic strains. Pest Manag Sci 68:684–691

Brent KJ, Hollomon DW (2007) Fungicide resistance: the assessment of risk, 2nd edn. CropLife International, Belgium, pp 1–52

Chapara V, Taylor RJ, Pasche JS, Gudmestad NC (2011) Competitive parasitic fitness of mefenoxam-sensitive and -resistant isolates of *Phytophthora erythroseptica* under fungicide selection pressure. Plant Dis 95:691–696

Cools HJ (2008) Are azole fungicides losing ground against Septoria wheat disease? Resistance mechanisms in *Mycosphaerella graminicola*. Pest Manag Sci 64:681–684

Corio-Costet M-F, Dufour M-C, Cigna J, Abadie P, Chen W-J (2010) Diversity and fitness of *Plasmopara viticola* isolates resistant to QoI fungicides. Eur J Plant Pathol 129:315–329

Hobbelen PHF, Paveley ND, Fraaije B, Lucas J, van den Bosch F (2011) Derivation and testing of a model to predict selection for fungicide resistance. Plant Pathol 60:304–313

Holmes G, Eckert J (1995) Relative fitness of imazalil-resistant and -sensitive biotypes of *Penicillium digitatum*. Plant Dis 79:1068–1073

Hu JH, Hong CX, Stromberg EL, Moorman GW (2008) Mefenoxam sensitivity and fitness analysis of *Phytophthora nicotianae* isolates from nurseries in Virginia, USA. Plant Pathol 57:728–736

Iacomi-Vasilescu B, Bataille-Simoneau N, Campion C, Dongo A, Laurent E, Serandat I, Hamon B, Simoneau P (2008) Effect of null mutations in the AbNIK1 gene on saprophytic and parasitic fitness of *Alternaria brassicicola* isolates highly resistant to dicarboximide fungicides. Plant Pathol 57:937–947

Kadish D, Cohen Y (1992) Overseasoning of metalaxyl-sensitive and metalaxyl-resistant isolates of *Phytophthora infestans* in potato tubers weeks after inoculation. Phytopathology 82:887–889

Karaoglanidis G, Thanassoulopoulos CC, Ioannidis PM (2001) Fitness of *Cercospora beticola* field isolates–resistant and–sensitive to demethylation inhibitor fungicides. Eur J Plant Pathol 107:337–347

Karaoglanidis G, Luo Y, Michailides T (2011) Competitive ability and fitness of *Alternaria alternata* isolates resistant to QoI fungicides. Plant Dis 95:178–182

Kim YK, Xiao CL (2011) Stability and fitness of pyraclostrobin- and boscalid-resistant phenotypes in field isolates of *Botrytis cinerea* from apple. Phytopathology 101:1385–1391

Ma B, Uddin W (2009) Fitness and competitive ability of an azoxystrobin-resistant G143A mutant of *Magnaporthe oryzae* from perennial ryegrass. Plant Dis 93:1044–1049

McDonald BA (2014) Using dynamic diversity to achieve durable disease resistance in agricultural ecosystems. Tropic Plant Pathol 39:191–196

Mikaberidze A, McDonald B, Bonhoeffer S (2014a) How to develop smarter host mixtures to control plant disease? Plant Pathol, in press, pre-print arXiv:14022788

Mikaberidze A, McDonald BA, Bonhoeffer S (2014b) Can high risk fungicides be used in mixtures without selecting for fungicide resistance? Phytopathology 104:324–331

Mundt CC (2002) Use of multiline cultivars and cultivar mixtures for disease management. Annu Rev Phytopathol 40:381–410

Peever TL, Milgroom MG (1994) Lack of correlation between fitness and resistance to sterol biosynthesis-inhibiting fungicides in *Pyrenophora teres*. Phytopathology 84:515

Peever T, Milgroom M (1995) Fungicide resistance – lessons for herbicide resistance management? Weed Technol 9:840

Shaw MW (2006) Is there such a thing as a fungicide resistance strategy? A modeller's perspective. Asp Appl Biol 78:37–43

Stukenbrock EH, McDonald BA (2008) The origins of plant pathogens in agro-ecosystems. Annu Rev Phytopathol 46:75–100

Suzuki F, Yamaguchi J, Koba A (2010) Changes in fungicide resistance frequency and population structure of *Pyricularia oryzae* after discontinuance of MBI-D fungicides. Plant Dis 94:329–334

Torriani SF, Brunner PC, McDonald BA, Sierotzki H (2009) QoI resistance emerged independently at least 4 times in European populations of *Mycosphaerella graminicola*. Pest Manag Sci 65:155–162

Webber J (1988) Effect of MBC fungicide tolerance on the fitness of *Ophiostoma ulmi*. Plant Pathol 37:217–224

Zhan J, McDonald BA (2013) Field-based experimental evolution of three cereal pathogens using a mark–release–recapture strategy. Plant Pathol 62:106–114

Zhan J, Stefanato F, McDonald BA (2006) Selection for increased cyproconazole tolerance in *Mycosphaerella graminicola* through local adaptation and in response to host. Mol Plant Pathol 7:259–268

Part II
Mechanisms of Resistance

Chapter 7
Anti-tubulin Agents

David H. Young

Abstract Commercial anti-tubulin fungicides belong to three structural classes: benzimidazoles, *N*-phenylcarbamates, and benzamides. These compounds inhibit microtubule assembly by binding to β-tubulin. Benzimidazoles are used primarily to control ascomycete pathogens and have encountered serious resistance problems and loss of efficacy in the field. The *N*-phenylcarbamate diethofencarb displays negatively correlated cross-resistance to benzimidazoles and was introduced to control benzimidazole-resistant strains of *Botrytis* and other fungi. However, this strategy led to selection of strains resistant to both chemistries. Resistance to benzimidazoles and *N*-phenylcarbamates in lab or field mutants is due primarily to mutations in β-tubulin, and positive or negative cross-resistance between these classes of compound is frequently observed depending on the particular mutation and compounds. The benzamide class of fungicides, which includes zoxamide and ethaboxam, is only used commercially to control oomycete diseases, and there are no reports to date of resistance in the field. Cross-resistance between the benzamide zoxamide, benzimidazoles, and diethofencarb in β-tubulin mutants from *Ascomycetes* and competitive binding assay data show that all three structural classes bind to the "colchicine" site on β-tubulin, a region targeted also by many experimental anticancer and herbicidal compounds. Recent structural studies of β-tubulin are providing new insights into binding of anti-tubulin fungicides and resistance mechanisms.

Keywords Tubulin • Fungicide • Resistance • Benzimidazole • *N*-phenylcarbamate • Benzamide • Carbendazim • Diethofencarb • Zoxamide

D.H. Young (✉)
Dow AgroSciences LLC, 9330 Zionsville Road, Bldg. 306, Indianapolis, IN 46268, USA
e-mail: dyoung@dow.com

H. Ishii, D.W. Hollomon (eds.), *Fungicide Resistance in Plant Pathogens*,
DOI 10.1007/978-4-431-55642-8_7

7.1 Introduction

Tubulin is a protein dimer, which consists of α- and β-subunits. It is the main component of microtubule filaments, which play a central role in nuclear division in all eukaryotic cells. The function of microtubules in nuclear division requires the reversible assembly of tubulin into microtubule polymers. A wide variety of chemicals have been shown to disrupt this process in cells either by inhibiting microtubule assembly or by stabilizing microtubules and preventing their disassembly. In addition to their use as fungicides, compounds which bind to tubulin or microtubules are used as herbicides and anticancer and anthelmintic drugs (Vaughn 2000; Hearn et al. 2006; Aguayo-Ortiz et al. 2013b).

Agricultural fungicides which act on tubulin (Fig. 7.1) belong to three classes: benzimidazoles, *N*-phenylcarbamates, and benzamides. Thiophanate fungicides are generally considered to belong to the benzimidazole class since they are profungicides which are converted to carbendazim. This chapter will summarize current

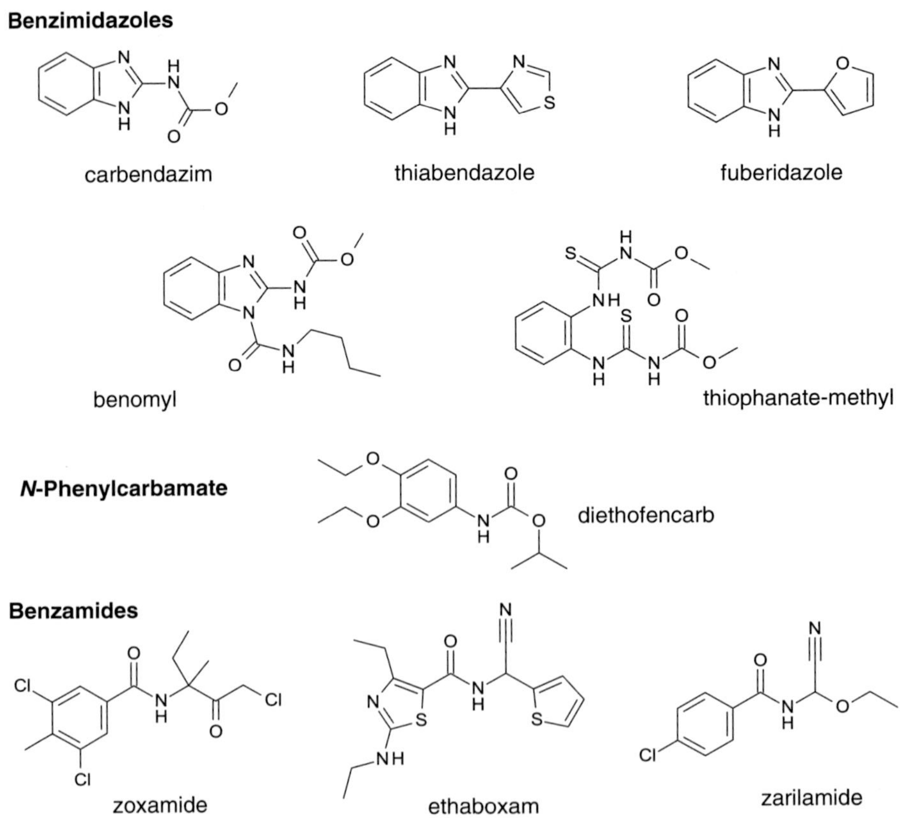

Fig. 7.1 Anti-tubulin fungicides

knowledge about the mode of action of anti-tubulin fungicides, the development of resistance, and mechanisms involved.

7.2 Benzimidazoles

Excellent in-depth reviews of the biological properties of benzimidazoles, their mode of action, and resistance development have been published (Davidse 1982, 1986; Davidse and Ishii 1995; Delp 1995), and the key features are summarized in this section. Benzimidazole fungicides were first introduced in the late 1960s as highly effective, systemic fungicides. They control a broad spectrum of ascomycete pathogens and some *Basidiomycetes* but are inactive against *Oomycetes*. Current commercial products include carbendazim, benomyl, thiabendazole, thiophanate-methyl, and fuberidazole. More than 100 fungi have developed resistance to benzimidazoles, and the speed with which resistance developed for different pathogens was greater for fungi with many life cycles per season (Fungicide Resistance Action Committee 2004). Shortly after their introduction, the development of resistance at a high level prompted intense interest in the mode of action of these fungicides and the mechanism of resistance. A combination of genetic and biochemical studies ultimately revealed that benzimidazoles block nuclear division by binding to β-tubulin and preventing microtubule assembly (reviewed by Davidse 1986; Davidse and Ishii 1995). Laboratory studies of resistant mutants, particularly in *Aspergillus nidulans* (Jung and Oakley 1990; Jung et al. 1992) and *Neurospora crassa* (Fujimura et al. 1992b, c; Koenraadt and Jones 1993), demonstrated a number of different mutations in β-tubulin which were capable of conferring resistance to benzimidazoles. Combined with evidence that binding of benzimidazoles to tubulin-containing protein extracts from resistant strains was lower than for sensitive strains (Davidse and Flach 1978), these studies led to the conclusion that the fungicidal effect of benzimidazoles results from binding to β-tubulin and that particular amino acid substitutions in the protein reduce binding and cause resistance.

Some of the mutations identified in laboratory mutants also occur in resistant field strains and lead to loss of pathogen control. These mutations occur in several codons, including codons 6, 50, 167, 198, 200, and 240. However, mutations E198A/G/K/Q and F200Y are the most prevalent (Fungicide Resistance Action Committee 2004; Koenraadt et al. 1992; Ma and Michailides 2005). In many cases, resistant mutants have been shown to persist in the population for many years even after benzimidazole use was discontinued (Leroux et al. 2005; Walker et al. 2013).

The majority of fungal plant pathogens have only one isotype of β-tubulin, but a few contain two isotypes. In cases where benzimidazole resistance has been documented in fungi with two isotypes, such as in *Fusarium graminearum* and *Colletotrichum* spp., resistance was found to result from mutation at codon 198 in only one tubulin isotype (Chen et al. 2009; Buhr and Dickman 1994; Maymon et al. 2006).

For benzimidazole-resistant field isolates, cross-resistance usually occurs between different benzimidazoles such that when resistance to one compound occurs, the entire class of benzimidazoles loses efficacy (Delp 1995). An exception to this rule has been demonstrated in laboratory studies for *A. nidulans* in which the A165V mutation in β-tubulin confers resistance to thiabendazole and supersensitivity to carbendazim (Jung and Oakley 1990). As expected for a target site-based resistance mechanism, the mutations which cause resistance to benzimidazoles do not affect sensitivity to fungicides which have a different mode of action.

While β-tubulin mutations are the principal cause of resistance development and loss of disease control among fungal populations in the field, other mechanisms may also affect sensitivity to benzimidazoles. In *Aspergillus nidulans*, an efflux mechanism involving a multidrug transporter protein was shown to affect sensitivity to carbendazim and thiabendazole (Andrade et al. 2000), but it is unclear whether such a mechanism affects sensitivity of plant pathogens to benzimidazoles in the field. *Colletotrichum acutatum* is inherently less sensitive to benzimidazoles than other *Colletotrichum* species. In some isolates, this low sensitivity appears to result from tubulin overexpression (Nakaune and Nakano 2007), whereas in other isolates, the substitution of glutamic acid at codon 198 by lysine has been implicated (Chung et al. 2006).

7.3 Diethofencarb

The discovery that certain *N*-phenylcarbamate herbicides were active toward benzimidazole-resistant fungi, yet inactive toward sensitive strains (Leroux and Gredt 1979), prompted interest in the potential of non-phytotoxic analogs to control resistant pathogens. This search led to the discovery and commercialization of diethofencarb (Takahashi et al. 1988), which was found to act by a similar mechanism to benzimidazoles (Fujimura et al. 1992a). This fungicide is unique among commercial fungicides with respect to its development specifically to control resistant strains based on negatively correlated cross-resistance.

The major commercial use of diethofencarb was to control benzimidazole-resistant *Botrytis cinerea* populations (Leroux 1995). The E198A mutation in *B. cinerea* is the primary cause of resistance to benzimidazoles in the field and is responsible for the highly resistant "Rb1" phenotype. Unfortunately, the use of diethofencarb in combination with carbendazim to control the Rb1 population soon led to the emergence of the "Rb2" population which shows resistance to both classes of fungicide and contains the F200Y mutation (Walker et al. 2013; Yarden and Katan 1993). Interestingly, the Rb1 and Rb2 phenotypes differ in their persistence after discontinuation of fungicide pressure. Whereas Rb1 strains were found to persist at high frequency in France many years after the use of diethofencarb-carbendazim mixtures was curtailed, the prevalence of Rb2 strains decreased quite rapidly (Walker et al. 2013).

The F200Y mutation, which confers resistance to both benzimidazoles and diethofencarb, and mutations at codon 198 have also been reported in other fungi (Albertini et al. 1999; Koenraadt et al. 1992). In addition to E198A, various other amino acid substitutions (E198A/E198G/E198K/E198N) occur, and these confer varying levels of sensitivity to the *N*-phenylcarbamate chemistry. The ability of a given mutation at codon 198 to confer sensitivity to *N*-phenylcarbamates also varies depending on the particular analog. In the case of diethofencarb, the mutations which confer sensitivity are E198A and E198G.

Although diethofencarb is the only fungicide to be commercialized for control of benzimidazole-resistant pathogens, other *N*-phenylcarbamates, *N*-phenylformamidoximes, benzamides, and diphenylamine also show negatively correlated cross-resistance toward benzimidazole-resistant fungi (Nakata et al. 1987; Leroux and Gredt 1979; Sholberg et al. 2005; Young and Slawecki 2005).

As in the case of benzimidazoles, genetic evidence for binding of diethofencarb and *N*-phenylformamidoximes to β-tubulin was complemented by binding studies which showed enhanced binding in extracts from benzimidazole-resistant fungi (Fujimura et al. 1992a; Ishii and Takeda 1989).

7.4 Benzamides

The benzamide class of anti-tubulin fungicides includes the commercial oomycete fungicides zoxamide (Egan et al. 1998) and ethaboxam (Kim et al. 2002) and the earlier experimental compound zarilamide (Heaney et al. 1988). All three compounds arrest nuclear division and disrupt the microtubule cytoskeleton (Uchida et al. 2005; Young 1991; Young and Slawecki 2001). Zoxamide and zarilamide have been shown by means of competitive binding assays to bind to a common site on β-tubulin (Young and Slawecki 2001). Ethaboxam is believed to bind to this same site based on its structural similarity to zarilamide (Young 2012) and its effects on microtubules at the cellular level (Uchida et al. 2005).

Zoxamide belongs to a series of α-haloketone-containing benzamides which originated from an exploratory synthesis effort around the herbicide pronamide, which acts on plant microtubules. Zoxamide is used commercially only to control oomycete diseases. However, α-haloketone-containing benzamides are also active against non-oomycete fungi, protozoan, plant, and mammalian cells (Young 2012). Although zoxamide is a mixture of two enantiomers, its biological activity is due to the S-enantiomer and involves a highly specific covalent binding to Cys-239 in β-tubulin (Young et al. 2006). This interaction enabled the development of cell-based radioligand binding assays in oomycete, plant, and mammalian cells which were used to identify other classes of anti-tubulin agents which bind to the same site (Young and Lewandowski 2000; Young and Slawecki 2001; Young et al. 2006). These include the benzimidazole nocodazole, *N*-phenylcarbamate and benzamide herbicides, colchicine, and various anticancer agents which bind to the colchicine site. Cys-239, with which zoxamide reacts, is also involved in covalent binding of some other colchicine site ligands, including the anticancer agents DCBT and T138067 and certain reactive colchicine derivatives (Bai et al. 1989, 2000; Shan et al. 1999).

The conclusion that anti-tubulin benzamides, benzimidazoles, and *N*-phenylcarbamates bind to a common site is also supported by the effect of β-tubulin mutations at codons 198 and 200. These mutations alter sensitivity to zoxamide as well as benzimidazoles and diethofencarb. Similar patterns of cross-resistance are found for zoxamide and diethofencarb in these mutants (Table 7.1).

Whereas benzimidazole fungicides and diethofencarb are classified by the Fungicide Resistance Action Committee as high risk for resistance development, zoxamide and ethaboxam are classified as low to medium risk. Since the commercial introduction of zoxamide in 2001 to control oomycete pathogens, there have been no reports of field resistance to benzamides, and their ability to control strains resistant to other uni-site fungicides such as phenylamides and carboxylic acid amides is a valuable feature in the oomycete market. The sharp contrast between the rapid emergence of benzimidazole- and diethofencarb-resistant fungal populations and the lack of field resistance to benzamides is likely determined by differences in the pathogens targeted. Since *Oomycetes* are diploid organisms and resistance mutations in β-tubulin are generally recessive, a high level of resistance would not be anticipated in a heterozygote containing a resistant allele (Young et al. 2001). In theory, sexual recombination involving two such heterozygotes could produce a highly resistant homozygous strain. However, the absence of field resistance to date suggests that such events are very rare. Consistent with the lack of observed resistance development in the field, early attempts in the laboratory to isolate mutants resistant to zarilamide and zoxamide in various *Phytophthora* species by adaptation or mutagenesis were unsuccessful (Eacott 1986; Young et al. 2001).

More recently, there have been reports of resistance in laboratory mutants. A 955-fold loss of sensitivity to zoxamide was reported for an isolate of *P. sylvaticum* following repeated transfer on zoxamide-amended medium (Martinez et al. 2005);

Table 7.1 Cross-resistance between zoxamide, benzimidazoles, and diethofencarb in fungal field isolates containing mutations in the β-tubulin gene

		Phenotype	
Mutation	Benzimidazole	Diethofencarb	Zoxamide
E198A	↓	↑	↑[a, b, c, d, e]
E198K	↓	↓	↓[c, e]
F200Y	↓	↓	↓[b, e]

↓ Reduced sensitivity or resistant
↑ Increased sensitivity
[a]*Pyrenopeziza brassicae* (Carter et al. 2013)
[b]*Colletotrichum gloeosporioides* (Deng et al. 2012)
[c]*Botrytis cinerea* (Malandrakis et al. 2011)
[d]*Monilinia laxa* (Malandrakis et al. 2012)
[e]*Venturia inaequalis* (Young and Slawecki 2005)

however, the mechanism involved is not known. In another study, UV-generated mutants of *P. capsici* showed >100-fold reduction in sensitivity to zoxamide; however, no mutations in the β-tubulin gene were detected (Yang et al. 2011). Further work will be needed to discover the mechanisms responsible for reduced sensitivity in such lab mutants and evaluate their possible relevance to potential resistance development in the field.

7.5 Insights from Structural Studies

Various binding sites exist for anti-tubulin agents. In studies of mammalian tubulin, agents which inhibit polymerization of microtubules most often bind to either the "colchicine site" or the "vinblastine site" on β-tubulin (Wilson and Jordan 1994). A distinct binding site on β-tubulin for the anticancer agent maytansine is also targeted by rhizoxin (Prota et al. 2014), a natural product with strong fungicidal activity (Ishii 1992). The anticancer natural product drug paclitaxel, which promotes microtubule polymerization, binds to yet another site on β-tubulin and demonstrates activity against oomycete fungi (Young et al. 1992). A class of synthetic tubulin polymerization promoters, represented by the triazolopyrimidine BAS600F, binds to a different site from paclitaxel (Zhang et al. 2007) and has been investigated extensively for fungicidal use by the agrochemical industry, although no compound has been commercialized (Lamberth et al. 2013). As discussed above, biochemical and mutation data support binding of benzimidazoles, benzamides, and diethofencarb to the colchicine site on β-tubulin. Thus, only one of several potential tubulin binding sites has so far been exploited commercially for controlling fungal pathogens.

Historically, it has proven difficult to crystallize tubulin to obtain detailed structural information on the colchicine site. However, the use of the phosphoprotein stathmin to stabilize mammalian tubulin by forming a complex with two dimers recently enabled the use of X-ray crystallography to provide structural details of the colchicine site and insight into the mechanism by which ligands inhibit microtubule assembly (Barbier et al. 2010; Dorleans et al. 2009; Gigant et al. 2000; Ravelli et al. 2004). Colchicine site ligands bind at the interface between the α- and β-tubulin subunits and appear to act by stabilizing a curved conformation which is not compatible with microtubule formation. In addition to fungi, benzimidazoles are active against nematodes and protozoal parasites, and some of the same mutations which confer resistance in fungal pathogens cause resistance in nematodes, notably F167Y, E198A, and F200Y. Homology models of nematode and protozoal β-tubulins have been generated based on the crystal structure of mammalian β-tubulin and used to study the binding of benzimidazoles. Amino acids at the mutation sites (F167, E198A, F200), and Cys-239 which is involved in the covalent binding of zoxamide, are located at the colchicine/benzimidazole binding site in these models (Aguayo-Ortiz et al. 2013a). It was proposed that the F167Y and F200Y resistance mutations reduce binding of benzimidazoles by increasing polarity in the binding pocket, whereas the E198A mutation prevents formation of a key H-bond interaction (Aguayo-Ortiz et al. 2013b).

Future structural studies will likely provide greater understanding of the mechanisms by which resistance mutations affect binding of different fungicides to the colchicine site and the structural basis for differential sensitivity between organisms to the various anti-tubulin fungicides. These insights could also lay the foundation for designing new anti-tubulin fungicides capable of controlling resistant fungal pathogens.

References

Aguayo-Ortiz R, Mendez-Lucio O, Medina-Franco JL, Castillo R, Yepez-Mulia L, Hernandez-Luis F, Hernandez-Campos A (2013a) Towards the identification of the binding site of benzimidazoles to β-tubulin of *Trichinella spiralis*: insights from computational and experimental data. J Mol Graph Model 41:12–19

Aguayo-Ortiz R, Mendez-Lucio O, Romo-Mancillas A, Castillo R, Yepez-Mulia L, Medina-Franco JL, Hernandez-Campos A (2013b) Molecular basis for benzimidazole resistance from a novel β-tubulin binding site model. J Mol Graph Model 45:26–37

Albertini C, Gredt M, Leroux P (1999) Mutations of the β-tubulin gene associated with different phenotypes of benzimidazole resistance in the cereal eyespot fungi *Tapesia yallundae* and *Tapesia acuformis*. Pestic Biochem Physiol 64:17–31

Andrade AC, Del Sorbo G, Van Nistelrooy JGM, De Waard MA (2000) The ABC transporter AtrB from *Aspergillus nidulans* mediates resistance to all major classes of fungicides and some natural toxic compounds. Microbiology 146:1987–1997

Bai RL, Duanmu C, Hamel E (1989) Mechanism of action of the antimitotic drug 2,4-dichlorobenzylthiocyanate. Biochim Biophys Acta 994:12–20

Bai RL, Covell DG, Pei XF, Ewell JB, Nguyen NY, Brossi A, Hamel E (2000) Mapping the binding site of colchicinoids on β-tubulin: 2-chloroacetyl-2-demethylthiocolchicine covalently reacts predominantly with cysteine 239 and secondarily with cysteine 354. J Biol Chem 275:40443–40452

Barbier P, Dorleans A, Devred F, Sanz L, Allegro D, Alfonso C, Knossow M, Peyrot V, Andreu JM (2010) Stathmin and interfacial microtubule inhibitors recognize a naturally curved conformation of tubulin dimers. J Biol Chem 285:31672–31681

Buhr TL, Dickman MB (1994) Isolation, characterization, and expression of a second beta-tubulin-encoding gene from *Colletotrichum gloeosporioides* f. sp. *aeschynomene*. Appl Environ Microbiol 60:4155–4159

Carter HE, Cools HJ, West JS, Shaw MW, Fraaije BA (2013) Detection and molecular characterisation of *Pyrenopeziza brassicae* isolates resistant to methyl benzimidazole carbamates. Pest Manag Sci 69:1040–1048

Chen C, Yu J, Bi C, Zhang Y, Xu J, Wang J, Zhou M (2009) Mutations in a β-tubulin confer resistance of *Gibberella zeae* to benzimidazole fungicides. Phytopathology 99:1403–1411

Chung W, Ishii H, Nishimura K, Fukaya M, Yano K, Kajitani Y (2006) Fungicide sensitivity and phylogenetic relationship of anthracnose fungi isolated from various fruit crops in Japan. Plant Dis 90:506–512

Davidse LC (1982) Benzimidazole compounds: selectivity and resistance. In: Dekker J, Georgopoulos SG (eds) Fungicide resistance in crop protection. Centre for Agricultural Publishing and Documentation, Wageningen, pp 60–70

Davidse LC (1986) Benzimidazole fungicides: mechanism of action and biological impact. Annu Rev Phytopathol 24:43–65

Davidse LC, Flach W (1978) Interaction of thiabendazole with fungal tubulin. Biochim Biophys Acta 543:82–90

Davidse LC, Ishii H (1995) Biochemical and molecular aspects of the mechanism of action of benzimidazoles, *N*-phenylcarbamates and *N*-phenylformamidoximes and the mechanism of resistance to these compounds in fungi. In: Lyr H (ed) Modern selective fungicides – properties, applications, mechanisms of action, 2nd edn. Gustav Fischer Verlag, Jena, pp 305–322

Delp CJ (1995) Benzimidazole and related fungicides. In: Lyr H (ed) Modern selective fungicides – properties, applications, mechanisms of action, 2nd edn. Gustav Fischer Verlag, Jena, pp 291–303

Deng W, Mei X, Yang M, Chen L, Bi Y, Cai M, Wang H, Puqian E, He X, Zhu S (2012) Correlation between sensitivity to zoxamide and the mutation of β-tubulin amino acid of *Colletotrichum gloeosporioides* and two *Phytophthora* spp. isolates. Nongyaoxue Xuebao 14:143–150

Dorleans A, Gigant B, Ravelli RBG, Mailliet P, Mikol V, Knossow M (2009) Variations in the colchicine-binding domain provide insight into the structural switch of tubulin. Proc Natl Acad Sci U S A 106:13775–13779

Eacott CJP (1986) Assessment of the risk of resistance to benzamide fungicides in *Phytophthora infestans*. Dissertation, University of London

Egan AR, Michelotti EL, Young DH, Wilson WJ, Mattioda H (1998) RH-7281: a novel fungicide for control of downy mildew and late blight. In: Proceedings of Brighton crop protection conference – pests and diseases, The British Crop Protection Council, Farnham, pp 335–342

Fujimura M, Kamakura T, Yamaguchi I (1992a) Action mechanism of diethofencarb to a benzimidazole-resistant mutant in *Neurospora crassa*. J Pestic Sci 17:237–242

Fujimura M, Kamakura T, Inoue H, Inoue S, Yamaguchi I (1992b) Sensitivity of *Neurospora crassa* to benzimidazoles and *N*-phenylcarbamates: effect of amino acid substitutions at position 198 in β-tubulin. Pestic Biochem Physiol 44:165–173

Fujimura M, Oeda K, Inoue H, Kato T (1992c) A single amino-acid substitution in the beta-tubulin gene of *Neurospora* confers both carbendazim resistance and diethofencarb sensitivity. Curr Genet 21:399–404

Fungicide Resistance Action Committee (2004) Benzimidazoles. http://www.frac.info/index.htm. Accessed 2 June 2014

Gigant B, Curmi PA, Martin-Barbey C, Charbaut E, Lachkar S, Lebeau L, Siavoshian S, Sobel A, Knossow M (2000) The 4 Å x-ray structure of a tubulin: stathmin-like domain complex. Cell 102:809–816

Heaney SP, Shephard MC, Crowley PJ, Shearing SJ (1988) ICIA0001: a novel benzamide fungicide. In: Proceedings of Brighton crop protection conference -- pests and diseases, The British Crop Protection Council, Thorntorn Heath, pp 551–558

Hearn BR, Shaw SJ, Myles DC (2006) Microtubule targeting agents. In: Taylor JB, Triggle DJ (eds) Comprehensive medicinal chemistry II, vol 7. Elsevier, New York, pp 81–110

Ishii H (1992) Target sites of tubulin-binding fungicides. In: Köller W (ed) Target sites of fungicide action. CRC Press, Boca Raton, pp 43–52

Ishii H, Takeda H (1989) Differential binding of a *N*-phenylformamidoxime compound in cell free extracts of benzimidazole-resistant and -sensitive isolates of *Venturia nashicola*, *Botrytis cinerea* and *Gibberella fujikuroi*. Neth J Plant Pathol 95(Suppl 1):99–108

Jung MK, Oakley BR (1990) Identification of an amino acid substitution in the *benA*, beta-tubulin gene of *Aspergillus nidulans* that confers thiabendazole resistance and benomyl supersensitivity. Cell Motil Cytoskel 17:87–94

Jung MK, Wilder IB, Oakley BR (1992) Amino acid alterations in the *benA* (beta-tubulin) gene of *Aspergillus nidulans* that confer benomyl resistance. Cell Motil Cytoskel 22:170–174

Kim DS, Lee YS, Chun SJ, Choi WB, Lee SW, Kim GT, Kang KG, Joe GH, Cho JH (2002) Ethaboxam: a new Oomycetes fungicide. In: Proceedings of Brighton crop protection conference – pests and diseases, The British Crop Protection Council, Farnham, pp 377–382

Koenraadt H, Jones AL (1993) Resistance to benomyl conferred by mutations in codon 198 or 200 of the beta-tubulin gene of *Neurospora crassa* and sensitivity to diethofencarb conferred by codon 198. Phytopathology 83:850–854

Koenraadt H, Somerville SC, Jones AL (1992) Characterizations of mutations in the beta-tubulin gene of benomyl-resistant field strains of *Venturia inaequalis* and other plant pathogenic fungi. Phytopathology 82:1348–1354

Lamberth C, Dumeunier R, Trah S, Wendeborn S, Godwin J, Schneiter P, Corran A (2013) Synthesis and fungicidal activity of tubulin polymerisation promoters. Part 3: imidazoles. Bioorg Med Chem 21:127–134

Leroux P (1995) Progress and problems in the control of *Botrytis cinerea* in grapevine. Pesticide Outlook 6:20–24

Leroux P, Gredt M (1979) Phenomena of negative cross resistance in *Botrytis cinerea* Pers. between benzimidazole fungicides and carbamate herbicides. Phytiatr Phytopharm 28:79–86

Leroux P, Gredt M, Walker AS, Moinard JM, Caron D (2005) Resistance of the wheat leaf blotch pathogen *Septoria tritici* to fungicides in France. In: Dehne HW, Gisi U, Kuck KH, Russell PE, Lyr H (eds) Modern fungicides and antifungal compounds IV. The British Crop Protection Council, Alton, pp 115–124

Ma Z, Michailides TJ (2005) Advances in understanding molecular mechanisms of fungicide resistance and molecular detection of resistant genotypes in phytopathogenic fungi. Crop Prot 24:853–863

Malandrakis A, Markoglou A, Ziogas B (2011) Molecular characterization of benzimidazole-resistant *B. cinerea* field isolates with reduced or enhanced sensitivity to zoxamide and diethofencarb. Pestic Biochem Physiol 99:118–124

Malandrakis AA, Markoglou AN, Ziogas BN (2012) PCR-RFLP detection of the E198A mutation conferring resistance to benzimidazoles in field isolates of *Monilinia laxa* from Greece. Crop Prot 39:11–17

Martinez C, Levesque CA, Belanger RR, Tweddell RJ (2005) Evaluation of fungicides for the control of carrot cavity spot. Pest Manag Sci 61:767–771

Maymon M, Zveibil A, Pivonia S, Minz D, Freeman S (2006) Identification and characterization of benomyl-resistant and -sensitive populations of *Colletotrichum gloeosporioides* from statice (*Limonium* spp.). Phytopathology 96:542–548

Nakata A, Sano S, Hashimoto S, Hayakawa K, Nishikawa H, Yasuda Y (1987) Negatively correlated cross-resistance to *N*-phenylformamidoximes in benzimidazole-resistant phytopathogenic fungi. Annu Phytopathol Soc Jpn 53:659–662

Nakaune R, Nakano M (2007) Benomyl resistance of *Colletotrichum acutatum* is caused by enhanced expression of β-tubulin 1 gene regulated by putative leucine zipper protein CaBEN1. Fungal Genet Biol 44:1324

Prota AE, Bargsten K, Diaz JF, Marsh M, Cuevas C, Liniger M, Neuhaus C, Andreu JM, Altmann K, Steinmetz MO (2014) A new tubulin-binding site and pharmacophore for microtubule-destabilizing anticancer drugs. Proc Natl Acad Sci U S A 111:13817–13821

Ravelli RBG, Gigant B, Curmi PA, Jourdain I, Lachkar S, Sobel A, Knossow M (2004) Insight into tubulin regulation from a complex with colchicine and a stathmin-like domain. Nature 428:198–202

Shan B, Medina JC, Santha E, Frankmoelle WP, Chou TC, Learned RM, Narbut MR, Stott D, Pengguang W, Jaen JC, Rosen T, Timmermans PBMWM, Beckman H (1999) Selective, covalent modification of β-tubulin residue Cys-239 by T138067, an antitumor agent with *in vivo* efficacy against multidrug-resistant tumors. Proc Natl Acad Sci U S A 96:5686–5691

Sholberg PL, Harlton C, Haag P, Levesque CA, O'Gorman D, Seifert K (2005) Benzimidazole and diphenylamine sensitivity and identity of *Penicillium* spp. that cause postharvest blue mold of apples using β-tubulin gene sequences. Postharv Biol Technol 36:41–49

Takahashi J, Nakamura S, Noguchi H, Kato T, Kamoshita K (1988) Fungicidal activity of *N*-phenylcarbamates against benzimidazole resistant fungi. J Pestic Sci 13:63–69

Uchida M, Roberson RW, Chun SJ, Kim DS (2005) *In vivo* effects of the fungicide ethaboxam on microtubule integrity in *Phytophthora infestans*. Pest Manag Sci 61:787–792

Vaughn KC (2000) Anticytoskeletal herbicides. In: Nick P (ed) Plant microtubules. Springer, Berlin/Heidelberg, pp 193–205

Walker AS, Micoud A, Rémuson F, Grosman J, Gredt M, Leroux P (2013) French vineyards provide information that opens ways for effective resistance management of *Botrytis cinerea* (grey mould). Pest Manag Sci 69:667–678

Wilson L, Jordan MA (1994) Pharmacological probes of microtubule function. In: Hyams JS, Lloyd CW (eds) Microtubules. Wiley-Liss, New York, pp 59–83

Yang B, Cui X, Lu X, Cai M, Liu X, Hao JJ (2011) Baseline sensitivity of natural population and resistance of mutants in *Phytophthora capsici* to zoxamide. Phytopathology 101:1104–1111

Yarden O, Katan T (1993) Mutations leading to substitutions at amino acids 198 and 200 of beta-tubulin that correlate with benomyl-resistance phenotypes of field strains of *Botrytis cinerea*. Phytopathology 83:1478–1483

Young DH (1991) Effects of zarilamide on microtubules and nuclear division in *Phytophthora capsici* and tobacco suspension-cultured cells. Pestic Biochem Physiol 40:149–161

Young DH (2012) Zoxamide: an antitubulin fungicide for the control of Oomycete pathogens. In: Kramer W, Schirmer U, Jeschke P, Witschel M (eds) Modern crop protection compounds, vol 2. Wiley-VCH, Weinheim, pp 739–748

Young DH, Lewandowski VT (2000) Covalent binding of the benzamide RH-4032 to tubulin in suspension-cultured tobacco cells and its application in a cell-based competitive-binding assay. Plant Physiol 124:115–124

Young DH, Slawecki RA (2001) Mode of action of zoxamide (RH-7281), a new Oomycete fungicide. Pestic Biochem Physiol 69:100–111

Young DH, Slawecki RA (2005) Cross resistance relationships between zoxamide, carbendazim and diethofencarb in field isolates of *Botrytis cinerea* and other fungi. In: Dehne HW, Gisi U, Kuck KH, Russell PE, Lyr H (eds) Modern fungicides and antifungal compounds IV. The British Crop Production Council, Alton, pp 125–131

Young DH, Michelotti EL, Swindell CS, Krauss NE (1992) Antifungal properties of taxol and various analogs. Experientia 48:882–885

Young DH, Spiewak SL, Slawecki RA (2001) Laboratory studies to assess the risk of development of resistance to zoxamide. Pest Manag Sci 57:1081–1087

Young DH, Rubio FM, Danis PO (2006) A radioligand binding assay for antitubulin activity in tumor cells. J Biomol Screen 11:82–89

Zhang N, Ayral-Kaloustian S, Nguyen T, Afragola J, Hernandez R, Lucas J, Gibbons J, Beyer C (2007) Synthesis and SAR of [1,2,4]triazolo[1,5-*a*]pyrimidines, a class of anticancer agents with a unique mechanism of tubulin inhibition. J Med Chem 50:319–327

Chapter 8
Respiration Inhibitors: Complex II

Gerd Stammler, Antje Wolf, Alice Glaettli, and Kristin Klappach

Abstract The succinate dehydrogenase inhibitors (SDHIs) are a highly attractive group of fungicides with regard to their history of use, chemical innovations, and spectrum of targeted diseases. They are the most modern, broadly effective fungicide group available to farmers for their disease control programs. As the group name indicates, the mode of action of SDHIs is the inhibition of succinate dehydrogenase in the respiration chain. Along with the broadening of their disease control palette, the structural complexity of the SDHIs has been increased. Nevertheless, the currently known SDHIs share common chemical features necessary for fungicidal activity, suggesting a very similar binding to the target as demonstrated here by three-dimensional alignment and computational docking experiments. Different target site mutations have been found in laboratory mutants and field isolates of various fungal species. Modeling studies with different target site mutations indicated that some of the observed target alterations conferring resistance to SDHIs have a direct impact on the binding behavior of SDHIs, whereas other mutations influence SDHI binding by long-range structural rearrangement in the transmembrane region of complex II. A diverse picture is seen regarding the effects of these mutations on the sensitivity toward various SDHIs in various pathogens and thus regarding cross-resistance. Some mutations cause a loss of sensitivity to all currently commercialized SDHIs, but there are also mutations where no complete cross-resistance can be found. Therefore, it can be stated that cross-resistance between SDHIs exists in general, but phenotypically exceptions are observed.

Keywords Binding • Carboxamides • Complex II • Fungicides • Resistance • SDHI • Succinate dehydrogenase • Target site mutation

---

G. Stammler (✉) • K. Klappach
Agricultural Center, BASF SE, Speyerer Strasse 2, D-67177 Limburgerhof, Germany
e-mail: gerd.stammler@basf.com

A. Wolf • A. Glaettli
Specialty Chemicals Research, BASF SE,
Carl-Bosch Strasse, D-67056 Ludwigshafen, Germany

H. Ishii, D.W. Hollomon (eds.), *Fungicide Resistance in Plant Pathogens*,
DOI 10.1007/978-4-431-55642-8_8

8.1 History of SDHIs

Before foundation of the Fungicide Resistance Action Committee (FRAC) Working Group for succinate dehydrogenase (complex II, succinate–ubiquinone reductase) inhibitors (SDHIs) in 2009, this class of compounds was known as carboxamides. After forming the Working Group for this mode of action (MoA), it was decided that the Working Group name should refer to the target rather than the chemistry. Both names, SDHIs and carboxamides, are still used, but the term "SDHIs" is replacing "carboxamides" more and more, and its use is recommended. Von Schmeling and Kulka (1966) described the systemic activity on basidiomycete fungi of carboxin and oxycarboxin. These two compounds were launched as seed disinfection and foliar spray agents in 1969 and 1975, respectively (von Schmeling and Kulka 1966). Thereafter, a number of structural analogues were introduced such as benodanil and fenfuram, both in 1974, followed by a range of SDHIs (e.g., mepronil, flutolanil, furametpyr, and thifluzamide) in the 1980s and 1990s, with a focus on *Rhizoctonia solani* in rice, turf, potatoes, and other crops. Despite the replacement of the 1,4-oxathiin ring in carboxin by a variety of alternative ring systems (e.g., phenyl, furan, pyrazole, or thiazole), the activity spectrum of these early SDHIs remained limited to *Basidiomycetes*. However, replacing the 1,4-oxathiin ring by a pyridine moiety and simultaneously introducing a phenyl group in the 2′ position of the anilide ring resulted in an SDHI with a considerable expansion of its disease control spectrum, to include various ascomycete fungi: boscalid, which has been launched in 2003. This was the first SDHI to control a broad range of pathogens in fruits, grapes, vegetables, and also arable crops such as cereals and canola (Stammler et al. 2007, 2008). Less than a decade later, the combination of boscalid as starting point and a competitive environment between research-based companies has expanded this second generation of SDHIs into a diverse group of compounds, and a number of new SDHI molecules have been or will be launched in all major agricultural crops. These reported new SDHIs are so far benzovindiflupyr, bixafen, fluopyram, fluxapyroxad, isofetamid, isopyrazam, penflufen, penthiopyrad, and sedaxane.

8.2 Mode of Action of SDHIs

Over the years, the structural complexity of SDHIs increased in parallel to the broadening of their disease control spectrum. Still, as a comparison of the chemical structures of some SDHIs shows (Fig. 8.1), they share a number of common features essential for fungicidal activity, such as the central amide moiety essential for hydrogen bond interactions in the ubiquinone binding site of SDH and the aromatic ring in the aniline part ensuring optimal hydrophobic contacts or π–π interactions to the binding site. Furthermore, most of the modern SDHIs (bixafen, boscalid, fluxapyroxad, isopyrazam, penthiopyrad, sedaxane) bear a nitrogen-containing

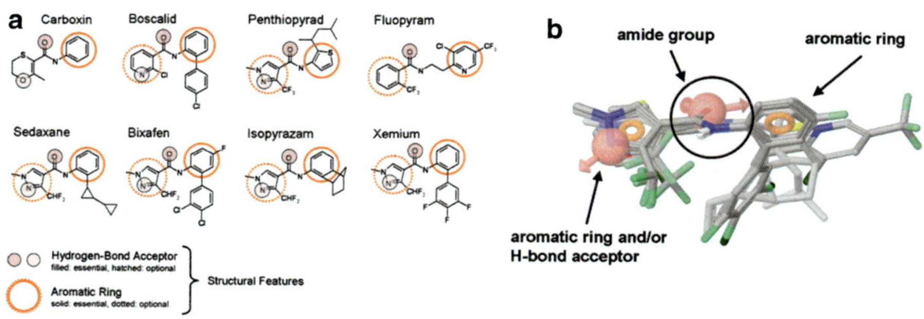

Fig. 8.1 Chemical structures of some SDHIs (**a**) and their structural alignment (**b**) illustrate common chemical features essential for fungicidal activity. The alignment suggests an identical binding mode at SDH

heterocycle (pyridine or pyrazole), which supposedly increases the binding affinity via π–π interactions and additional hydrogen bonding (through the aromatic nitrogen) to the binding site. The alignment of their three-dimensional structures demonstrates that the common chemical features discussed above superimpose very well, suggesting an identical binding mode at SDH. Furthermore, the three-dimensional alignment shows that the left-hand part of SDHIs, the carboxylic acid building block, appears to be structurally quite conserved, whereas the right-hand part, the aniline building block, is more variable.

Early mode of action studies with carboxin showed profound effects on respiration (Mathre 1970; Ragsdale and Sisler 1970). Further investigations pointed at complex II of the mitochondrial respiration chain as a possible target of SDHIs (Mathre 1971; Ulrich and Mathre 1972), which has been confirmed by the detection of mutations in the SDH gene in carboxin-resistant strains of *Ustilago maydis* (Georgopoulos et al. 1972). Complex II is a membrane-anchored protein and represents a link between the mitochondrial respiration and the Krebs cycle (Cecchini 2003). It consists of four subunits: a flavoprotein (SDH-A) catalyzing the oxidation of succinate to fumarate, an iron–sulfur protein (SDH-B) containing the three iron–sulfur clusters responsible for the electron transfer from succinate to ubiquinone, and two membrane anchor subunits (C and D) with the heme b located between two antiparallel helices of SDH-C and SDH-D (Fig. 8.2).

The question of the mechanism of inhibition and site of action of SDHIs has been addressed by various studies (Kuhn 1984). Electron paramagnetic resonance (EPR) spectroscopy studies proposed that carboxin interrupts the electron transfer between the [3Fe–4S] cluster and the ubiquinone (Ackrell et al. 1977). Photoaffinity labeling studies revealed binding of azidocarboxin to the subunit C and D (Ramsay et al. 1981). Both findings point to the ubiquinone binding pocket (Q-site) as a possible site of action, which has been later confirmed by X-ray studies on complex II (Yankovskaya et al. 2003; Huang et al. 2006).

The ubiquinone binding site is a hydrophobic pocket formed by residues of the subunits B, C, and D and is highly conserved throughout a range of organisms

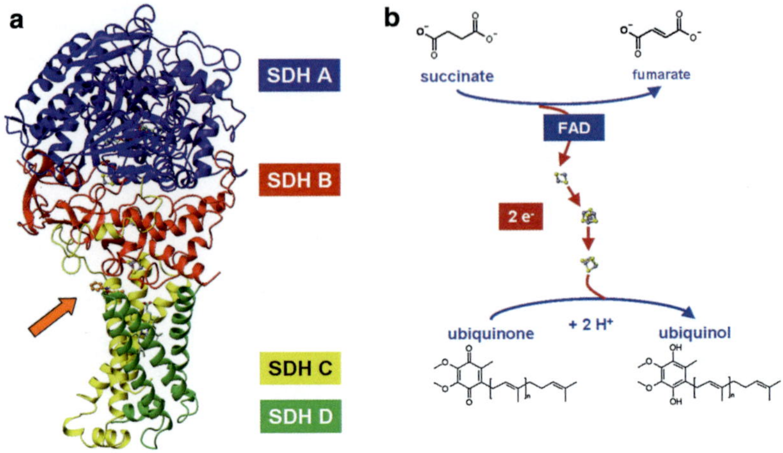

Fig. 8.2 Homology model of SDH from *B. cinerea* with subunits A (flavoprotein, *blue*), B (iron–sulfur protein, *red*), C, and D (membrane anchors, *yellow* and *green*) (**a**). The *orange arrow* points at the Q-site. SDH catalyzes the oxidation of succinate to fumarate and transports the released electrons from the flavin via the three iron–sulfur clusters to the Q-site, where the reduction of ubiquinone to ubiquinol takes place (**b**). Technical details of the homology model construction are described elsewhere (Glaettli et al. 2009)

(Horsefield et al. 2004). Docking experiments suggest an identical binding mode for different SDHIs, forming hydrogen bonds via the central amide moiety to the highly conserved tryptophan of subunit B and tyrosine of subunit C (Fig. 8.3, Panel A). They also show that SDHIs bind deeper into the Q-site than ubiquinone: The carboxylic acid building block directly interacts with the histidine at the rear end of the pocket (Fig. 8.3, Panel B).

8.3 Mechanisms of Resistance to SDHIs

Carboxin and other active ingredients of the first-generation SDHIs were among the first substances subjected to the then newly established scientific discipline of fungicide resistance research in the mid-1970s. Despite the fact that its target pathogens were considered to have a low resistance risk based on their biology and history of resistance development, the first cases of resistance to carboxin, in, *e.g.*, corn smut and chrysanthemum rust, appeared 5–7 years after market introduction (Abiko et al. 1977; Ben-Yephet et al. 1975; Georgopoulos et al. 1975; Leroux and Berthier 1988). Thus, the modes of resistance, fitness of resistant individuals, the role of target site mutations, and the inheritance of these mutations were intensively investigated. Studies showed that single-site mutations were responsible for the loss of fungicide efficacy (Keon et al. 1991; Skinner et al. 1998; Matsson et al. 1998; Matsson and Hederstedt 2001; Ito et al. 2004). With investigations of SDHI

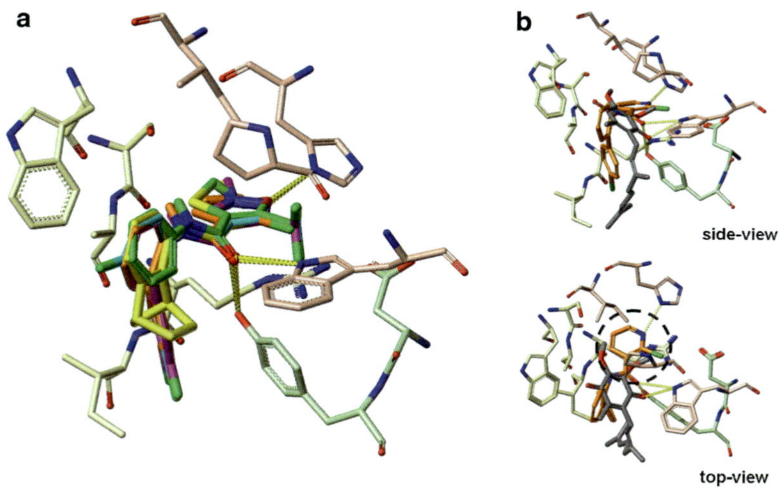

Fig. 8.3 Ubiquinone binding site with carboxin (*green*), boscalid (*orange*), bixafen (*cyan*), isopyrazam (*yellow*), and fluxapyroxad (*magenta*) bound as suggested by computational docking experiments (**a**). Comparison of the binding modes of ubiquinone (*grey*) and SDHIs represented by boscalid (*orange*), *side view* and *top view*, showing that SDHIs bind deeper into the Q-site than ubiquinone itself (**b**). Computational details on the docking experiments are reported elsewhere (Glaettli et al. 2009)

resistance moving forward after the introduction of broad-spectrum SDHIs, a complex picture is forming. Several mutations in the target protein at different positions in three SDH subunits B, C, and D were detected in field isolates of some plant pathogens such as *Botrytis cinerea* (Stammler et al. 2007; Veloukas et al. 2011), *Corynespora cassiicola* (Miyamoto et al. 2009, 2010a), *Alternaria alternata* (Avenot and Michaelidis 2007), *Alternaria solani* (Miles et al. 2013), *Didymella bryoniae* (Avenot et al. 2012; Fernandez-Ortuno et al. 2012), *Podosphaera xanthii* (Miyamoto et al. 2010b), and *Sclerotinia sclerotiorum* (Glaettli et al. 2009) and in laboratory mutants of *Zymoseptoria tritici* (=*Mycosphaerella graminicola*) (Skinner et al. 1998; Stammler et al. 2010; Fraaije et al. 2012; Scalliet et al. 2012). Different mutations were found at one location (*e.g.*, B-P225L/F/T or B-H272Y/R/L/V in *B. cinerea*) and at different locations in different subunits (*e.g.*, B-H277Y, C-N75S, C-G79R, C-H134R, C-S135R, D-D124N/E, D-H143R, D-D145G in *Pyrenophora teres*, Stammler et al. 2014). The numbering of mutations needs some explanation. Several of these amino acids, which have been found to be exchanged in different species, are homologous, but have different numbers, since the number of amino acids ahead in the protein may vary in different species. Homologous mutations found in several species but with a different numbering are described in Table 8.1 (Table 8.2). This diversity of mutations complicates the interpretation of sensitivity findings in an unprecedented manner. There are mutations with amino acid exchanges leading to more or less complete resistance to all SDHIs, but some exchanges have also been detected which affect sensitivity to different SDHIs differently. Mutations at homologous sites may even have different effects in different fungal species (e.g., mutations at P220 in *Z. tritici* and at the homologous site P225 in *B. cinerea* as described by Scalliet et al. 2012). A general cross-resistance can be postulated for SDHIs, even if the effects may be somewhat different in some cases. However, there are exceptions, where no cross-resistance seems to be present (Ishii et al. 2011; Veloukas et al. 2013, Fig. 8.4). In addition, uses of specific SDHIs also influence which mutation will occur in various fungal species (Table 8.3), and variability in the frequency of mutations is therefore a result not only of the fungal species but also of the use of various SDHIs.

Some amino acid exchanges are part of the binding site, with explainable effects on SDHI binding, but some exchanges are so far away from the binding site that a direct influence on SDHI binding can be excluded. The impact of the mutation on the resistance level is not correlated with its proximity to the binding site, and exchanges even at one position can cause different resistance factors to various SDHIs (e.g., H272Y/R/L/V in *B. cinerea*; for H272Y and H272L, see also Fig. 8.4). Locating mutations causing SDHI resistance in the three-dimensional homology model of SDH (Fig. 8.5) shows that some amino acid exchanges found in *B. cinerea* isolates are situated in direct proximity to the ubiquinone binding site (Q-site) of SDH-B. Proline at position 225 is an integral part of the Q-site, contributing to SDHI binding through hydrophobic contacts. The mutation of proline into an amino acid with a bulkier side chain such as phenylalanine and leucine, or into a slightly more polar residue such as threonine, could result in a decreased binding affinity for SDHIs. The observed mutations of the histidine at position 272, with its

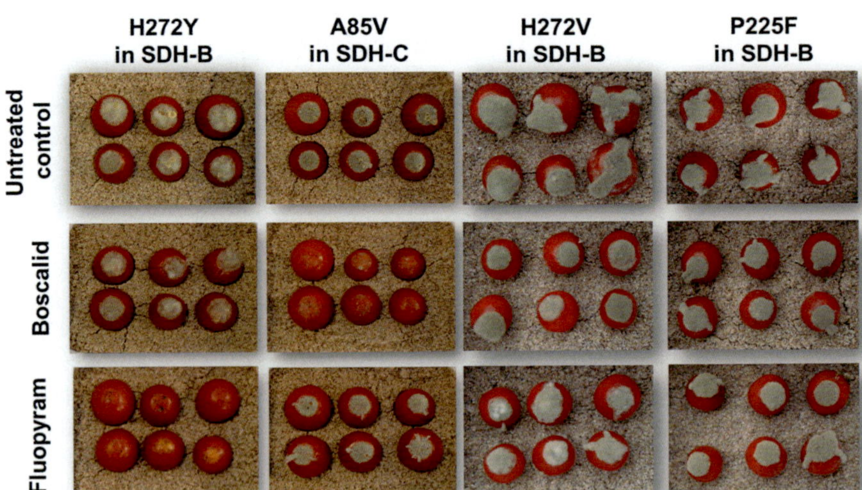

Fig. 8.4 Infection of tomato fruits with four different strains of *B. cinerea* (with B-H272Y, B-H272V, B-P225F, and C-A85V) on untreated cherry tomatoes and treated with two different SDHIs 1 day before inoculation (Linck 2013). Cross-resistance for these 2 SDHIs is found for B-H272V and B-P225F but not for B-H272Y and C-A85V

Table 8.1 List of cases of SDHI-resistant fungal plant pathogen species, their origin, and mutations found conferring SDHI resistance

Species name		Reported from host	Origin	Resistance mechanism (subunit-mutation)
Ustilago maydis	a	(Laboratory)	Lab	B-H257L
Aspergillus oryzae	b	(Laboratory)	Lab	B-H249Y/L/N, C-T90I, D-D124E
Zymoseptoria tritici	c	(Laboratory)	Lab	B-N225I, B-H267Y/R/L, B-I269V, C-A84V, C-H152R, C-T79I, C-N86K, C-G90R, D-H129E, and several others
Zymoseptoria tritici	d	Wheat	Field	B-N225T, C-T79N, C-W80S, C-N86S
Pyrenophora teres	e	Barley	Field	B-H277Y, C-N75S, C-G79R, C-H134R, C-S135R, D-D124N/E, D-H134R, D-D145G
Botrytis cinerea	f	Various	Field	B-P225L/T/F, B-H272Y/R/L/V, B-N230I, D-H132R, C-A85V
Botrytis elliptica	g	Lilies	Field	B-H272Y/R
Alternaria alternata	h	Pistachio	Field	B-H277Y/R, C-H134R, D-D123E, D-H133R
Alternaria solani	i	Potatoes	Field	B-H277Y/R, D-H133R
Corynespora cassiicola	j	Cucurbits	Field	B-H278Y/R, C-S73P, D-S89P, D-G109V
Didymella bryoniae	k	Cucurbits	Field	B-H277R/Y
Podosphaera xanthii	l	Cucurbits	Field	B-H->Y (homologous to H272 in *B. cinerea*)
Sclerotinia sclerotiorum	m	Oilseed rape	Field	B-H273Y, C-H146R, D-H132R
Stemphylium vesicarium	n	Asparagus	Field	B-P225L, H272Y/R
Venturia inaequalis	o	Apple	Field	C-H151R

Letter codes given for species are used in Table 8.2. Table reflects the list published on the FRAC webpage (status July 2014) with some updates based on BASF, unpublished data

Table 8.2 Amino acids in the different SDH subunits which were found to be exchanged and to influence SDHI sensitivity (first column)

Amino acid	Homologous positions
B-Proline	220c, 225 f, n
B-Histidine	257a, 249b, 267c, 273m, 277e, h, I, k, 272f, g, n, 278j
B-Asparagine	225c, d, 230f
C-Alanine	84c, 85f, also homologous position to S73 in *C. cassiicola*
C-Asparagine	86c, d, 75e
C-Glycine	90c, 79e
C-Histidine	145c, 134e, h, 146m
C-Histidine	152c, 151o
D-Asparagine acid	124e, 123h
D-Histidine	118c, 132f, m, 134e, 133h

Homologous position (number) in different species (for letter codes, see Table 8.1) is given in the second column. For example, P220 in *Z. tritici* is homologous to P225 *in B. cinerea* and P225 in *S. vesicarium*

Table 8.3 Frequency of *B. cinerea* strains with B-H272Y and B-P225L, without chemical control and after four applications of two different SDHIs, in a strawberry trial, with sampling after last application

Mutation	Untreated (%)	SDHI 1 (%)	SDHI 2 (%)
B-P225L	<2	3	56
B-H272Y	<2	92	29

Analysis of mutations in strains was done by pyrosequencing

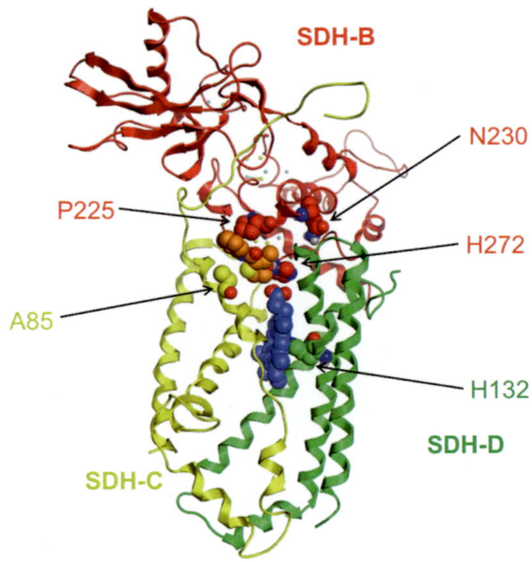

Fig. 8.5 SDH-B, SDH-C, and SDH-D model for *B. cinerea* with localization of sites where mutations have been found conferring SDHI resistance in *B. cinerea*. The heme b group is shown in *blue*. The inhibitor carboxin (shown in *orange*) was added to indicate the binding site. Modeling was made with the X-ray structure of *Gallus gallus* and the gene sequence of *B. cinerea*

side chain located at the furthest point from the opening of the Q-site, will also have a direct impact on the SDHI binding affinity, as SDHIs bind deeper into the Q-site than ubiquinone and are in direct contact to H272 via hydrogen bonding (Horsefield et al. 2006; Huang et al. 2006; Ruprecht et al. 2009; see also Fig. 8.3). The observed difference in the binding mode between ubiquinone and SDHIs could explain why in almost all fungi that developed resistant strains, amino acid exchanges are found at this particular histidine. Furthermore, it is noteworthy that both amino acids, H272 and P225, are in close vicinity to the [3Fe–4S] cluster. This could alter the reduction potential of the [3Fe–4S] cluster and consequently affect the electron transfer from succinate to ubiquinone. In addition to the mutations in the SDH-B, the iron-coordinating histidines in SDH-C and SDH-D are hot spots of mutations for SDHI resistance. The histidine in SDH-C (*e.g.*, C-H134 in *A. alternata* and *P. teres*) is highly conserved and located about 12–13 Å from the binding site. It is involved in the iron coordination of heme b. A mutation at this position may result in some structural rearrangement, indirectly affecting the topology of the binding site and hence the binding affinity of SDHIs. A comparable structural effect could be expected from the second axial histidine coordinating the iron of heme b (*e.g.*, D-H133R in *A. alternata,* D-H134R in *P. teres*). To date, the exact role of heme b in the electron transfer is still unclear and a matter of scientific debate (Horsefield et al. 2004; Oyedotun et al. 2007; Maklashina et al. 2010). However, a few findings appear to be confirmed by several studies: The heme is not an essential component of the catalytic mechanism of SDH, as shown by several site-directed mutagenesis experiments of both the axial histidines in *Escherichia coli* or *Saccharomyces cerevisiae* (Oyedotun et al. 2007; Tran et al. 2007; Maklashina et al. 2010). Nevertheless, it seems to play a pivotal role in providing a lower-energy pathway for the electron transfer from the [3Fe–4S] center to ubiquinone and as such contributes to the maintenance of a high rate of catalysis (Anderson et al. 2005; Tran et al. 2007). There is also strong evidence that heme b plays a crucial role in the structural stabilization of the enzyme (Nakamura et al. 1996; Maklashina et al. 2001; Tran et al. 2007; Maklashina et al. 2010), whereas its previously discussed function in reactive oxygen species (ROS) suppression (Yankovskaya et al. 2003) has not been confirmed in recent studies (Tran et al. 2007; Maklashina et al. 2010). Based on these results, various structural and functional consequences of the amino acid exchanges of the histidines coordinating the iron of heme b to arginine are conceivable: As arginine is not known to be able to act as a metal-coordinating residue (Dokmanic et al. 2007), this amino acid exchange probably leads to a loss of the sixth coordination partner for the heme iron. While in the case of *E. coli* an alternative histidine residue nearby (C-H30) is supposed to coordinate the heme iron instead (Tran et al. 2007), no such option seems plausible for the fungi described here, based on sequence comparisons (data not shown). Given that no residue nearby is able to function as a surrogate ligand, the low-spin hexa-coordinated heme is expected to be converted into a high-spin penta-coordinated heme which will affect its reduction potential and spectral properties (Tran et al. 2007; Maklashina et al. 2010). Another option would be the loss of the heme b group. The consequent decrease in structural stability could be compensated by hydrogen bonding between subunits C and D, as suggested for histidine-to-tyrosine mutations in *E. coli* and *S. cerevisiae* (Maklashina et al. 2010). To address these hypotheses, experimental investigation of the structural and functional consequences of the iron-coordinating histidine exchanges to arginine for fungal SDH would be necessary and is certainly of enormous interest. Although not directly involved in iron coordination, C-G90 is located in close proximity to the heme b group. A mutation to the much larger amino acid arginine, as found in *Z. tritici* (C-G90R) and *P. teres* (C-G79R), likely leads to spatial rearrangements which might not allow for a correct positioning or even lead to the loss of the heme b molecule. In *Z. tritici* and *V. inaequalis*, another histidine at the deeper part of the Q-site was found to be mutated: C-H152 (C-H151 in *V. inaequalis*). An exchange at this position certainly interferes with SDHI binding – either directly or by influencing the conformation of neighboring residues of the binding pocket, e.g., H272 of the SDH-B chain.

The wild-type sequence of a target protein is thought to represent the optimum, whose evolution is "created" under conditions without fungicide selection pressure. It could be speculated that any change, *e.g.*, in the form of single amino acid exchanges, may cause penalties for enzyme activity and eventually for the fungus. And in fact, a lower fitness of isolates with a reduced sensitivity to different modes of action has been described in various studies (Fraile et al. 1986; Raposo et al. 2000; Billard et al. 2012), and such fitness costs are obviously also relevant for the field, as the frequency of resistance decreased in time periods with no or reduced selection pressure (Johnson et al. 1994; Genet et al. 2006). Reduced enzyme activity was also shown for SDH with different mutations in different organisms (Li et al. 2006; Scalliet et al. 2012), and reduced fitness and less competitiveness in dual inoculation tests were recently described by Veloukas et al. (2014) for different SDH mutants of *B. cinerea.*

The high number of mutations which cause SDHI resistance complicates the development of genetic assays for sensitivity monitoring. Different mutations can occur in a specific codon, but also different codons in different subunits can cause SDHI resistance. It can also be expected that combinations of mutations might occur under high selection pressure. A further complicating and uncertain factor is the occurrence of mutations in field isolates. An important question is which mutation(s) will occur in the field under selection pressure of different SDHIs. A number of different mutations have been described for laboratory mutants of *Z. tritici* (Skinner et al. 1998; Stammler et al. 2010; Fraaije et al. 2012; Scalliet et al. 2012), but the first mutations reported for field isolates of *Z. tritici* were not identified in lab mutants: C-T79N, C-N86S (BASF data, www.frac.info) and B-N225T, C-W80S (www.frac.info). Therefore, it is advisable to monitor SDHI sensitivity with in vivo or in vitro assays and in the case of reduced sensitivity to identify the responsible mutations in the target gene in a number of isolates. If nearly all resistant field isolates carry a limited number of mutations, genetic assays can be developed, as it has been done for *B. cinerea, S. vesicarium*, or *P. teres* (BASF, unpublished data). Main SDHI selection pressure in the last years has been caused by a limited number of SDHIs. With the launch of new SDHIs, the frequency of mutations might change, or even other, so far undetected and unknown mutations or combinations of mutations might occur. Therefore, concomitant "classical" sensitivity tests should be run in monitoring programs at least for a number of isolates and from time to time, in order to ensure that established genetic assays are still valid, *i.e.*, that the "right" codons are analyzed and new mutations are included in the molecular genetic monitoring.

References

Abiko K, Kishi K, Yoshioka A (1977) Occurrence of oxycarboxin-tolerant isolates of *Puccinia horiana* P. Hennings in Japan. Ann Phytopathol Soc Jpn 43:145–150

Ackrell BAC, Kearney EB, Coles CJ, Singer TP, Beinert H, Wan YP, Folkers K (1977) Kinetics of the reoxidation of succinate dehydrogenase. Arch Biochem Biophys 182:107–117

Anderson RF, Hille R, Shinde SS, Cecchini G (2005) Electron transfer within complex II – succinate: ubiquinone oxidoreductase of *Escherichia coli*. J Biol Chem 280:33331–33337

Avenot HF, Michaelidis TJ (2007) Resistance to boscalid fungicide in *Alternaria alternata* isolates from pistachio in California. Plant Dis 91:1345–1350

Avenot HF, Thomas A, Gitaitis RD, Langston DB, Stevenson KL (2012) Molecular characterization of boscalid- and penthiopyrad-resistant isolates *of Didymella bryoniae* and assessment of their sensitivity to fluopyram. Pest Manag Sci 68:645–651

Ben-Yephet Y, Dinoor A, Henis Y (1975) The physiological basis of carboxin sensitivity and tolerance in *Ustilago hordei*. Phytopathology 65:936–942

Billard A, Fillinger S, Leroux P, Lachaise H, Beffa R, Debieu D (2012) Strong resistance to the fungicide fenhexamid entails a fitness cost in *Botrytis cinerea*, as shown by comparisons of isogenic strains. Pest Manag Sci 68:684–691

Cecchini G (2003) Function and structure of complex II of the respiratory chain. Annu Rev Biochem 72:77–109

Dokmanic I, Sikic M, Tomic S (2007) Metals in proteins: correlation between the metal-ion type, coordination number and the amino-acid residues involved in the coordination. Acta Crystallogr D64:257–263

Fernandez-Ortuno D, Chen F, Schnabel G (2012) Resistance to pyraclostrobin and boscalid in *Botrytis cinerea* isolates from strawberry fields in the Carolinas. Plant Dis 96:1198–1203

Fraaije BA, Bayon C, Atkins S, Cools HJ, Lucas JH, Fraaije MW (2012) Risk assessment studies on succinate dehydrogenase inhibitors, the new weapons in the battle to control Septoria leaf blotch in wheat. Mol Plant Pathol 13:263–275

Fraile A, Alonso A, Sagasta EM (1986) Some characteristics of *Botrytis cinerea* isolates tolerant to procymidone. Plant Pathol 35:82–85

Genet JL, Jaworska G, Deparis F (2006) Effect of dose rate and mixtures of fungicides on selection for QoI resistance in populations of *Plasmopara viticola*. Pest Manag Sci 62:188–194

Georgopoulos SG, Alexandri E, Chrysayi M (1972) Genetic evidence for the action of oxathiin and thiazole derivatives on the succinic dehydrogenase system of *Ustilago maydis* mitochondria. J Bacteriol 110:809–817

Georgopoulos SG, Chrysayi M, White GA (1975) Carboxin resistance in the haploid, the heterozygous diploid, and the plant-parasitic dicaryotic phase of *Ustilago maydis*. Pestic Biochem Physiol 5:543–551

Glaettli A, Stammler G, Schlehuber S (2009) Mutations in the target proteins of succinate-dehydrogenase inhibitors (SDHI) and 14α-demethylase inhibitors (DMI) conferring changes in the sensitivity – structural insights from molecular modelling. In: 9th international conference on plant diseases, Tours, France, pp 670–681

Horsefield R, Iwata S, Byrne B (2004) Complex II from a structural perspective. Curr Protein Pept Sci 5:107–118

Horsefield R, Yankovskaya V, Sexton G, Whittingham W, Shiomi K, Omura S, Byrne B, Checchini G, Iwata S (2006) Structural and computational analysis of the quinone-binding site of complex II (succinate-ubiquinone oxidoreductase). J Biol Chem 281:7309–7316

Huang L, Sun G, Cobessi D, Wang AC, Shen JT, Tung EY, Anderson VE, Berry EA (2006) 3-Nitropropionic acid is a suicide inhibitor of mitochondrial respiration that, upon oxidation by complex II, forms a covalent adduct with a catalytic base arginine in the active site of the enzyme. J Biol Chem 281:5965–5972

Ishii H, Miyamoto T, Ushio S, Kakishima M (2011) Lack of cross-resistance to a novel succinate dehydrogenase inhibitor, fluopyram, in highly boscalid-resistant isolates of *Corynespora cassiicola* and *Podosphaera xanthii*. Pest Manag Sci 67:474–482

Ito Y, Muraguchi H, Seshime Y, Oita S, Yanagi SO (2004) Flutolanil and carboxin resistance in *Coprinus cinereus* conferred by a mutation in the cytochrome b560 subunit of succinate dehydrogenase complex (Complex II). Mol Genet Genomics 272:328–335

Johnson KB, Sawyer TL, Powelson ML (1994) Frequency of benzimidazole- and dicarboximide-resistant strains of *Botrytis cinerea* in western Oregon small fruit and snap bean plantings. Plant Dis 78:572–577

Keon JP, White GA, Hargreaves JA (1991) Isolation, characterization and sequence of a gene conferring resistance to the systemic fungicide carboxin from the maize smut pathogen, *Ustilago nuda*. Curr Genet 19:475–481

Kuhn PJ (1984) Mode of action of carboxamides. Br Mycol Soc Sym Ser 9:155–183

Leroux P, Berthier G (1988) Resistance to carboxin and fenfuram in *Ustilago nuda* (Jens) Rostr, the causal agent of barley loose smut. Crop Prot 7:16–19

Li J, Zhou M, Li H, Chen C, Wang J, Zhang Y (2006) A study on the molecular mechanism of resistance to amicarthiazol in *Xanthomonas campestris* pv. *citri*. Pest Manag Sci 62:440–445

Linck H (2013) Method development for the efficacy evaluation of biological control agents against *Botrytis cinerea.* Master thesis, University Hohenheim

Maklashina E, Rothery RA, Weiner JH, Cecchini G (2001) Retention of heme in axial ligand mutants of succinate-ubiquinone oxidoreductase (complex II) form *Escherichia coli.* J Biol Chem 276:18968–18976

Maklashina E, Rajagukguk S, McIntire WS, Cecchini G (2010) Mutation of the heme axial ligand of *Escherichia coli* succinate-quinone reductase: Implications for heme ligation in mitochondrial complex II from yeast. Biochim Biophys Acta 1797:747–754

Mathre DE (1970) Mode of action of oxathiin systemic fungicides. I. Effect of carboxin and oxycarboxin on the general metabolism of several basidiomycetes. Phytopathology 60:1464–1469

Mathre DE (1971) Mode of action of oxathiin systemic fungicides. III. Effect on mitochondrial activities. Pestic Biochem Physiol 1:216–224

Matsson M, Hederstedt L (2001) The carboxin-binding site on *Paracoccus denitrificans* succinate: quinone reductase identified by mutations. J Bioenerg Biomembr 33:99–105

Matsson M, Ackrell BAC, Cochran B, Hederstedt L (1998) Carboxin resistance in *Paracoccus denitrificans* conferred by a mutation in the membrane anchor of succinate: quinone reductase (complex II*)*. Arch Mircobiol 170:27–37

Miles TD, Miles LA, Fairchild KL, Wharton PS (2013) Screening and characterization of resistance to succinate dehydrogenase inhibitors in *Alternaria solani*. Plant Pathol 53:155–164

Miyamoto T, Ishii H, Seko T, Kobori S, Tomita Y (2009) Occurrence of *Corynespora cassiicola* isolates resistant to boscalid on cucumber in Ibaraki Prefecture, Japan. Plant Pathol 58:1144–1151

Miyamoto T, Ishii H, Stammler G, Koch A, Ogawara T, Tomita T, Fountaine J, Ushio T, Seko T, Kobori S (2010a) Distribution and molecular characterization of *Corynespora cassiicola* strains resistant to boscalid. Plant Pathol 59:873–881

Miyamoto T, Ishii H, Tomita T (2010b) Occurrence of boscalid resistance in cucumber powdery mildew in Japan and molecular characterization of the iron-sulfur protein of succinate dehydrogenase of the causal fungus. J Gen Plant Pathol 76:261–267

Nakamura K, Yamaki M, Sarada M, Nakayama S, Vibat CRT, Gennis RB, Nakayashiki T, Inokuchi H, Kojima S (1996) Two hydrophobic subunits are essential for the heme b ligation and functional assembly of complex II (succinate-ubiquinone oxidoreductase) from *Escherichia coli.* J Biol Chem 271:521–527

Oyedotun KS, Sit CS, Lemire BD (2007) The *Saccharomyces cerevisiae* succinate dehydrogenase does not require heme for ubiquinone reduction. Biochim Biophys Acta 1767:1436–1445

Ragsdale NN, Sisler HD (1970) Metabolic effects related to fungitoxicity of carboxin. Phytopathology 60:1422–1427

Ramsay RR, Ackrell BAC, Coles CJ, Singer TP, White GA, Thorn GD (1981) Reaction site of carboxanilides and of thenoyltrifluoroacetone in complex II. Proc Natl Acad Sci U S A 78:825–828

Raposo R, Gomez V, Urrutia T, Melgarejo P (2000) Fitness of *Botrytis cinerea* associated with dicarboximide resistance. Phytopathology 90:1246–1249

Ruprecht J, Yankovskaya V, Maklashina E, Iwata S, Cecchini G (2009) Structure of *Escherichia coli* succinate:quinone oxidoreductase with an occupied and empty quinone-binding site. J Biol Chem 284:29836–29846

Scalliet G, Bowler J, Luksch T, Kirchhofer-Allan L, Steinhauer D, Ward K, Niklaus M, Verras A, Csukai M, Daina A (2012) Mutagenesis and functional studies with succinate dehydrogenase inhibitors in the wheat pathogen *Mycosphaerella graminicola*. PLoS One 7, e35429

Skinner W, Bailey A, Renwick A, Keon J, Gurr S, Hargreaves J (1998) A single amino-acid substitution in the iron-sulphur protein subunit of succinate dehydrogenase determines resistance to carboxin in *Mycosphaerella graminicola*. Curr Genet 34:393–398

Stammler G, Brix HD, Glaettli A, Semar M, Schoefl U (2007) Biological properties of the carboxamide boscalid including recent studies on its mode of action. In: Proceedings of BCPC XVI international plant protection congress, Alton. British Crop Protection Council Publications, Hampshire, pp 16–21

Stammler G, Brix HD, Nave B, Gold R, Schoefl U (2008) Studies on the biological performance of boscalid and its mode of action. In: Dehne HW et al (eds) Modern fungicides and antifungal compounds V. 15th international Reinhardsbrunn symposium 2007, pp 45–51

Stammler G, Glaettli A, Koch A (2010) Mutations in the target protein conferring resistance to SDHI fungicides. In: Dehne HW et al (eds) Modern fungicides and antifungal compounds V. 16th international Reinhardsbrunn symposium 2010, pp 195–198

Stammler G, Rehfus A, Prochnow J, Bryson R, Strobel D (2014) New findings on the development of insensitive isolates of *Pyrenophora teres* towards SDHI fungicides. Julius-Kühn-Archiv 447:568

Tran QM, Rothery RA, Maklashina E, Cecchini G, Weiner JH (2007) *Escherichia coli* succinate dehydrogenase variant lacking the heme b. Proc Natl Acad Sci U S A 104:18007–18012

Ulrich JT, Mathre DE (1972) Mode of action of oxathiin systemic fungicides. V. Effect on electron transport system of *Ustilago maydis* and *Saccharomyces cerevisiae*. J Bacteriol 110:628–632

Veloukas T, Leroch M, Hahn M, Karaoglanidis GS (2011) Detection and molecular characterization of boscalid-resistant *Botrytis cinerea* isolates from strawberry. Plant Dis 95:1302–1307

Veloukas T, Markoglou AN, Karaoglanidis GS (2013) Differential effect of *Sdh*B gene mutations on the sensitivity to SDHI fungicides in *Botrytis cinerea*. Plant Dis 97:118–122

Veloukas T, Kalogeropoulou P, Markoglou A, Karaoglanidis G (2014) Fitness and competitive ability of *Botrytis cinerea* field-isolates with dual resistance to SDHI and QoI fungicides, associated with several *sdh*B and the *cyt*b G143A mutations. Phytopathology 104:347–356

von Schmeling B, Kulka G (1966) Systemic fungicidal activity of 1,4-oxathiin derivatives. Science 152:659–660

Yankovskaya V, Horsefield R, Törnroth S, Luna-Chavez C, Miyoshi H, Léger C, Byrne B, Ceccini G, Iwata S (2003) Architecture of succinate dehydrogenase and reactive oxygen species generation. Science 299:700–704

Chapter 9
Respiration Inhibitors: Complex III

Helge Sierotzki

Abstract Respiration inhibitor fungicides are highly successful in controlling a wide range of plant pathogens. The biggest group is the QoI fungicides possessing activity toward many major pathogens within *Deuteromycetes*, *Ascomycetes*, *Basidiomycetes*, and *Oomycetes*. The inhibitor binding to the target pocket is well investigated, and also the resistance mechanism is exhaustively characterized. Resistance to QoIs is based on a point mutation, G143A, as main mechanism; also F129L and G137R can cause resistance but with minor relevance. Resistance is widespread in many plant pathogens and evolved rather quickly in high-risk pathogen/host systems, like powdery mildew on cereals or downy mildew on grapes. In pathogens considered as low risk for resistance development, resistance has developed less rapidly and only recently after more than 10 years of usage, such as in *Rhizoctonia solani* on rice. In rust pathogens, resistance to QoIs has not developed until today based on the presence of a bi2 intron after position 143 in the *cyt b* gene. The QiI and the recently launched QoSI fungicides are active against oomycete pathogens. QiIs bind to a different, inner heme pocket of the cytochrome bc1 complex. Confirmed field resistance has been observed so far only in *Phytophthora capsici*; it is based on an unknown mutation. QoSIs bind to the distal area of the Qo heme pocket of the cytochrome b complex and are not cross-resistant to QoI fungicides. Resistance has not yet been observed in field populations. Respiration inhibitors are still one of the most frequently used fungicide groups for controlling plant diseases. Their success in the market is based on the spectrum of activity, the beneficial effects on plant physiology, and also the strict and well-accepted use recommendations.

Keywords Respiration inhibitors • QoI • QiI • QoSI • Mitochondria • *Oomycetes* • *Ascomycetes* • *Basidiomycetes* • *Deuteromycetes* • Cytochrome b gene • G143A mutation • F129L mutation • Intron • Alternative respiration • Stress tolerance • Quantitative PCR technologies

---

H. Sierotzki (✉)
Syngenta Crop Protection AG, Schaffhauserstrasse, 4332 Stein, Switzerland
e-mail: helge.sierotzki@syngenta.com

H. Ishii, D.W. Hollomon (eds.), *Fungicide Resistance in Plant Pathogens*,
DOI 10.1007/978-4-431-55642-8_9

9.1 Introduction

Among agricultural fungicides, respiration inhibitors are the second most important class of chemicals after the DMI (azole) fungicides. The mitochondrial respiration in fungal and oomycete plant pathogens is basically identical to all other eukaryotic organisms with several major enzyme complexes and active centers. They bind to the natural substrates for electron transport and also serve as binding pockets for inhibitors. Important fungicides in the respiration chain are currently known for complexes II (succinate dehydrogenase) and III (cytochrome bc1 complex) as well as for ATPase inhibition and uncoupling. Focusing on complex III inhibitors, fungicides from three different binding classes are currently on the market: QoI, QoSI, and QiI fungicides. Qo and Qi are two distinct pockets within the cytochrome b complex, the ubiquinol oxidizing outer and the ubiquinone reducing inner pocket, respectively, whereas QoSIs bind slightly different into the Qo pocket than QoIs. The QoI fungicides have a very broad spectrum and strong activity with a wide range of agronomical applications against many pathogens from all taxonomic groups, whereas QiIs are mainly active against oomycete pathogens. The intensive usage of QoI fungicides has led to ample resistance evolution in many pathogen species. The main resistance mechanism is based on several mutations in the target site gene, the cytochrome b (*cyt b*) gene. Additional mechanisms have been described only occasionally (Brandt et al. 1993; Fernández-Ortuño et al. 2008). Mutations leading to QoI resistance include G143A, F129L, and G137R in *cyt b* gene. However, the G143A mutation is the most important mechanism and has the strongest effect on fungicide activity in field populations. Resistance evolution depends mainly on pathogen biology and the molecular configuration of the *cyt b* gene. Among the first pathogens having acquired resistance was *Blumeria graminis* in Northern Germany (Reschke 1999), whereas the last one, most recent pathogen, after more than 13 years of QoI usage, was *Rhizoctonia solani* on rice in the USA (Olaya et al. 2012). In certain pathogen groups, no resistance evolved to date; among them are the rust fungi; the mechanism of this lack of resistance was elucidated to be due to an intron directly after the position G143 in the *cyt b* gene. Resistance to QiI and to QoS fungicides is not confirmed for field isolates of any pathogen species; however, it is assumed to potentially appear in the future. Despite ample resistance in many pathogen species, QoI fungicides are still in use. The broad spectrum activity and the beneficial effects on plant physiology make them still important tools for modern agronomy. However, the frequent and widespread applications of QoIs exert a high and continuous selection pressure for resistance on target (and nontarget) plant pathogens. Recommendations for sustainable usage have been developed by FRAC (www.frac.info); they need to be enforced by all users and official bodies.

9.2 Diversity of Compounds

Fungicides inhibiting respiration at complex III are classified into three different modes of action groups. The most important one is the QoI group, also called "strobilurins" including 20 compounds subdivided into 9 chemical classes. Azoxystrobin, coumoxystrobin, enoxastrobin, flufenoxystrobin, picoxystrobin, and pyraoxystrobin belong to the methoxy-acrylates; mandestrobin to the methoxy-acetamides; pyraclostrobin, pyrametostrobin, and triclopyricarb to the methoxy-carbamates; kresoxim-methyl and trifloxystrobin to the oximino-acetates; and dimoxystrobin, fenaminostrobin, metominostrobin, and orysastrobin to the oximino-acetamides, whereas famoxadone, fluoxastrobin, fenamidone, and pyribencarb belong to the oxazolidine-diones, dihydro-dioxazines, imidazolinones, and benzyl-carbamates, respectively (www.frac.info). The chemical history of the QoIs has been summarized exhaustively by Bartlett et al. (2002) and Sauter (2012). All compounds have a toxophore moiety that contains a carbonyl oxygen, with similar structural appearance. The toxophore is linked to a more or less hydrophobic tail (Sauter 2012).

The QiI compounds are classified into two chemical groups, the cyano-imidazoles and sulfamoyl-triazoles with one compound each, cyazofamid and amisulbrom, respectively (Sauter 2012). The most recent molecule, ametoctradin, also binds to the QoI pocket of the cytochrome bc1 enzyme complex, but in a different manner than the other QoIs mentioned above. The fungicide ametoctradin is classified as QoS inhibitor belonging to the chemical class of triazolo-pyrimidylamine (Gold et al. 2009). QiI and QoSI compounds are so far registered only against *Peronosporales* within oomycete pathogens, controlling downy mildews and late blight.

9.3 History

Prior to the first market introduction of QoI fungicides in 1996, extensive research has been made on cytochrome b inhibitors (Brasseur et al. 1996) leading to the discovery of strobilurins as fungicides in 1977 by Anke (Anke 1995). Compounds like kresoxim-methyl and azoxystrobin have been developed and patented shortly afterward in the mid-1980s (Sauter et al. 1995). Although major QoI fungicides were introduced in the late 1990s and the beginning of the new millennium, still in recent years new compounds are registered such as orazostrobin, enoxastrobin, mandestrobin, and pyrametostrobin. Today, QoI fungicides are formulated in several hundreds of products and used in more than 80 crops to control pathogens in more than 70 countries. QoIs are one of the few fungicide classes that can control pathogen species of all fungal classes such as *Ascomycetes*, *Deuteromycetes*, and *Basidiomycetes* and also of the *Oomycetes*. This is due to the general inhibition of a central step in the primary metabolism, the electron flow in the respiration chain at the Qo center of the cytochrome *bc*1 complex in mitochondria. Interestingly, the sensitivity of fungal and oomycete pathogens on the one hand and the intrinsic insensitivity of plants and mammals on the other hand originate from the metabolic instability of QoI fungicides in the latter organisms rather than from differences in the binding pocket of the target enzyme (Sauter 2012). Accordingly, human HeLa cultures were not inhibited by the compounds. The natural lead compounds produced by basidiomycete fungi like *Strobilurus tenacellus* are instable toward UV light and needed chemical modification for their stabilization (Sauter 2012).

Two QiI fungicides were commercially introduced in 2001 (cyazofamid) and in 2003 (amisulbrom) (Mitani et al. 2001, Förch et al. 2007). The compounds are active exclusively against oomycete pathogens like downy mildews and late blight. The same activity spectrum is also true for the QoSI compound ametoctradin, discovered in 2004 and officially announced at the BCPC Congress in 2009 (Gold et al. 2009).

9.4 Market Aspects

In 2010, the eight most important QoI fungicides had a total sales volume of about 2790 mUS$ (McDougall 2012). In 2012, about 28 % of the fungicide market was covered by QoI fungicides. The sales volume is forecasted to grow in the near future to about 3.5 bnUS$ (pers. communication Luc Henry, Syngenta internal data). Therefore, the use intensity and consequently the selection pressure for resistance toward QoIs will still increase in the coming years, especially in Asia, where most of the new QoIs were announced. The QiIs and QoSIs are estimated to cover around 0.5 % of the global fungicide market, within the oomycetes about 7 % (Syngenta estimation).

Besides the activity against fungal and oomycete plant pathogens, additional effects have been reported for QoIs such as the induction of systemic acquired resistance (Herms et al. 2002) and physiological stimulation of plants leading to greening effects, healthier and more vigorous plants, and higher stress tolerance (Ruske et al. 2003; Venacio et al. 2003; Nason et al. 2007). It is assumed that these applications will contribute significantly to the current and future sales of QoIs. However, the majority of today's QoI applications are still driven by fungicidal control of plant pathogens; the broad activity spectrum of QoIs is the major reason for the increased usage of this fungicide class (Fernández-Ortuño et al. 2010).

QoIs can control several diseases in one single crop. In cereals, they control diseases like rusts, *Septoria* leaf blotch, powdery mildews, net blotch, and *Rhynchosporium* leaf spot. Also in other crops like grapes, potato, vegetables, soybean, rice, and small fruits, QoIs can control simultaneously several major diseases, e.g., downy and powdery mildews. However, QoIs can vary quite significantly in their activity spectra according to their intrinsic activity. Interestingly, the usage of some QoIs has expanded in recent years to new crops such as soybean and maize.

They remain major mixing partners in many fungicide products either for spectrum and resistance management reasons or also based on their positive effects on plant physiology.

9.5 QoI Fungicides

9.5.1 Mode of Binding

The mode of action of QoI fungicides has been elucidated already in 1981 by Becker et al. (1981) based on studies with myxothiazol and some natural compounds such as strobilurin A and oudemansins. The inhibitors interfere with the electron transfer at the Qo site (outer quinol oxidation site) of the cytochrome *bc*1 complex of the mitochondrial respiration chain. The resulting blocking of electron flow leads to an energy deficiency due to a lack of ATP in fungal cells. Mitochondrial respiration is important in all life stages of fungi, which are highly energy dependent, such as spore germination or zoospore movement. Protein co-crystallization revealed the mode of binding of QoI compounds into the ubiquitin binding pocket of the cytochrome *b* (Fig. 9.1; Link et al. 2003). All QoI fungicides bind to the Glu 272 of the ef helix of the cytochrome b enzyme, based on the toxophore containing a carbonyl oxygen moiety (except for famoxadone and fenamidone). Other parts of the toxophore seem to be close to positions Ala 128 and Phe 129 of cytochrome b. These

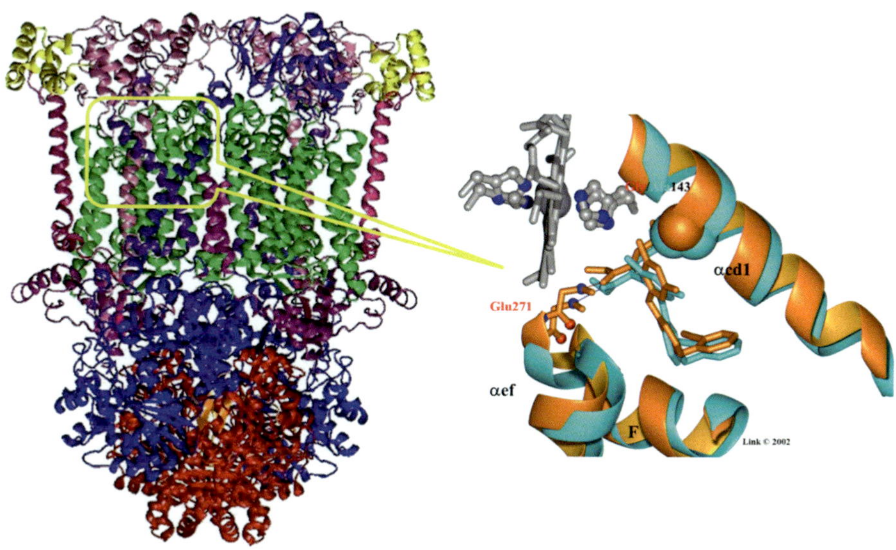

Fig. 9.1 Detailed cytochrome bc1 structure with binding pocket for QoI molecules (e.g., azoxystrobin). *Left panel* shows the overall enzyme complex and *right panel* the binding pocket with enzyme helix. The changes due to the G143A mutation are indicated in blue (Link et al. 2003)

positions within the ubiquitin binding pocket are about 6 Å away from the heme bL (Fig. 9.1; Link et al. 2003). The so-called benzene linker of QoI fungicides was introduced into the molecules as light stabilizing element and is approaching the αcd 1 helix at position Gly143. The G143A mutation is not involved in enzyme activity, whereas the F129L is closer to the heme and is potentially involved in electron transfer. These mechanistic features might explain the different frequencies of the mutations and their impact on sensitivity. The G143A mutation is by far the most frequent mechanism of resistance to QoI fungicides in plant pathogens, followed by the F129L and other mutations (see below). Resistance in some pathogens has also been described to be based on mechanisms outside mutations in the cytochrome b gene (e.g., Fernández-Ortuño et al. 2008).

9.5.2 Resistance Mechanisms

Mutant analysis in several different organisms, ranging from bacteria to mice, revealed many mutations possibly involved in reduced binding of Qo inhibitors, such as myxothiazol (Brasseur et al. 1996). The mutant analysis revealed amino acid changes in resistant phenotypes in two hot spots in the *cyt b* gene (aa numbers 120–150 and aa numbers 250–290). The three-dimensional models and crystal structures of the cytochrome b complex suggested that these fragments form the ubiquitin pocket (Brasseur et al. 1996; Crofts 2004). Many mutations in the *cyt b* gene fragment do not have a strong effect on resistance, but some impair the enzyme activity. Early experiments with yeast assessing the resistance risk for QoI fungicides failed to produce mutants, suggesting initially a medium-risk classification for resistance to QoIs. In addition, mutagenesis experiments with *Zymoseptoria tritici* did not produce stable target site mutants *in planta* (Ziogas et al. 1997). The resistant in vitro phenotypes were mediated by overexpression of alternative oxidase, bypassing the cytochrome *b*. However, this mechanism is not able to cause disease management problems in the field (Joseph-Horne and Hollomon 2000). Three mutations have been detected in plant pathogenic fungi leading to QoI resistance in field populations. The change of glycine to alanine at position 143 in the *cyt b* gene (G143A) has the strongest effect and is also most widely spread in terms of number of affected pathogen species as well as frequency in populations. The mutation leads to a steric interference with the linker part of QoI molecules resulting in markedly reduced binding and therefore conferring strong resistance (Fig. 9.1). The resistance factors can be up to several hundreds depending on the species and significantly impact field performance of solo QoI fungicides. The change of phenylalanine to leucine at position 129 (F129L) is close to the heme bL and interferes with the toxophore of QoI fungicides. The resistance factors are normally much lower; the mutation is assumed to have a stronger effect on fitness of resistant isolates compared to the G143A mutation (Esser et al. 2004; Sierotzki et al. 2007). However, the F129L mutation can affect the performance of solo QoI fungicides, especially when occurring at high frequency in populations, for example, in

Alternaria solani (Pasche et al. 2002). The exchange of glycine by arginine at position 137 (G137R) is of minor importance. It is not well understood how it affects binding; it mediates only low to moderate resistance and most probably has the biggest fitness penalty.

Interestingly, not all pathogen species treated by frequent and repeated QoI applications over many years have developed resistance. The most prominent group is the rust fungi (Grasso et al. 2006), but also in other species no QoI resistance based on G143 mutation has been detected (see below). Comparing the mitochondrial cytochrome *b* gene structure in different pathogen species showed a huge diversity in number, size, and position of introns. In some species, no intron is present (e.g., *Blumeria graminis*), whereas in others up to 10 introns can be found, some of which can be very long (up to several kb's each, e.g., in *Venturia inaequalis*). The positions of introns are different among species, but one specific intron, bi2, has proven to significantly affect the development of QoI resistance (Fig. 9.2; Grasso et al. 2006). It is a type I self-splicing intron located between the codons at positions 143 and 144 in several pathogen species. If present at this position in *cyt b* gene, the G143A mutation most likely does not evolve in nature because it would lead to a disordered splicing of the pre-messenger RNA, which would result in nonfunctional

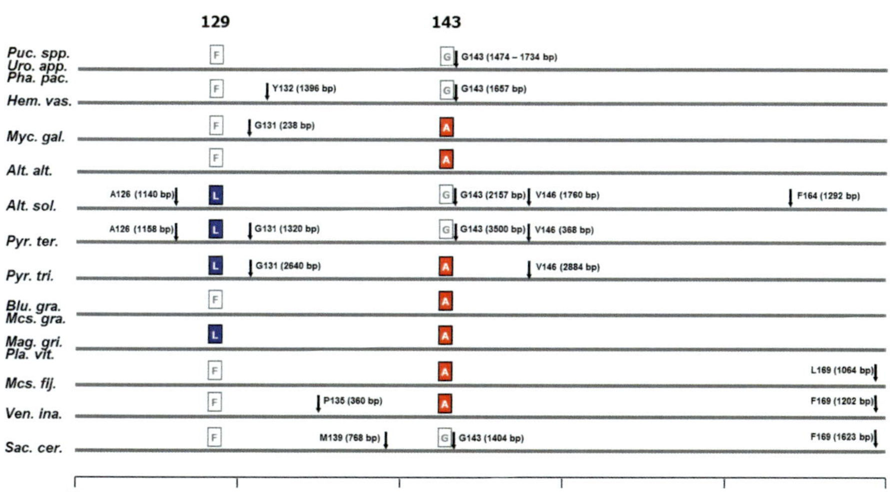

Fig. 9.2 Cytochrome *b* gene structure in different plant pathogen species, in the yeast *Saccharomyces cerevisiae* (*Sac. cer.*), and in the strobilurin-producing basidiomycete *Mycena galopoda* (*Myc. gal.*). *Lines* indicate the nucleotide sequence; in *boxes* are the position and amino acids involved in QoI resistance; *vertical arrows* show intron positions (amino acid position and length). *Puc. spp.* includes *Puccinia recondita* f. sp. *tritici*, *P. graminis* f. sp. *tritici*, *P. striiformis* f. sp. *tritici*, *P. coronata* f. sp. *avenae*, *P. hordei*, *P. recondita* f. sp. *secalis*, *P. sorghi*, and *P. horiana*; *Uro. app.*, *Uromyces appendiculatus*; *Pha. pac.*, *Phakopsora pachyrhizi*; *Hem. vas.*, *Hemileia vastatrix*; *Myc. gal.*, *Mycena galopoda*; *Alt. alt.*, *Alternaria alternata*; *Alt. sol.*, *Alternaria solani*; *Blu. gra.*, *Blumeria graminis*; *Mag. gri.*, *Magnaporthe grisea*; *Mcs. fij.*, *Mycosphaerella fijiensis*; *Mcs. gra.*, *Mycosphaerella graminicola (Zymoseptoria tritici)*; *Ven. ina.*, *Venturia inaequalis*; *Pla. vit.*, *Plasmopara viticola*

proteins and finally a lower level of cytochrome *b* and respiratory activity (Vallières et al. 2011). The intron has to be located within the 4 to 6 bases after position 143 to prevent the mutation to occur (Lazowska 1980). An intron directly after codon 143 has been identified in several plant pathogenic fungal species (Grasso et al. 2006; Miessner and Stammler 2010; Miessner et al. 2011; Stammler 2012; Stammler et al. 2013) and is claimed to be responsible for the low resistance risk toward QoI fungicides. The following pathogen species have been verified to possess an intron directly after position 143 in the *cyt b* gene and did not develop resistance (based on G143A) to QoI fungicides so far: *Alternaria solani*, *Bipolaris maydis*, *Cercospora zeae-maydis*, *Cochliobolus carbonum*, *Guignardia bidwellii*, *Guignardia citricarpa*, *Monilinia laxa*, *Setosphaeria turcica*, and all rust fungi such as *Hemileia vastatrix*, *Phakopsora pachyrhizi*, *Puccinia coronata*, *P. graminis*, *P. hordei*, *P. horiana*, *P. secalis*, *P. sorghi*, *P. striiformis*, *P. triticina* (*recondita*), and *Uromyces appendiculatus*.

In a few pathogens, the intron-exon structure of the *cyt b* gene is polymorphic. *Saccharomyces cerevisiae* (Bonjardin et al. 1996) and *Botrytis cinerea* (Banno et al. 2009) populations contain individuals with or without an intron at position 143, whereas in *Alternaria solani* populations the polymorphisms are at position 126 (Leiminger et al. 2013). However, two additional mutations have been identified in plant pathogenic fungi: the F129L mutation that has been detected in *Alternaria solani* and *Pyrenophora teres*, both possessing an intron at position 143 in the *cyt b* gene, and the G137R mutation in isolates of *Pyrenophora teres* from Ireland (www.frac.info). Independent of which of the three mutations is present, cross-resistance to all members of the QoI fungicide class exists for the vast majority of isolates. This also includes compounds which do not obviously fit to the typical QoI kind of "molecule shape," such as famoxadone, fenamidone, and pyribencarb. The cross-resistance behavior can be nicely demonstrated when comparing the median EC_{50} values of sensitive and resistant isolates of *Zymoseptoria tritici* for two intrinsically different QoI fungicides such as azoxystrobin and pyraclostrobin (Fig. 9.3). The presence of the G143A mutation always correlates well with resistance toward QoI fungicides. The combination of the bioassay with the molecular test allows setting a clear threshold between sensitive and resistant phenotypes among isolates even if the results of the bioassay might be not so conclusive.

The first resistant isolates of *Venturia inaequalis* detected in Swiss orchards did not contain any mutations in the *cyt b* gene (Steinfeld et al. 2002). The mechanism has never been fully elucidated; however, it may be due to a differential organization of the mitochondrial genome organization leading to an upregulation of the respiration rate (Brandt et al. 1993). Reduced QoI sensitivity without the presence of mutations in *cyt b* gene was found in isolates of *V. inaequalis* also in Northern Germany few years after market introduction of kresoxim-methyl; the resistance mechanism was claimed as being an increased metabolism of QoIs by enhanced esterase activity (Jabs et al. 2001). However, later on, the development of resistance in *V. inaequalis* populations was based always on the presence of G143A which was so dominant that all other mechanisms were of minor importance. There are only rare additional cases of reduced QoI sensitivity without the presence of mutations in the

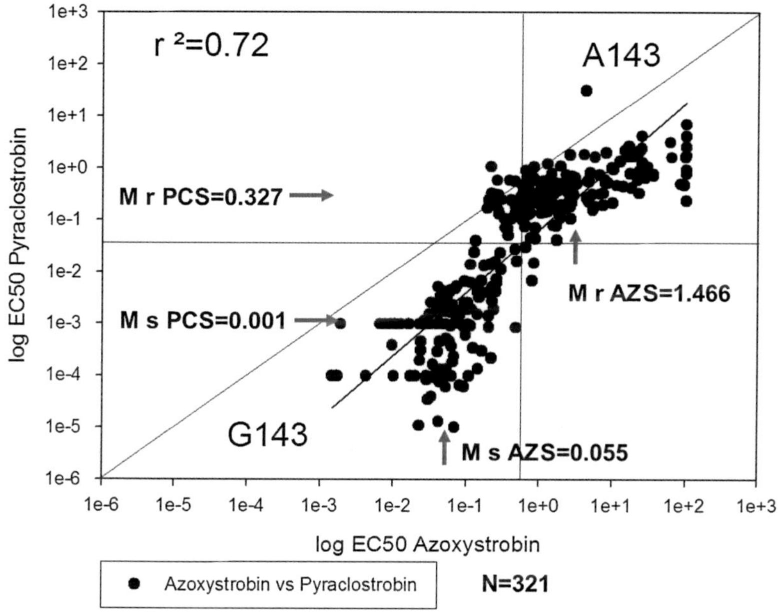

Fig. 9.3 Cross-resistance between QoI fungicides for *Zymoseptoria tritici* field isolates. *PCS* pyraclostrobin, *AZS* azoxystrobin. *M* indicates the median values for each fungicide and sensitivity group (sensitive or resistant). The *horizontal* and *vertical lines* indicate the threshold EC_{50} value for the G143A mutants. The *diagonal lines* indicate the theoretical and actual ($r^2=0.72$) regression, respectively, for identical activity (EC_{50}) against isolates ($N=321$)

cytochrome *b* gene (e.g., *Podosphaera fusca*, Fernández-Ortuño et al. 2008, *Botrytis cinerea*, Ishii et al. 2009).

The knowledge of the molecular target for the QoI fungicides and the coding gene sequence enable the design of molecular test methods to quantitatively determine the appearance and frequency of mutations in field populations (Fraaije et al. 2002; Sierotzki and Gisi 2003; Stammler et al. 2006; Semar et al. 2007; Gisi and Sierotzki 2008). The information gained from molecular tests, in most cases quantitative PCR technologies and pyrosequencing, complements the results generated by bioassays in a wide range of pathogens. Quantitative molecular DNA-based assays provide an estimate of resistance frequencies present in a field, area, region, or country and are especially strong when comparing different years or different treatments with each other. By comparing EC_{50} valucs with the presence or absence of mutations, thresholds for resistance in populations can be set. For the G143A mutation, this was relatively easy since the sensitivity difference between sensitive and resistant strains was huge (more than a factor of 100), because the resistance selection process is disruptive. Other mutations, such as the F129L, causing lower resistance factors, lead to a more or less continuous sensitivity distribution. In such cases, the molecular technique was especially helpful to monitor the spread of resistant isolates (Sierotzki et al. 2007).

Molecular technologies allow the testing of bulked samples and are in most cases very sensitive and reliable (e.g., Dufour et al. 2011).

9.5.3 Evolution of Resistance

The emergence of QoI resistance was/is highly dependent on the pathogen species. Resistance based on the G143A mutation was observed in powdery mildews of cereals and cucurbits, downy mildews of grapes and vegetables, and black sigatoka on banana already 2 years after the commercial use of QoIs (Reschke 1999; Heaney et al. 2000; Sierotzki et al. 2000a, b). Thereafter, many other pathogens developed resistance toward QoIs, either based on G143A, such as *Alternaria alternata*, *Venturia inaequalis*, *Erysiphe necator*, and *Ramularia collo-cygni* (Ma et al. 2003; Wilcox et al. 2003; Matusinsky et al. 2010; Fountaine et al. 2011; Baudoin et al. 2008), or based on F129L, such as *Alternaria solani* and *Pyrenophora teres*. In some pathogens, such as *Pyricularia oryzae* and *Pyrenophora tritici-repentis*, both mutations can be detected. Recently, additional pathogens being affected by QoI resistance have been reported, such as *Cercospora sojina* (Zeng et al. 2015), *Rhizoctonia solani*, and *Fulvia fulva*. Until today, 42 pathogen species have been reported to have developed resistance to QoI fungicides (Table 9.1). Resistance occurred in all three major fungal groups, i.e., *Ascomycetes*, *Deuteromycetes*, and *Basidiomycetes*, as well as in *Oomycetes*. However, the only basidiomycete pathogen with QoI resistance is *R. solani* AG-1 IA isolated from rice, detected in the USA in 2011 (www.FRAC.info). Resistance based on G143A seems to be stable in plant pathogen populations and does not necessarily induce a significant fitness penalty. However, a slow decrease of mutation frequency in *Plasmopara viticola* (downy mildew of grapes) has been reported after significant reductions in QoI usage (Genet et al. 2006; Toffolati et al. 2008). For some pathogens, such as *Rhynchosporium secalis* and *Phytophthora infestans*, the reasons for the absence of resistance are not clear; for others, such as the rust fungi, a molecular mechanism was described (see above).

Since an extensive database of *cyt b* gene sequences was available already in the 1990s, it was rather straightforward to investigate the *cyt b* gene in several plant pathogens, such as *Blumeria graminis* f. sp. *tritici*, *Plasmopara viticola*, and *Mycosphaerella fijiensis* (Gisi et al. 2000; Sierotzki et al. 2000a, b). In most QoI-resistant isolates of the mentioned plant pathogen species, mutations in the *cyt b* gene have been found: the most prominent and frequent was G143A, but also F129L was detected early, e.g., in *P. viticola* (Gisi et al. 2002), *A. solani* (Pasche et al. 2002), and *Magnaporthe grisea* (Avila-Adame and Köller 2003). In pathogens where both G143A and F129L have been found, G143A became quickly dominant in populations and overcame other mechanisms as in *P. viticola* (Sierotzki et al. 2005), *Pyricularia grisea* (Kim et al. 2003), *Zymoseptoria tritici* (Lucas and Fraaije 2008; Kildea et al. 2010), and *Pyrenophora tritici-repentis* (www.frac.info). The *cyt b* gene is located in the mitochondrial genome and is inherited cytoplasmatically implying that resistance is donated uniparentally during sexual recombination.

Table 9.1 List of QoI-resistant plant pathogen species (for references, see www.frac.info)

Species name	Common name	Host	Geographical distribution	Type of resistance	Years until R detection	Comments
Alternaria alternata	Early blight	Potato and tomato	EU	G143A		Abundant
Alternaria alternata, *Alternaria tenuissima*, *Alternaria arborescens*	*Alternaria* blight	Pistachio	USA	G143A	5	Abundant
Alternaria mali	*Alternaria* blotch	Apple	USA	G143A		?
Alternaria solani	Early blight	Potato	USA, EU	F129L	4	Abundant
Blumeria graminis f. sp. *tritici* and f. sp. *hordei*	Powdery mildew	Wheat and Barley	EU	G143A	2	Abundant
Botrytis cinerea	Gray mold	Strawberries, small fruits, etc.	EU, Japan, USA	G143A, 10 % with intron	8	Abundant
Cercospora beticola	*Cercospora* leaf spot	Sugar beet	USA	G143A		Increasing importance
Cercospora sojina	Frogeye spot	Soybean	USA	G143A	12	Limited occurrence
Cladosporium carpophilum	Scab	Almond	USA	?		?
Colletotrichum graminicola	Anthracnose	Turfgrass	USA	G143A	5	Abundant
Corynespora cassiicola	*Corynespora* leaf spot	Cucumber	Japan	G143A	6	Abundant
Didymella bryoniae	Gummy stem blight	Cucurbits	USA	G143A	3	Abundant
Erysiphe necator	Powdery mildew	Grape	USA, EU	G143A	5	Abundant
Glomerella cingulata (= *Colletotrichum gloeosporioides*)	Anthracnose	Strawberries, small fruits, etc.	Japan	G143A		Abundant?
Microdochium majus	Snow mold, ear blight	Wheat	EU, Japan	G143A	5	Abundant
Microdochium nivale	Snow mold, ear blight	Wheat	EU, Japan	G143A	5	Abundant
Monilinia fructicola	Brown rot	Peach	USA	G143A	12	?

(continued)

Table 9.1 (continued)

Species name	Common name	Host	Geographical distribution	Type of resistance	Years until R detection	Comments
Mycosphaerella fijiensis	Black sigatoka	Banana	Central and South America, Africa, and Asia	G143A	1	Abundant
Mycosphaerella musicola	Yellow sigatoka	Banana	South America, Australia	G143A		?
Mycovellosiella nattrassii	Leaf mold	Eggplant	Japan	G143A		?
Passalora fulva	Leaf mold	Tomato	Japan	F129L		?
Pestalotiopsis longiseta	Gray blight	Tea	Japan	G143A and F129L		?
Phaeosphaeria nodorum	*Septoria nodorum* leaf blotch	Wheat	EU	G143A		Importance unclear
Plasmopara viticola	Downy mildew	Grape	EU, Japan	G143A and F129L	2	Abundant
Podosphaera fusca	Powdery mildew	Cucurbits	EU, Japan	No mutations detected		?
Pseudoperonospora cubensis	Downy mildew	Cucurbits	EU, Asia	G143A	2	Abundant
Pyrenophora teres	Net blotch	Barley	EU	F129L	5	Abundant
Pyrenophora tritici-repentis	Tan spot	Wheat	EU	G143A, F129L and G137R	5	Abundant
Pyricularia grisea (*Magnaporthe oryzae*)	Gray leaf spot	Turfgrass	USA	G143A and F129L	4	Abundant
	Blast	Rice	Japan	G143A	12	Abundant
Pythium aphanidermatum	*Pythium* blight	Turfgrass	USA	F129L	4	Abundant
Ramularia areola	Grey mildew	Cotton	Brazil	?		Abundant
Ramularia collo-cygni	*Ramularia*	Barley	EU	G143A	4	Abundant

Rhizoctonia solani AG-1 IA	Sheath blight	Rice	USA	F129L	13	Limited occurrence
Rhynchosporium secalis	Scald	Barley	EU	G143A	10	Limited occurrence
Sphaerotheca fuliginea (*Podosphaera xanthii*)	Powdery mildew	Cucurbits	EU, Asia	G143A and other mech.	1	Abundant
Stemphylium vesicarium	Brown spot	Asparagus and pear	EU	G143A	8	?
Venturia inaequalis	Scab	Apple	EU, USA	G143A	4	Abundant
Venturia pirina	Scab	Pear	USA	G143A		?
Zymoseptoria tritici	*Septoria* leaf blotch	Wheat	EU	G143A and F129L	4	Abundant

Anisogamous heterothallic species, like many *Ascomycetes*, inherit cytoplasm and mitochondria by the ascogonium-forming partner ("maternally"), whereas the antheridium-forming partner donates only the nucleus. The segregation for individual crosses should then be 0:1; however, in all species with ample sexual recombination and a mating type equilibrium, the segregation in populations will be 1:1. This kind of inheritance is assumed for most ascomycete plant pathogens such as *Venturia inaequalis* (Steinfeld et al. 2002). However, for isogamous homothallic species, both partners can donate mitochondria. Ascospores of a specific perithecium possess mitochondria from one parent only, but different perithecia vary leading to a 1:1 segregation for each cross. This type of inheritance has been shown for *Blumeria graminis* f. sp. *tritici* (Robinson et al. 2002). Ample reports are available showing that insufficient mitochondria (heteroplasmic stage with different genomes) are removed from cells faster than expected by stochastic probability (Kües and Casselton 1992). However, it seems that in some plant pathogens heteroplasmy is tolerated at least for a short time (Ishii et al. 2007). The underlying mechanisms are not clear since mitochondrial leakage, spontaneous back mutation, or low-frequency survival of sensitive mitochondria in a primarily resistant environment has not been clearly demonstrated. Similar phenomena were attributed to the gradual loss of resistance in *Podosphaera xanthii* over 3 years without QoI selection pressure (Ishii et al. 2007). However, in most pathogen species, mitochondrial heteroplasmy is obviously not stable since selection for functional mitochondria within cells is very high and supported by targeted mitochondrial apoptosis of nonfunctional mitochondria (Kües and Casselton 1992). This process ensures that only fully functional mitochondria are present in cells allowing cells to adapt faster to changes than just by stochastic redistribution during mitosis.

9.5.4 Examples

In *Zymoseptoria tritici*, first detection of a very low frequency of G143A was in 2001 (Fraaije et al. 2003). In 2002, QoI resistance was detected by bioassay and quantitative real-time PCR at five different locations in Europe. The main mutation was G143A, although in two Irish isolates, F129L have been reported (Lucas and Fraaije 2008; Kildea et al. 2010). The simultaneous emergence of resistance in five regions (Ireland, North England, Northern France, Denmark, and Northern Germany) in 2002 was based on independent selection, as shown by comparing the mitochondrial genome of isolates; the mutation occurred in different "haplotypes" simultaneously (Torriani et al. 2008), indicating that the selection process for QoI resistance between 1996 and 2002 was similar in these regions of Europe. Unfortunately, resistance frequency in Ireland was obviously already so high in 2003 that performance of solo QoI fungicides was significantly affected. Despite immediate implementation of a strict resistance management (use of mixtures only and limitation of number of applications), resistance frequencies increased quite rapidly, especially in the UK and Ireland, followed by Northern France and Northern

Germany. Resistance development in Southern parts of Germany and France was slower, most probably due to lower levels of epidemics and therefore lower number of applications (lower selection pressure). As a consequence of increased QoI resistance in *Z. tritici*, QoI usage in wheat dropped from over 50 % in 2004 to less than 30 % in the following years (Fig. 9.4).

In *Pyrenophora tritici-repentis*, causing tan spot on wheat, all three mutations G143A, F129L, and G137R have been found, offering the opportunity to study the effects of different mutations and their impact on resistance development in a single species (Sierotzki et al. 2007). The first resistant isolates, possessing F129L, were detected in Sweden in 2003; 1 year later, a low frequency of G143A was recorded in Denmark and Germany. A low frequency of G137R was also present, but disappeared again. Nowadays, G143A is dominating the populations in Europe (www.frac.info). The resistance factors for G143A isolates were much higher than for F129L and G137R. The latter mutations seem to impose higher fitness penalties than the G143A mutation. Disease produced by artificially inoculating plants in greenhouse experiments was well controlled by solo QoI applications when individuals carried no mutation or the F129L or the G137R mutations, whereas

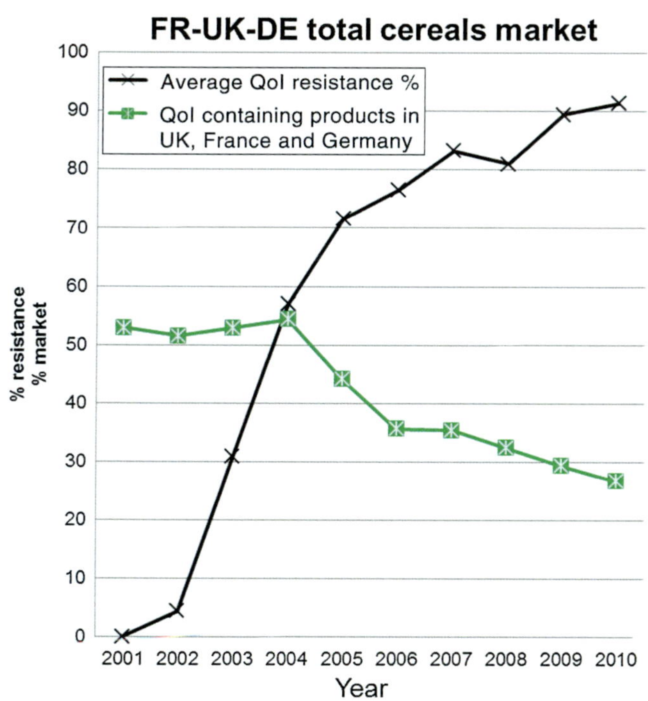

Fig. 9.4 Evolution of QoI resistance in *Zymoseptoria tritici* field populations measured as frequency of G143A mutation in bulked samples (average for the UK, France, and Germany) and market share of QoI-containing fungicides in the same countries between 2001 and 2010 (resistance and market data Syngenta internal information)

isolates carrying the G143A mutation were hardly controlled even at full label rates (Stammler et al. 2006, BASF internal data). An interesting comparison can be drawn between *P. tritici-repentis* (on wheat) and *P. teres*, causing net blotch on barley, since in the latter species the G143A mutation has not been found until today (intron present at position 143 in the cytochrome *b* gene as described above). In *P. teres*, F129L and G137R evolved almost in the same time period as in *P. tritici-repentis*. F129L can be present in European *P. teres* populations quite frequently (details on www.frac.info, QoI Working Group), while G137R is restricted to single sites in Ireland. Field trials showed that *P. teres* populations dominated by F129L can still be well controlled by QoI fungicides (Semar et al. 2007).

In *Plasmopara viticola*, resistance was detected for the first time in 1999 in Brazil and developed rapidly after widespread usage of QoIs in France and Italy (Gisi et al. 2002). Molecular tests revealed that G143A and F129L mutations were present in the populations; however, already in 2003, F129L frequency became very low, and monitoring for this mutation was stopped. However, G143A further spread in most vine-growing countries. In both *P. viticola* and *Podosphaera xanthii*, resistance was observed to decline in field populations when QoI treatments were stopped (Sierotzki et al. 2005; Genet et al. 2006). The decline occurred within five generations under greenhouse and within three seasons under field conditions, and reappearance could be delayed by using fungicide mixtures. These observations have been supported by studies performed in Italy measuring the G143A frequency and QoI resistance of oospores collected at the end of the growing season in different vineyards with different treatment regimes. When QoI fungicides were applied in mixtures or not used at all, the frequency of QoI resistance was significantly lower than in vineyards treated with solo QoIs (Toffolati et al. 2008). The mechanisms leading to this behavior have never been elucidated, but migration of sensitive isolates, reverse mutation, or heteroplasmy might contribute to dilute resistant populations. However, when solo QoIs were applied again, resistance quickly reappeared.

Ramularia collo-cygni is an important, widespread, and regularly occurring pathogen on barley. Described already in 1893 in Italy, the pathogen was not causing any obvious epidemics in barley field for a long time. However, in the last 20 years, frequent occurrence has been reported, and clear association to leaf spots on barley has been made. Recently, it was recognized that QoI fungicide does not provide sufficient control of this disease (Oxley and Hunter 2005). Fungicide sensitivity monitoring revealed that QoI resistance based on G143A was already fairly widespread (Fountaine and Fraaije 2009; Matusinsky et al. 2010; Fountaine et al. 2011). The reason for the recent increase of this pathogen is not fully understood; however, resistance evolution might have contributed and obviously developed largely unrecognized until it has reached rather high levels.

Rhizoctonia solani AG (anastomosis group)-1 IA isolates resistant to QoI fungicides were detected in 2011 in rice fields in Acadia Parish, Louisiana, USA, and have affected disease control in some fields. However, the resistance factors were rather low (around 40) still allowing a QoI dose response when isolates were tested under lab conditions (Olaya et al. 2012). In earlier studies with *R. solani*, no intron has been detected in the *cyt b* gene directly after position 143. This would indicate that G143A and/or F129L (or other mutations) may cause resistance to QoI fungicides (Grasso et al. 2006). Recent sequencing analyses of resistant *R. solani* isolates revealed the presence of F129L in the *cyt b* gene (Syngenta internal data). Resistance monitoring has been initiated in rice and soybean fields in the USA in 2011 and 2012 showing that resistance can be detected in close vicinity around the initial detection spots and spread is so far limited (Olaya 2012). QoI fungicides have been used since many years in this area to control diseases in the two alternating crops soybean and rice. Each crop normally receives only one to two fungicide applications per season; however, the *Rhizoctonia* populations were under continuous selection over many years. The resistance evolution, therefore, has to be considered as rather slow, confirming the classification of *Rhizoctonia* as a low-risk pathogen for resistance evolution (www.frac.info). It is not known why in *Rhizoctonia* the presumable less effective mutation F129L was selected and not G143A and whether this is only a transient stage before other, stronger mutations will replace F129L. Additionally, the question raises whether resistance might evolve soon also in other AGs of *R. solani*. The AG-1 IA isolates are known as aerial *Rhizoctonia* infecting the upper parts of plants, whereas most other pathogenic AGs seem to remain in the soil or close to the soil surface. It is not known if the origin of the inoculum would have an impact on resistance evolution.

9.5.5 Recommendations

The use recommendations for QoI-containing fungicides need to cover pathogen species that have already developed resistance but also those which represent still sensitive targets for the large QoI activity spectrum. Use recommendations especially include appropriate dosages, timing intervals, and mixing partners. The overall resistance risk (fungicide x pathogen x agronomy; see www.frac.info) for QoIs ranges from very high (e.g., for *Plasmopara viticola* treated several times under high disease pressure) to moderate to low (e.g., seed treatment against *Rhizoctonia*). In many crops, the number of applications of QoI-containing products per season is considerably restricted, and, wherever possible, mixtures or strict alternations are advised. The mixing or alternating partners need to be active at the used dosage against the current field population. The current QoI use recommendations agreed by the Working Group of the Fungicide Resistance Action Committee (FRAC) are published on www.frac.info, QoI Working Group, and are yearly updated. Additionally, also country and regional recommendation of FRAC exists, which should be within the "global" FRAC guidelines. The recommendations include diseases on cereals (all diseases), vine (powdery and downy mildew), pome fruits (scab), potato (late and early blight), soybean (all diseases, but mainly rust), sugar beet (*Cercospora* leaf spot), greenhouse vegetables, and related multiple-spray crops. For rice and banana, specific guidelines are given by specific groups within FRAC or others.

9.6 QiI Fungicides

9.6.1 *Mode of Binding*

The biochemical mode of action of cyazofamid has been investigated by mycelial growth assays (in the presence of SHAM), measurements of respiration in intact mycelia, electron transfer in mitochondria, and activity of the cytochrome bc1 enzyme complex (Mitani et al. 2001a). The experiments showed that the inhibition occurs specifically at the Qi site of the cytochrome b in oomycete pathogens; however, there was no activity in *Botrytis cinerea*, *Saccharomyces cerevisiae*, rat liver, and potato tuber tissue. Cyazofamid is active against *Phytophthora infestans*, *Pseudoperonospora cubensis*, and other oomycete pathogens (Mitani et al. 2001b). Cyazofamid provides mainly strong preventative activity, but also residual, moderate translaminar, and some curative activities were reported.

9.6.2 *Resistance Mechanisms*

The two heme centers in the cytochrome b (Qo and Qi) are sufficiently separated from each other to allow molecules to specifically bind to one or the other pocket. As a consequence, a different set of mutations has been reported to affect the activity of inhibitors of the two classes (Brasseur et al. 1996). The Qo inhibitors are affected by mutations at amino acid positions 106–336 and 279–336 of the cytochrome b gene (see above), whereas the Qi inhibitors are impacted by mutations at positions 33–52 and 209–252.

9.6.3 *Evolution of Resistance*

Mutations in the Qi binding pocket have been described to lead to Qi resistance in several organisms (Brasseur et al. 1996). However, resistance in agronomically relevant oomycete pathogens has rarely been reported. In *Phytophthora capsici*, resistance has been found in the USA on watermelon (Kousik and Keinath 2008). Until today, resistance to cyazofamid has not been correlated to any mutations, and the reports remain unconfirmed. Also, reports on reduced sensitivity in *Plasmopara viticola* (Note technique commune maladies de la vigne en France 2014, web page access: http://www.vignevin-sudouest.com/cartes/bulletins/documents/notetechniquecommuneVigne2014.pdf) were not followed up and could not be confirmed. Therefore, this information is not further considered in this article.

9.6.4 Recommendations

Isolates *P. capsici* resistant to QiIs are not cross-resistant to QoIs and QoSIs, suggesting to use them in mixture as resistance management tool for each other. The resistance risk for the highly specific single-site QiIs is assumed to be medium to high, and therefore, resistance management guidelines should be developed. However, mixtures between QoIs and QiIs may not be advisable due to the high risk for both fungicide classes. In addition, QiIs are active against oomycete pathogens, of which most are medium- to high-risk pathogens. Therefore, it is advisable to restrict the number of applications.

9.7 QoSI Fungicides

9.7.1 Mode of Binding

The Qo pocket of the cytochrome bc1 enzyme complex is well investigated, and various inhibitors are known to bind to this pocket (Brasseur et al. 1996). Some of these inhibitors were described to bind in a somewhat different manner reflected by their specific interaction with certain residues. Especially for myxothiazol and stigmatellin, a different binding has been recognized (Brasseur et al. 1996; Vallières et al. 2011). The binding characteristics for ametoctradin have been elucidated by mutant analysis, including G143A, F129L, G137R, G37V, and alternative oxidase (AOX) overexpression. Only AOX overexpression slightly affected the activity of the QoSI inhibitor (Schiffer et al. personal communication, Pest Manag Sci). Overexpression of AOX leads to a strong fitness penalty of isolates and is most probably not able to mediate field resistance (Joseph-Horne and Hollomon 2000). Comparative analysis of the heme absorbance pattern in the presence of the inhibitors revealed a strong similarity to stigmatellin binding; the predicted docking suggests that ametoctradin and stigmatellin bind to the Rieske protein by forming a hydrogen bond to amino acid H181. The inhibitors either interact directly or indirectly with the carboxylate group at E272 of cytochrome b. Anyway, QoSI compounds occupy the binding pocket in a different way compared to QoIs by not approaching the G143A position (Schiffer et al. personal communication, Pest Manag Sci; Fehr et al. 2015; Xiaolei et al. 2015).

9.7.2 Resistance Mechanisms

Sensitivity monitoring for *Plasmopara viticola* and *Phytophthora infestans* did not indicate any signs of resistance in field populations, and no target site mutations were identified (BASF data). Some strains of *P. viticola* with reduced sensitivity were detected for which the mechanism was described as AOX overexpression (Schiffer et al. personal communication, Pest Manag Sci). Mutagenesis experiments in *P. infestans* did not yield any resistant isolates (Gold et al. 2011).

9.7.3 Recommendations

No resistance has been reported so far for QoSIs. However, use recommendations have been developed including limitation of the number of applications, use of mixtures, alternation, and preventive use (Gold et al. 2011).

9.8 Summary and Outlook

Respiration inhibitors are one of the most successful fungicide classes, especially the QoIs. Despite ample and high frequencies of resistance and elevated resistance risk, these fungicides are still very important for plant disease control throughout the world. The broad spectrum and the advantageous toxicological and ecotoxicological profiles allow the usage of these fungicides in a wide range of agronomic situations (Bartlett et al. 2002). Additionally, the positive effects on plant growth make QoI fungicide very attractive for farmers. The QiI and QoSI fungicides are active exclusively against oomycete pathogens; resistance has not been confirmed in field populations, but resistance management is mandatory.

QoIs, QoSIs, and QiIs are single-site fungicides inhibiting respiration at the Qo and Qi site, respectively, of the cytochrome *bc*1 complex. Resistance for QoIs is induced by single point mutations in the *cyt b* gene. Additional mechanisms play a minor role. Resistance levels observed for the G143A mutation are high and have a significant impact on disease control. The F129L and G137R mutations induce lower resistance factors and therefore do not necessarily affect field performance of QoI-containing fungicides. In some pathogens, resistance evolution was very fast, such as in powdery mildews of cereals and cucurbits, downy mildew of grapes, or black sigatoka of bananas, whereas in other pathogens the evolution was much later and slower, e.g., in *Erysiphe necator* and *Rhizoctonia solani.* For some pathogens, resistance is still limited to particular areas and/or is not further spreading or disappeared (e.g., in *Rhynchosporium secalis*), or resistance did not occur at all, e.g., in *Phytophthora infestans*, *Sclerotinia homoeocarpa*, *S. sclerotiorum*, *Leptosphaeria maculans*, *L. biglobosa*, *Fusarium* spp., and rusts. Rust fungi and some ascomycete species possess a special molecular mechanism based on the presence of an intron in the *cyt b* gene preventing the occurrence of the G143A mutation. More and more pathogens are reported to also contain the bi2 intron at position 143 in the *cyt b* gene and therefore make resistance development unlikely. However, in about 40 pathogen species, resistance to QoI have been found and characterized. Not in all of them resistance leads to a complete loss of QoI activity, especially not in regions where these pathogens cause moderate disease problems. All QoI fungicides are in a cross-resistance group among themselves, but no cross-resistance exists to QiIs and QoSIs. In some pathogen species, QoI resistance is not stable possibly due to reduced enzyme activity or mitochondrial heteroplasmy, where temporary mixtures of sensitive and resistant mitochondria may be present in individual isolates. After releasing QoI selection pressure, the sensitivity may reestablish, as it has been shown for *Plasmopara viticola*. Resistance to QiIs is not widespread currently; it was detected only in *Phytophthora capsici* and not confirmed in *P. viticola*, whereas no resistance to QoSIs is reported so far.

The industrial partners producing and selling QoIs have agreed on a series of use recommendations which have been published as FRAC guidelines under www.frac.info. In general, QoI fungicides are considered as high-risk inhibitors for inducing resistance. The modulation of the overall resistance risk depends in addition on the biology and behavior of the pathogen and the agronomic risk factors. Use recommendations for QoI products have to be adapted to each host/pathogen system. The main elements are the restriction of the number of applications per season, the use of fungicide mixtures, or alternation and appropriate timing and intervals of applications. The use recommendations are summarized on www.frac.info.

The respiration chain offers a range of binding pockets. Until today, the Qo pocket has been investigated in detail; however, the QiI and recently also the QoSI fungicides indicate that additional binding sites are possible targets. Especially broadening the activity spectrum of QiIs and QoSIs beyond oomycetes may be an interesting approach. Recent patent applications raise hope that novel fungicides may be found targeting alternative binding pockets within the cytochrome b complex (complex III) and adding a new mode of action able to control all types of pathogen isolates in the most important crops (Carson et al. 2009).

References

Anke T (1995) The antifungal strobilurins and their possible ecological role. Can J Bot 73:940–945

Avila-Adame C, Köller W (2003) Impact of alternative respiration and target-site mutations on responses of germinating conidia of *Magnaporthe grisea* to Qo-inhibiting fungicides. Pest Manag Sci 59:303–309

Banno S, Ymashita K, Fukumor F, Okada K, Uekusa H, Takagaki M, Kimura M, Fujimura M (2009) Characterization of QoI resistance in Botrytis cinerea and identification of two types of mitochondrial cytochrome b gene. Plant Pathol 58:120–129

Bartlett DW, Godwin JR, Clough JM, Hall AA, Hamer M, Parr-Dobrzanski B (2002) The strobilurin fungicides. Pest Manag Sci 58:649–662

Baudoin A, Olaya G, Delmotte F, Colcol JF, Sierotzki H (2008) QoI resistance of *Plasmopara viticola* and *Erysiphe necator* in the Mid-Atlantic United States. Plant Management Network, Plant Health Progress, 11 February 2008

Becker WF, von Jagow G, Anke T, Steglich W (1981) Oudemansin, strobilurin A, strobilurin B and myxothiazol: new inhibitors of the bc1 segment of the respiratory chain with an E-ß-methoxyacrylate system as common structural element. FEBS Lett 132:329–333

Bonjardin CA, Pereira LS, Nobrega FG (1996) Analysis of exon and intron mutants in the cytochrome b mitochondrial gene of *Saccharomyces cerevisiae*. Curr Genet 30:200–205

Brandt U, Haase U, Schägger H, von Jagow G (1993) Speziesspezifität und Wirkmechanismus der Strobilurine. In: Wege zu neuen Produkten und Verfahren der Biotechnologie. Verlag Chemie, Weinheim, pp 27–38 (in German)

Brasseur G, Saribas AS, Daldal F (1996) A compilation of mutations located in the cytochrome b subunit of the bacterial and mitochondrial bc1 complex. Biochim Biophys Acta 1275:61–69

Carson Ch, Klittich CJR, Owen WJ, Schobert, Ch, Young DH (2009) Methods to control QoI-resistant fungal pathogens. Patent: US 20090306142 A1, published: 10.12.2009. Dow Agrosciences

Crofts RA (2004) The cytochrome bc1 complex: function in the context of structure. Annu Rev Physiol 66:689–733

Dufour MC, Fontaine S, Montarrya J, Corio-Costeta MF (2011) Assessment of fungicide resistance and pathogen diversity in *Erysiphe necator* using quantitative real-time PCR assays. Pest Manag Sci 67:60–69

Esser L, Quinn B, Li YF, Zhang M, Elberry M, Yu L, Yu CA, Xia D (2004) Crystallographic studies of quinol oxidation site inhibitors: a modified classification of inhibitors for the cytochrome bc1 complex. J Mol Biol 341:281–302

Fernández-Ortuño D, Torés JA, De Vicente A, Pérez-García A (2008) Field resistance to QoI fungicides in *Podosphaera fusca* is not supported by typical mutations in the mitochondrial cytochrome b gene. Pest Manag Sci 64:694–702

Fernández-Ortuño D, Torés JA, De Vicente A, Pérez-García A (2010) The QoI fungicides, the rise and fall of a successful class of agricultural fungicides. In: Carisse O (ed) Fungicides. ISBN: 978-953-307-266-1. InTech, available from: http://www.intechopen.com/books/fungicides/the-qoifungicides

Fehr M, Wolf A, Stammler G (2015) Binding of the respiratory chain inhibitor ametoctradin of the mitochondrial bc1 complex. Pest Manag Sci. Apr 27.doi: 10.1002/ps.403 (Epub ahead of print)

Förch M, Kessel G, Spits H, Hasunuma N (2007) Baseline sensitivity of *Phytophthora infestans* lifecycle components to NC–224 20SC (amisulbrom). Tenth Workshop of an European network for development of an integrated control strategy of potato late blight. Bologna, Italy, PPO-Special Report no. 12:337–344

Fountaine JM, Burnett FJ, Fraaije BA (2011) Development of QoI resistance in *Ramularia collo-cygni* populations. In: Dehne HW, Deising HB, Gisi U, Kuck K-H, Russell P, Lyr H (eds) Modern fungicides and antifungal compounds VI. DPG Spectrum Phytomedizin, Braunschweig, pp 199–202

Fountaine JM, Fraaije BA (2009) Development of QoI resistant alleles in populations of *Ramularia collo-cygni*. Asp Appl Biol 92:123–126

Fraaije BA, Butters JA, Coelho JM, Jones DR, Hollomon DW (2002) Following the dynamics of strobilurin resistance in *Blumeria graminis* f. sp. *tritici* using quantitative allele-specific real-time PCR measurements with the fluorescent dye SYBR Green I. Plant Pathol 51:45–54

Fraaije BA, Lucas JA, Clark WS, Burnett FJ (2003) QoI resistance development in populations of cereal pathogens in the UK. In: Proceedings of the BCPC Intr Congr, Crop Science and Technology, vol 2. The British Crop Protection Council, Alton, Hampshire, pp 689–694

FRAC (2013) www.frac.info. Accessed 1 May 2013

Genet JL, Jaworska G, Deparis F (2006) Effect of dose rate and mixtures of fungicides on selection for QoI resistance in populations of *Plasmopara viticola*. Pest Manag Sci 62:188–194

Gisi U, Sierotzki H (2008) Molecular and genetic aspects of fungicide resistance in plant pathogens. In: Dehne HW, Gisi U, Kuck KH, Russell PE, Lyr H (eds) Modern fungicides and antifungal compounds IV. DPG, Bonn, pp 53–61

Gisi U, Chin KM, Knapova G, Küng Färber R, Mohr U, Parisi S, Sierotzki H, Steinfeld U (2000) Recent developments in elucidating modes of resistance to phenylamide, DMI and strobilurin fungicides. Crop Prot 19:863–872

Gisi U, Sierotzki H, Cook A, McCaffery A (2002) Mechanisms influencing the evolution of resistance to Qo inhibitor fungicides. Pest Manag Sci 58:859–867

Gold RE, Scherer M, Rether J, Speakman J, Navé B, Levy T, Storer R, Marris D (2009) Initium – an innovative fungicide. Proc BCPC Congr, Glasgow, pp 1–6

Gold RE, Schiffer H, Speakman J, Stammler G, Klappach K, Brix HD, Schlehuber S (2011) Initium – a new innovative fungicide for the control of Oomycetes in speciality crops. In: Dehne HW, Deising HB, Gisi U, Kuck K-H, Russell P, Lyr H (eds) Modern fungicides and antifungal compounds VI. DPG Spectrum Phytomedizin, Braunschweig, pp 55–61

Grasso V, Palermo S, Sierotzki H, Garibaldi A, Gisi U (2006) Cytochrome b gene structure and consequences for resistance to Qo inhibitor fungicides in plant pathogens. Pest Manag Sci 62:465–472

Heaney SP, Hall AA, Davies SA, Olaya G (2000) Resistance to fungicides in the QoI-STAR cross-resistance group: current perspectives. Brighton Crop Prot Conf: 755–762

Herms S, Seehaus K, Koehle H, Conrath U (2002) A strobilurin fungicide enhances the resistance of tobacco against tobacco mosaic virus and *Pseudomonas syringae* pv *tabaci*. Plant Physiol 130:120–127

Ishii H, Yano K, Date H, Furuta A, Sagehashi Y, Yamaguchi T, Sugiyama T, Nishimura K, Hasama W (2007) Molecular characterization and diagnosis of QoI resistance in cucumber and eggplant pathogens. Phytopathology 97:1458–1466

Ishii H, Foutaine L, Chung W-H, Kansako M, Nishimura K, Takahashi K, Oshima M (2009) Characterisation of QoI-resistant field isolates of *Botrytis cinerea* from citrus and strawberry. Pest Manag Sci 65:916–922

Jabs T, Cronshaw K, Freund A (2001) New strobilurin resistance mechanism in apple scab (*Venturia inaequalis*). Phytomedizin 31:15–16

Joseph-Horne T, Hollomon DW (2000) Functional diversity within the mitochondrial electron transport chain of plant pathogenic fungi. Pest Manag Sci 56:24–30

Kildea S, Dunne B, Mullins E, Cooke LR, Mercer PC, O'Sullivan E (2010) Pyraclostrobin reduces germ tube growth of QoI-resistant *Mycosphaerella graminicola* pycnidiospores and the severity of *Septoria tritici* blotch on winter wheat. Plant Pathol 59:1091–1098

Kim YS, Dixon EW, Vincelli P, Farman ML (2003) Field resistance to strobilurin (QoI) fungicides in *Pyricularia grisea* caused by mutations in the mitochondrial cytochrome b gene. Phytopathology 93:891–900

Kousik CS, Keinath AP (2008) First report of insensitivity to cyazofamid among isolates of *Phytophthora capsici* from the Southeastern United States. Plant Dis 92:979

Kües U, Casselton LA (1992) Fungal mating type genes – regulators of sexual development. Mycol Res 96:993–1006

Lazowska J (1980) Sequence of introns and flanking exons in wild type and box3 mutants of cytochrome b reveals an interlaced splicing protein coded by an intron. Cell 22:333–348

Leiminger JH, Adolf B, Hausladen H (2013) Occurrence of the F129L mutation in Alternaria solani populations in Germany in response to QoI application, and its effect on sensitivity. Plant Pathol 63(3):640–650

Link T, Iwata M, Björkman J, van der Spoel D, Stocker A, Iwata S (2003) Molecular modeling of inhibitors at Qi and Qo sites in cytochrome bc1 complex. In: Voss G, Ramos G (eds) Chemistry of crop protection. Wiley-VCH Weinheim, Germany

Lucas JA, Fraaije BA (2008) QoI resistance in *Mycosphaerella graminicola*: what have we learned so far? In: Dehne HW, Deising HB, Gisi U, Kuck K-H, Russell P, Lyr H (eds) Modern fungicides and antifungal compounds V. DPG Spectrum Phytomedizin, Braunschweig, pp 71–77

Ma Z, Felts D, Michailides TJ (2003) Resistance to azoxystrobin in *Alternaria* isolates from pistachio in California. Pestic Biochem Physiol 77:66–74

Matusinsky P, Svobodava-Leisova L, Marik P, Stemberkova L, Hanusova M, Minarikova V, Vysohlidova M, Spitzer T (2010) Frequency of a mutant allele of cytochrome b conferring resistance to QoI fungicides in the Czech population of *Ramularia collo-cygni*. J Plant Dis Prot 117:248–252

McDougall (2012) http://www.croplife.org/view_document.aspx?docId=2478

Miessner S, Stammler G (2010) *Monilinia laxa, M. fructigena* and *M fructicola*: risk estimation of resistance to QoI fungicides and identification of species with cytochrome b gene sequences. J Plant Dis Protec 117:162–167

Miessner S, Mann W, Stammler G (2011) *Guignardia bidwellii*, the causal agent of black rot of grapevine has a low risk for QoI resistance. J Plant Dis Prot 118:51–53

Mitani S, Araki S, Takii Y, Ohshima T, Matsuo N, Miyoshi H (2001a) The biochemical mode of action of the novel selective fungicide cyazofamid: Specific inhibition of the mitochondrial complex III in *Pythium spinosum*. Pestic Biochem Physiol 71:107–115

Mitani S, Araki S, Ymaguchi T, Takii Y, Ohshima T, Matsuo N (2001b) Biological properties of the novel fungicide cyazofamid against *Phytophthora infestans* on tomato and *Pseudoperonospora cubensis* on cucumber. Pest Manag Sci 58:139–145

Nason MA, Farrar J, Bartlett D (2007) Strobilurin fungicides induce changes in photosynthetic gas exchange that do not improve water use efficiency of plants grown under conditions of water stress. Pest Manag Sci 63:1191–1200

Olaya G, Buitrago C, Pearsaul D, Sierotzki H, Tally A (2012) Detection of resistance to QoI fungicides in *Rhizoctonia solani* isolates from rice. Phytopathology 102:S4.88 (Abstr)

Oxley SJP, Hunter AE (2005) Appropriate fungicide doses on winter barley: producing dose–response data for a decision guide. HGCA Project Report No 366

Pasche JS, Wharam CM, Gudmestad NC (2002) Shift in sensitivity of *Alternaria solani* (potato early blight) to strobilurin fungicides. Proc BCPC Conf Pests Dis, 841–846

Reschke M (1999) Strobis in Gefahr. DLG-Mitteilungen 1:51 (in German)

Robinson HL, Ridout CJ, Sierotzki H, Gisi U, Brown JK (2002) Isogamous, hermaphroditic inheritance of mitochondrion-encoded resistance to Qo inhibitor fungicides in *Blumeria graminis* f. sp. *tritici*. Fungal Genet Biol 36:98–106

Ruske RE, Gooding MJ, Jones SA (2003) The effects of triazole and strobilurin fungicide programmes on nitrogen uptake, partitioning, remobilization and grain N accumulation in winter wheat cultivars. J Agric Sci 140:395–407

Sauter H (2012) Strobilurins and other complex III inhibitors. In: Krämer W, Schirmer U (eds) Modern crop protection compounds. Wiley-VCH, Weinheim, pp 584–627

Sauter H, Ammermann E, Benoit R, Brand R, Gold RE, Grammenos W, Köhle H, Lorent G, Müller B, Röhl F, Schirmer U, Speakman JB, Wenderoth B, Wingert H (1995) Mitochondrial respiration as a target for antifungals: lessons from research on strobilurins. In: Dixon GK, Copping LG, Hollomon DW (eds) Antifungal agents- discovery and mode of action. BIOS Scientific Publishers, Oxford, pp 173–192

Semar M, Strobel D, Koch A, Klappach K, Stammler G (2007) Field efficacy of pyraclostrobin against populations of *Pyrenophora teres* containing the F129L mutation in the cytochrome b gene. J Plant Dis Prot 114:117–119

Sierotzki H, Gisi U (2003) Molecular diagnostics for fungicide resistance in plant pathogens. In: Voss G, Ramos G (eds) Chemistry of crop protection. Wiley-VCH, Weinheim, pp 71–88

Sierotzki H, Frey R, Wullschleger J, Palermo S, Karlin S, Godwin J, Gisi U (2007) Cytochrome b gene sequence and structure of *Pyrenophora teres* and *P. tritici-repentis* and implications for QoI resistance. Pest Manag Sci 63:225–233

Sierotzki H, Kraus N, Assemat P, Stanger C, Cleere S, Windass J, Gisi U (2005) Evaluation of resistance to QoI fungicides in *Plasmopara viticola* populations in Europe. In: Dehne HW, Gisi U, Kuck KH, Russell PE, Lyr H (eds) Modern fungicides and antifungal compounds IV. BCPC, Alton, pp 73–80

Sierotzki H, Parisi S, Steinfeld U, Tenzer I, Poirey S, Gisi U (2000a) Mode of resistance to respiration inhibitors at the cytochrome bc1 enzyme complex of *Mycosphaerella fijiensis* field isolates. Pestic Sci 56:833–841

Sierotzki H, Wullschleger J, Gisi U (2000b) Point-mutation in cytochrome b gene conferring resistance to strobilurin fungicides in *Erysiphe graminis* f. sp. *tritici* field isolates. Pestic Biochem Physiol 68:107–112

Stammler G (2012) Resistance risk of corn pathogens to QoI fungicides. Outlook Pest Manag 23:211–214

Stammler G, Schutte GC, Speakman J, Miessner S, Crous PW (2013) *Phyllosticta* species on citrus: risk estimation of resistance to QoI fungicides and identification of species with cytochrome *b* gene sequences. Crop Prot 48:6–12

Stammler G, Strobel D, Semar M, Klappach K (2006) Diagnostics of fungicide resistance and relevance of laboratory data for the field. Aspects Appl Biol 78: 29–36. In: Fungicide resistance: are we winning the battle but losing the war?

Steinfeld U, Sierotzki H, Parisi S, Gisi U (2002) Comparison of resistance mechanisms to strobilurin fungicides in *Venturia inaequalis*. In: Dehne HW, Gisi U, Kuck KH, Russel PE, Lyr H (eds) Modern fungicides and antifungal compounds III. AgroConcept, Bonn, pp 167–176

Toffolati SL, Prandato M, Serrati L, Sierotzki H, Gisi U, Vercesi A (2008) Monitoring QoI resistance in *Plasmopara viticola* oospore populations. In: Dehne HW, Gisi U, Kuck KH, Russell PE, Lyr H (eds) Modern fungicides and antifungal compounds IV. DPG, Bonn, pp 79–84

Torriani SFF, Brunner PC, McDonald BA, Sierotzki H (2008) QoI resistance emerged independently at least 4 times in European populations of *Mycosphaerella graminicola*. Pest Manag Sci 65:155–162

Vallières C, Trouillard C, Dujardin G, Meunier B (2011) Deleterious effect of the Qo inhibitor compound resistance-conferring mutation G143A in the intron-containing cytochrome *b* gene and mechanisms for bypassing it. Appl Environ Microbiol 77:2088–2093

Venacio WS, Rodrigues MAT, Begliomini E, de Souza NL (2003) Physiological effects of strobilurin fungicides on plants. Ponta Grossa 9:59–68

Wilcox WF, Burr JA, Riegel DG, Wong FP (2003) Practical resistance to QoI fungicides in New York populations of *Uncinula necator* associated with quantitative shifts in pathogen sensitivities. Phytopathology 93:S90 (abstract)

Xiaolei Z, Mengmeng Z, Jingling L, Jingming G, Guangfu Y (2015) Ametoctradin is a potent Qo site inhibitor of the mitochondrial respiration complex III. J Agric Food Chem 63:3377–3386

Zeng F, Zhang G, Olaya G, Wullschleger J, Sierotzki H, Ming R, Bradley CA (2015) Characterization of quinone outside inhibitor fungicide resistance in Cercospora sojina and development of diagnostic tools for its identification. Plant Disease 99(4):544–550

Ziogas BN, Baldwin BC, Young JE (1997) Alternative respiration: a biochemical mechanism of resistance to azoxystrobin (ICIA 5504) in *Septoria tritici*. Pestic Sci 50:28–34

Chapter 10
Oomycete Fungicides: Phenylamides, Quinone Outside Inhibitors, and Carboxylic Acid Amides

Ulrich Gisi and Helge Sierotzki

Abstract Oomycetes are "fungal-like" protista that are phylogenetically distinct from fungi, they are diploid, and cell walls contain cellulose. Oomycetes comprise major plant pathogens within the orders Saprolegniales, Pythiales, Peronosporales ("downy mildews" including species of *Plasmopara*, *Phytophthora*, *Peronospora*, *Pseudoperonospora*, *Bremia*), and Sclerosporales. Control of oomycete diseases relies mainly on chemical measures using products within 16 different chemical groups, among which the phenylamides (PAs), quinone outside inhibitors (QoIs), carboxylic acid amides (CAAs), and multisite inhibitors are most widely used. However, resistance evolved against most single-site inhibitors in many oomycete pathogen species. Resistance against PAs (inhibitors of RNA polymerase I) was detected in most oomycete pathogen species; the molecular mechanism was associated recently with the presence of the Y382F mutation in the *RNApolI* gene. QoIs inhibit electron transport in complex III of the respiration chain; two major mutations, G143A (in *Plasmopara viticola*) and F129L (in *Pythium* spp.), have been identified in the cytochrome b, *cyt b* gene coding for resistance. However, no resistant isolates were observed in *Phytophthora* species. CAAs inhibit cellulose synthase; several mutations, G1105S/V (in *P. viticola*) and G1105V/W (in *Pseudoperonospora cubensis*), were detected in the cellulose synthase, *CesA3* gene of CAA-resistant field isolates. In *Phytophthora infestans and P. capsici*, no CAA-resistant field isolates were discovered, whereas in artificial mutants, the mutations G1105A/V and V1109L/M were found. The L1109 and M1109 amino acid configurations are responsible for the intrinsic insensitivity to CAAs of the entire genus *Pythium* and all oomycetes outside Peronosporales. Resistance is inherited by one recessive nuclear gene (*CesA3*) for CAAs, one maternal mitochondrial gene (*cytb*) for QoIs, and mainly one semidominant nuclear gene for PAs. The consequences of

U. Gisi (✉)
University of Basel, Basel, Switzerland

Syngenta Crop Protection AG, Stein, Switzerland
e-mail: u.gisi@bluewin.ch

H. Sierotzki
Syngenta Crop Protection AG, Schaffhauserstrasse, 4332 Stein, Switzerland
e-mail: helge.sierotzki@syngenta.com

H. Ishii, D.W. Hollomon (eds.), *Fungicide Resistance in Plant Pathogens*,
DOI 10.1007/978-4-431-55642-8_10

the molecular mechanisms for evolution and stability of resistance are discussed in the light of oomycete pathogen biology.

Keywords Carboxylic acid amides • Fungicides • Fungicide resistance • Oomycetes • Peronosporales • Phenylamides • *Phytophthora* • *Plasmopara* • *Pseudoperonospora* • Quinone outside inhibitors

10.1 Introduction

10.1.1 Biology of Oomycetes

Oomycetes are, together with Chytridiomycetes and Phaeophyta (brown algae), a group of organisms called Chromista or Mastigomycota or recently Straminipila. Together with Plasmodiophoromycota and Myxomycota, they form the "fungal-like protista" within the kingdom Protista and are clearly separate from Fungi, Prokaryota (bacteria), Plantae, and Animalia (Whittaker's five kingdoms). The differentiation of oomycetes from fungi is based on many important characteristics in their life cycle like ploidity, cell wall and cell membrane composition, metabolic pathways, and sensitivity to a range of inhibitors. Oomycetes are diploid, have cellulose as fibrillar cell wall component, are mostly sterol auxotrophic, produce lysine along the DAP (diaminopimelic acid) pathway, have hyphae with no (or few) septa, and are insensitive to DMI (sterol demethylation inhibitor) fungicides and sensitive to phenylamide (PA) and carboxylic acid amide (CAA) fungicides (Table 10.1). Oomycetes embrace major orders like Saprolegniales, Pythiales, Albuginales, Peronosporales, and Sclerosporales; most species are important plant pathogens (with the Peronosporaceae, the downy mildews, being biotrophic, "obligate" pathogens), but some are pathogenic to fish (e.g., *Saprolegnia* spp.) or have also strong saprotrophic activity (e.g., many *Pythium* spp.). Recently, oomycetes have been regrouped based on a novel molecular marker, the amino acid configuration at position 1109 in the cellulose synthase, *CesA3* gene sequence, showing a close relatedness of *Phytophthora* spp. with the "classical" downy mildew clade, and a clear distinction from Pythiales among which two subgroups can be distinguished (the "*aphanidermatum*" and the "*ultimum*" clade) (Fig. 10.1) (Blum et al. 2012). This regrouping is also reflected in a different sensitivity of *Phytophthora* and *Pythium* species, the former being sensitive and the latter insensitive (sometimes also defined as tolerant) to CAA fungicides (Table 10.1, more details in Sect. 10.4).

In terms of market size for anti-oomycete products, the most important foliar pathogen species are *Phytophthora infestans* (tomato and potato late blight), *Plasmopara viticola* (grape downy mildew), *Pseudoperonospora cubensis* (cucurbit downy mildew), *Bremia lactucae* (lettuce downy mildew), and several *Peronospora* spp. (downy mildews) on various crops like peas, *Brassica*, and tobacco (Hermann and Gisi 2012). In addition, many soilborne oomycetes including species of the

Table 10.1 Characteristics of oomycetes and fungi

Characteristics	Oomycetes	Fungi (zygo-, asco-, basidiomycetes)
Ploidity	Diploid (2*n*) (rarely also polyploid), as plants and animals; Mendelian segregation	Haploid (*n*)
Cell wall, fibrillar fraction	Cellulose (1-4-β-bound) (10–30 % of cell wall), as in plants	Chitin, chitosan (1-4-β-bound), as in insects
Cell wall, amorphous fraction	1-3/1-6-β-glucans (70–90 % of cell wall)	1-3/1-6-β-glucans
Cell membrane, functional sterol	Cholesterol, as in animals; sterol auxotrophy	Ergosterol
Lysine biosynthesis	DAP-pathway[a], as in plants	AAP-pathway[a]
Hyphae (septa)	No (few) septa	Regular septation
Asexual spore stages	Zoospores (and sporangia) for dissemination, chlamydospores for survival, conidia (*Peronospora*)	Conidia (and related forms) for dissemination
Sexual spore stages	Oospores	Zygo-, asco-, basidiospores
Sensitivity to:		
Pimaricin (antibiotic)	Tolerant	Mostly sensitive
MBCs (benomyl)	Tolerant	Mostly sensitive (resistant)
DMIs	Tolerant	Mostly sensitive (resistant)
Phenylamides	Sensitive (resistant)	Tolerant
CAAs	Sensitive (resistant) for Peronosporales, tolerant for other oomycetes	Tolerant
OSBPIs[b]	Sensitive (resistant) (except *Pythium*, tolerant)	Tolerant
QoIs	Mostly sensitive (resistant)	Mostly sensitive (resistant)

[a]*DAP* Diaminopimelic acid, *AAP* amino adipic acid
[b]*OSBPIs* Oxysterol-binding protein inhibitors

genus *Pythium* (e.g., on corn, soybean, vegetables, turf, ornamentals) and *Phytophthora*, e.g., *P. cinnamomi*, *P. nicotianae*, *P. erythroseptica*, and *P. capsici*, cause key diseases against which chemical control measures are required, mainly by soil application. Also seed-borne oomycetes are known to cause serious damage in several crops such as *Plasmopara halstedii* on sunflower and *Sclerospora*, *Peronosclerospora*, and *Sclerophthora* spp. on corn and millet, against which seed treatment with several compounds is a common practice.

In several oomycetes, sexual reproduction has been reported either as integral part of the disease cycle (e.g., in *P. viticola*) or as geographic specialization (e.g., in *P. infestans* in Northern Europe or Mexico). As a result, genetic diversity of populations can increase, also affecting the frequency and distribution of different pathotypes (with different virulence factors) and genotypes with different sensitivities to

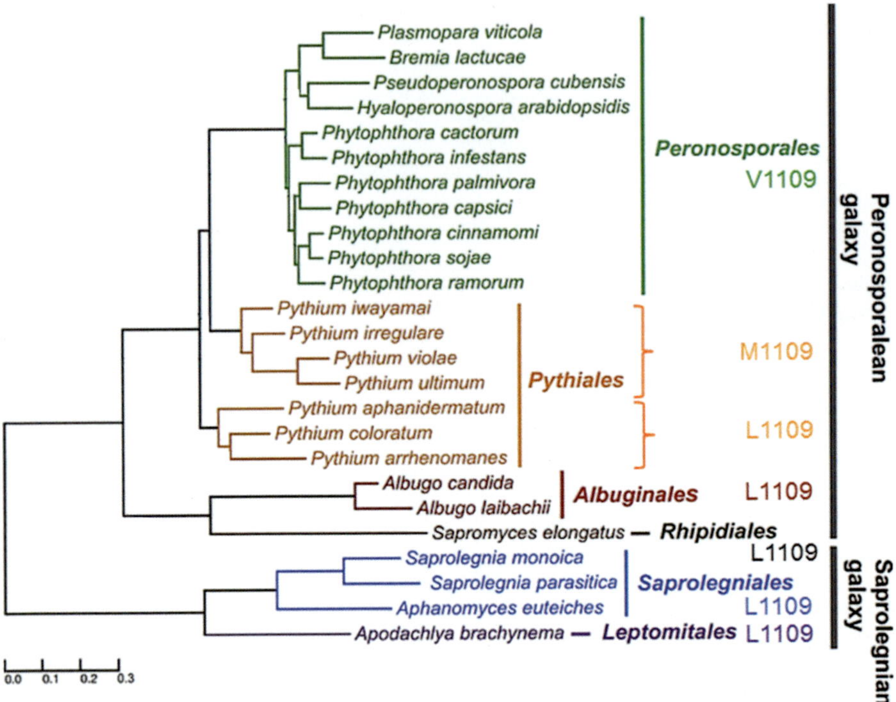

Fig. 10.1 Relatedness among oomycetes based on *CesA3* gene and amino acid configurations at position 1109, V/M/L1009 (After Blum et al. 2012). *CesA3 ce*llulose *s*ynthase gene 3, *V* valine, *M* methionine, *L* leucine

fungicides (sensitive, intermediate, resistant individuals, see PA chapter Sect. 10.2) or homozygote and heterozygote individuals (see CAA chapter Sect. 10.3), providing in many sites a rather stable equilibrium between different genotypes (alleles, subpopulations) (Gisi and Cohen 1996; Gisi et al. 2011). However, clonal populations can dominate local populations for a certain time period until new genotypes may appear through migration and import with infected plant material as it is known for *P. infestans* in many European countries (Hermann and Gisi 2012). Sexual recombination requires always two mating types (called A1 and A2 in *Phytophthora* spp.) which can be unified in one individual (homothallic species, e.g., *Phytophthora cactorum, P. erythroseptica, P. megasperma, P. palmivora*) or separate in two individuals (heterothallic species, e.g., *P. capsici, P. cinnamomi, P. infestans, P. nicotianae, P. parasitica*). In some heterothallic species, the sex ratio (frequency of the two mating types in field populations) is often close to 1:1 (e.g., in *P. capsici* and also in *Plasmopara viticola*, P1 and P2, visible only if single sporangiophore isolates are collected from single lesions; Scherer and Gisi 2006), whereas in others, the two mating types can be locally, globally, or temporarily quite separated (e.g., in *P. infestans*) resulting in isolated clonal populations. Sexual behavior and frequency of sexual recombination in the disease cycle of oomycete plant pathogens strongly influence the evolution and stability of fungicide resistance (see below).

10.1.2 Oomycete Fungicides

A total of 16 chemical groups with different modes of action are available for oomycete control: The single-site inhibitors such as phenylamides (PAs), quinone outside inhibitors (QoIs), and carboxylic acid amides (CAAs), as well as several multisite inhibitors (mostly inhibiting sulfhydration) including dithiocarbamates (e.g., mancozeb), phthalimides (e.g., folpet), chlorothalonil, copper formulations, and thiram (described in more detail below, Table 10.2). In addition, several groups with smaller market shares and some with unknown mode of action are available such as cymoxanil, fosetyl-Al, fluazinam (uncoupler), quinone inside respiration inhibitors (QiIs, like cyazofamid), fluopicolide (inhibiting spectrin delocalization), and ethaboxam (inhibiting beta-tubulin assembly) (Table 10.2). Also, older products are still in use against oomycete diseases such as hymexazol (heteroaromatics), fentin acetate (organo-tins), etridiazole (thiadiazoles), propamocarb (carbamates), and

Table 10.2 Major fungicide groups and key active ingredients (sorted according to market size in 2011) available for oomycete control, FRAC codes, and resistance risk

Fungicide groups and key active ingredients	FRAC codes[a]	Sales 2011 in mUS$[b]	Resistance risk[a]
Phenylamides	4 (A1)	405	High
Metalaxyl/mefenoxam; oxadixyl; benalaxyl/kiralaxyl			
QoI fungicides	11 (C3)	n.a.[c]	High
Azoxystrobin; fenamidone; famoxadone			
Multisites	(M1–M5)	n.a.[c]	Low
For example, mancozeb; chlorothalonil; copper			
CAA fungicides	40 (H5)	<275	Moderate
Dimethomorph; flumorph; iprovalicarb; benthiavalicarb; mandipropamid			
Cyanoacetamide oximes (cymoxanil)	27 (U)	145	Moderate
Dinitroanilines (fluazinam)	29 (C5)	145	Moderate
Phosphonates (fosetyl-Al)	33 (U)	130	Low
QiI fungicides	21 (C4)	<80	Medium to high
Cyazofamid; amisulbrom			
Benzamides (fluopicolide)	43 (B5)	45	Moderate
Benzamides/carboxamides	22 (B3)	<40	Low
Ethaboxam; zoxamide			

[a]Nomenclature according to FRAC Mode of Action Code List 2014, www.frac.info; Kuck et al. (2012)
[b]Figures according to Phillips McDougall Product Directory V13 (2012)
[c]QoIs and multisites are broad-spectrum fungicides including activity against fungi and oomycetes; total market size for each fungicide group > 2 bn US$; no separation of oomycete market fraction available (n.a.)

the plant defense inducer acibenzolar-S-methyl, but also novel compound such as ametoctradin (quinone inhibitor at unknown site, QoSI) became available recently. Almost all chemical groups are used for foliar treatments, PAs, fosetyl-Al, CAAs, propamocarb, QoIs, QiIs, and fluopicolide also for soil treatments, whereas hymexazol and etridiazole are used exclusively for soil applications. A smaller group of compounds are available especially for seed treatments including PAs, QoIs, QiIs, cymoxanil, hymexazol, and acibenzolar-S-methyl.

10.2 Phenylamide (PA) Fungicides

10.2.1 Mode of Action

Seven PA molecules have been introduced between 1977 and 2007: metalaxyl and benalaxyl including the corresponding active isomers (mefenoxam = metalaxyl-M and kiralaxyl = benalaxyl-M, respectively) oxadixyl, furalaxyl, and ofurace, the latter two having been withdrawn from the market (Hermann and Gisi 2012). PAs inhibit the polymerization of ribosomal RNA biosynthesis in oomycetes (Fisher and Hayes 1982). Davidse (1995) showed that metalaxyl affects the polymerase I complex of rRNA synthesis in *Phytophthora megasperma*, which was considered as primary site of action. PAs inhibit several life stages of oomycetes, mainly hyphal growth, haustoria, and sporangia formation (Schwinn and Staub 1995), whereas no effect on zoospore release or motility and cystospore germination was observed. PAs are intrinsically highly active and specifically control plant pathogens within the oomycetes including downy mildews of the Peronosporales and Sclerosporales as well as most members of the Pythiales and also Saprolegniales (Gisi 2002). The long-lasting preventative and broad activity within oomycetes (including also seed- and soilborne pathogens such as *Pythium* and *Phytophthora* spp.), the high mobility in plants, curative potential, and excellent crop safety make the PAs a valuable tool for disease management in many crops (Müller and Gisi 2012). They are used mostly as co-formulated products with multisite fungicides or unrelated single-site inhibitors to broaden the activity spectrum beyond oomycetes and to manage resistance development.

10.2.2 Resistance Mechanism

Although many investigations on the mode of action and mechanism of resistance to PAs have been undertaken over the last 30 years, the putative resistance gene(s) and site of mutation(s) in the genome have not been elucidated until recently (Müller and Gisi 2012). Davidse (1988) assumed that a mutation in the RNA polymerase gene is responsible for resistance because endogenous RNA polymerase activity of isolated nuclei of *P. megasperma* and *P. infestans* was highly susceptible to metalaxyl when gained from sensitive but not from resistant individuals. This hypothesis was confirmed recently by Whisson et al. (2011) who identified a specific genomic region in a resistant *P. infestans* isolate encoding the large subunit of RNA polymerase I that, when transferred by transformation into a sensitive isolate, led to a resistant individual. The responsible mutation conferring resistance was identified as Y382F in the *RNApolI* gene, which showed an association with resistance to mefenoxam in about 85 % of field isolates (Randall et al. 2014). Thus, it is not completely clear whether an additional mutation in another gene, associated or not with Y382F or additional mechanisms, is involved in PA resistance as hypothesized after the identification of MEX I and MEX II loci by Judelson and Roberts (1999).

Mendelian segregation of PA resistance in F1 and F2 progeny was observed in crossing studies with *P. infestans*, suggesting that PA resistance is monogenic (Shattock 1986, 1988) and is based on one incompletely dominant gene (Shaw and Shattock 1991). Studies with *P. capsici* (Lucas et al. 1990), *P. sojae* (Bhat et al. 1993), and *B. lactucae* (Crute and Harrison 1988) also suggested that resistance to metalaxyl is controlled by a single incompletely dominant gene. However, Fabritius et al. (1997) and Abu-El Samen et al. (2005) observed a continuous segregation pattern in F1 progeny of *P. infestans* and *P. erythroseptica* when metalaxyl-resistant and metalaxyl-sensitive parents were crossed, suggesting that a semidominant locus together with several minor loci may be involved in PA resistance. Crossing experiments with sensitive and resistant field isolates of both *P. infestans* and *P. viticola* (Gisi and Sierotzki 2008) yielded F1 progeny isolates that were exclusively intermediate in sensitivity. The F2 progeny of *P. viticola* deviated to some degree from the expected Mendelian distribution (s:i:r = 1:2.7:2, instead of 1:2:1; Fig. 10.2 Gisi et al. 2007), indicating that multiple mechanisms may be involved in PA resistance in this pathogen. Further research is required to elucidate the molecular mechanisms involved in PA resistance.

10.2.3 Resistance Evolution in P. infestans

Resistant isolates of *P. infestans* were found in Israel and Europe 2 to 3 years after first introduction of metalaxyl (Gisi and Cohen 1996) and subsequently in most countries with intensive potato and tomato cropping around the globe (FRAC List of Resistant Pathogenic Organisms 2013, www.frac.info). The evolutionary forces driving PA resistance are mutations, migration, selection, and recombination. The mutation frequency is generally believed not to be linked to the use of fungicides. In fact, PA-resistant isolates of *P. infestans* were detected in a strain collection established long before the introduction of PAs (Dagget et al. 1993) and probably existed at very low frequencies in most *P. infestans* populations worldwide. While many oomycetes (e.g., *P. cubensis*, *Peronospora* spp.) can be disseminated hundreds of kilometers by aerosol movement containing sporangia (conidia), others (e.g., *P. viticola*) migrate only locally by splash dispersal of sporangia. In *P. infestans,* sporangia can be disseminated in rain droplets for hundreds of meters, but more importantly,

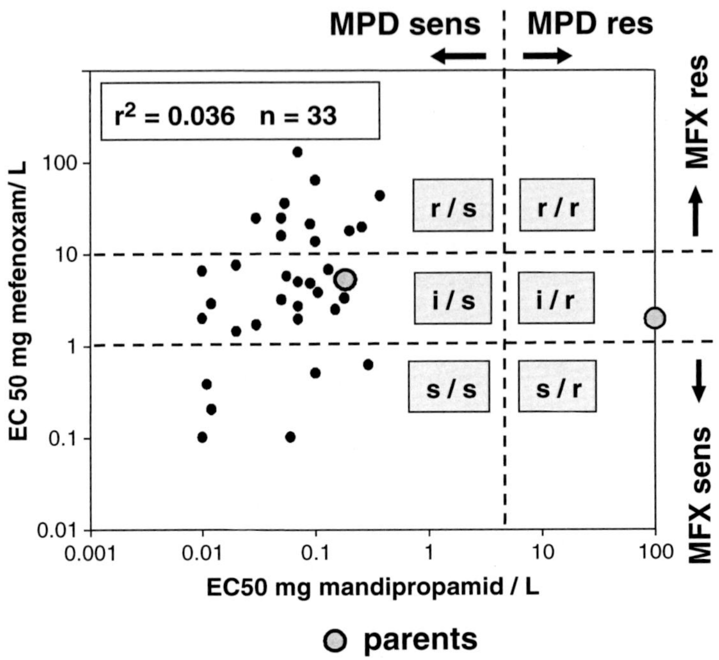

Fig. 10.2 Segregation of sensitivity/resistance (EC 50) to mandipropamid (MPD) in F1 and to mefenoxam (MFX) in F2 progeny of *Plasmopara viticola* (After Gisi et al. 2007)

the pathogen is distributed globally by shipment of infected plant material (containing mycelium) such as tubers (potato), seeds, and seedlings (tomato). Thus, migration can bring new genotypes including PA-resistant individuals to new locations followed by local selection potentially leading to dominance of new clones.

Important population changes in *P. infestans* have been observed repeatedly in the USA and Canada, where the originally sensitive A1 clones (e.g., US1) were displaced in the 1990s by waves of PA resistant, mostly A2 clones (US6, US7, US8, US11, US17) (Fry et al. 1993; Goodwin et al. 1998), followed between 2002 and 2009 by new waves of PA sensitive clones (US22, US23, US24) (Hu et al. 2012). In 2009, about 60 % of the *P. infestans* isolates were sensitive US22 genotypes. These changes are believed to be caused only partly by the use of PAs, because in parallel, also mating-type distribution has changed. However, the origin and driving forces for these migrations are largely unknown.

In only a few places (e.g., Mexico, Northern Europe), *P. infestans* populations are genetically diverse (non-clonal) (Yuen and Andersson 2013). Since about 2004, major changes in *P. infestans* populations were observed in Western and Central Europe, with an increase of more aggressive strains, genotypic differentiation, and a switch from A1 to A2 mating-type dominance. Sensitive, intermediate, and resistant phenotypes were present in both A1 and A2 mating types, so no genetic linkage existed between mating type and PA resistance (Gisi et al. 2011). When isolates collected in 1997 were compared to those from 2007 by molecular marker techniques (SSR, simple sequence repeats, Lees et al. 2006; Li et al. 2013), new genotypes appeared in many locations resulting in an increase of A2 mating types from almost zero to 50–90 % (Gisi et al. 2011). By coincidence, most of these A2 isolates (about 75 %, Syngenta internal data) were PA resistant and rather aggressive, subsequently referred to as "blue 13" or 13_A2 lineages (Lees et al. 2006; Cooke et al. 2012). Interestingly, the European *P. infestans* populations have changed again since 2008: The frequency of PA-resistant isolates has dropped from about 50 % (2008/09) to 30 % (2010/11) and 10 % (2012) (Table 10.3). In addition, a new genotype emerged recently, referred to as "green 33," mostly PA sensitive and reaching frequencies in the Netherlands in 2011 of 20 % (G. Kessel, Wageningen, pers. communication). The appearance of new SSR genotypes and changes in resistance frequencies in European *P. infestans* populations may be caused by (I) sexual recombination – although probably not frequent in most parts of Europe – (II) import of infected plant material from "producer countries," e.g., infected potato tubers and solanaceous seedling plants (e.g., tomatoes); and (III) selection of strains by local solanaceous weeds (e.g., black night shade; Grönberg et al. 2012), certain potato and tomato cultivars (Lebreton et al. 1999), climatic conditions, and PA use pattern (Gisi et al. 2011). Although the "blue 13" clones dominated populations in Western Europe for some years, they declined again in the last few years. One reason for this behavior may be their competitive disadvantage during sexual recombination, because many "blue 13" strains are triploid/trisomic (Hamed and Gisi 2013). At locations where the *P. infestans* disease cycle also includes sexual recombination,

Table 10.3 Sensitivity to mefenoxam of *Phytophthora infestans* bulk isolates[a] collected from potato fields in Europe (NL, F, A, CH, D, DK, UK, S, RU)[b] between 2004 and 2013

		Sensitivity[c]		
Year	Number of isolates	%S	%I	%R
2004	40	65	15	20
2005	57	47	6	47
2006	95	26	22	52
2007	124	11	15	74
2008	147	31	16	53
2009	28	39	11	50
2010	82	61	6	33
2011	85	57	15	28
2012	43	75	16	9
2013	Few[a]	nd	nd	nd

[a]Data from Syngenta internal sensitivity monitoring (unpublished). In 2013, only few samples were received, no reliable data available (nd)

[b]*NL* Netherlands, *F* France, *A* Austria, *CH* Switzerland, *D* Germany, *DK* Denmark, *UK* United Kingdom, *S* Sweden, *RU* Russia

[c]Thresholds: sensitive (S): $EC_{50} \leq 0.1$ mg/l; intermediate (I): $0.1 > EC_{50} \leq 10$ mg/l; resistant (R): $EC_{50} > 10$ mg/l

e.g., in Southern Sweden, the common European clones (e.g., "blue 13" strains) did not dominate local populations (Widmark et al. 2007).

As a result of selection, PA-resistant isolates of *P. infestans* may be present in field populations at rather high proportions at the end of the season, but epidemics may start with mainly PA-sensitive populations at the beginning of the following season (Gisi and Cohen 1996). The increase in frequency of resistant isolates during the epidemics was observed in a recent study again in the Netherlands (in Hermann et al. 2015). Also Ivors and Gudmestad observed mainly PA-sensitive isolates in *P. infestans* on solanaceous vegetables at season start but an increase of resistant isolates during the season (in Hermann et al. 2015). In many locations, the proportion of PA-resistant *P. infestans* isolates increased during the season also in potato fields which were not treated with PA products (Gisi and Cohen 1996). Thus, PA selection cannot be the only reason for this phenomenon. In the absence of PA-fungicide treatment, several authors (e.g., Kadish et al. 1990) observed a faster progress of epidemics imposed by PA-resistant than by PA-sensitive *P. infestans* isolates suggesting an equally high or greater pathogenic fitness of PA-resistant isolates. Probably, as a result of this behavior, PA-resistant isolates also colonized tuber tissue faster than sensitive ones (Montarry et al. 2007) resulting in a lower overwintering potential (lower survival fitness) (Kadish and Cohen 1992). This observation might be another reason why resistance frequencies in the primary inoculum at the beginning of the next season can be lower than at the end of the previous season (Gisi and Cohen 1996). However, higher pathogenic fitness in resistant isolates may not be genetically linked to PA resistance but more likely to a different genetic background in the new races (genotypes). Higher pathogenic fitness was also observed for PA-resistant isolates in *P. erythroseptica* (Taylor et al. 2006; Chapara et al. 2011).

10.2.4 Resistance Evolution in **P. viticola, P. cubensis,** *and* **B. lactucae**

Resistant isolates of *P. viticola*, a pathogen undergoing sexual recombination under European conditions every year, were detected already in 1981 in vineyards that have been treated continuously with PAs (Staub and Sozzi 1981) but also at sites without PA treatment (Bosshard and Schüepp 1983). Since then, resistance was detected in many vineyards in many countries around the globe (Gisi 2002; Gisi and Sierotzki 2008). In general, the proportion of sensitive isolates declines during the epidemics every year (Gisi and Sierotzki 2008). During the 2006–2013 period, about 5–25 % of isolates collected in European vineyards were resistant to mefenoxam, whereas 15–40 % showed intermediate sensitivity (Table 10.4) supporting that sexual recombination occurs regularly. A rather high and stable proportion (40–80 %) of isolates was sensitive to PAs during the last years although with large regional variations and although an average of one PA treatment per season is made. No PA-resistant oosporic isolates were found in Italy between 2007 and 2009

Table 10.4 Sensitivity to PAs (mefenoxam), QoIs (azoxystrobin), and CAAs (mandipropamid) of *Plasmopara viticola* bulk isolates[a] collected from vineyards in Europe (F, A, CH, D, I, H, E, P, SL)[b] between 2006 and 2013

		Phenylamides[c]			QoIs[d]		CAAs[e]
Year	Number of isolates	%S	%I	%R	%R (F)	%R (E)	% R
2006	98	64	18	18	48	fs[a]	31
2007	123	79	13	8	69	32	19
2008	135	81	13	6	71	30	23
2009	32	50	38	12	66	36	30
2010	117	43	30	27	53	26	38
2011	57	75	27	8	44	fs[a]	38
2012	63	13	21	66[f]	39	27	38
2013	180	24	52	24	nd[a]	fs[a]	25

[a]Data from Syngenta internal sensitivity monitoring (unpublished). Only few samples (fs) were received in some cases, no reliable data available (nd)

[b]*F* France, *A* Austria, *CH* Switzerland, *D* Germany, *I* Italy, *H* Hungary, *E* Spain, *P* Portugal, *SL* Slovenia

[c]Thresholds PAs: sensitive (S): $EC_{50} \leq 10$ mg/l; intermediate (I): $10 > EC_{50} \leq 100$ mg/l; resistant (R): $EC_{50} > 100$ mg/l

[d]Threshold QoIs: Resistance expressed as frequency of G143A in bulk sample: sensitive (S): G143A ≤ 10 %; resistant (R): 10–100 %. (F): France; (E) Spain

[e]Threshold CAAs: Discriminatory dose of mandipropamid in bioassay is 6.25 mg/l

[f]Resistance frequency unexpectedly high possibly based on biased sampling

(Toffolatti et al. 2011). The more or less Mendelian segregation of PA resistance in sexual populations, as it is typical for *P. viticola* (heterothallic pathogen with two mating types, P1 and P2, intimately unified in single lesions; Scherer and Gisi 2006) and some other oomycete pathogens (e.g., the homothallic *P. erythroseptica*), has an important consequence for resistance evolution: It will always generate isolates with intermediate resistance (which can be controlled adequately by full PA rates at short intervals) and will bring back sensitivity to unselected populations (theoretically 25 % in F2 progeny). However, resistance will genetically not disappear.

Resistance to PAs in the downy mildew of cucurbits, caused by *P. cubensis*, was described already 2 years after introduction of metalaxyl in Israel (Reuveni et al. 1980) and thereafter in several countries at varying frequencies (Lebeda and Cohen 2012). Epidemics can be quite heavy (e.g., in greenhouses) but may fluctuate in open fields from year to year. Recently, the disease occurred more frequently especially on cucurbits in the USA (McGrath 2012) but also elsewhere. A broad sensitivity distribution from highly sensitive to fully PA resistant was found in 49 isolates collected during 2008–2011 in Europe, Israel, and the USA (Hermann et al. 2015). However, the majority of isolates from Israel and the USA were resistant. Also, isolates collected in China during 2011–2012 and Vietnam in 2013 were mostly resistant to mefenoxam (Syngenta unpublished data). Although varying frequencies of PA-resistant isolates were detected in Czech Republic between 2005 and 2009, a large part of the populations was still sensitive (Lebeda and Cohen 2012).

Populations of the downy mildew pathogen on lettuce, *B. lactucae*, mostly contain many races, and sexual reproduction is believed to be frequent in European populations (Crute et al. 1994). Resistance to PAs was widespread already in the 1990s, and sexual outcrosses were often intermediate in sensitivity to PAs. However, in the USA (especially California), populations were reported to be mostly clonal: All 134 isolates collected in 1999 and 2000 in California were resistant to metalaxyl (Brown et al. 2004). However, recent isolates collected from the same area, when tested against mefenoxam, were sensitive (S.T. Koike, personal communication to McGrath 2012). Whether these isolates emerged through sexual recombination or migration from fields or regions with sensitive populations is largely unknown.

10.2.5 Resistance Evolution in Seed- and Soilborne Oomycetes

PAs are widely used as seed treatment or soil application against seed- and soilborne oomycetes. Although resistance was discovered many years ago (FRAC List of Resistant Pathogenic Organisms 2013, www.frac.info), the frequency of PA resistance in *Pythium* spp. seems to remain rather limited and lower than in foliar pathogens, but many cases of resistance have been reported (Table 10.5). In a large survey over 4 years (1997–2000) in 16 states of the USA and two Canadian provinces, Taylor et al. (2002) found about 4 % of *P. ultimum* isolates from potato tubers to be resistant to mefenoxam. Lu et al. (2012) analyzed 169 isolates of *Pythium* spp. associated with cavity spot in carrots in California and Michigan: 93 % were sensitive and 7 % resistant to mefenoxam (Table 10.5). In ornamentals grown under greenhouse conditions, resistance to PAs in *Pythium* was more frequent probably because of a more intensive PA use: In Pennsylvania, up to 62 % of *P. aphanidermatum* and *P. irregulare* were resistant to mefenoxam (Moorman et al. 2002), and a similarly high proportion of isolates was resistant in the Syngenta internal sensitivity monitoring in 2011 (Table 10.5).

In *Phytophthora cinnamomi*, a root pathogen in numerous woody plant species, resistance is known to occur since long (FRAC List of Resistant Pathogenic Organisms 2013, www.frac.info), but the majority of isolates collected from ornamental nurseries in Virginia were sensitive to mefenoxam (Hu et al. 2010, Table 10.5).

The population structure of *Phytophthora nicotianae*, causal agent of black shank in tobacco, was investigated recently by Parkunan et al. (2010) in Virginia: No PA-resistant isolates were found (Table 10.5); 2 % showed an intermediate response. These results confirm earlier findings from other tobacco-growing states in the USA (Csinos and Bertrand 1994; Gallup et al. 2009), where populations were mostly sensitive to PAs.

In a large survey across the USA, the frequency of PA resistance in *Phytophthora erythroseptica*, causing pink rot in potato tubers, ranged between 3 and 36 % (Taylor et al. 2002; Table 10.5). Resistant isolates seem to be highly fit and competitive even in the absence of PA selection pressure (Taylor et al. 2006; Porter et al. 2007; Chapara et al. 2011). In recent studies, higher frequencies of PA-resistant isolates

Table 10.5 Resistance to mefenoxam in soilborne *Pythium* and *Phytophthora* species in the USA (recent selected references)

Pathogen	Crop	Years	No. of isolates	Percent resistance	Reference
Pythium spp.	Carrot	2010	169	7	Lu et al. (2012)
Pythium spp.	Ornamentals	1996–2001	29	12–62	Moorman et al. (2002)
Pythium spp.	Ornamentals	2010–2011	20	60	Olaya (2011)[a]
Phytophthora cinnamomi	Ornamentals	1986–2004	65	15	Hu et al. (2010)
Phytophthora nicotianae	Tobacco	2006–2008	217	0	Parkunan et al. (2010)
Phytophthora erythroseptica	Potato	1997–2000	805	3–36	Taylor et al. (2002)
		2001–2002	224	70	Porter et al. (2007)
Phytophthora capsici	Diverse (see text)		498	13	Lamour and Hausbeck (2000)
			150	59	Parra and Ristaino (2001)
			53	51	French-Monar et al. (2006)
			120	7	Keinath (2007)
			75	63	Café-Filho and Ristaino (2008)
		2007	257	0–65	Dunn et al. (2010)
			75	10	Olaya (2009)[b]

[a]Data from Syngenta internal sensitivity monitoring 2011 (unpublished)

[b]Data from Syngenta internal sensitivity monitoring 2009 (unpublished)

were found in the USA (Porter et al. 2007; Chapara et al. 2010; Table 10.5). Although resistance seems to be widespread in some counties of the USA, it may stabilize over the years mainly due to the fact that *P. erythroseptica* is a homothallic species undergoing frequent sexual reproduction. However, solo PA applications should be avoided or strictly integrated into a solid resistance management program (including also mixtures with unrelated products).

Several authors investigated PA resistance in *Phytophthora capsici*, an important soil- and leaf-borne pathogen especially in vegetables such as zucchini, pepper, squash, water melon, pumpkin, and tomato. Frequency of resistance to PAs seems to be highly variable ranging from 0 to 65 % depending on year, crop, and geographic origin (Table 10.5). The pathogen undergoes frequent sexual recombination (sex ratio in field populations nearly 1:1); thus, highly diverse but also clonal populations can be found (Dunn et al. 2010). In a study conducted by Syngenta (Olaya 2009, personal communication), most isolates of *P. capsici* collected from various vegetables in different states of the USA (South Carolina, Florida, California) were sensitive to mefenoxam, but some few intermediate and about 10 % resistant isolates were found (Table 10.5). Similarly, low proportions were found also by Keinath (2007) in South Carolina and by Lamour and Hausbeck (2000) in Michigan. However, higher proportions were discovered in New Jersey, North Carolina, and Florida by other authors (Table 10.5) probably as a response to more intensive PA uses.

Also in many other important soilborne and foliar *Phytophthora* species, PA-resistant isolates have been detected (e.g., *P. cactorum, P. megasperma, P. melonis, P. porri, P. ramorum*) (FRAC List of Resistant Pathogenic Organisms 2013, www.frac.info), although no data on changes in populations are available. Resistance evolution in seed-borne oomycete pathogens seems to be less frequent, because selection pressure on seed- and soilborne populations are lower (only one application, rapid product dilution) than in foliar pathogen populations, although fungicide concentrations on the seed surface are rather high (Gisi 2014). PA resistance has been detected in the seed-borne sunflower downy mildew, *Plasmopara halstedii*, several years back (FRAC List of Resistant Pathogenic Organisms 2013, www.frac.info). Also, in the seed transmitted *Peronospora viciae* on peas, PA resistance emerged in New Zealand (Falloon et al. 2000). In some cases, it is not clear whether resistance was really selected by seed treatment or by previous soil drench or foliar applications at the same site where treated seeds are planted later (e.g., resistant *Pythium* spp. isolates collected from soil planted with PA-treated corn seeds, Syngenta internal information).

10.2.6 Resistance Management for PAs as Recommended by FRAC

The strict implementation of anti-resistance guidelines defined by the PA-FRAC group (Working Group until 1996, Expert Forum since then) and local adaptation by the involved companies and regulatory bodies, together with the availability of new modes of actions for chemical control of oomycete pathogens (Table 10.2) allowing more diversified application programs, have significantly contributed to cope with PA resistance over the past three decades. Foliar applications of PAs are always done in mixture with contact fungicides such as mancozeb or chlorothalonil. In general, product performance against foliar diseases remained adequate if resistant individuals were not dominating the populations (Staub 1994; Wicks et al. 1994). For the control of soilborne diseases (e.g., potato pink rot), suitable mixing partners are actually lacking, and the use of solo PAs does not give adequate disease control if PA-resistant populations prevail (Taylor et al. 2011).

The use recommendations for PA-based products issued by FRAC (2013) have remained unchanged since 1997. They are intended as general guidelines that have to be adapted to the respective pathosystem, disease pressure, local conditions (e.g., climate), fungicide use history, and resistance levels. Major elements of a sound resistance management program for foliar applications include a preventive (not eradicative) use pattern of PAs always in mixture with unrelated effective (residual) products used at an interval not exceeding 14 days (or less when disease pressure is high). The number of PA applications per season is limited (two to four per crop and year or less if resistance is already established); they should be integrated in a spray program positioned early season or during the period of active vegetative growth of the crop. When solo formulations are made available for soil use, strategies must be implemented which prevent any possibilities for foliar applications. In many regions or countries, more restrictive regulations or use recommendations have been put in place to avoid further buildup of resistance.

10.2.7 Summary and Outlook for PA Fungicides

The unique properties of PA fungicides with long-lasting preventive activity, high mobility in plants, curative potential, and broad activity spectrum against all major oomycete pathogens (including members of Peronosporales, Sclerosporales, and Pythiales), combined with good crop safety, make PA products a valuable tool for disease management in many crops. PAs are believed to inhibit RNA polymerase I during RNA biosynthesis. However, the molecular basis of resistance and the involved mutations have not been characterized in detail. Uses include foliar, soil, and seed treatment applications. For foliar and seed treatment, PAs are used as co-formulated products with contact fungicides or other unrelated compounds to broaden the activity spectrum beyond oomycetes and to manage resistance. Resistant strains appeared in key pathogens like *P. cubensis*, *P. infestans*, and *P. viticola* as early as 2–3 years after introduction of PAs and subsequently also in other crop pathogens across the globe and are widespread especially in foliar oomycete pathogens. Inheritance of PA resistance has been described as basically Mendelian and monogenic with intermediates in F1 progeny and a proportion of 1:2:1 of sensitive, intermediate, and resistant outcrosses in the F2 generation. Over the last 30 years, resistance has stabilized in most locations due to restriction of PA uses and reappearance of sensitivity through sexual recombination and introduction (migration) of new sensitive clonal lineages.

Development and spread of resistance in soil- and seed-borne oomycetes seem to be significantly slower and more patchy than for foliar pathogens. After intense use of solo PA applications, rather high but variable frequencies of resistance were found in pathogens like *Pythium* spp., *P. capsici*, and *P. erythroseptica* requiring adapted use strategies including product mixtures (when possible) and rotation of different modes of action. Although quite high proportions of PA resistance have evolved in *P. erythroseptica* after soil uses of solo PAs in potato in the USA, sensitivity to mefenoxam has generally returned in *P. infestans* in the same crop suggesting that (I) PA translocation from the soil to leaf tissue may be negligible for the selection process in *P. infestans* (due to spatial and timing separation) and (II) independent of PA use pattern, new sensitive genotypes have been imported from other places. Even with a broad range of new fungicides available for oomycete control, PAs remain standard fungicides with high intrinsic activity against foliar and seed- and soilborne pathogens including members of the Pythiales. If appropriate strategies for resistance management are maintained and adapted to the current local situations, PAs will remain an important tool for chemical control of oomycete diseases. However, sexual recombination, migration, and appearance of new genotypes in local populations will continuously affect sensitivity to PAs requiring a repeated adaptation of control strategies.

10.3 Quinone Outside Respiration Inhibitors (QoIs)

10.3.1 Mode of Action and Mechanisms of Resistance

Within QoIs, only few molecules are active also against oomycete pathogen, the main representatives being azoxystrobin, famoxadone, and fenamidone. QoI fungicides are inhibitors of mitochondrial respiration; they interrupt electron transport in cytochrome b (complex III) by binding to the Qo site, the ubiquinol oxidizing pocket which is located at the positive, outer side of mitochondrial membranes (Gisi et al. 2002). The cytochrome b (*cyt b*) gene, encoding the target protein for QoI binding, is located in the mitochondrial genome. Long before the introduction of agricultural QoIs, resistance to similar molecules (e.g., myxothiazol) was described by medical scientists as being based on several mutations in the *cyt b* gene in a range of organisms (Di Rago et al. 1989; Geier et al. 1992; Degli-Esposti et al. 1993; Brasseur et al. 1996). However, it was not known which mutations would appear in plant pathogens. In 2000, Sierotzki et al. (2000a, b) detected the G143A substitution (exchange of glycine by alanine at position 143) for the first time in QoI-resistant isolates of plant pathogens, the banana black sigatoka pathogen, *Mycosphaerella fijiensis*, and the wheat powdery mildew, *Blumeria (Erysiphe) graminis* f. sp. *tritici*. This substitution is based on a single nucleotide polymorphism (SNP) in the triplet at position 143 from GGT to GCT in the *cyt b* gene. It was detected in resistant isolates of many important plant pathogen species including *Plasmopara viticola* and *Pseudoperonospora cubensis* (Heaney et al. 2000, FRAC List of Resistant Pathogenic Organisms 2013, www.FRAC.info). The mutation is associated with high levels of resistance (high RF values) or "complete" resistance which leads to a complete loss of disease control if QoIs are used as solo products.

A second mutation, F129L (exchange of phenylalanine by leucine at position 129), was discovered in resistant isolates of additional pathogen species such as *P. viticola* (Sierotzki et al. 2005) and *Pythium aphanidermatum* (Olaya et al. 2003) resulting in "partial," less pronounced resistance leading to reduced disease control. In *P. viticola*, the F129L mutation was detected in 2002 in few individuals but was much less frequent (less than 2 % in population) as compared to the G143A mutation and was never found together with G143A in the same individual; in recent populations, F129L frequency is very low (Sierotzki et al. 2005). In *Phytophthora infestans, Bremia lactucae,* and *Peronospora* spp. and in all rust species (e.g., *Puccinia, Uromyces, Phakopsora, Hemileia*), no resistant isolates (and no mutations) were detected so far. For rusts, the lack of resistance (based on G143A) has been elucidated recently: In the *cyt b* gene, an intron is present between aa positions

143 and 144 which has to be removed for correct transcription and translation by splicing. If mutated from GGT to GCT at position 143, splicing will not occur resulting in a nonfunctional cytochrome b which is lethal (Grasso et al. 2006). The reason for the absence of mutations in *P. infestans* is not known. More details on mode of action and mechanisms of resistance in pathogens outside oomycetes are described in chapter "Complex III respiration inhibitors" by H. Sierotzki in this book.

Based on the mitochondrial origin of QoI resistance, a maternal inheritance of resistance can be postulated, i.e., a segregation of 0:1 or 1:0 in single crosses, depending on whether resistance is located in the male or female parent. When a sensitive and a QoI-resistant single sporangiophore parental isolate (P1 and P2) of *P. viticola* were co-inoculated on grape leaves and F1 progeny isolates collected from germinating oospores, sensitive and resistant offsprings emerged. Surprisingly, segregation of resistance was not 0:1, but r:s = 8:23 (or ~1:3) (Blum and Gisi 2008). Possible reasons for the unexpected segregation might be mitochondrial leakage or irregularities in the mating process. The first possibility is rather unlikely, nobody described the existence of stable heteroplasmic stages in any plant pathogens, while the second hypothesis was described for *P. infestans* and named femaleness (Judelson 1997). Assuming segregation of QoI resistance would follow a near to 0:1 pattern in single crosses and 1:1 in populations (assuming the P1:P2 ratio is about 1:1), then the evolution of resistance under QoI selection is expected to be rather quick and stable in sexual populations. In fact, the G143A mutation was very rare in unselected populations of *P. viticola* prior to the use of QoIs, but was quickly selected through the continuous use of these fungicides (Gisi and Sierotzki 2008). In addition, *P. viticola* is a pathogen with a high rate of sexual recombination resulting in high genetic diversity and many different genotypes every season. On the other side, migration rate is rather low (dispersal mainly within few meters) resulting in local epidemics. Thus, resistance evolution might be driven not only by fungicide selection but also by "local" epidemiological processes.

10.3.2 Resistance Evolution in P. viticola *and* P. cubensis

QoI resistance (based on G143A) in *P. viticola* populations in Europe was first detected in 2000 (Heaney et al. 2000; Gullino et al. 2004; Sierotzki et al. 2005) and evolved quickly with a rapid increase of resistant isolates reaching mean frequencies in 2003 of about 75 % in France and about 30 % in the Northwest of Spain (Galicia) (Gisi and Sierotzki 2008). Mean frequencies of 30 % (range 0–70 %) were observed in the Bordeaux area and 33 % (range 0.01–77 %) in the Champagne region by Corio-Costet et al. (2008). In the North of Italy (collected results in Toffolatti and Vercesi 2012) and in Switzerland, frequencies in 2002 were high only in certain area, whereas in Portugal, Germany, and Austria, they were low (Sierotzki et al. 2008). An increase of resistance (percent of resistant oospores and G143A in field populations) was observed in Northern Italy and Puglia especially when several solo QoI applications per season were made against both grape powdery and downy mildews; however, a decline was noticed at many sites when mixtures with unrelated fungicides were used or when QoI treatments were abandoned completely (Toffolatti et al. 2007; Toffolatti and Vercesi 2012). In general, resistance frequencies in European *P. viticola* populations remained more or less stable over the last years (Table 10.4); they are rather high at some locations (e.g., Czech Republic, Veneto in Italy); moderate in France, Germany (Pfalz, Württemberg), Spain, Slovakia, and Hungary; and low in Romania, Greece, and Bulgaria (QoI FRAC working group minutes, 2013, www.frac.info). Since the collected leaf samples with sporulating lesions always represented a bulk of isolates, the measured frequency of resistance (A143 allele in Q-PCR test) represents the average proportion of resistance in the population of a specific vineyard. However, if single sporangiophore isolates were picked from lesions, they were always either completely sensitive (100 % G143 allele) or completely resistant (100 % A143 allele); heteroplasmic stages were never detected (Sierotzki et al. 2008).

QoI-resistant *P. cubensis* isolates (with G143A mutation) were first detected in Japan in 1999–2000 by Takeda et al. (1999) and Heaney et al. (2000) and confirmed by Ishii et al. (2001, 2002). Soon afterwards, solo QoI applications were stopped or no longer recommended against cucumber downy mildew. However, resistance frequencies remained high at many locations (Lebeda and Cohen 2012), especially at the East Coast of the USA (QoI FRAC working group minutes, 2013, www.frac.info), although no systematic investigations over several years were undertaken. In some Japanese conditions, resistance declined after few years without QoI applications (Ishii et al. 2001). No sensitivity information is available for *B. lactucae*.

10.3.3 Sensitivity in **Phytophthora** *and* **Pythium** *spp.*

It is an interesting and also surprising finding that QoI-resistant *P. infestans* isolates have never been detected and no mutations were identified (neither G143A nor F129L) although QoI-containing products have been used in potato since more than 10 years. It has been hypothesized that the intrinsic activity of QoIs (e.g., azoxystrobin) or their use intensity may not be high enough for selection of resistant individuals (which is rather unlikely) or that *P. infestans* has some molecular control systems eliminating mutations from the mitochondrial genome rather quickly, especially when cells would become heteroplasmic. However, no experimental evidence has been generated to support this hypothesis. Also, in other *Phytophthora* spp., QoI-resistant isolates (and mutations) seem to be absent (no information available in literature). In contrast, the F129L mutation was discovered in QoI-resistant isolates of *Pythium aphanidermatum* collected from turf (Olaya et al. 2003), but their frequency was rather low.

10.3.4 Resistance Management for QoI Fungicides as Recommended by FRAC

For downy mildew control in grapes (*P. viticola*), a maximum of three preventive QoI fungicide-containing sprays per vine crop or a maximum of 33 % of the total number of applications can be applied. QoIs have to be used only in mixture with effective partners from different cross-resistance groups; no solo products against grape downy mildew are recommended. Because of the high frequency of resistant *P. viticola* isolates in certain French vineyards, the use of QoI products is no longer recommended by officials. Special resistance management precautions should be taken, if both grape downy and powdery mildews have to be controlled at the same location. When QoIs are applied alone to control late blight in potato (*P. infestans*), do not exceed one spray out of three with a maximum of three sprays per crop and do not use more than two consecutive applications. Where QoI fungicide products are applied in mixtures (co-formulations or tank mixtures), do not exceed 50 % of the total number of sprays or a maximum of six QoI fungicide applications whichever is the lower. Do not use more than three consecutive QoI fungicide-containing sprays.

10.3.5 Summary and Outlook for QoI Fungicides

QoI fungicides are active against a broad range of diseases; some of them like azoxystrobin, famoxadone, and fenamidone control also oomycete pathogens, especially *P. viticola, P. cubensis, P. infestans*, and some *Pythium* species. QoIs are respiration inhibitors binding to the quinone outside membrane pocket of cytochrome b in complex III of the respiration chain. Resistance based on the G143A mutation in the mitochondrial *cytb* gene evolved shortly after product introduction mainly in *P. viticola* and *P. cubensis*, but not in *P. infestans*. At few locations, also the F129L mutation was discovered in *P. viticola* and some *Pythium* species. After sexual recombination, resistance is inherited maternally (through mitochondria) resulting in a 0:1 or 1:0 segregation in the F1 progeny depending on whether resistance is located in the male or female parent; in unselected populations, resistance segregates in a s:r = 1:1 pattern (if mating-type distribution is close to 1:1). Therefore, the evolution of resistance in sexual populations under QoI selection is rather quick and stable. In grape and cucumber downy mildew populations, resistance frequencies have become rather high at certain locations; as a consequence, QoIs have to be applied only in mixture with unrelated partner fungicides, and the number of applications per season and crop is limited.

10.4 Carboxylic Acid Amide Fungicides (CAAs)

10.4.1 Mode of Action

The carboxylic acid amide fungicides include three subclasses: cinnamic acid amides (dimethomorph, flumorph, pyrimorph), valinamide carbamates (benthiavalicarb, iprovalicarb, valiphenalate), and mandelic acid amides (mandipropamid) (Gisi et al. 2012) exhibiting cross-resistance among all group members as was demonstrated for sexual F2 outcrosses and the vast majority of the tested field isolates of *Plasmopara viticola* (Gisi et al. 2007). CAAs exhibit specific activity against most oomycetes including pathogens of the Peronosporales such as *P. viticola* in grapes; *B. lactucae* on lettuce; *Peronospora* spp. on tobacco, pea, onion and vegetables; *P. cubensis* on cucurbits; and several *Phytophthora* species on many crops such as potato, tomato, pepper, and pineapple. However, the entire genus *Pythium* and other oomycetes outside Peronosporales are insensitive (tolerant), as are all other pathogens outside the oomycetes. Recent studies performed with mandipropamid contributed to elucidate the mode of action for CAA fungicides in *P. infestans*. The incorporation of ^{14}C-labeled glucose into the β-1,4 glucan (cellulose) fraction of cell walls of germinating cystospores was inhibited in the presence of mandipropamid (Blum et al. 2010a). Gene sequencing of artificially generated mutants of *P. infestans* which were resistant to mandipropamid revealed an amino acid substitution in the cellulose synthase, *CesA3* gene at position 1105 from glycine to alanine or valine, G1105A/V. In addition, the transformation and expression of a mutated *CesA3* allele in a sensitive *P. infestans* isolate resulted in a CAA-resistant phenotype (Blum et al. 2010a). Thus, cellulose synthase can be postulated as the primary target enzyme for CAA activity.

10.4.2 Resistance Mechanism

In sexual crosses made with sensitive (homozygous AA) and CAA-resistant (homozygous aa) single sporangiophore isolates of *P. viticola*, segregation of resistance was studied (Gisi et al. 2007). All F1 progeny isolates were sensitive to CAAs (Fig. 10.2) representing a segregation pattern of s:r = 1:0. When F1 progeny isolates (siblings) were crossed, the segregation in the F2 progeny was about 9:1 (s:r) (Gisi et al. 2007) and in a later study about 4:1 (Fig. 10.3; Blum et al. 2010b). The latter segregation pattern suggests that resistance to CAAs is inherited by one recessive nuclear gene. In the F2 progeny, resistance co-segregated for all tested CAAs (mandipropamid, dimethomorph, iprovalicarb) resulting in cross-resistance among all CAAs (Gisi et al. 2007). Since intrinsic activity of single CAAs can vary considerably (e.g., dimethomorph is about ten times less active than mandipropamid against certain isolates; Gisi et al. 2007), special care has to be taken in analyzing the cross-resistance correlation for the entire populations.

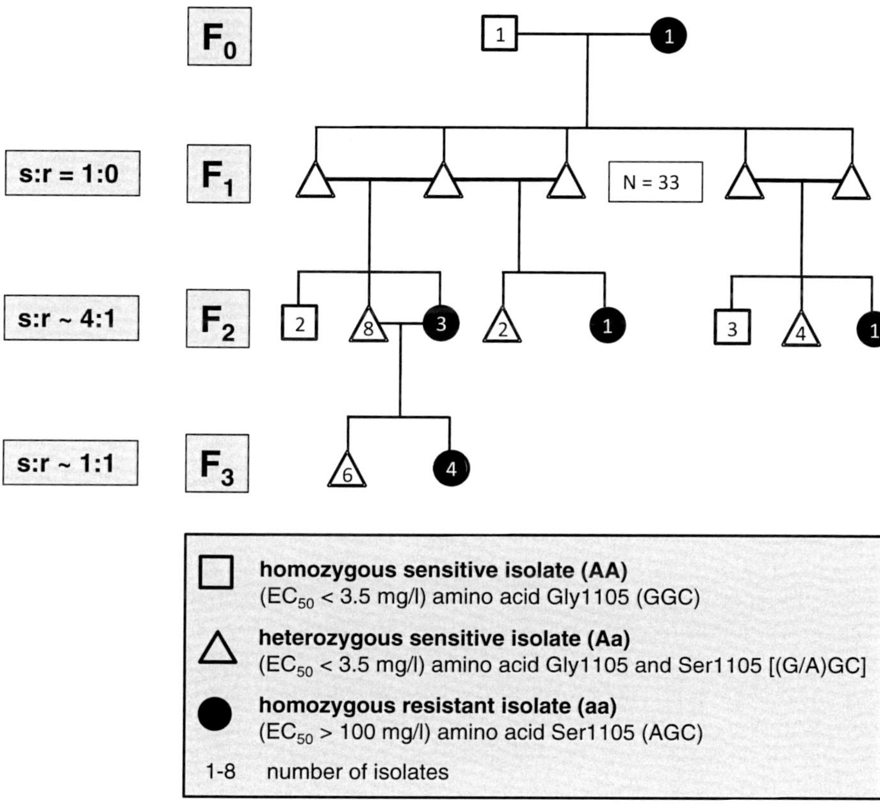

Fig. 10.3 Co-segregation of CAA resistance (EC50 > 100 mg/l) and G1105S mutation in *CesA3* gene in F1, F2, and F3 progeny of *Plasmopara viticola* generated from an initial cross of homozygous sensitive and resistant parents (F0) (After Blum et al. 2010b)

Sequencing of the four cellulose synthase genes, *CesA1, CesA2, CesA3*, and *CesA4* in CAA-sensitive and CAA-resistant single sporangiophore isolates of *P. viticola*, revealed several single nucleotide polymorphisms (SNPs) affecting the amino acid configurations of the protein (Blum et al. 2010b). SNP inheritance in F1-, F2-, and F3-progeny isolates revealed resistance to be correlated with one specific SNP located in the *CesA3* gene: Only if present in both alleles, this SNP led to the substitution of glycine for serine at position 1105 (G1105S), thus conferring CAA resistance (Blum et al. 2010b). The G1105S mutation co-segregated in all F2- and F3-progeny isolates with the resistant phenotype (Fig. 10.3). The results demonstrate that one recessive mutation in the *CesA3* gene causes inheritable resistance to CAAs in *P. viticola* and confirm cellulose synthase to be the target enzyme for CAA activity. More recent studies revealed that in most CAA-resistant *P. viticola* field isolates, G1105S and rarely also G1105V (glycine for valine) were present (Sierotzki et al. 2011). In *P. cubensis*, resistance was associated with different amino acid exchanges at the same position: G1105V (glycine for valine; Israelian isolates mainly from cucumber) or G1105W (glycine for tryptophane, US isolates from cucumber, watermelon, zucchini, cantaloupe); these isolates exhibited cross-resistance to all CAAs (Blum et al. 2011; Sierotzki et al. 2011).

Surprisingly, CAA-resistant field isolates have not been discovered so far in *Phytophthora* species including *P. infestans* and *P. capsici*, although CAAs (especially dimethomorph) have been used in potato and several vegetables since almost 20 years. On the other hand, artificial mutants of *P. infestans* and *P. capsici* have been generated by EMS and UV treatment by several authors (e.g., Rubin et al. 2008; Blum et al. 2010a; Chen et al. 2011) yielding the mutations G1105A or G1105V in *CesA3*, but also two mutations at position 1109 have been identified, V1109L and V1109M; V1009L was found also in *P. melonis* (Chen et al. 2012). Artificial mutants are useful tools to study CAA mode of action (Blum et al. 2010a) and assess resistance risk (Rubin et al. 2008), but it is very speculative to predict which mutations will evolve in field populations. In contrast to *P. infestans*, artificial mutants can be generated quite easily for *P. capsici*, e.g., through mass selection on fungicide-amended agar (Pang et al. 2013), for CAAs but also for other modes of action like fluopicolide (Lu et al. 2011). Recent observations have shown that all investigated species of the Peronosporales (and probably also of the Sclerosporales) have a valine at position 1109 and are intrinsically sensitive to CAAs, whereas in all other orders of oomycetes, species carry a leucine at position 1109, L1109 (or a methionine M1109 for the *Pythium ultimum* species clade) (Fig. 10.1). These two amino acid configurations are coding for the intrinsic insensitivity (tolerance) to CAAs of all oomycetes outside Peronosporales (e.g., all *Pythium* spp.; Blum and Gisi 2012; Blum et al. 2012).

10.4.3 Resistance Evolution in **P. viticola, P. cubensis,** *and* **B. lactucae**

Resistance to CAAs in *P. viticola* field populations has already been reported in 1994, shortly after the introduction of dimethomorph in France (Chabane et al. 1996). As a consequence, intensive sensitivity monitoring was done across European vineyards by several companies; resistant isolates were repeatedly detected mainly in certain grape-growing regions of France, Italy, Switzerland, and Germany. Resistance gradually increased between 2005 and 2010 in these countries but seems to be stable since then at a moderate level (Table 10.4); it is low in other regions and completely absent in some European countries (e.g., Portugal). Also, resistance in *P. cubensis* was reported at several trial sites and in commercial fields of the Western coast in the USA and in some places of Israel and China (CAA FRAC working group, www.frac.info; Zhu et al. 2007). No signs of resistance evolution in *B. lactucae* field populations were reported so far.

10.4.4 Resistance Risk Assessment and Management for CAA Fungicides

Resistance risk (inherent risk) has to be assessed separately for each fungicide – pathogen species combination, and should also include local climatic and agronomic conditions (agronomic risk). As a consequence, adapted use strategies can be designed to delay resistance development (management risk). For CAAs and *P. viticola*, the following factors contribute to an increased resistance risk: the presence of CAA-resistant isolates in recent field populations, stability of CAA-resistant isolates, inheritable resistance (in phenotype) after sexual recombination, and presence of G1105S in the target gene (*CesA3*) of resistant isolates (Gisi et al. 2007; Blum et al. 2010b). However, other factors contribute to a decrease of resistance risk such as rather slow spread of CAA resistance in field populations (over many years and limited to some European countries and regions; CAA FRAC working group, www.frac.info), decline of resistance in absence of product use (Gisi and Sierotzki 2008), recessive nature of CAA resistance, and resistance (and G1105S) expressed only in homozygous isolates (both alleles are mutated; Blum et al. 2010b). Therefore, the intrinsic risk and extent of resistance for CAAs in *P. viticola* was classified by FRAC as moderate. As a consequence, CAAs are recommended only in mixtures with multisite fungicides or other effective non-cross-resistant partner fungicides. In Europe, a maximum of four treatments (three when high disease pressure) during one season may contain a CAA fungicide which should be used in a preventative manner. A moderate resistance risk can be attributed also to *P. cubensis*.

No resistant isolates have been detected in field populations of *P. infestans*, although some CAAs are in use since more than 15 years. In many investigations, artificial mutants were produced but in most cases did not express stable resistance or were less fit, and no resistant individuals were detected upon enforced selection (Cohen et al. 2007; Rubin et al. 2008). No heterozygous isolates (at position 1105 in *CesA3* gene) have been detected so far in field populations (Syngenta internal data). Therefore, resistance risk for CAAs was classified as low to medium for *P. infestans*, and CAAs (especially mandipropamid) can be recommended as solo products for the control of late blight. However, as a resistance management precaution, the number of CAA applications per season is limited to no more than 50 % of the treatments against *P. infestans*. Alternation with products of a different mode of action group should be considered (CAA FRAC working group recommendations; www.frac.info).

10.4.5 Summary and Outlook for CAA Fungicides

CAA fungicides are inhibitors of cellulose synthase, they are active against all pathogen species within Peronosporales, whereas Pythiales (and all other oomycetes orders) are intrinsically insensitive (tolerant). Resistance to CAAs has evolved in several European countries in *P. viticola* and, however, in only a few countries (the USA, Israel, China) in *P. cubensis*. No resistance has been detected in *P. infestans* and *B. lactucae*. Inheritance of CAA resistance is based on one recessive nuclear gene. Resistance is based mainly on the G1105S mutation in the *CesA3* gene; the phenotype (resistance) and genotype (mutation) always co-segregate during sexual recombination. CAA resistance is expressed in homozygous isolates only; heterozygous isolates are sensitive. Thus, CAA resistance is recessive. The resistance risk is classified as low to moderate for *P. infestans* and moderate for *P. viticola* and *P. cubensis*. For the control of grape downy mildew, CAAs are recommended only in mixture with other effective, non-cross-resistant fungicides, and the number of applications per season is limited.

Interestingly, resistance in *P. infestans* has evolved for one class of fungicides only, the phenylamides, whereas in *P. viticola*, all major single-site fungicides are confronted (to different degrees) with resistance problems (mainly PAs, CAAs, QoIs, cymoxanil). The reasons for this differential behavior of *P. infestans* are largely unknown but may be related to rather rare sexual recombination (in most parts of the world), frequent chromosomal abnormalities and heterokaryosis (Catal et al. 2010), and quite high frequencies of triploid (or trisomic) individuals in (European) field populations with some fitness penalties during sexual recombination (Hamed and Gisi 2013).

References

Abu-El Samen FM, Oberoi K, Taylor RJ, Secor GA, Gudmestad NC (2005) Inheritance of mefenoxam resistance in selfed populations of the homothallic oomycete *Phytophthora erythroseptica* (Pethybr.), cause of pink rot of potato. Am J Potato Res 82:105–115

Bhat RG, McBlain BA, Schmitthenner AF (1993) The inheritance of resistance to metalaxyl and to fluorophenylalanine in matings of homothallic *Phytophthora sojae*. Mycol Res 97:865–870

Blum M, Gisi U (2008) Inheritance of fungicide resistance in *Plasmopara viticola*. In: Dehne HW, Gisi U, Kuck KH, Russell PE, Lyr H (eds) Modern fungicides and antifungal compounds V, Proceedings 15th international Reinhardsbrunn symposium. DPG Selbstverlag, Braunschweig, pp 101–104

Blum M, Gisi U (2012) Insights into the molecular mechanism of tolerance to Carboxylic Acid Amide (CAA) fungicides in *Pythium aphanidermatum*. Pest Manag Sci 68:1171–1183

Blum M, Boehler M, Randall E, Young V, Csukai M, Kraus S, Moulin F, Scalliet G, Avrova AO, Whisson SC, Fonné-Pfister R (2010a) Mandipropamid targets the cellulose synthase-like PiCesA3 to inhibit cell wall biosynthesis in the oomycete plant pathogen *Phytophthora infestans*. Mol Plant Pathol 11:227–243

Blum M, Waldner M, Gisi U (2010b) A single point mutation in the novel *PvCesA3* gene confers resistance to the carboxylic acid amide fungicide mandipropamid in *Plasmopara viticola*. Fungal Genet Biol 47:499–510

Blum M, Waldner M, Olaya G, Cohen Y, Gisi U, Sierotzki H (2011) Resistance mechanism to carboxylic acid amide (CAA) fungicides in the cucurbit downy mildew pathogen *Pseudoperonospora cubensis*. Pest Manag Sci 67:1211–1214

Blum M, Gamper HA, Waldner M, Sierotzki H, Gisi U (2012) The cellulose synthase 3 (CesA3) gene of oomycetes: structure phylogeny and influence on sensitivity to carboxylic acid amide (CAA) fungicides. Fungal Biol 116:529–542

Bosshard E, Schüepp H (1983) Variability of selected strains of *Plasmopara viticola* with respect to their metalaxyl sensitivity under field conditions. Z Pflkrank Pflschutz 90:449–459 (in German)

Brasseur G, Saribas AS, Daldal F (1996) A compilation of mutations located in the cytochrome b subunit of the bacterial and mitochondrial bc1 complex. Biochim Biophys Acta 1275:61–69

Brown S, Koike ST, Ochoa OE, Laemmlen F, Michelmore RW (2004) Insensitivity to the fungicide fosetyl-aluminium in California isolates of the lettuce downy mildew pathogen *Bremia lactucae*. Plant Dis 88:502–508

Café-Filho AC, Ristaino JB (2008) Fitness of isolates of *Phytophthora capsici* resistant to mefenoxam from squash and pepper fields in North Carolina. Plant Dis 92:1439–1443

Catal M, King L, Tumbalam P, Wiriyajitsomboon P, Kirk WW, Adams GC (2010) Heterokaryotic nuclear conditions and a heterogeneous nuclear population are observed by flow cytometry in *Phytophthora infestans*. Cytometry Part A 77A:769–775

Chabane K, Leroux P, Maia N, Bompeix G (1996) Resistance to dimethomorph in laboratory mutants of *Phytophthora parasitica*. In: Lyr H, Russell PE, Sisler HD, Lyr H (eds) Modern fungicides and antifungal compounds I, Proceedings 11th international Reinhardsbrunn symposium. Intercept, Andover, pp 387–391

Chapara V, Taylor RJ, Pasche JS, Gudmestad NC (2010) Prevalence of mefenoxam resistance among *Phytophthora erythroseptica* (Pethybridge) isolates in Minnesota and North Dakota. Am J Potato Res 87:521–530

Chapara V, Taylor RJ, Pasche J, Gudmestad NC (2011) Competitive parasitic fitness of mefenoxam-sensitive and -resistant isolates of *Phytophthora erythroseptica* under fungicide selection pressure. Plant Dis 95:691–696

Chen L, Wang Q, Lu X, Zhu S, Liu X (2011) Semidominant mutations in *CesA3* leading to resistance to CAA fungicides in *Phytophthora capsici*. Phytopathology 101:S34 (abstract)

Chen L, Zhu S, Lu X, Pang Z, Cai M, Liu X (2012) Assessing the risk that *Phytophthora melonis* can develop a point mutation (V1109L) in *CesA3* conferring resistance to carboxylic acid amide fungicides. PLoS ONE 7(7), e42069. doi:10.1371/journal.pone.0042069

Cohen Y, Rubin E, Hadad T, Gotlieb D, Sierotzki H, Gisi U (2007) Sensitivity of *Phytophthora infestans* to mandipropamid and the effect of enforced selection pressure in the field. Plant Pathol 56:836–842

Cooke DEL, Cano LM, Raffaele S, Bain RA, Cooke LR, Etherington GJ, Deahl KL, Farrer RA, Gilroy EM, Goss EM, Grünwald NJ, Hein I, MacLean D, McNicol JW, Randall E, Oliva RF, Pel MA, Shaw DS, Squires JN, Taylor MC, Vleeshouwers VGAA, Birch PRJ, Lees AK, Kamoun S (2012) Genome analyses of an aggressive and invasive lineage of the Irish potato famine pathogen. PLoS Pathog 8(10), e1002940. doi:10.1371/journal.ppat.1002940

Corio-Costet MF, Martinez F, Delmotte F, Douence L, Richart-Cervera S, Chen WJ (2008) Resistance of *Plasmopara viticola* to QoI fungicides: origin and diversity. In: Dehne HW, Gisi U, Kuck KH, Russell PE, Lyr H (eds) Modern fungicides and antifungal compounds V, Proceedings 15th international Reinhardsbrunn symposium. DPG Selbstverlag, Braunschweig, pp 107–112

Crute IR, Harrison JM (1988) Studies on the inheritance of resistance to metalaxyl in *Bremia lactucae* and on the stability and fitness of field isolates. Plant Pathol 37:231–250

Crute IR, Gordon PL, Moss NA (1994) Variation for response to phenylamides in UK populations of *Bremia lactucae* (lettuce downy mildew) and *Peronospora parasitica* (brassica downy mildew). In: Heaney S, Slawson D, Hollomon DW, Smith M, Russell PE, Parry DW (eds) Fungicide resistance. British Crop Protection Council, Surrey, pp 155–162

Csinos AS, Bertrand PF (1994) Distribution of *Phytophthora parasitica* var. *nicotianae* races and their sensitivity to metalaxyl in Georgia. Plant Dis 78:471–474

Dagget SS, Goetz E, Therrien CD (1993) Phenotypic changes in populations of *Phytophthora infestans* from eastern Germany. Phytopathology 83:319–323

Davidse LC (1988) Phenylamide fungicides: mechanism of action and resistance. In: Delp CJ (ed) Fungicide resistance in North America. APS Press, St Paul, pp 63–65

Davidse LC (1995) Phenylamide fungicides: biochemical action and resistance. In: Lyr H (ed) Modern selective fungicides, 2nd edn. Gustav Fischer, Jena, pp 347–354

Degli-Esposti M, De Fries S, Crimi M, Ghelli A, Patarnello T, Meyer A (1993) Mitochondrial cytochrome b: evolution and structure of the protein. Biochim Biophys Acta 1143:243–271

Di Rago JP, Coppée JY, Colson AM (1989) Molecular basis for resistance to myxothiazol, mucidin (strobilurin A) and stigmatellin. J Biol Chem 264:14543–14548

Dunn AR, Milgroom MG, Meitz JC, McLeod A, Fry WE, McGrath MT, Dillard HR, Smart CD (2010) Population structure and resistance to mefenoxam of *Phytophthora capsici* in New York State. Plant Dis 94:1461–1468

Fabritius AL, Shattock RC, Judelson HS (1997) Genetic analysis of metalaxyl insensitivity loci in *Phytophthora infestans* using linked DNA markers. Phytopathology 87:1034–1040

Falloon RE, Follas GB, Butler RC, Goulden DS (2000) Resistance in *Peronospora viciae* to phenylamide fungicides: reduced efficacy of seed treatments of pea (*Pisum sativum*) and assessment of alternatives. Crop Prot 19:313–325

Fisher DJ, Hayes AL (1982) Mode of action of the fungicides furalaxyl, metalaxyl and ofurace. Pestic Sci 13:330–339

FRAC (2014) Several documents on www.frac.info

French-Monar RD, Jones JB, Roberts PD (2006) Characterization of *Phytophthora capsici* associated with roots of weeds on Florida vegetable farms. Plant Dis 90:345–350

Fry WE, Goodwin SB, Dyer AT, Matuszak JM, Drenth A, Tooley PW, Sujkowski LS, Koh YJ, Cohen BA, Spielman LJ, Deahl KL, Inglis DA, Sandlan KP (1993) Historical and recent migrations of *Phytophthora infestans*: chronology, pathways, and implications. Plant Dis 84:731–735

Gallup CA, Ivors KL, Greene MD, Lannon KR, Shew HD (2009) Population structure of *Phytophthora nicotianae*, the causal agent of black shank, in North Carolina tobacco fields. North Carolina State University, Raleigh

Geier BM, Schägger H, Brandt U, Colson AM, von Jagow G (1992) Point mutation in cytochrome b of yeast ubihydroquinone-cytochrome-oxidoreductase causing myxothiazol resistance and facilitated dissociation of the iron-sulfur subunit. Eur J Biochem 208:375–380

Gisi U (2002) Chemical control of downy mildews. In: Spencer PTN, Gisi U, Lebeda A (eds) Advances in downy mildew research. Kluwer Academic Publishers, Dordrecht, pp 119–159

Gisi U (2014) Assessment of selection and resistance risk for DMI fungicides in *Aspergillus fumigatus* in agriculture and medicine: a critical review. Pest Manag Sci 70:352–364

Gisi U, Cohen Y (1996) Resistance to phenylamide fungicides: a case study with *Phytophthora infestans* involving mating type and race structure. Annu Rev Phytopathol 34:549–572

Gisi U, Sierotzki H (2008) Fungicide mode of action and resistance in downy mildews. Eur J Plant Pathol 122:157–167

Gisi U, Sierotzki H, Cook A, Mc Caffery A (2002) Mechanisms influencing the evolution of resistance to Qo inhibitor fungicides, Pest Manag Sci 58:859-867

Gisi U, Waldner M, Kraus N, Dubuis PH, Sierotzki H (2007) Inheritance of resistance to carboxylic acid amide (CAA) fungicides in *Plasmopara viticola*. Plant Pathol 56:199–208

Gisi U, Walder F, Resheat-Eini Z, Edel D, Sierotzki H (2011) Change of genotype, sensitivity and aggressiveness in *Phytophthora infestans* isolates collected in European countries in 1997, 2006 and 2007. J Phytopathol 159:223–232

Gisi U, Lamberth C, Mehl A, Seitz T (2012) Carboxylic Acid Amide (CAA) fungicides. In: Krämer W, Schirmer U, Jeschke P, Witschel M (eds) Modern crop protection compounds, 2nd edn. Wiley-VCH, Weinheim, pp 807–830

Goodwin SB, Smart CD, Sandrock RW, Deahl KL, Punja ZK, Fry WE (1998) Genetic change within populations of *Phytophthora infestans* in the United States and Canada during 1994 to 1996: role of migration and recombination. Phytopathology 85:478–479

Grasso V, Palermo S, Sierotzki H, Garibaldi A, Gisi U (2006) Cytochrome b gene structure and consequences for resistance to QoI fungicides in plant pathogens. Pest Manag Sci 62:465–472

Grönberg L, Andersson B, Yuen J (2012) Can weed hosts increase aggressiveness of *Phytophthora infestans* on potato? Phytopathology 102:429–433

Gullino ML, Gilardi G, Tinivella F, Garibaldi A (2004) Observations on the behaviour of different populations of *Plasmopara viticola* resistant to QoI fungicides in Italian vineyards. Phytopathol Medit 43:341–350

Hamed BH, Gisi U (2013) Generation of pathogenic F1 progeny from crosses of *Phytophthora infestans* isolates differing in ploidy. Plant Pathol 62:708–718

Heaney SP, Hall AA, Davies SA, Olaya G (2000) Resistance to fungicides in the QoI-STAR cross-resistance group: current perspectives. Proc Brighton Crop Protect Conf 2:755–762

Hermann D, Gisi U (2012) Fungicide resistance in Oomycetes with special reference to *Phytophthora infestans* and phenylamides. In: Thind TS (ed) Fungicide resistance in crop protection. CAB International, Oxfordshire, pp 133–140

Hermann D, McKenzie D, Cohen Y, Gisi U (2015) Phenylamides: market trends and resistance evolution for important Oomycete pathogens 35 years after the first product introduction. In: Stevenson KL, Wyenandt CA, McGrath ML (eds) Fungicide resistance development in North America. APS Press, St. Paul (in press)

Hu J, Hong C, Stromberg EL, Moorman GW (2010) Mefenoxam sensitivity in *Phytophthora cinnamomi* isolates. Plant Dis 94:39–44

Hu CH, Perez FG, Donahoo R, McLeod A, Myers K, Ivors K, Secor G, Roberts PD, Deahl KL, Fry WE, Ristaino JB (2012) Recent genotypes of *Phytophthora infestans* in the Eastern United States reveal clonal populations and reappearance of mefenoxam sensitivity. Plant Dis 96:1323–1330

Ishii H, Fraaije BA, Sugiyama T, Noguchi K, Nishimura K, Takeda T, Amano T, Hollomon DW (2001) Occurrence and molecular characterization of strobilurin resistance in cucumber powdery mildew and downy mildew. Phytopathology 91:1166–1171

Ishii H, Sugiyama T, Nishimura K, Ishikawa Y (2002) Strobilurin resistance in cucumber pathogens: persistence and molecular diagnosis of resistance. In: Dehne HW, Gisi U, Kuck KH, Russell PE, Lyr H (eds) Modern fungicides and antifungal compounds III, Proceedings 13th international Reinhardsbrunn symposium. Agro Concept, Bonn, pp 149–159

Judelson HS (1997) Expression and inheritance of sexual preference and selfing propensity in *Phytophthora infestans*. Fungal Genet Biol 21:188–197

Judelson HS, Roberts S (1999) Multiple loci determining insensitivity to phenylamide fungicides on *Phytophthora infestans*. Phytopathology 89:754–760

Kadish D, Cohen Y (1992) Overseasoning of metalaxyl-sensitive and metalaxyl-resistant isolates of *Phytophthora infestans* in potato tubers. Phytopathology 82:887–889

Kadish D, Grinberger M, Cohen Y (1990) Fitness of metalaxyl-sensitive and metalaxyl-resistant isolates of *Phytophthora infestans* on susceptible and resistant potato cultivars. Phytopathology 80:200–205

Keinath AP (2007) Sensitivity of populations of *Phytophthora capsici* from South Carolina to mefenoxam, dimethomorph, zoxamide, and cymoxanil. Plant Dis 91:743–748

Kuck KH, Leadbeater AJ, Gisi U (2012) FRAC mode of action classification and resistance risk of fungicides. In: Krämer W, Schirmer U, Jeschke P, Witschel M (eds) Modern crop protection compounds, 2nd edn. Wiley-VCH, Weinheim, pp 539–557

Lamour KH, Hausbeck MK (2000) Mefenoxam insensitivity and the sexual stage of *Phytophthora capsici* in Michigan cucurbit fields. Phytopathology 90:396–400

Lebeda A, Cohen Y (2012) Fungicide resistance in *Pseudoperonospora cubensis*, the causal pathogen of cucurbit downy mildew. In: Thind TS (ed) Fungicide resistance in crop protection: risk and management. CABI, Wallingford, pp 44–63

Lebreton L, Lucas JM, Andrivon D (1999) Aggressiveness and competitive fitness of *Phytophthora infestans* isolates collected from potato and tomato in France. Phytopathology 89:679–686

Lees AK, Wattier R, Shaw DS, Sullivan L, Williams NA, Cooke DEL (2006) Novel microsatellite markers for the analysis of *Phytophthora infestans* populations. Plant Pathol 55:311–319

Li Y, Cooke DEL, Jacobsen E, van der Lee T (2013) Efficient multiplex simple sequence repeat genotyping of the oomycete plant pathogen *Phytophthora infestans*. J Microbiol Methods 92:316–322

Lu XH, Hausbeck MK, Liu XL, Hao JJ (2011) Wild type sensitivity and mutation analysis for resistance risk to fluopicolide in *Phytophthora capsici*. Plant Dis 95:1535–1541

Lu XH, Davis RM, Livingston S, Nunez J, Hao J (2012) Fungicide sensitivity of *Pythium* spp. associated with cavity spot of carrot in California and Michigan. Plant Dis 96:384–388

Lucas JA, Greer G, Oudemans PV, Coffey MD (1990) Fungicide sensitivity in somatic hybrids of *Phytophthora capsici* by protoplast fusion. Physiol Mol Plant Pathol 36:175–187

McGrath MT (2012) Challenge of fungicide resistance and anti-resistance strategies in managing vegetable diseases in the USA. In: Thind ST (ed) Fungicide resistance in crop protection: risk and management. CABI, Wallingford, pp 191–207

Montarry J, Corbiere R, Andrivon D (2007) Is there a trade-off between aggressiveness and over-winter survival in *Phytophthora infestans*? Funct Ecol 21:603–610

Moorman GW, Kang S, Geiser DM, Kim SH (2002) Identification and characterization of *Pythium* species associated with greenhouse floral crops in Pennsylvania. Plant Dis 86:1227–1231

Müller U, Gisi U (2012) Newest aspects of nucleic acid synthesis inhibitors: metalaxyl-M. In: Krämer W, Schirmer U, Jeschke P, Witschel M (eds) Modern crop protection compounds, 2nd edn. Wiley-VCH, Weinheim, pp 901–908

Olaya G, Cleere S, Stanger C, Burbidge J, Hall A, Windass J (2003) A novel potential target site and QoI fungicide resistance mechanism in *Pythium aphanidermatum*. Phytopathology 93(6) Supplement: S67 (abstract)

Pang Z, Shao J, Chen L, Lu X, Hu J, Qin Z, Liu X (2013) Resistance to the novel fungicide pyrimorph in *Phytophthora capsici*: risk assessment and detection of point mutations in *CesA3* that confer resistance. PLoS ONE 8(2), e56513. doi:10.1371/journal.pone.0056513

Parkunan V, Johnson CS, Bowman BC, Hong CX (2010) Population structure, mating type, and mefenoxam sensitivity of *Phytophthora nicotianae* in Virginia tobacco fields. Plant Dis 94:1361–1365

Parra G, Ristaino JB (2001) Resistance to mefenoxam and metalaxyl among field isolates of *Phytophthora capsici* causing *Phytophthora* blight of bell pepper. Plant Dis 85:1069–1075

Phillips McDougall (2012) AgriReference reports, release 6.0, product directory V13, www.phillipsmcdougall.com

Porter LD, Miller JS, Nolte P, Price WJ (2007) *In vitro* somatic growth and reproduction of phenylamide-resistant and -sensitive isolates of *Phytophthora erythroseptica* from infected potato tubers in Idaho. Plant Pathol 56:492–499

Randall E, Young V, Sierotzki H, Scalliet G, Birch PRJ, Cooke DEL, Csukai M, Whisson SC (2014) Sequence diversity in the large subunit of RNA polymerase I contributes to mefenoxam insensitivity in *Phytophthora infestans*. Mol Plant Pathol. doi:10.1111/mpp.12124

Reuveni M, Eyal H, Cohen Y (1980) Development of resistance to metalaxyl in *Pseudoperonospora cubensis*. Plant Dis 64:1108–1109

Rubin A(E), Gotlieb D, Gisi U, Cohen Y (2008) Mutagenesis of *Phytophthora infestans* for resistance against carboxylic acid amide and phenylamide fungicides. Plant Dis 92:675–683

Scherer E, Gisi U (2006) Characterization of genotype and mating type in European isolates of *Plasmopara viticola*. J Phytopathol 154:489–495

Schwinn F, Staub T (1995) Oomycete fungicides: phenylamides and other fungicides against Oomycetes. In: Lyr H (ed) Modern selective fungicides, 2nd edn. Gustav Fischer, Jena, pp 323–346

Shattock RC (1986) Inheritance of metalaxyl resistance in the potato late blight fungus. In: Proceedings Brighton crop protection conference, vol 2. British Crop Protection Council, Thornton Heath, Surrey, pp 555–560

Shattock RC (1988) Studies on inheritance of resistance to metalaxyl in *Phytophthora infestans*. Plant Pathol 37:4–11

Shaw DS, Shattock RC (1991) Genetics of *Phytophthora infestans*: the Mendelian approach. In: Lucas JA, Shattock RC, Shaw DS, Cooke LR (eds) Phytophthora. Cambridge University Press, Cambridge, UK, pp 218–230

Sierotzki H, Parisi S, Steinfeld U, Tenzer I, Poirey S, Gisi U (2000a) Mode of resistance to respiration inhibitors at the cytochrome bc1 enzyme complex of *Mycosphaerella fijiensis* field isolates. Pest Manag Sci 56:833–841

Sierotzki H, Wullschleger J, Gisi U (2000b) Point-mutation in cytochrome b gene conferring resistance to strobilurin fungicides in *Erysiphe graminis* f.sp. *tritici* field isolates. Pestic Biochem Physiol 68:107–112

Sierotzki H, Kraus N, Assemat P, Stanger C, Cleere S, Windass J, Gisi U (2005) Evolution of resistance to QoI fungicides in *Plasmopara viticola* populations in Europe. In: Dehne HW, Gisi U, Kuck KH, Russell PE, Lyr H (eds) Modern fungicides and antifungal compounds IV, Proceedings 14th international Reinhardsbrunn symposium. BCPC, Alton, pp 73–80

Sierotzki H, Kraus N, Pepin S, Fernandes N, Gisi U (2008) Dynamics of QoI resistance in *Plasmopara viticola*. In: Dehne HW, Gisi U, Kuck KH, Russell PE, Lyr H (eds) Modern fungicides and antifungal compounds V, Proceedings 15th international Reinhardsbrunn symposium. DPG Selbstverlag, Braunschweig, pp 151–157

Sierotzki H, Blum M, Olaya G, Waldner-Zulauf M, Wullschleger J, Cohen Y, Gisi U (2011) Sensitivity to CAA fungicides and frequency of mutations in cellulose synthase *(CesA3)* gene of oomycete pathogen populations. In: Dehne HW, Deising HB, Gisi U, Kuck KH, Russell PE, Lyr H (eds) Modern fungicides and antifungal compounds VI, Proceedings 16th international Reinhardsbrunn symposium. DPG Selbstverlag, Braunschweig, pp 103–110

Staub T (1994) Early experiences with phenylamide resistance and lessons for continued successful use. In: Heaney S, Slawson D, Hollomon DW, Smith M, Russell PE, Parry DW (eds) Fungicide resistance. British Crop Protection Council, Farnham, pp 131–138

Staub T, Sozzi D (1981) Résistance au métalaxyl en pratique et les consequences pour son utilisation. Phytiatr Phytopharm 30:283–291 (in French)

Takeda T, Kawagoe Y, Uchida K, Fuji M, Amano T (1999) The appearance of resistant isolates to strobilurins. Ann Phytopathol Soc Jpn 65:655 (Japanese abstract)

Taylor RJ, Salas B, Secor GA, Rivera V, Gudmestad NC (2002) Sensitivity of North American isolates of *Phytophthora erythroseptica* and *Pythium ultimum* to mefenoxam (metalaxyl). Plant Dis 86:797–802

Taylor RJ, Pasche JS, Gudmestad NC (2006) Biological significance of mefenoxam resistance in *Phytophthora erythroseptica* and its implications for the management of pink rot of potato. Plant Dis 90:927–934

Taylor RJ, Pasche JS, Gudmestad NC (2011) Effect of application method and rate on residual efficacy of mefenoxam and phosphorous acid fungicides in the control of pink rot of potato. Plant Dis 95:997–1006

Toffolatti SL, Vercesi A (2012) QoI resistance in *Plasmopara viticola* in Italy: evolution and management strategies. In: Thind TS (ed) Fungicide resistance in crop protection. CAB International, Oxfordshire, pp 172–183

Toffolatti SL, Serrati L, Sierotzki H, Gisi U, Vercesi A (2007) Assessment of QoI resistance in *Plasmopara viticola* oospores. Pest Manag Sci 63:194–201

Toffolatti SL, Serrati L, Vercesi A (2011) CAA, phenylamide and QoI resistance assessment in *Plasmopara viticola* oospores. In: Dehne HW, Deising HB, Gisi U, Kuck KH, Russell PE, Lyr H (eds) Modern fungicides and antifungal compounds VI, Proceedings 16th international Reinhardsbrunn symposium. DPG Selbstverlag, Braunschweig, pp 267–272

Whisson SC, Fonné-Pfister R, Csukai M (2011) Molecular approaches to elucidate pathways and sites of fungicide resistance in Oomycetes. In: Dehne HW, Deising HB, Gisi U, Kuck KH, Russell PE, Lyr H (eds) Modern fungicides and antifungal compounds VI, Proceedings 16th international Reinhardsbrunn symposium. DPG Selbstverlag, Braunschweig, pp 91–102

Wicks TG, Hall B, Pezzanati P (1994) Fungicidal control of metalaxyl-insensitive strains of *Bremia lactucae* on lettuce. Crop Prot 13:617–624

Widmark AK, Andersson B, Cassel-Lundhagen A, Sandström M, Yuen JE (2007) *Phytophthora infestans* in a single field in southwest Sweden early in spring: symptoms, spatial distribution and genotypic variation. Plant Pathol 56:573–579

Yuen JE, Andersson B (2013) What is the evidence for sexual reproduction of *Phytophthora infestans* in Europe? Plant Pathol 62:485–491

Zhu SS, Liu XL, Wang Y, Wu XH, Liu PF, Li JQ, Yuan SK, Si NG (2007) Resistance of *Pseudoperonospora cubensis* to flumorph on cucumber in plastic houses. Plant Pathol 56:967–975

Chapter 11
Melanin Biosynthesis Inhibitors

Makiichi Takagaki

Abstract Carpropamid was the first commercial fungicide from the scytalone dehydratase inhibitors of melanin biosynthesis group (MBI-D). It has been widely used in Japan as the chemical agent for nursery-box treatment against leaf blast of paddy rice since 1998. Further MBI-Ds, diclocymet and fenoxanil, were launched in 2000 and 2001, respectively. In 2001, failure of control of rice blast was reported in southern part of Japan, where MBI-Ds have been used since 1998. The inhibitory activity of carpropamid on scytalone dehydratase (SDH) extracted from a carpropamid-resistant strain of *Magnaporthe oryzae* was dramatically reduced in comparison with that on SDH extracted from the sensitive strain. A single-point mutation (G to A) located at the upstream region (233 bp downstream from the ATG codon) resulting in a one-amino-acid substitution (valine [GTG] 75 to methionine [ATG]: V75M) was found in the resistant strain. To examine whether the V75M mutation is the primary reason for decreasing the sensitivity of SDH to carpropamid, the SDH cDNAs of both the sensitive and the resistant strain were cloned into a GST-fused protein expression vector system. The recombinant SDHs of both strains exhibited the same sensitivities to carpropamid as those extracted from the mycelia of the respective strains. These data clearly revealed that the V75M mutation causes the low sensitivities of the SDHs of the carpropamid-resistant strains and strongly suggest that the V75M mutation confers resistance of these strains to carpropamid.

Keywords Carpropamid • Scytalone dehydratase • Fungicide resistance • Point mutation • MBI-D fungicides • *Magnaporthe oryzae* • Melanin • Rice blast

---

M. Takagaki (✉)
Kumiai Chemical Industry Co. Ltd.,
Marketing Section, 1-4-26, Ikenohata, Taitoh-ku, Tokyo 110-8782, Japan
e-mail: m-takagaki@kumiai-chem.co.jp

H. Ishii, D.W. Hollomon (eds.), *Fungicide Resistance in Plant Pathogens*,
DOI 10.1007/978-4-431-55642-8_11

11.1 Introduction

Fungal melanin is a black substance wherein dihydroxynaphthalene has been oxidatively polymerized. It is synthesized through the polyketide pathway which includes fusion of five isoprenyl units, two steps of reduction, two steps of dehydration, and polymerization of 1,8-dihydroxynaphthalene (Bell et al. 1976a, b). The melanin biosynthesis pathways have been resolved not only in *Magnaporthe oryzae* but also in plant pathogenic fungi such as *Verticillium*, *Colletotrichum*, *Cochliobolus*, and *Alternaria* (Bell and Wheeler 1986). Although *Cochliobolus heterostrophus* and *Alternaria alternata* require no melanin in the infectious process into the host plants, the biosynthesis of melanin is essentially required in *M. oryzae* and *C. lagenarium* (Bell et al. 1976b; Woloshuk et al. 1980; Okuno et al. 1983; Bell and Wheeler 1986). Therefore, melanin biosynthesis inhibitors have attracted attention in developing rice blast fungicides.

The scytalone dehydratase inhibitors in melanin biosynthesis (MBI-D), carpropamid, diclocymet, and fenoxanil, block the dehydration reactions from scytalone to 1,3,8-trihydroxynaphthalene and from vermelone to 1,8-dihydroxynaphthalene (Tsuji et al. 1997; Kurahashi et al. 1998; Motoyama et al. 1998b). In contrast, tricyclazole, pyroquilon, and fthalide, which were developed in the 1970s, as hydroxynaphthalene reductase inhibitors in melanin biosynthesis (MBI-R), block the reduction reactions from 1,3,6,8-tetrahydroxynaphthalene to scytalone and from 1,3,8-trihydroxynaphthalene to vermelone (Fig. 11.1) (Tokousbalides and Sisler 1978, 1979; Woloshuk et al. 1981; Yamaguchi et al. 1982; Wheeler and Greenblatt 1988).

Carpropamid is a long-acting agent for controlling rice blast showing a systemic action (Kurahashi et al. 1997). It has been used mainly for nursery-box treatment and contributes to saving labor and lessening application frequency since 1998 in Japan. Further MBI-Ds, diclocymet and fenoxanil, were launched in 2000 and 2001, respectively. In 2001, however, insufficient performance of carpropamid was reported in southern part of Japan, where MBI-Ds have been used since 1998 (Yamaguchi et al. 2002). Isolates of *M. oryzae* from this area showed decreased sensitivity to carpropamid in vivo (So et al. 2002). It has also been shown that these isolates displayed cross-resistance to the fungicides: diclocymet and fenoxanil, which are also categorized in the MBI-D class (Sawada et al. 2004). So far no resistant strains to MBI-R have been reported, although it has been used for over 40 years. Therefore, the mechanism of resistance of the MBI-D resistant strain was of particular interest.

11.2 Mechanism of Resistance

When the speed of metabolic degradation of carpropamid in the MBI-D resistant strain was compared with that in the sensitive strain, no significant difference was observed (Takagaki et al. 2002). In an inoculation test using young seedlings,

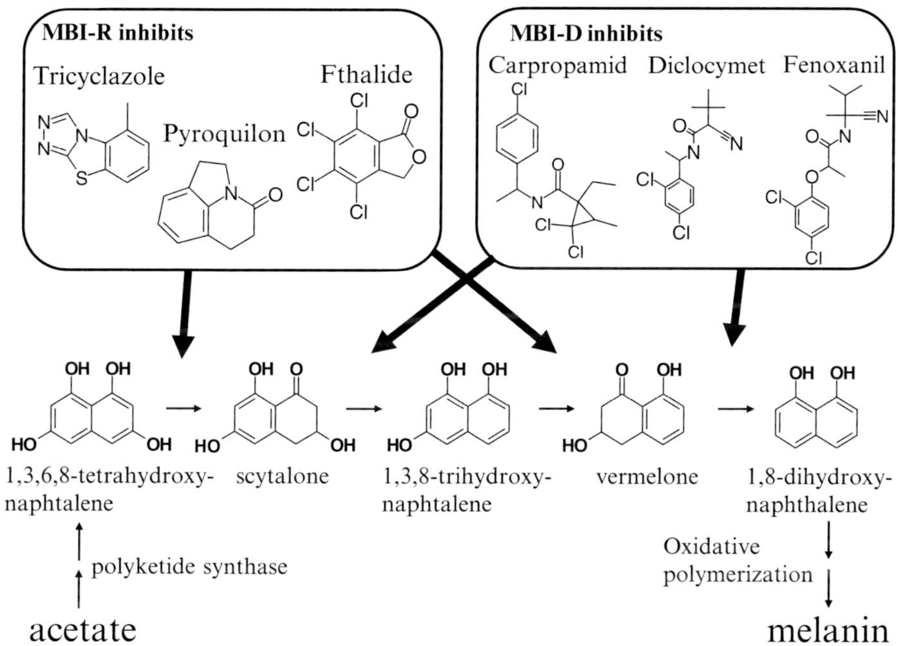

Fig. 11.1 Mode of action of melanin biosynthesis inhibitors (MBIs)

Table 11.1 Sensitivity of MBI-D resistant strain to carpropamid and tricyclazole in foliar spray tests

	Efficacy %		
	Carpropamid (mg liter^{-1})		Tricyclazole (mg liter^{-1})
Isolate	1	10	10
R (A14)	1.6	18.6	100
S (SS35)	100	100	100

R resistant, *S* sensitive

tricyclazole (MBI-R) showed an effect on the MBI-D resistant strain comparable with the effect on the sensitive strain, but carpropamid showed an obviously reduced efficacy (Table 11.1). On the basis of these results, it was supposed that the MBI-D resistant strain might have a target-site mutation in the scytalone dehydratase (SDH) gene.

The inhibitory activity of carpropamid on SDH prepared from the MBI-D resistant strain was examined. The inhibitory activity of carpropamid on SDH extracted from the resistant strain was dramatically reduced in comparison with the SDH extracted from the sensitive strain (Fig. 11.2). To conclude whether or not the affinity between SDH and carpropamid was lowered by amino-acid substitution in the SDH protein, the SDH gene was analyzed. Comparison of the sequences of both strains revealed a single-point mutation (G to A), which was distinctive for the resistant strain. This mutation, which locates in the upstream region (233 bp downstream from the ATG codon), results in a one-amino-acid substitution (valine [GTG] 75 to methionine [ATG]: V75M) (Fig. 11.3). The recombinant SDHs of the sensitive and the resistant strain exhibited the same sensitivities to carpropamid as those extracted from the respective mycelia (Fig. 11.4). These data clearly showed that the V75M mutation causes the low sensitivities of the SDHs of the carpropamid-resistant strains and strongly suggested that the V75M mutation confers resistance of these strains to carpropamid.

Since carpropamid competitively inhibits SDH activity at an extremely low concentration relative to the enzyme concentration, it has been suggested that carpropamid is a tight-binding competitive inhibitor of the SDH (Motoyama et al. 1998c). Cryogenic X-ray structural analysis has been carried out by crystallizing the SDH of *M. oryzae* and the structure of an SDH-carpropamid complex was analyzed. As a result, a function model has been provided, wherein carpropamid completely fits into the hydrophobic active center pocket in the SDH. This would completely block

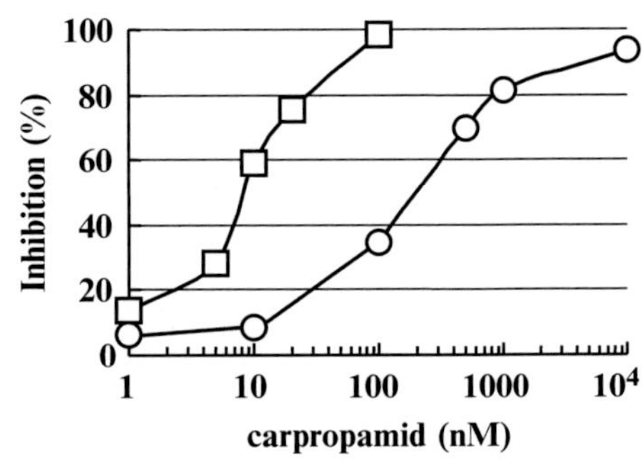

Fig. 11.2 Inhibition of SDHs extracted from the strains, SS35 and A14, by carpropamid. □, SDH from sensitive strain, SS35; ○, SDH from resistant strain, A14

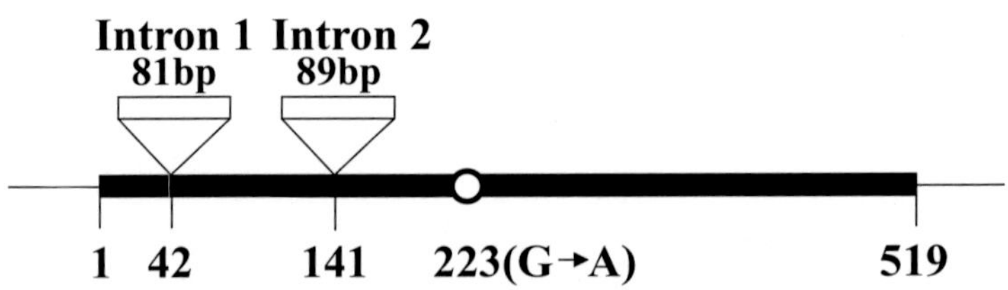

a) TTCTCAGGTGAGCATAATATCCCCCTCCAAAAAGAAAATAGCGGTGAAGCCACCA ACGACAGTACCGCTGACCCTAATTCCCCTCCAGACTAC

b) CTGCGCGTATGTTCCGCCCTGCCATGTTTATTTTTACTTTCCCACACCAAATCCA GACTTTAACAGCGACGACCAAAAAAAAAAAAAAAAAAACAGATTGAC

Fig. 11.3 Map of SDH cDNA sequence and location of introns and the point mutation (indicated with a *dot*) in the resistant rice blast fungus. Intron1 and intron2 are inserted after the 42nd and the 141st nucleotide, respectively. (**a**) The nucleotide sequence of intron1 (*underlined*); (**b**) the nucleotide sequence of intron 2 (*underlined*)

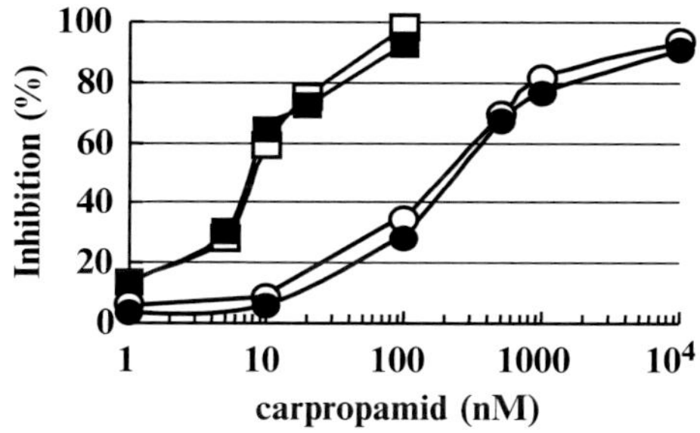

Fig. 11.4 Comparison of SDHs inhibition by carpropamid between SDHs from mycelia and GST-fused SDHs expressed in *Escherichia coli*. □, SDH prepared from mycelia of sensitive strain, SS35; ■, GST-fused SDH expressed from the SS35 cDNA; ○, SDH prepared from mycelia of resistant strain, A14; ●, GST-fused SDH expressed from the A14 cDNA

Y30, Y50, H85, and H110, which are directly required for the catalytic enzymatic activity, and S129 and N131, required for binding the substrate (Lundqvist et al. 1994; Nakasako et al. 1998). To analyze the detailed function of the SDH, enzymes carrying mutations in amino acids at a site directly relating to the catalytic site, the substrate-binding site, the C-terminal region, and around them were constructed. In each case, the catalytic efficiency was largely reduced (Motoyama et al. 1998a, c). In particular, the C-terminal region serves as a cap covering carpropamid. Therefore, it has been demonstrated that this region is important not only for binding to carpropamid but also for the enzymatic reaction or maintaining the enzyme structure (Motoyama et al. 1998a, c). In this study, it could be shown that the amino-acid sequence of the MBI-D resistant strain carries a mutation of valine 75 to methionine (V75M). This mutation site, which is located at the C-terminal region, is considered important for binding to the MBI-D and the substrate or the enzymatic function. In particular, the enzymatic efficiency was largely reduced in a model test wherein valine at the 75-position was substituted by alanine (Motoyama et al. 1998a, c). It is therefore considered that the V75M mutation reduces exclusively the binding force to the MBI-D while sustaining the binding force to the substrate, hence causing the resistance to the MBI-D.

References

Bell AA, Puhalla JE, Tolmsoff WJ, Stipanovic RD (1976a) Use of mutants to establish (+)-scytalone as an intermediate in melanin biosynthesis by *Verticillium dahliae*. Can J Microbiol 22:787–799

Bell AA, Stipanovic RD, Puhalla JE (1976b) Pentaketide metabolites of *Verticillium dahliae* – identification of (+)-scytalone as a natural precursor to melanin. Tetrahedron 32:1353–1356

Bell AA, Wheeler MH (1986) Biosynthesis and functions of fungal melanins. Annu Rev Phytopathol 24:411–451

Kurahashi Y, Sakawa S, Kinbara T, Tanaka K, Kagabu S (1997) Biological activity of carpropamid (KTU3616): a new fungicide for rice blast disease. J Pestic Sci 22:108–112

Kurahashi Y, Araki Y, Kinbara T, Pontzen R, Yamaguchi I (1998) Intermediates accumulation and inhibition sites of carpropamid in the melanin biosynthesis pathway of *Pyricularia oryzae*. J Pestic Sci 23:22–28

Lundqvist T, Rice J, Hodge CN, Basarab GS, Pierce J, Lindqvist Y (1994) Crystal structure of scytalone dehydratase – a disease determinant of the rice pathogen, *Magnaporthe grisea*. Structure 2:937–944

Motoyama T, Imanishi K, Yamaguchi I (1998a) cDNA cloning, expression, and mutagenesis of scytalone dehydratase needed for pathogenicity of the rice blast fungus, *Pyricularia oryzae*. Biosci Biotechnol Biochem 62:564–566

Motoyama T, Imanishi K, Kinbara T, Kurahashi Y, Yamaguchi I (1998b) Inhibition of scytalone dehydratase in melanin biosynthesis by carpropamid, a novel rice blast controlling agent. J Pestic Sci 23:58–61

Motoyama T, Nakasako M, Yamaguchi I (1998c) Molecular action mechanism of a new melanin biosynthesis inhibitor. In: Lyr H et al (eds) Modern fungicides and antifungal compounds II. Intercept, Andover, pp 111–119

Nakasako M, Motoyama T, Kurahashi Y, Yamaguchi I (1998) Cryogenic X-ray crystal structure analysis for the complex of scytalone dehydratase of a rice blast fungus and its tight-binding inhibitor, carpropamid: the structural basis of tight-binding inhibition. Biochemistry 37:9931–9939

Okuno T, Matsuura K, Furusawa I (1983) Recovery of appressorial penetration by some melanin precursors in *Pyricularia oryzae* treated with tricyclazole and in a melanin deficient mutant. J Pestic Sci 8:357–360

Sawada H, Sugihara M, Takagaki M, Nagayama K (2004) Monitoring and characterization study of *Magnaporthe grisea* showing decreased sensitivity to scytalone dehydratase inhibitors. Pest Manag Sci 60:777–785

So K, Fuji M, Iwabuchi H, Kanayama M, Yamaguchi J (2002) Effects of various fungicides against less carpropamid-sensitive rice blast fungus isolated from the northwest area in Saga Prefecture. Jpn J Phytopathol 68:262 (Japanese abstr)

Takagaki M, Shimizu T, Miura I, Sawada H, So K, Nagayama K (2002) Mechanism of less-sensitive *Magnaporthe grisea* to carpropamid and the method for testing the sensitivity. Jpn J Phytopathol 68:262 (Japanese abstr)

Tokousbalides MC, Sisler HD (1978) Effect of tricyclazole on growth and secondary metabolism in *Pyricularia oryzae*. Pestic Biochem Physiol 8:26–32

Tokousbalides MC, Sisler HD (1979) Site of inhibition by tricyclazole in the melanin biosynthesis pathway of *Verticillium dahliae*. Pestic Biochem Physiol 11:64–73

Tsuji G, Takeda T, Furusawa I, Horino O, Kubo Y (1997) Carpropamid, an anti-rice blast fungicide, inhibits scytalone dehydratase activity and appressorial penetration in *Colletotrichum lagenarium*. Pestic Biochem Physiol 57:211–219

Wheeler MH, Greenblatt GA (1988) The inhibition of melanin biosynthetic reactions in *Pyricularia oryzae* by compounds that prevent rice blast disease. Exp Mycol 12:151–160

Woloshuk CP, Sisler HD, Tokousbalides MC, Dutky SR (1980) Melanin biosynthesis in *Pyricularia oryzae* : Site of tricyclazole inhibition and pathogenicity of melanin-deficient mutants. Pestic Biochem Physiol 14:256–264

Woloshuk CP, Wolkow PM, Sisler HD (1981) The effect of three fungicides, specific for the control of rice blast disease, on the growth and melanin biosynthesis by *Pyricularia oryzae* Cav. Pestic Sci 12:86–90

Yamaguchi I, Sekido S, Misato T (1982) The effect of nonfungicidal anti-blast chemicals on the melanin biosynthesis and infection by *Pyricularia oryzae*. J Pestic Sci 7:523–529

Yamaguchi J, Kuchiki F, Hirayae K, So K (2002) Decreased effect of carpropamid for rice blast control in the west north area of Saga Prefecture in 2001. Jpn J Phytopathol 68:261 (Japanese abstr)

Chapter 12
Histidine Kinase Inhibitors

Makoto Fujimura, Shinpei Banno, Akihiko Ichiishi, and Fumiyasu Fukumori

Abstract Dicarboximides and phenylpyrroles have been mainly used to control diseases caused by fungal strains that belong to the genera *Botrytis*, *Sclerotinia*, *Monilinia*, and *Alternaria*. Both types of fungicides overactivate Hog-like mitogen-activated protein kinases in the osmotic signal transduction pathway and result in cell death. Cross-resistance among dicarboximides, phenylpyrroles, and aromatic hydrocarbons has been observed in most laboratory *Botrytis cinerea*-resistant mutants, which are generally hyperosmotic sensitive. However, such resistant strains have rarely been isolated from the fields. All dicarboximide-resistant field isolates contained point mutations in a putative osmosensor histidine kinase *BcOS1/Daf1*, did not show cross-resistance to phenylpyrroles, and were insensitive to osmotic stress. In contrast, *Alternaria* field-resistant strains carried various mutations, including null mutations, in their osmosensor histidine kinase genes. The introduction of several new fungicides against *B. cinerea*, such as anilinopyrimidine fungicides, fenhexamid, QoIs, and succinate dehydrogenase inhibitors, reduced the use of dicarboximides, thereby reducing the populations of dicarboximide-resistant strains. However, several types of multidrug resistance strains, in which efflux pumps are activated, have emerged. Gain-of-function mutations of the transcription factor Mrr1, which leads to an overexpression of the ATP-binding cassette transporter *AtrB*, confers reduced sensitivities to some fungicides, including fludioxonil and cyprodinil. In addition, strains that overexpress the major facilitator superfamily transporter mfsM2 by promoter rearrangements lead to reduced sensitivities to iprodione, fenhexamid, and cyprodinil. Therefore, in addition to target modifications of BcOS1, multidrug resistance caused by the overexpression of drug transporters is another resistance mechanism in *B. cinerea* against dicarboximides and phenylpyrroles.

M. Fujimura (✉) • A. Ichiishi • F. Fukumori
Faculty of Life Sciences, Toyo University, Ora, Gunma 374-0193, Japan
e-mail: mfujimura@toyo.jp

S. Banno
Faculty of Life Sciences, Toyo University, Ora, Gunma 374-0193, Japan

Yamanashi Experiment Station, Japan Plant Protection Association, Minami, Yamanashi 405-0042, Japan

H. Ishii, D.W. Hollomon (eds.), *Fungicide Resistance in Plant Pathogens*,
DOI 10.1007/978-4-431-55642-8_12

Keywords Histidine kinase • MAP kinase • Osmotic stress • Multidrug resistance • ABC transporter • MFS transporter

12.1 Introduction

Leading compounds of dicarboximide fungicides were discovered in the early 1970s by Sumitomo Chemical Co., Ltd. (Fujinami et al. 1971). In the mid-1970s, dicarboximides, such as iprodione, procymidone, vinclozolin, and chlozolinate, were commercially available and widely used to control diseases caused by the taxonomically related pathogens belonging to the genera *Botrytis*, *Sclerotinia*, *Monilinia*, and *Alternaria* (Pommer and Lorenz 1995). Formerly, benzimidazoles were used extensively for preventing grapevine gray mold, but the occurrence of its resistant strains became a serious problem, and, therefore, dicarboximides superseded benzimidazoles. After several years of dicarboximide use in European countries, dicarboximide-resistant *B. cinerea* strains were frequently isolated from crop fields. Notably, *B. cinerea* dicarboximide-resistant strains were often resistant to benzimidazoles (Leroux and Clerjeau 1985; Beever et al. 1989; Locke and Fletcher 1988). Dicarboximide-resistant strains have been found in many areas where the chemicals were used continuously. Most dicarboximide-resistant isolates show moderate levels of resistance, whereas laboratory mutants generally show higher levels. The discontinuance of dicarboximide applications to the field led to a decrease in populations of the resistant strains, suggesting a reduced fitness of dicarboximide-resistant strains in the natural habitat (Raposo et al. 2000; Walker et al. 2013).

Phenylpyrrole fungicides, such as fenpiclonil and fludioxonil, are sported from an antifungal antibiotic pyrrolnitrin that is produced by several *Pseudomonas* spp. (Gehmann et al. 1990). Although phenylpyrroles are not chemically related to dicarboximide, their antifungal spectra are similar (Leroux et al. 1992). Antifungal activities of phenylpyrroles are generally stronger than those of dicarboximides. Therefore, phenylpyrrole fungicides have been applied not only on the foliage but also on the seeds of cereals and fruits postharvest. In most cases, laboratory-isolated *B. cinerea* dicarboximide-resistant mutants showed cross-resistance to phenylpyrroles and aromatic hydrocarbons, such as PCNB and tolclofos-methyl; therefore, these fungicides were considered to have a similar mode of action (Leroux et al. 1992). However, field-isolated dicarboximide-resistant strains did not show cross-resistance to phenylpyrroles. Thus, the Fungicide Resistance Action Committee's (FRAC) estimates of the intrinsic risk for resistance evolution to dicarboximides (FRAC code 2) and phenylpyrroles (FRAC code 12) were medium to high and low to medium, respectively. In fact, there are a limited number of reports of fludioxonil-resistant field isolates.

In this chapter, we focus on the mode of action of these fungicides on the model fungus *Neurospora crassa* and the resistance mechanisms in field isolates of *B. cinerea* and other plant pathogens.

12.2 Mode of Action of Histidine Kinase Inhibitors in *Neurospora crassa* and Other Fungi

The mode of action and the resistance mechanism of dicarboximides and phenylpyrroles were not clear for a long time, despite extensive studies. Although dicarboximides do not affect respiration or primary metabolisms related to DNA and RNA synthesis, the treated *B. cinerea* hyphae immediately stopped elongation, increased the number of septa, and then swelled, which leads to cell burst (Hisada and Kawase 1977 and Hisada et al. 1978). Phenylpyrroles and aromatic hydrocarbons also cause similar morphological alterations (Leroux et al. 1992). Many effects of these chemicals on various cellular functions, such as mitosis, lipid metabolism, and reactive oxygen species production, have been reported (Edlich and Lyr 1995). Dicarboximides could induce the membrane lipid peroxidation by interfering with flavin-containing enzymes. As the antioxidant α-tocopheryl acetate and the cytochrome P-450 inhibitor piperonyl butoxide are known to reduce the fungitoxicity of dicarboximides and phenylpyrroles in *B. cinerea* (Leroux et al. 1992), the peroxidation of the membrane lipid was initially considered the primary mode of action of these fungicides. However, it was determined that treatments of fenpiclonil, a phenylpyrrole, lead to the accumulation of polyols in *Fusarium sulphureum* (Jespers et al. 1993 and Jespers and De Waard 1995) and that fludioxonil induced the glycerol synthesis in *N. crassa* (Pillonel and Meyer 1997).

The mode of action of histidine kinase (HK) inhibitors has been elucidated mainly in the model fungus *N. crassa*. The fungus accumulated glycerol in response to osmotic stress, and several mutants, such as *os-1*, *os-2*, *os-4*, and *os-5*, were isolated as osmotic-sensitive mutants (Mays 1969). These mutants show resistance to dicarboximides, phenylpyrroles, and aromatic hydrocarbons (Beever and Byrde 1982; Grindle and Temple 1982; Fujimura et al. 2000). Fludioxonil was found to inhibit a type III-like protein kinase in *N. crassa* (Pillonel and Meyer 1997). Genetic analyses revealed that the *os-1* gene encodes a group III HK (Schumacher et al. 1997; Alex et al. 1996), and the *os-4*, *os-5*, and *os-2* genes encode a MAPKK kinase, a MAPK kinase, and a MAP kinase, respectively, which forms the MAP kinase cascade (Fujimura et al. 2003; Zhang et al. 2002). The HK and three kinase proteins are components of the osmotic signal transduction (OS) pathway, suggesting that both dicarboximides and phenylpyrroles interfere with the OS pathway.

The OS pathway of *N. crassa* consists of the upstream His-Asp phosphorelay system and the downstream MAP kinase cascade (Fig. 12.1). The His-Asp phosphorelay system (also called a two-component system) is used to sense and respond to changes in many different environmental conditions in various organisms, such as bacteria, fungi, slime molds, and plants, but is absent in mammals

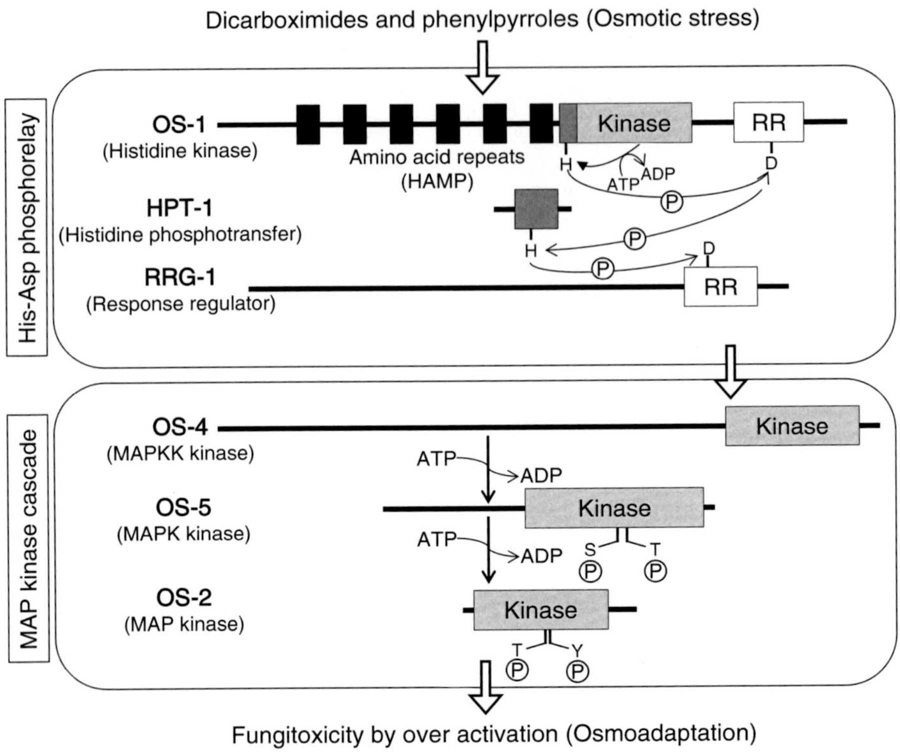

Fig. 12.1 The osmotic signal transduction pathway in *Neurospora crassa*. This pathway consists of the upstream His-Asp phosphorelay system and the downstream MAP kinase cascade. Domain structures shown in boxes are HAMP domain (*black*), histidine phosphotransfer (*dark gray*), kinase (*light gray*), and response regulator (*white*). The amino acids that can be phosphorylated: histidine (H), aspartic acid (D), serine (S), threonine (T), and tyrosine (Y). Null mutants of all components, except *hpt-1*, show high levels of resistance to dicarboximides and phenylpyrroles and increased sensitivities to osmotic stress

(Schaller et al. 2011). In most prokaryote systems, the HK domain and the response regulator domain are located in separate proteins, and in some cases, the response regulator protein has a DNA-binding domain that acts as a transcription factor, allowing the target gene expression level to adapt to environmental change(s). In contrast, eukaryotes often use hybrid HKs that contain the kinase domain and the response regulator domain on one protein. In the *Neurospora* OS pathway, the hybrid HK OS-1 probably acts as an osmosensor and transmits signals to the response regulator RRG-1 *via* the histidine phosphotransfer protein HPT-1. The OS-1 protein, which belongs to the group III HK, has six HAMP (HK-adenylate cyclase-methyl-accepting chemotaxis protein-phosphatase) domains in its N-terminal half region and the response regulator domain in its C-terminal region. The unique six HAMP repeats (approximately 90 amino acids each) probably sense the change in the osmotic condition, and its conformational change regulates the kinase activity in the catalytic domain. When the catalytic kinase domain of OS-1 is activated, OS-1 is autophosphorylated at the conserved histidine residue, and this phosphoryl group is in turn transferred to the aspartate residue of the response regulator domain. The phosphoryl group on the response regulator domain is then shuttled from OS-1 to the conserved histidine residue on HPT-1 and subsequently to the terminal response regulator protein RRG-1. The response regulator RRG-1 regulates the MAP kinase cascade consisting of OS-4, OS-5, and OS-2. OS-2 is phosphorylated by OS-5 *via* OS-4 in response to osmotic stress. Activated OS-2 regulates transcription factors, such as ATF-1, and consequently regulates the target gene expression. The *N. crassa* OS pathway regulates various genes including those for glycerol synthesis (Noguchi et al. 2007; Watanabe et al. 2007; Yamashita et al. 2007, 2008; Kamei et al. 2013). Like *os* mutants, the response regulator *rrg-1* mutant shows both the osmotic sensitivity and the fungicide resistance (Jones et al. 2007). In contrast, the histidine phosphotransferase HPT-1 is essential for cell survival unless any of the downstream MAP kinase genes are defective (Banno et al. 2007). Null mutants of most components of the OS pathway acquire resistance to dicarboximide and phenylpyrrole fungicides. These results imply that these fungicides cause cell death *via* improper activation of the OS-2 MAP kinase in *N. crassa*. In fact, both fungicides stimulate the phosphorylation of the OS-2MAP kinase in the wild-type strain but not in *os-1*, *rrg-1*, *os-4*, and *os-5* mutants (Yoshimi et al. 2005; Noguchi et al. 2007). Constitutive activation of the OS pathway by these fungicides leads to glycerol accumulation and cell swelling without an osmolarity increase (Fujimura et al. 2000). Different from some other site-specific fungicides, such as ergosterol biosynthesis inhibitors, benzimidazoles, QoIs, and succinate dehydrogenase inhibitors (SDHIs), the HK inhibitors are unique because they act as the activators of signal transduction pathways. Thus, any loss-of-function mutations in *os-1*, *rrg-1*, *os-4*, *os-5*, or *os-2* genes in the OS pathway confer resistance to these fungicides. This is probably a reason that the resistant mutants have been easily obtained in laboratory conditions in various fungi and that most of them show the osmotic sensitivity.

The whole-genome analyses in fungi revealed that HKs could be categorized into 11 classes (Catlett et al. 2003). *N. crassa* has 11 HK genes, but yeasts have smaller numbers of HKs; only one HK (Sln1) in *Saccharomyces cerevisiae*, three in *Schizosaccharomyces pombe*, and three in *Candida albicans*. Among these yeasts, *S. cerevisiae* and *S. pombe*, which lack group III HKs, are insensitive to dicarboximide and phenylpyrrole (Okada et al. 2005). In contrast, *C. albicans* has a group III HK, Canik1, and is sensitive to these fungicides. However, deletion mutants in this gene are resistant to them (Ochiai et al. 2002). The high-osmolarity glycerol response (HOG) pathway of *S. cerevisiae*, which consists of the His-Asp system (Sln1-Ypd1-Ssk1) and the MAK kinase cascade (Ssk2/22-Pbs2-Hog1), is the most well-characterized signal transduction pathway in fungi (Maeda et al. 1994). The HOG pathway is essentially very similar to the OS pathway of *N. crassa*. The osmotic-sensitive phenotype of *hog1* mutants in *S. cerevisiae* can be complemented by the *N. crassa os-2* gene, indicating the functional similarities of these two pathways. The deletion of the osmosensor *SLN1* gene of *S. cerevisiae* is lethal owing to an improper activation of the HOG pathway (Maeda et al. 1994). Unlike *S. cerevisiae*, *SLN1* orthologs are commonly dispensable in filamentous fungi (Furukawa et al. 2002), probably because of the presence of multiple HK genes. OS-1 HK (group III) is structurally very different from Sln1 (group VI) in the N-terminal region. Sln1 has two transmembrane domains that are responsible for its membrane localization. In contrast, as OS-1 homologs lack the transmembrane domain, they are predicted to be soluble proteins. Moreover, they have a unique N-terminal region consisting of HAMP repeats. The involvement of group III HKs in sensitivities to dicarboximides and phenylpyrroles has been confirmed by the fact that their heterozygous expression, such as *Magnaporthe grisea* Hik1, *Alternaria brassicicola* AbNik1, and *C. albicans* CaNik1, in *S. cerevisiae* conferred fungicide sensitivity to the host cells (Motoyama et al. 2005b; Dongo et al. 2009; Meena et al. 2010).

The OS pathway has also been characterized in various fungi, including *Fusarium graminearum* (Ochiai et al. 2007; Van Thuat et al. 2012), *M. grisea* (Motoyama et al. 2005a), *Aspergillus nidulans* (Furukawa et al. 2005; Abe et al. 2009), *Cochliobolus heterostrophus* (Yoshimi et al. 2004 and 2005; Izumitsu et al. 2007 and 2009), *Colletotrichum lagenarium* (Kojima et al. 2004), and *B. cinerea* (Segmuller et al. 2007; Liu et al. 2008). Although the function and the regulation of OS pathways are diverse among fungi, this pathway is essential for susceptibility to dicarboximides and phenylpyrroles in plant pathogens. For example, deletion mutants of *F. graminearum*, a causal agent of head blight of wheat, show quite similar phenotypes to those of *N. crassa*: *FgOS1* (HK), *FgOS4* (MAPKK kinase), *FgOS5* (MAPK kinase), and *FgOS2* (MAP kinase) mutants show the osmotic sensitivity and fludioxonil resistance (Ochiai et al. 2007). Interestingly, this signal transduction pathway is also involved in the regulation of mycotoxin production in *F. graminearum*, suggesting various roles of the OS pathway in filamentous fungi.

Another type of signaling pathway is involved in the fungicide susceptibility of the phytopathogenic fungus *Ustilago maydis*. A serine (threonine) protein kinase and a cAMP signal transduction pathway mediate resistance to dicarboximide and aromatic hydrocarbon fungicides in *U. maydis* (Orth et al. 1995; Ramesh et al. 2001).

12.3 Resistance Mechanism of Histidine Kinase Inhibitors in Plant Pathogens

12.3.1 **Botrytis cinerea** *(Teleomorph* **Botryotinia fuckeliana***)*

Although, resistance mechanisms against HK inhibitors have been studied for various fungi, the most attention has been paid to that of *B. cinerea* because of its agricultural and economic importance. Dicarboximide fungicides have been used to protect various crops from gray mold, which is caused by the fungus. Dicarboximide-resistant *B. cinerea* isolates have spread worldwide, and genetic analyses have demonstrated that the resistance is conferred by changes in a single locus, designated *Daf1* (Faretra and Pollastro 1991). A sequence analysis of the *BcOS1* gene, which encodes an OS-1-like group III HK, revealed that the dicarboximide-resistant phenotype was associated with amino acid substitutions in the BcOS1 protein (Leroux et al. 2002; Cui et al. 2004; Oshima et al. 2002, 2006), indicating that *Daf1* is the same as *BcOS1* (also called *Bos1*). Similar to other group III HKs, BcOS1 has six HAMP domains, an HK domain, and a response regulator domain. Although several SNPs are found in the *BcOS1* gene, initial studies revealed that five types of point mutations in two specific amino acid residues conferred resistance to dicarboximides (Table 12.1). Among them, three types of resistant strains had substitutions of

Table 12.1 Mutations associated with dicarboximide and phenylpyrrole resistance in *Botrytis* and other plant pathogens

Mutation	Position[a]	Dc[b]	PP[b]	OS[c]	References[d]
Botrytis cinerea					
I365S	2nd AA	MR	S	–	1, 2, 3, 4, 5, 6
I365N	2nd AA	MR	S	–	3, 4, 5, 6
I365R	2nd AA	MR	S	–	3, 4
I365S and Q369P	2nd AA	MR	S	–	3, 4, 5, 6
Q369P and N373S	2nd AA	MR	S	–	2, 4
V368F, Q369H, T447S	2nd + 3rd AA	MR	S	–	2
Knockout[e]	–	HR	HR	S	7
Alternaria alternata (AaHK1)					
M246 (11 bp insertion)	1st AA	HR	HR	MS	8
A642 (4 bp deletion)	5th AA	HR	HR	MS	8
Alternaria brassicicola (AbNIK1)					
M495 (flame shift)	4th AA	HR	HR	S	9
W634*	5th AA	HR	HR	S	9
E753K	H-Kinase	HR	HR	MS	9
Q988*	H-Kinase	HR	HR	S	9
Knockout[e]	–	HR	HR	S	10
Alternaria longipes (AlHK1)					
G420D	3rd AA	HR	HR	S	11
582–689 (deletion)	5th–6thAA	HR	HR	S	11
Knockout[e]	–	HR	HR	S	12
Stemphylium vesicarium (SvHK1)					
F267L	1st AA	LR	S	–	13
L290S	1st AA	MR	S	–	13
T765R	H-kinase	HR	HR	–	13
Q777R	H-kinase	HR	HR	–	13

[a]Position of resistant mutation in *BcOS1*: *AA* amino acid repeat domain, *H-kinase* histidine kinase domain

[b]Sensitivity to dicarboximides (Dc) and phenylpyrroles (PP): *HR* highly resistant, *MR* moderately resistant, *LR* low resistant, *S* sensitive

[c]Osmotic sensitivity: *S* sensitive, *MS* moderately sensitive, – normal

[d]1 Oshima et al. (2002); 2 Oshima et al. (2006); 3 Leroux et al. (2002); 4 Cui et al. (2004); 5 Ma et al. (2007); 6 Grabke et al. (2014); 7 Viaud et al. (2006); 8 Dry et al. (2004); 9 Avenot et al. (2005); 10 Iacomi-Vasilescu et al. (2008); 11 Luo et al. (2008); 12 Luo et al. (2013); 13 Alberoni et al. (2010)

[e]Knockout: null mutant by gene disruption

the isoleucine at BcOS1 position 365 to serine (I365S), asparagine (I365N), or arginine (I365R). The other two resistant strains had an amino acid change of the glutamine at position 369 to proline (Q369P) or histidine (Q369H). Resistant strains with BcOS1^{I365S} are frequently isolated worldwide, which suggests that it should be the major dicarboximide-resistant mutation. The strains were isolated from various Japanese fields; however, mutations with either Q369P or Q369H were isolated from restricted regions, and no isolates with mutations I365N or I365R were identified (Oshima et al. 2006). In New Zealand, most resistant strains have BcOS1^{I365S}, but some strains have other mutations, namely, I365N or I365R. Recently, Grabke et al. (2014) reported that dicarboximide-resistant isolates from North Carolina and South Carolina blackberry fields possessed three types of amino acid changes in BcOS1 (I365S, I365N, and Q369P+N373S). In other words, dicarboximide-resistant mutations in field isolates could be restricted to two amino acid positions in BcOS1, I365, and Q369. These dicarboximide-resistant mutations are located in the less conserved terminal region of the second repeat in BcOS1 HK (Fig. 12.2). These amino acid changes confer moderate resistance to dicarboximides but do not associate with fludioxonil resistance and/or the osmotic sensitivity phenotypes. *B. cinerea* strains with BcOS1 amino acid changes other than these two might not tolerate the environmental pressure. Laboratory-induced dicarboximide-resistant mutations are also located within the *BcOS1* gene (Cui et al. 2002; Fillinger et al. 2012), and resistant mutations are found, in most cases, in the second repeat region of BcOS1 but in other repeats as well (Table 12.1). Most show higher levels of resistance to dicarboximides and phenylpyrroles and the osmotic sensitivity. Like some other filamentous fungi, BcOS1 has six repeats of ca. 90 amino acids, each of which contains a HAMP domain (50 amino acids) (Fig. 12.2). The HAMP motif has two amphiphilic helices (AS1 and AS2) joined by a nonhelical connector segment. Each of the two helices that compose a typical heptad repeat with hydrophobic residues is considered to be facing each other (Fillinger et al. 2012). In addition, HAMP domains have been shown to interact with each other to form a dimer. The major dicarboximide-resistant mutations, I365S and Q369P, are located at the terminal region of the second HAMP domain. In contrast, amino acid changes within the HAMP domain that affect helix structure confer higher resistance to dicarboximides

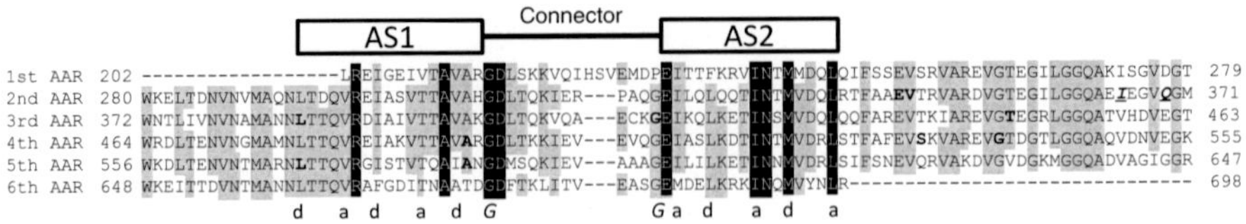

Fig. 12.2 Multiple alignments of amino acid repeat domains of the BcOS1 histidine kinase of *Botrytis cinerea*. Portions of the typical two amphipathic helices (AS1 and AS2) of the HAMP structure and the connector are indicated above the amino acid sequence. Hydrophobic residues of the helices are denoted by letter "a" and "d" to show their heptad (a–g) periodicity. "*G*" corresponds to the conserved glycine residues of the connector motif. Sites of amino acid substitutions found in laboratory-induced resistant mutants are written in boldface, and those found in field-resistant isolates are shown in boldface and italics with underline (I365 and Q369)

and phenylpyrroles associated with the osmotic sensitivity, similar to phenotypes of *BcOS1* deletion mutants (Viaud et al. 2006; Fillinger et al. 2012). Although the function of the HAMP domain remains unclear, these results suggest that the coiled-coil structure of the HAMP domain repeats might be important for the regulation of the kinase catalytic activity of BcOS1. More recently, a yeast two-hybrid assay of the HAMP domains present in group III HKs in the osmotolerant and halotolerant yeast *Debaryomyces hansenii* revealed that each HAMP domain selectively interacts with other HAMP domains (Meena et al. 2010). It is proposed that the fourth HAMP domain, which strongly interacts with the fifth HAMP domain, is the most important, and alternative interaction among the HAMP domains regulates the activity of DhNik1p, similar to an "on-off switch" for osmosensing by DhNik1p.

12.3.2 **Alternaria** ***spp.***

Resistance mechanisms against HK inhibitors have also been studied in some *Alternaria* spp., such as *Alternaria alternata* (Dry et al. 2004), *A. brassicicola* (Avenot et al. 2005), and *A. longipes* (Luo et al. 2008, 2013). *A. alternata* causes leaf spots and blights on a wide range of host plants, including cotton, sunflower, kiwifruit, citrus, tomato, bean, cherry, and apple. The development of iprodione-resistant *A. alternata* strains grown on several crops, such as passionfruit, tangelos, cherries, and apples, has been reported. Dicarboximide fungicides have been used on passionfruit fields to control *A. alternata*, which causes leaf and fruit spots that result in premature defoliation and fruit drop; however, within 5 years of its use, the development of resistant strains had significantly reduced the effectiveness of dicarboximide fungicides in Australia (Hutton 1988). These strains are highly resistant to dicarboximides and phenylpyrroles and moderately sensitive to osmotic stress. A sequence analysis of the group III HK *AaHK1* genes of two representative resistant isolates indicated that, distinct from *B. cinerea*, they contained null mutations: one had an insertion (11 bp) and the other had a deletion (4 bp) in the HK gene, leading to premature termination of the HK protein (Dry et al. 2004). These field-resistant strains are pathogenic to detached passionfruit, suggesting that the lack of AaHK is not essential for virulence in *A. alternata.*

A. brassicicola is the causal agent of brassica dark leaf spot on most *Brassica* species, including cabbage, broccoli, canola, and mustard. This seed-borne fungal pathogen is responsible for considerable financial losses to growers. *A. brassicicola* dicarboximide-resistant field strains were isolated from *Raphanus sativus* and *Brassica oleracea* in France and Romania (Avenot et al. 2005). They were also highly resistant to dicarboximides and phenylpyrroles and sensitive to osmotic stress. Various mutations have been found in its group III HK AbNIK1 (Table 12.1); nonsense mutations within the fifth AA repeat and HK domain and a frameshift at the fourth AA repeat domain both lead to the impaired translation of HK (Avenot et al. 2005). A missense mutation, which causes a single amino acid substitution, E753K, which is located in the H-box of the HK domain, may also result in the loss of HK activity. These results indicate that most field-resistant *A. brassicicola* strains lack the AbNIK1 function, whereas some resistant strains have no mutations within the *AbNIK1* gene. These AbNIK1-null mutants have no defects in mycelial growth, sporulation, or conidial germination under standard conditions and are infectious to radish. However, they show significant growth defects under osmotic stress and were weak in competitiveness (Iacomi-Vasilescu et al. 2008).

Studies on the HK inhibitor-resistant field mutants of *A. longipes*, a pathogen that causes tobacco brown spot disease, have shown that they are either deletion mutants of a group III HK gene, *AlHK1*, or single amino acid change (G420D) mutants in the third amino acid repeat of AlHK1. Both of these field isolates, as well as *AlHK1* deletion mutants by homologous recombination, were highly resistant to dicarboximides and phenylpyrroles and were sensitive to osmotic stress (Luo et al. 2008; 2013). Like other *Alternaria* spp. field HK inhibitor-resistant strains, *AlHK1-null* mutants show osmotic sensitivity. Interestingly, despite being osmotic sensitive, these null mutants had a stronger pathogenicity to the host plant than their parental strain, suggesting AlHK1 is a negative factor of virulence in *A. longipes.*

12.3.3 Stemphylium vesicarium

Since the late 1970s, brown spot on pear, which is caused by *S. vesicarium* (teleomorph *Pleospora allii*), has been one of the most important fungal diseases in Europe. Three types of dicarboximide-resistant strains (R1, R2, and S+) have been isolated from the fields (Alberoni et al. 2005; 2010). The most frequent phenotype, R1, is highly resistant to procymidone and moderately resistant to iprodione, vinclozolin, and chlozolinate. In contrast, the R2 phenotype (highly resistant to all dicarboximides) and S+ phenotype (slightly resistant to procymidone and iprodione) are very rare. Cross-resistance between dicarboximides and phenylpyrroles is observed only in the R2 phenotype strains. Amino acid substitutions, L290S (R1) and F267L (S+) in the first amino acid repeat of the group III HK, have been identified. For the R2 isolates, the substitutions T765R or Q777R were located within the HK domain (Alberoni et al. 2010).

12.3.4 Others

Cochliobolus heterostrophus dicarboximide-resistant mutants isolated under laboratory conditions have been characterized in detail (Yoshimi et al. 2004; Izumitsu et al. 2009). Three genes, *Dic1*, *Dic2*, and *Dic3*, that confer osmotic adaptation and fungicidal sensitivity were identified. *Dic1* and *Dic2* encode a group III HK and Skn7-response regulator, respectively. Although the *Dic1* phenotype is pleiotropic, null mutants and some strains with missense mutations in the N-terminal repeat region are highly resistant to dicarboximides and phenylpyrroles and highly sensitive to osmotic stress. In contrast, a single amino acid change within the kinase domain, or the regulator domain, altered the sensitivity to osmotic stress and conferred moderate resistance to the fungicides. Although the *Dic3* gene has not been identified yet, ChSsk1 (an RRG-1-like response regulator), BmSsk2 (OS-4-like MAPKK kinase), BmPbs2 (OS-5-like MAPK kinase), and BmHog1 (OS-2-like MAP kinase) are involved in the fungicidal sensitivity of *C. heterostrophus* (Yoshimi et al. 2004; Izumitsu et al. 2009). Additionally, null mutants of group III HK genes have been isolated and characterized in several plant pathogens, such as *Monilinia fructicola* (Ma et al. 2006), *Sclerotinia sclerotiorum* (Duan et al. 2013), and *M. grisea* (Motoyama et al. 2005a). The phenotypes of fungicide resistance and osmotic sensitivity are common in various fungi; however, the effects of group III HK deletions on pathogenicity to the host plants may differ in each fungus. As BcOS1 is important for the normal macroconidiation and full virulence in *B. cinerea,* null mutants were significantly reduced in their ability to infect host plants (Viaud et al. 2006). *S. sclerotiorum*, the Shk1 HK mutant, showed a significant reduction in vegetative hyphal growth and was unable to produce sclerotia (Duan et al. 2013). In contrast, the disruption of *HIK1* in *M. grisea* caused no defect in cell growth on normal media and in pathogenicity to rice plants (Motoyama et al. 2005a). In *A. longipes*, AlNik1 HK negatively regulates pathogenicity because its deletion mutants showed stronger pathogenicity than the parental strains (Luo et al. 2013). These differences are reflected in the fitness cost of field-resistant strains.

12.4 Multidrug Resistances in *B. cinerea*

Many different types of mechanisms can contribute to the development of resistance to drug chemicals: modification of the target site, upregulation of the target protein, enzyme-catalyzed drug degradation, development of alternative pathways, and overexpression of efflux pumps. Among them, target-site mutations, which might reduce the binding affinity of the fungicide, is the major resistance mechanism to agricultural fungicides. For example, specific mutations in β-tubulin, cytochrome *b*, and succinate dehydrogenase confer resistance to benzimidazoles, QoIs, and SDHIs, respectively, in *B. cinerea* and other plant pathogens. As described above, the major dicarboximide and/or phenylpyrrole resistance mechanism in plant pathogens is the modification of group III HK. However, a recent survey revealed that the multiple drug resistance (MDR) caused by the increase of drug efflux activity is involved in fungal resistance to HK inhibitors in field isolates (Kretschmer et al. 2009; Mernke et al. 2011). In general, target-site modifications lead to higher levels of specific resistance to a single class of fungicides. In contrast, the MDR mediated by active efflux pumps, such as ATP-binding cassettes (ABCs) and major facilitator superfamily (MFS) transporters, is lower than that in the target modifications; however, it shows a broad-range resistance to structurally unrelated chemicals. *B. cinerea* populations with MDR phenotypes of resistance to chemically unrelated fungicides have been observed at high frequencies in French and German vineyards (Mernke et al. 2011; Walker et al. 2013).

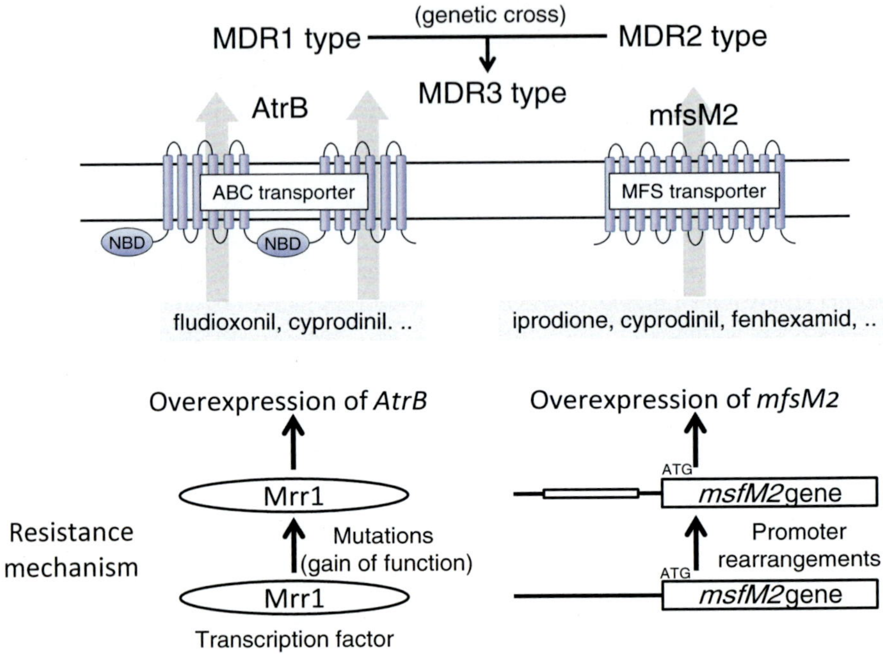

Fig. 12.3 Molecular and genetic mechanisms of multidrug resistance (MDR) in *Botrytis cinerea*. Three multidrug resistance phenotypes, MDR1, MDR2, and MDR3, appear in the European *B. cinerea* population. Mutations in the transcription factor gene *mrr1* lead to an overexpression of the ATP-binding cassette (ABC) transporter gene *atrB* in MDR1 strains. In contrast, the overexpression of the major facilitator superfamily (MFS) transporter gene *mfsM2*, caused by the unique promoter rearrangement, is responsible for the MDR2 phenotype. MDR3 strains, which may be yielded by crosses between MDR1 and MDR2 strains, have both mutations

B. cinerea MDR strains can be classified into three groups, MDR1, MDR2, and MDR3, according to their phenotypes (Fig. 12.3). MDR1 and MDR2 strains were originally identified as AniR2 and AniR3 strains, respectively, because of their reduced sensitivity to anilinopyrimidines, such as cyprodinil, pyrimethanil, and mepanipyrim (Chapeland et al. 1999; Leroux et al. 1999). MDR1 strains show partial resistance to fludioxonil, cyprodinil, and tolnaftate. MDR2 strains show partial resistance to iprodione, cyprodinil, fenhexamid, tolnaftate, and cycloheximide. However, tolnaftate and cycloheximide have not been used in agriculture. MDR3 strains have spectra with the combined fungicide resistance of MDR1 and MDR2. MDR1 phenotypes are correlated with the gain-of-function point mutations in the Zn(II)2Cys6 zinc cluster transcription factor Mrr1, leading to the overexpression of an ABC transporter, AtrB. The expression of *AtrB* was induced by fludioxonil in fludioxonil-sensitive strains, but MDR1 strains constitutively overexpressed *AtrB* (Kretschmer et al. 2009). Deletion mutants of the *Mrr1* gene lacked the ability to induce the *AtrB* gene and showed increased sensitivities to fludioxonil and cyprodinil. In contrast, MDR2 strains constitutively overexpressed another MSF transporter gene, *mfsM2* (major facilitator superfamily transporter involved in MDR2), which is caused by a unique promoter rearrangement in the *mfsM2* gene (Mernke et al. 2011). A sensitive strain, in which *mfsM2* was artificially overexpressed, showed an MDR2-like phenotype, in contrast, a deletion of the *mfsM2* gene in an MDR2 strain resulted in a lost MDR2 phenotype. Both mutations found in MRD1 strains and MDR2 strains were detected in MDR3 strains, meaning the constitutive overexpression of two types of transporter *AtrB* and *mfsM2* genes.

More recently, a stronger MDR1 phenotype strain, MDR1h, has become widespread in German strawberry fields (Leroch et al. 2013). MDR1h shows more resistance to fludioxonil and cyprodinil than the MDR1 strains. The MDR1h phenotype is correlated with a higher expression level of the *AtrB* gene. A 3-bp deletion mutation, causing ΔL497 in transcription factor-Mrr1, is correlated with MDR1h. In the high-risk pathogen *Botrytis*, various fungicides are used repeatedly in rotation to control gray mold. These situations may result in the development of multidrug resistance populations in *B. cinerea*.

References

Abe K, Furukawa K, Fujioka T, Hagiwara D, Maeda H, Marui J, Mizutani O, Takahashi T, Yoshimi A, Yamagata Y, Gomi K, Hasegawa F (2009) Novel industrial applications of *Aspergillus oryzae* genomics. In: Machida M, Gomi K (eds) *Aspergillus*: molecular biology and genomics. Caister Academic Press, Washington, DC, pp 199–227

Alberoni G, Collina M, Pancaldi D, Brunelli A (2005) Resistance to dicarboximide fungicides in *Stemphylium vesicarium* of Italian pear orchards. Eur J Plant Pathol 113(2):211–219

Alberoni G, Collina M, Lanen C, Leroux P, Brunelli A (2010) Field strains of *Stemphylium vesicarium* with a resistance to dicarboximide fungicides correlated with changes in a two-component histidine kinase. Eur J Plant Pathol 128:171–184

Alex LA, Borkovich KA, Simon MI (1996) Hyphal development in *Neurospora crassa*: involvement of a two-component histidine kinase. Proc Natl Acad Sci U S A 93(8):3416–3421

Avenot H, Simoneau P, Iacomi-Vasilescu B, Bataille-Simoneau N (2005) Characterization of mutations in the two-component histidine kinase gene *AbNIK1* from *Alternaria brassicicola* that confer high dicarboximide and phenylpyrrole resistance. Curr Genet 47(4):234–243

Banno S, Noguchi R, Yamashita K, Fukumori F, Kimura M, Yamaguchi I, Fujimura M (2007) Roles of putative His-to-Asp signaling modules HPT-1 and RRG-2, on viability and sensitivity to osmotic and oxidative stresses in *Neurospora crassa*. Curr Genet 51(3):197–208

Beever RE, Byrde RJW (1982) Resistance to the dicarboximide fungicides. In: Dekker J, Georgopoulos SG (eds) Fungicide resistance in crop protection. Pudoc, Wageningen, the Netherlands, pp 101–117

Beever RE, Laracy EP, Pak HA (1989) Strains of *Botrytis cinerea* resistant to dicarboximide and benzimidazole fungicides in New Zealand vineyards. Plant Pathol 38(3):427–437

Catlett NL, Yoder OC, Turgeon BG (2003) Whole-genome analysis of two-component signal transduction genes in fungal pathogens. Eukaryot Cell 2(6):1151–1161

Chapeland F, Fritz R, Lanen C, Gredt M, Leroux P (1999) Inheritance and mechanisms of resistance to anilinopyrimidine fungicides in *Botrytis cinerea* (*Botryotinia fuckeliana*). Pestic Biochem Physiol 64(2):85–100

Cui W, Beever RE, Parkes SL, Weeds PL, Templeton MD (2002) An osmosensing histidine kinase mediates dicarboximide fungicide resistance in *Botryotinia fuckeliana* (*Botrytis cinerea*). Fungal Genet Biol 36(3):187–198

Cui W, Beever RE, Parkes SL, Templeton MD (2004) Evolution of an osmosensing histidine kinase in field strains of *Botryotinia fuckeliana* (*Botrytis cinerea*) in response to dicarboximide fungicide usage. Phytopathology 94(10):1129–1135

Dongo A, Bataille-Simoneau N, Campion C, Guillemette T, Hamon B, Iacomi-Vasilescu B, Katz L, Simoneau P (2009) The group III two-component histidine kinase of filamentous fungi is involved in the fungicidal activity of the bacterial polyketide ambruticin. Appl Environ Microbiol 75(1):127–134

Dry IB, Yuan KH, Hutton DG (2004) Dicarboximide resistance in field isolates of *Alternaria alternata* is mediated by a mutation in a two-component histidine kinase gene. Fungal Genet Biol 41(1):102–108

Duan Y, Ge C, Liu S, Wang J, Zhou M (2013) A two-component histidine kinase Shk1 controls stress response, sclerotial formation and fungicide resistance in *Sclerotinia sclerotiorum*. Mol Plant Pathol 14(7):708–718

Edlich W, Lyr H (1995) Mechanism of action of dicarboximide fungicides. In: Lyr H (ed) Modern selective fungicides: properties, applications, and mechanisms of action. Spektrum Akademischer Verlag, New York, pp 119–131

Faretra F, Pollastro S (1991) Genetic basis of resistance to benzimidazole and dicarboximide fungicides in *Botryotinia fuckeliana* (*Botrytis cinerea*). Mycol Res 95(8):943–951

Fillinger S, Ajouz S, Nicot PC, Leroux P, Bardin M (2012) Functional and structural comparison of pyrrolnitrin- and iprodione-induced modifications in the class III histidine-kinase Bos1 of *Botrytis cinerea*. PLoS One 7(8), e42520

Fujimura M, Ochiai N, Ichiishi A, Usami R, Horikoshi K, Yamaguchi I (2000) Sensitivity to phenylpyrrole fungicides and abnormal glycerol accumulation in *os* and *cut* mutant strains of *Neurospora crassa*. J Pestic Sci 25:31–35

Fujimura M, Ochiai N, Oshima M, Motoyama T, Ichiishi A, Usami R, Horikoshi K, Yamaguchi I (2003) Putative homologs of SSK22 MAPKK kinase and PBS2 MAPK kinase of *Saccharomyces cerevisiae* encoded by *os-4* and *os-5* genes for osmotic sensitivity and fungicide resistance in *Neurospora crassa*. Biosci Biotechnol Biochem 67(1):186–191

Fujinami A, Ozaki T, Nodera K, Tanaka K (1971) Studies on biological activity of cyclic imide compounds. Part II. Antimicrobial activity of 1-phenylpyrrolidine-2,5-diones and related compounds. Agric Biol Chem 35:1707–1719

Furukawa K, Katsuno Y, Urao T, Yabe T, Yamada-Okabe T, Yamada-Okabe H, Yamagata Y, Abe K, Nakajima T (2002) Isolation and functional analysis of a gene, *tcsB*, encoding a transmembrane hybrid-type histidine kinase from *Aspergillus nidulans*. Appl Environ Microbiol 68(11):5304–5310

Furukawa K, Hoshi Y, Maeda T, Nakajima T, Abe K (2005) *Aspergillus nidulans* HOG pathway is activated only by two-component signalling pathway in response to osmotic stress. Mol Microbiol 56(5):1246–1261

Gehmann K, Nyfeler R, Leadbeater A, Nevill D, Sozzi D (1990) CGA 173506: a new phenylpyrrole fungicide for broad-spectrum disease control. In: Proceedings of the Brighton crop protection conference pests diseases 2. British Crop Protection Council, Farnham, pp 339–406

Grabke A, Fernandez-Ortuno D, Amiri A, Li X, Peres NA, Smith P, Schnabel G (2014) Characterization of iprodione resistance in *Botrytis cinerea* from strawberry and blackberry. Phytopathology 104(4):396–402

Grindle M, Temple W (1982) Fungicide resistance of *os* mutants of *Neurospora crassa*. Neurospora Newsl 29:16–17

Hisada Y, Kawase Y (1977) Morphological studies of antifungal action of N-(3′, 5′-dichlorophenyl)-1, 2-dimethylcyclopropane-1, 2-dicarboximide on *Botrytis cinerea*. Ann Phytopathol Soc Jpn 43:151–158

Hisada Y, Kato T, Kawase Y (1978) Mechanism of antifungal action of procymidone in *Botrytis cinerea*. Ann Phytopathol Soc Jpn 44(4):509–518

Hutton DG (1988) The appearance of dicarboximide resistance in *Alternaria alternata* in passionfruit in south-east Queensland. Australas Plant Pathol 17(2):34–36

Iacomi-Vasilescu B, Bataille-Simoneau N, Campion C, Dongo A, Laurent E, Serandat I, Hamon B, Simoneau P (2008) Effect of null mutations in the *AbNIK1* gene on saprophytic and parasitic fitness of *Alternaria brassicicola* isolates highly resistant to dicarboximide fungicides. Plant Pathol 57:937–947

Izumitsu K, Yoshimi A, Tanaka C (2007) Two-component response regulators Ssk1p and Skn7p additively regulate high-osmolarity adaptation and fungicide sensitivity in *Cochliobolus heterostrophus*. Eukaryot Cell 6(2):171–181

Izumitsu K, Yoshimi A, Hamada S, Morita A, Saitoh Y, Tanaka C (2009) *Dic2* and *Dic3* loci confer osmotic adaptation and fungicidal sensitivity independent of the HOG pathway in *Cochliobolus heterostrophus*. Mycol Res 113(Pt 10):1208–1215

Jespers ABK, De Waard MA (1995) Effect of fenpiclonil on phosphorylation of glucose in *Fusarium sulphureum*. Pestic Sci 44(2):167–175

Jespers ABK, Davidse LC, Dewaard MA (1993) Biochemical effects of the phenylpyrrole fungicide fenpiclonil in *Fusarium sulphureum* (Schlecht). Pestic Biochem Physiol 45(2):116–129

Jones CA, Greer-Phillips SE, Borkovich KA (2007) The response regulator RRG-1 functions upstream of a mitogen-activated protein kinase pathway impacting asexual development, female fertility, osmotic stress, and fungicide resistance in *Neurospora crassa*. Mol Biol Cell 18(6):2123–2136

Kamei M, Takahashi M, Banno S, Ichiishi A, Fukumori F, Fujimura M (2013) Regulation and physiological function of MAP kinase and cAMP-PKA pathways. In: Kasbekar D, McCluskey K (eds) Neurospora: genomics and molecular biology. Horizon Scientific Press/Caister Academic Press, Norwich, pp 171–192

Kojima K, Takano Y, Yoshimi A, Tanaka C, Kikuchi T, Okuno T (2004) Fungicide activity through activation of a fungal signalling pathway. Mol Microbiol 53(6):1785–1796

Kretschmer M, Leroch M, Mosbach A, Walker AS, Fillinger S, Mernke D, Schoonbeek HJ, Pradier JM, Leroux P, De Waard MA, Hahn M (2009) Fungicide-driven evolution and molecular basis of multidrug resistance in field populations of the grey mould fungus *Botrytis cinerea*. PLoS Pathog 5(12):e1000696

Leroch M, Plesken C, Weber RW, Kauff F, Scalliet G, Hahn M (2013) Gray mold populations in German strawberry fields are resistant to multiple fungicides and dominated by a novel clade closely related to *Botrytis cinerea*. Appl Environ Microbiol 79(1):159–167

Leroux P, Clerjeau M (1985) Resistance of *Botrytis cinerea* Pers. and *Plasmopara viticola* (Berk. & Curt.) Berl. and de Toni to fungicides in French vineyards. Crop Prot 4:137–160

Leroux P, Lanen C, Fritz R (1992) Similarities in the antifungal activities of fenpiclonil, iprodione and tolclofos-methyl against *Botrytis cinerea* and *Fusarium nivale*. Pestic Sci 36:255–261

Leroux P, Chapeland F, Desbrosses D, Gredt M (1999) Patterns of cross-resistance to fungicides in *Botryotinia fuckeliana* (*Botrytis cinerea*) isolates from French vineyards. Crop Prot 18(10):687–697

Leroux P, Fritz R, Debieu D, Albertini C, Lanen C, Bach J, Gredt M, Chapeland F (2002) Mechanisms of resistance to fungicides in field strains of *Botrytis cinerea*. Pest Manag Sci 58(9):876–888

Liu W, Leroux P, Fillinger S (2008) The HOG1-like MAP kinase Sak1 of *Botrytis cinerea* is negatively regulated by the upstream histidine kinase Bos1 and is not involved in dicarboximide- and phenylpyrrole-resistance. Fungal Genet Biol 45(7):1062–1074

Locke T, Fletcher JT (1988) Incidence of benomyl and iprodione resistance in isolates of *Botrytis cinerea* in tomato crops in England and Wales in 1986. Plant Pathol 37(3):381–384

Luo Y, Yang J, Zhu M, Yan J, Mo M, Zhqng K (2008) Characterization of mutations in *AlHK1* gene from *Alternaria longipes*: implication of limited function of two-component histidine kinase on conferring dicarboximide resistance. J Microbiol Biotechnol 18(1):15–22

Luo YY, Yang JK, Zhu ML, Liu CJ, Li HY, Lu ZB, Pan WZ, Zhang ZH, Bi W, Zhang KQ (2013) The group III two-component histidine kinase AlHK1 is involved in fungicides resistance, osmosensitivity, spore production and impacts negatively pathogenicity in *Alternaria longipes*. Curr Microbiol 64(5):449–456

Ma Z, Luo Y, Michailides T (2006) Molecular characterization of the two-component histidine kinase gene from *Monilinia fructicola*. Pest Manag Sci 62(10):991–998

Ma Z, Yan L, Luo Y, Michailides TJ (2007) Sequence variation in the two-component histidine kinase gene of *Botrytis cinerea* associated with resistance to dicarboximide fungicides. Pestic Biochem Physiol 88(3):300–306

Maeda T, Wurgler-Murphy SM, Saito H (1994) A two-component system that regulates an osmosensing MAP kinase cascade in yeast. Nature 369(6477):242–245

Mays L (1969) Isolation, characterization, and genetic analysis of osmotic mutants of *Neurospora crassa*. Genetics 63:781–794

Meena N, Kaur H, Mondal AK (2010) Interactions among HAMP domain repeats act as an osmosensing molecular switch in group III hybrid histidine kinases from fungi. J Biol Chem 285(16):12121–12132

Mernke D, Dahm S, Walker AS, Laleve A, Fillinger S, Leroch M, Hahn M (2011) Two promoter rearrangements in a drug efflux transporter gene are responsible for the appearance and spread of multidrug resistance phenotype MDR2 in *Botrytis cinerea* isolates in French and German vineyards. Phytopathology 101(10):1176–1183

Motoyama T, Kadokura K, Ohira T, Ichiishi A, Fujimura M, Yamaguchi I, Kudo T (2005a) A two-component histidine kinase of the rice blast fungus is involved in osmotic stress response and fungicide action. Fungal Genet Biol 42(3):200–212

Motoyama T, Ohira T, Kadokura K, Ichiishi A, Fujimura M, Yamaguchi I, Kudo T (2005b) An Os-1 family histidine kinase from a filamentous fungus confers fungicide-sensitivity to yeast. Curr Genet 47(5):298–306

Noguchi R, Banno S, Ichikawa R, Fukumori F, Ichiishi A, Kimura M, Yamaguchi I, Fujimura M (2007) Identification of OS-2 MAP kinase-dependent genes induced in response to osmotic stress, antifungal agent fludioxonil, and heat shock in *Neurospora crassa*. Fungal Genet Biol 44(3):208–218

Ochiai N, Fujimura M, Oshima M, Motoyama T, Ichiishi A, Yamada-Okabe H, Yamaguchi I (2002) Effects of iprodione and fludioxonil on glycerol synthesis and hyphal development in *Candida albicans*. Biosci Biotechnol Biochem 66(10):2209–2215

Ochiai N, Tokai T, Nishiuchi T, Takahashi-Ando N, Fujimura M, Kimura M (2007) Involvement of the osmosensor histidine kinase and osmotic stress-activated protein kinases in the regulation of secondary metabolism in *Fusarium graminearum*. Biochem Biophys Res Commun 363(3):639–644

Okada A, Banno S, Ichiishi A, Kimura M, Yamaguchi I, Fujimura M (2005) Pyrrolnitrin interferes with osmotic signal transduction pathway in *Neurospora crassa*. J Pestic Sci 30:378–383

Orth AB, Rzhetskaya M, Pell EJ, Tien M (1995) A serine (threonine) protein kinase confers fungicide resistance in the phytopathogenic fungus *Ustilago maydis*. Appl Environ Microbiol 61(6):2341–2345

Oshima M, Fujimura M, Banno S, Hashimoto C, Motoyama T, Ichiishi A, Yamaguchi I (2002) A point mutation in the two-component histidine kinase *BcOS-1* gene confers dicarboximide resistance in field isolates of *Botrytis cinerea*. Phytopathology 92(1):75–80

Oshima M, Banno S, Okada K, Takeuchi T, Kimura M, Ichiishi A, Yamaguchi I, Fujimura M (2006) Survey of mutations of a histidine kinase gene *BcOS1* in dicarboximide-resistant field isolates of *Botrytis cinerea*. J Gen Plant Pathol 72(1):65–73

Pillonel C, Meyer T (1997) Effect of phenylpyrroles on glycerol accumulation and protein kinase activity of *Neurospora crassa*. Pestic Sci 49(3):229–236

Pommer E, Lorenz G (1995) Dicarboximide fungicides. In: Lyr H (ed) Modern selective fungicides – properties, applications, mechanisms of action. Gustav Fischer Verlag, New York, pp 99–118

Ramesh MA, Laidlaw RD, Durrenberger F, Orth AB, Kronstad JW (2001) The cAMP signal transduction pathway mediates resistance to dicarboximide and aromatic hydrocarbon fungicides in *Ustilago maydis*. Fungal Genet Biol 32(3):183–193

Raposo R, Gomez V, Urrutia T, Melgarejo P (2000) Fitness of *Botrytis cinerea* associated with dicarboximide resistance. Phytopathology 90(11):1246–1249

Schaller GE, Shiu SH, Armitage JP (2011) Two-component systems and their co-option for eukaryotic signal transduction. Curr Biol 21(9):R320–R330

Schumacher MM, Enderlin CS, Selitrennikoff CP (1997) The *osmotic-1* locus of *Neurospora crassa* encodes a putative histidine kinase similar to osmosensors of bacteria and yeast. Curr Microbiol 34(6):340–347

Segmuller N, Ellendorf U, Tudzynski B, Tudzynski P (2007) BcSAK1, a stress-activated mitogen-activated protein kinase, is involved in vegetative differentiation and pathogenicity in *Botrytis cinerea*. Eukaryot Cell 6(2):211–221

Van Thuat N, Schafer W, Bormann J (2012) The stress-activated protein kinase FgOS-2 is a key regulator in the life cycle of the cereal pathogen *Fusarium graminearum*. Mol Plant Microbe Interact 25(9):1142–1156

Viaud M, Fillinger S, Liu W, Polepalli JS, Le Pecheur P, Kunduru AR, Leroux P, Legendre L (2006) A class III histidine kinase acts as a novel virulence factor in *Botrytis cinerea*. Mol Plant Microbe Interact 19(9):1042–1050

Walker AS, Micoud A, Remuson F, Grosman J, Gredt M, Leroux P (2013) French vineyards provide information that opens ways for effective resistance management of *Botrytis cinerea* (grey mould). Pest Manag Sci 69(6):667–678

Watanabe S, Yamashita K, Ochiai N, Fukumori F, Ichiishi A, Kimura M, Fujimura M (2007) OS-2 mitogen activated protein kinase regulates the clock-controlled gene *ccg-1* in *Neurospora crassa*. Biosci Biotechnol Biochem 71(11):2856–2859

Yamashita K, Shiozawa A, Banno S, Fukumori F, Ichiishi A, Kimura M, Fujimura M (2007) Involvement of OS-2 MAP kinase in regulation of the large-subunit catalases CAT-1 and CAT-3 in *Neurospora crassa*. Genes Genet Syst 82(4):301–310

Yamashita K, Shiozawa A, Watanabe S, Fukumori F, Kimura M, Fujimura M (2008) ATF-1 transcription factor regulates the expression of *ccg-1* and *cat-1* genes in response to fludioxonil under OS-2 MAP kinase in *Neurospora crassa*. Fungal Genet Biol 45(12):1562–1569

Yoshimi A, Tsuda M, Tanaka C (2004) Cloning and characterization of the histidine kinase gene *Dic1* from *Cochliobolus heterostrophus* that confers dicarboximide resistance and osmotic adaptation. Mol Genet Genomics 271(2):228–236

Yoshimi A, Kojima K, Takano Y, Tanaka C (2005) Group III histidine kinase is a positive regulator of Hog1-type mitogen-activated protein kinase in filamentous fungi. Eukaryot Cell 4(11):1820–1828

Zhang Y, Lamm R, Pillonel C, Lam S, Xu JR (2002) Osmoregulation and fungicide resistance: the *Neurospora crassa os-2* gene encodes a HOG1 mitogen-activated protein kinase homologue. Appl Environ Microbiol 68(2):532–538

Chapter 13
Sterol Biosynthesis Inhibitors: C14 Demethylation (DMIs)

Basil Nicholas Ziogas and Anastasios Andreas Malandrakis

Abstract DMIs are a commercially very successful group of systemic fungicides targeting cell membrane integrity by inhibiting C14 demethylation during sterol formation. Regardless of their site-specific mode of action, resistance development evolved in a typical stepwise manner leading to a prolonged effectiveness for more than four decades and rendering those fungicides a unique paradigm. A number of resistance mechanisms acting individually or in combination have been identified over the years to be associated with decreased sensitivity to these compounds in certain pathogen populations in the field including target-site modification, target gene (*cyp51*) overexpression, increased efflux, and multiple paralogues of the target gene. Various mutations in the coding region of the *cyp51* gene usually confer different levels or no resistance to different members of DMIs contrary to ones leading to target gene overexpression which decrease the sensitivity of isolates to all members of the DMI group. Increased efflux mediated by drug transporters belonging to the ABC or MFS transporter families results in resistance to all DMI members as well as to other unrelated fungicides. Pleiotropic effects of resistance mutations on fitness parameters, variable levels, and lack of cross-resistance between members of the DMI group probably account for the delayed resistance development and can be partly attributed to a polygenic control of resistance. These unique characteristics and the large number of registered members make DMI indispensable components of control programs against important plant pathogens.

Keywords DMI resistance • Fungicide resistance • *cyp51* paralogues • SBIs • EBIs • Fungicide efflux • *cyp51* mutations • *cyp51* overexpression • C14-demethylase

B.N. Ziogas (✉) • A.A. Malandrakis
Laboratory of Pesticide Science, Agricultural University of Athens, 118 55 Votanikos, Athens, Greece
e-mail: ziv@aua.gr

H. Ishii, D.W. Hollomon (eds.), *Fungicide Resistance in Plant Pathogens*,
DOI 10.1007/978-4-431-55642-8_13

13.1 Introduction

Sterols are essential constitutes of fungal cell membranes regulating their stability and permeability, and, therefore, sterol biosynthesis is a very important target for fungal growth inhibition. Ergosterol is the main sterol of most fungi from the classes of Ascomycetes, Basidiomycetes, and Fungi imperfecti with a number of exceptions from powdery mildew, rust, and apple scab fungi which have other sterols such as ergosta-5,29(28)-dienol and Δ^7-sterols as their main sterol (Loeffler et al. 1984; Pontzen et al. 1990; Koeller 1992). The ergosterol biosynthesis inhibitors (EBIs), or sterol biosynthesis inhibitors (SBIs), due to their broad spectrum of fungicidal activity and systemic properties are valuable fungicides in agriculture for the control of important fungal diseases in a wide range of crops and for the control of important mycoses in humans and animals.

The SBIs appeared in agricultural practice since the second half of the 1960s when a number of fungicides from the chemical groups of imidazoles, morpholines, pyrimidines, pyridines, piperazines, and triazoles and the more recently introduced piperidines, hydroxyanilides, and spiroketalamines were patented by different agrochemical companies. Biochemical studies using radiolabeled acetate by a number of research groups (Ragsdale and Sisler 1972; Sisler et al. 1983; Kato et al. 1974; Buchenauer 1977; Baloch et al. 1984; Kerkenaar 1990; Ziogas et al. 1991) have shown that these fungicides interfere with the fungal sterol biosynthesis pathway, mainly at the steps catalyzed by C14-demethylase, Δ14-reductase, C4-demethylase, Δ8,7-isomerase, and squalene epoxidase (Fig. 13.1).

DMI fungicides, being the largest member of the group of SBIs, comprise an important group of commercially successful compounds (belonging to the chemical classes of triazoles, imidazoles, piperazines, pyrimidines, and pyridines) registered for the control of plant diseases caused by Ascomycetes, Basidiomycetes and Fungi imperfecti. The imidazole and triazole derivatives, known as azoles, have been proven to be the most important DMI fungicides based on the number of compounds used in agricultural practice. Since 1970, more than 40 DMI molecules have been introduced in agriculture to control a large number of plant pathogens, dominating the agricultural fungicide market (Table 13.1).

DMIs inhibit the sterol C14-demethylation step during the process of sterol formation in higher fungi (Fig. 13.1). The lanosterol C14 demethylation is mediated by a mixed-function oxygenase cytochrome P450, which is a kind of hemoprotein (Fig. 13.2). DMIs bind to the heme iron of cytochrome P450 with a nitrogen atom and inhibit the O_2 binding and its transfer to lanosterol C14-methyl group, which is the main step in lanosterol C14-demethylation process. However, a recent study of the way of DMIs binding to *Mycosphaerella graminicola cyp51* suggests for prothioconazole an inhibition of the enzyme in a manner other than direct coordination to the heme (Parker et al. 2011).

DMI fungicides despite their extensive use to control a wide range of plant pathogens for the past four decades and in spite of their site-specific mode of action, which generally favors resistance development, still retain most of their effectiveness.

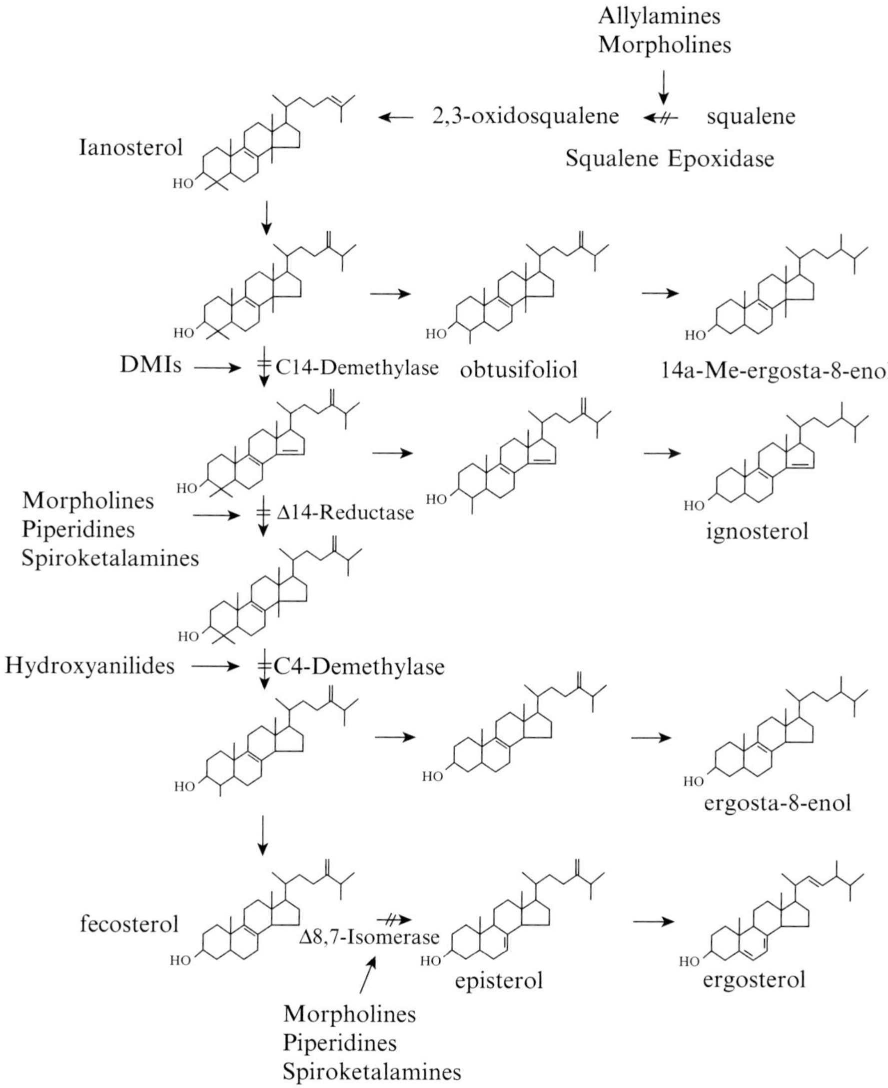

Fig. 13.1 Ergosterol biosynthesis and sites blocked by SBIs (From Ziogas and Markoglou 2010)

The polygenic control of resistance coupled with the pleiotropic effects of resistance mutations on phytopathogenic fitness has been suggested to be responsible for this prolonged effectiveness and delayed resistance development (Fuchs and De Waard 1982; Georgopoulos 1987; Kalamarakis et al. 1991). Eventually, however, the intensive use of DMIs in agriculture has led to a stepwise manner resistance development leading to practical control problems in certain pathogen populations starting from the 1980s (Hollomon et al. 1984; Brent and Hollomon 1988, 1998; Hollomon 1993; Ma and Michailidis 2005).

Table 13.1 DMI (C14-demethylase inhibiting) fungicides presented by chemical group and year of introduction

Chemical group	Common name and the year of introduction
Piperazines	Triforine (1969)
Pyridines	Pyrifenox (1986), buthiobate (1975), pyrisoxazole (2011–2012)
Pyrimidines	Fenarimol (1975), nuarimol (1975)
Imidazoles	Imazalil (1972), prochloraz (1977), fenapanil (1978), triflumizole (1982), pefurazoate (1990), oxpoconazole (2000)
Triazoles	Triadimefon (1973), fluotrimazole (1973), myclobutanil (1977)
	Triadimenol (1977), bitertanol (1978), diclobutrazol (1979), etaconazole (1979), propiconazole (1979), azaconazole (1983)
	Diniconazole (1983), flusilazole (1983), flutriafol (1983), penconazole (1983), cyproconazole (1986) hexaconazole (1986), tebuconazole (1986), difenoconazole (1988), fenbuconazole (1988), imibenconazole (1988), tetraconazole (1988), furconazole (1988), bromuconazole (1990), epoxiconazole (1990), fluquinconazole (1992), metconazole (1992), triticonazole (1992), ipconazole (1994),
	Simeconazole (2001), prothioconazole (2002)

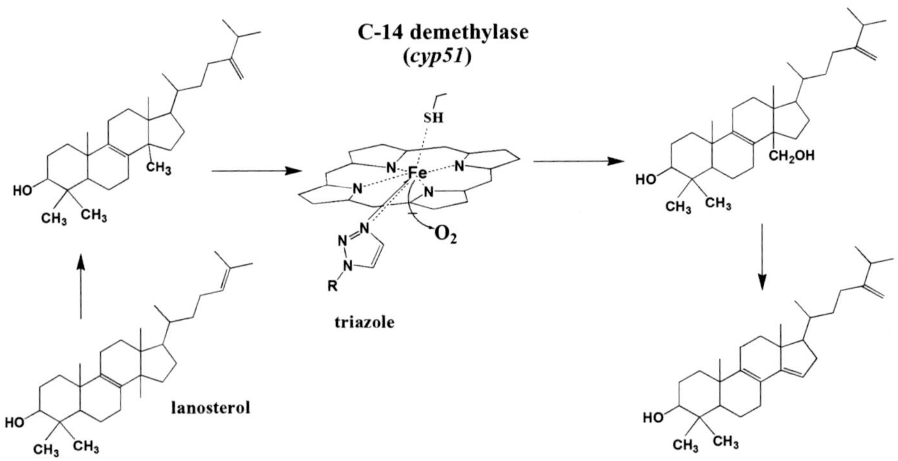

Fig. 13.2 Process of C14 lanosterol demethylation and its inhibition by DMIs

13.2 Resistance Mechanisms

Studies on biochemical mechanisms involved in DMI resistance have reported a variety of models as mechanisms for acquired resistance to DMIs: (a) survival of mutated cells with abnormal C14-methyl sterols in their plasma membranes (deficiency in C14 demethylation), (b) decreased intracellular fungicide accumulation mediated by an increased energy-dependent fungicide efflux, (c) target-site modifications in the C14-demethylase (*CYP51*) gene resulting in decreased affinity of fungicides, (d) overexpression of the target gene during ergosterol formation, and (e)

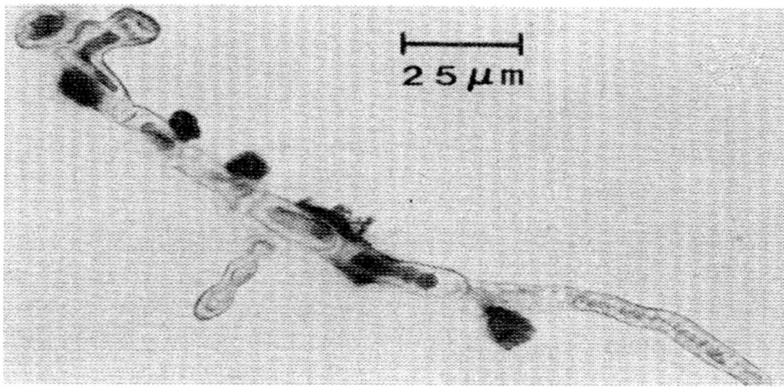

Fig. 13.3 Microphotograph of hyphae of *Monilinia fructicola* developed in 48 h from a conidium on agar medium containing 0.2 μg/ml fenarimol. Note contents extruding from hyphal cells (From Sisler et al. 1983)

multiple *CYP51s*, which seem to reduce the fitness costs associated with the changes in the structure of C14-demethylase or overexpression of encoding gene and increase the spectrum of fungicide resistance. The available data from a number of biochemical and molecular studies on DMI resistance seem to support the hypothesis that a combination of the above resistance mechanisms could contribute to the decrease in field efficacy of DMIs.

13.2.1 Survival with C14-Methyl Sterols

The deficiency in the sterol C14 demethylation has been demonstrated to be a functional DMI-resistance mechanism in the case of budding fungi such as *Ustilago maydis* and the yeasts *Candida albicans* and *Saccharomyces cerevisiae* (Trocha et al. 1977; Pierce et al. 1978; Walsh and Sisler 1982) but not in filamentous phytopathogenic fungi. Apparently, mutations blocking C14 demethylation or earlier steps of ergosterol formation are lethal, since the C14-methyl sterols produced do not support sufficiently stable and functional lipoprotein membranes in filamentous fungi (Ziogas et al. 1983). This fact is demonstrated by a delayed bursting of cells observed in several filamentous fungi at low concentrations of C14-demethylation inhibitors (Fig. 13.3).

13.2.2 Increased Energy-Dependent Fungicide Efflux

Membrane-bound drug transporters have the ability to secrete antifungal compounds to the extracellular space of mycelial cells and prevent the accumulation of the fungicide at the fungal target site (Fig. 13.4). Increased function of this efflux

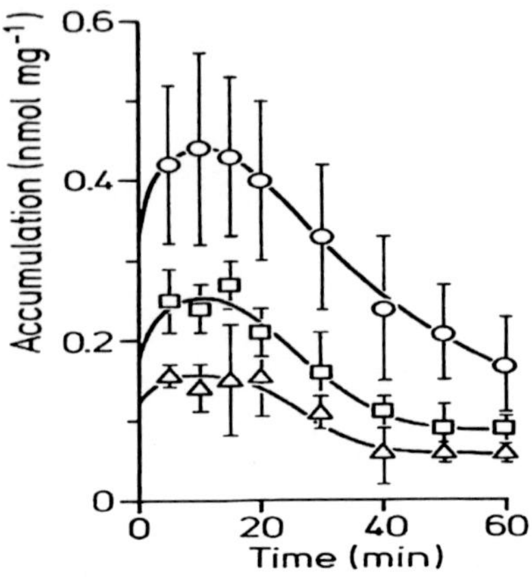

Fig. 13.4 Accumulation of 14C fenarimol (10 μM) by mycelium of wild-type strain No 38 (○) and fenarimol-resistant isolates FN-377 (□) and FN-76 (Δ) of *Nectria haematococca* var. *cucurbitae* (From Kalamarakis et al. 1991)

pump mediated by the drug transporters leads to the development of multidrug resistance (MDR) due to the limited specificity of these transporters in secreting xenobiotics (De Waard et al. 2006).

The most important drug transporter families involved in MDR are the ATP-binding cassette (ABC) and major facilitator superfamily (MFS) transporters (Del Sorbo et al. 2000). Overexpression of genes encoding fungicide efflux pumps is the most frequently recognized mechanism of azole resistance in human pathogens leading to clinical failures of disease control (Akins 2005; Prasad and Kapoor 2005). It should be noted though that in the majority of plant pathogenic fungi, MDR resistance is connected with low saprophytic fitness and virulence of mutant strains which means that such mutants may not be readily established in the field (De Waard et al. 2006).

However, resistance to DMIs based on an increased energy-dependent efflux has been supported in a number of filamentous fungi such as *Aspergillus nidulans*, *Penicillium italicum*, *Penicillium digitatum*, *Monilinia fructicola*, *Nectria haematococca* var. *cucurbitae*, *Botrytis cinerea*, *Sclerotinia homeocarpa*, *Mycosphaerella graminicola*, *Pyrenophora tritici*, and *Oculimacula yallundae* (Table 13.2).

A number of genes encoding multidrug resistance and the capacity to export DMIs outside of the mycelial cells have been identified including *AtrB*, *AtrD*, and *AtrG* in *A. nidulans*; *BcatrB*, *BcatrD*, and *Bcmfs1* in *B. cinerea*; *MgAtr1*-*MgAtr5* and *MgMfs1* in *M. graminicola*; *PdPMR1* and *PdPMR5* in *P. digitatum*; and *ShatrD* in *S. homeocarpa* (Table 13.2). Genes encoding drug efflux pumps conferring DMI resistance were also identified in other *Aspergillus* species, such as *Apmdr* in *A. parasiticus*, *Afumdr* in *A. fumigatus*, and *Afmdr1* in *A. flavus* (Table 13.2).

Of particular interest is the use of well-confirmed blockers or modulators of ABC transporters to fight the fungicide multidrug resistance in plant pathogens resulting from increased function of fungicide efflux pumps (De Waard et al. 2006). The use of MDR reversal agents to block the expression of ABC transporters is a very promising strategy to increase the fungicide accumulation in mycelia of mutant strains by inactivating the MDR resistance mechanism (De Waard and Van Nistelroy 1982, 1984; Kalamarakis et al. 1991; Hayashi et al. 2003).

Table 13.2 Reported fungicide efflux resistance mechanism in DMI-resistant fungal isolates

Pathogen	Origin	Gene involved (name/family)	References
Pyrenophora tritici	Lab		Reimann and Deising (2005)
Botrytis cinerea	Lab	*BcatrD*/ABC	Hayashi et al. (2001)
Botrytis cinerea	Lab	*Bcmfs1*/MFS	Hayashi et al. (2002)
Botrytis cinerea	Lab	BcatrK/ABC	Nakajima et al. (2001)
Aspergillus parasiticus	Lab	*Mdr*/ABC	Doukas et al. (2012)
Aspergillus nidulans	Lab	*AtrBp*/ABC	Andrade et al. (2000) and Do Nascimento et al. (2002)
Aspergillus fumigatus	Lab	*Afmdr*	Nascimento et al. (2003)
Aspergillus flavus	Lab	*Aflmdr*	Tobin et al. (1997)
Venturia inaequalis	Lab		Vijaya-Palani and Lalithakumari (1999)
Nectria haematococca var. *cucurbitae*	Lab	*fen-1, fen-7, fen-9*	Kalamarakis et al. (1991)
Botrytis cinerea	Field	*AtrB*/ABC	Kretschmer et al. (2009)
Botrytis cinerea	Field	*mfsM2*/MFS	Kretschmer et al. (2009)
Penicillium digitatum	Field	*PMR1, PMR5/ABC*	Nakaune et al. (1998, 2002)
Penicillium digitatum	Field	*PMR5*/ABC	Sanchez-Torres and Tuset (2011)
Penicillium italicum	Field		Guan et al. (1992)
Oculimacula yallundae	Field		Leroux et al. (2013)
Sclerotinia homeocarpa	Field	*ShatrD*/ABC	Hulvey et al. (2012)
Mycosphaerella graminicola	Field	*MgAtr*/ABC	Zwiers et al. (2002) and Leroux and Walker (2011)
Monilinia fructicola	Field	*MfABC 1/ABC*	Luo and Schnabel (2008)

13.2.3 Target Site Modification

Before the development of molecular methodology in fungicide research, a structural change in the cytochrome P450-dependent C14-demethylase as biochemical mechanism of resistance to certain DMI fungicides in plant pathogenic fungi was proposed after electrophoretic profile studies of microsomal fractions comparing a wild-type and a triadimenol-resistant (*tri*-1) mutant strain on *N. haematococca* var. *cucurbitae* with a chromosomal major gene mutation specific for resistance to diastereoisomer triadimenol A (Kalamarakis et al. 1989; Demopoulos and Ziogas 1994). A major difference was observed between the two strains regarding the heme dissociation from the apoprotein of cytochrome P450 during electrophoresis (Fig. 13.5). The instability of C14-demethylase in the mutant strain was explained by a change in the heme binding to the protein moiety of the enzyme caused by the *tri*-1 chromosomal mutation.

The rapid development over about the past 20 years of PCR-based diagnostic methods for the identification of mutations causing resistance gave the possibility to identify a number of point mutations in the *cyp51* gene of C14-demethylase (Table 13.3). The *cyp51* point mutations, in contrast to mutations causing resistance

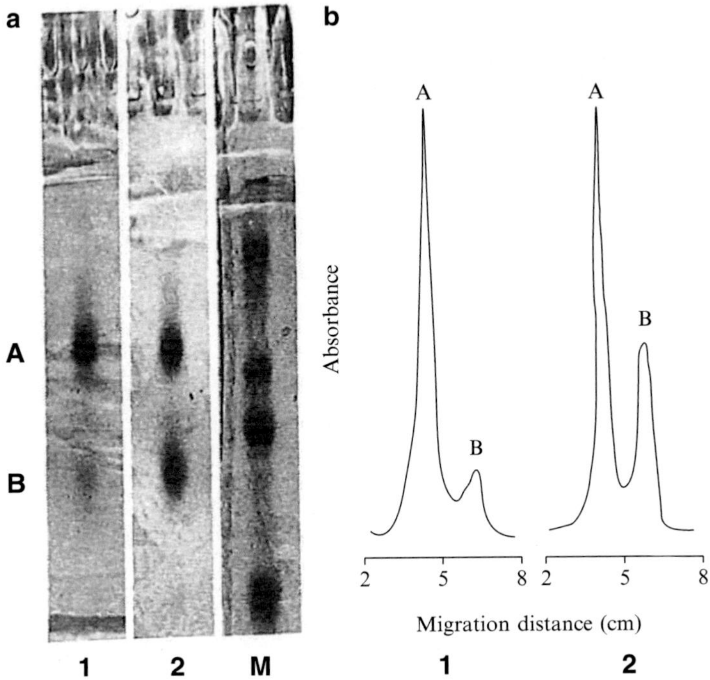

Fig. 13.5 Electrophoresis of micromosomal protein preparations from a wild-type (*1*) and a triadimenol-resistant (*tri*-1) mutant strain (*2*) of *N. haematococca* var. *cucurbita*e on SDS-polyacrylamide gels. (**a**) A photographic representation of the hemoprotein profile from each type of microsomal preparation. (**b**) Scans of the hemoprotein bands A and B of the gels shown in (**a**) (From Demopoulos and Ziogas 1994)

to other single-site inhibitors, affect in most of the cases the sensitivity to specific or certain members of the DMIs. Specific amino acid modifications in *cyp51* protein could be neutral, positive, or negative regarding the fungitoxicity of different DMIs. Therefore, the *cyp51* variants are differentially selected by the different DMIs providing an option to manage DMI-resistance development by alterations or combinations of different members of the DMI group.

The most frequently reported *cyp51* amino acid alteration seems to be at the residue Y136 or equivalent residues Y134 and Y137 in both human and plant pathogenic fungi (Becher and Wirsel 2012). The substitution of tyrosine (Y) by phenylalanine (F) at the above codon positions has been reported in field isolates of grape and cereal powdery mildews resistant to triadimenol and propiconazole (Delye et al. 1997, 1998; Wyand and Brown 2005; Kuck and Mehl 2004), in *Mycosphaerella* species resistant to propiconazole (Cañas-Gutierrez et al. 2009; Chong et al. 2011; Leroux and Walker 2011; Cools and Fraaije 2012) and in *Puccinia triticina* resistant to epoxiconazole (Stammler et al. 2009). Furthermore, additional amino acid substitutions in DMI-resistant isolates were frequently reported in *Blumeria graminis* f. sp. *hordei*, *Cercospora beticola*, *M. graminicola*,

Table 13.3 Reported mutations at *cyp51* gene in DMI-resistant fungal isolates

Pathogen	Strain origin	Amino acid substitution/ deletion	References
Aspergillus flavus	Lab	K197N, Y132N, T469S, D282E, M288L, H399P, D411N, T454P, T486P	Krishnan-Natesan et al. (2008)
Aspergillus fumigatus	Lab	G54W	Diaz-Guerra et al. (2003) and Maan et al. (2003)
Aspergillus parasiticus	Lab	G54W	Doukas et al. (2012)
Blumeria graminis f. sp. *hordei*	Field	Y136F, K147Q	Delye et al. (1998) and Wyand and Brown (2005)
Blumeria graminis f. sp. *tritici*	Field	Y136F, S79T, K175N	Kuck and Mehl (2004) and Wyand and Brown (2005)
Cercospora beticola	Field	E297K, I330T, P384S	Nikou et al. (2009)
Erysiphe necator	Field	Y136F	Delye et al. (1997)
Monilinia fructicola	Lab	Y136F	Chen et al. (2012)
Mycosphaerella fijiensis	Field	Y136F, A313G, A381G, Y461D, G462A, Y463D/H/N	Cañas-Gutierrez et al. (2009) and Chong et al. (2011)
Mycosphaerella graminicola	Field	L50S, D107V, D134G, V136A,V136C/G, Y137F,M145L, N178S, S188N, S208T, N284H, H303Y, A311G, G312A, A379G, I381V, A410T, G412A, Y459C, Y459D/N/S/P, G460D, Y461D/H/S, ΔY459, ΔG460, V490L, G510C, N513K, S524T	Cools and Fraaije (2012)
Phakopsora pachyrhizi	Field	Y131F, Y131H, K142R, F120L, I145F, I475T	Schmitz et al. (2014)
Puccinia triticina	Field	Y134F	Stammler et al. (2009)
Pyrenopeziza brassicae	Field	G460S, S508T	Carter et al. (2014)

M. fijiensis, *Phakopsora pachyrhizi*, and *A. parasiticus* (Table 13.3). The replacement of phenylalanine by leucine at position 180 of the *cyp51* gene was suggested to be responsible for the natural resistance of *Tapesia acuformis* to triazoles (Albertini et al 2003). Moreover, the deletion of two amino acids (Y459/G460) within the target-encoding *cyp51* gene has been shown to play an important role in the inhibitor binding and enzyme function and was associated with azole resistance in *Mycosphaerella* species (Leroux et al. 2007; Cañas-Gutierrez et al. 2009; Mullins et al. 2011). However, the involvement of all these amino acid substitutions in the DMI resistance is not clear. It is possible that some of the multiple mutations could synergistically affect the affinity of the fungicides at the *cyp51* binding site while some of them have been shown to compensate for the fitness costs of specific targetsite mutations in *M. graminicola* (Cools et al. 2013).

13.2.4 Overexpression of C14-Demethylase

The development of quantitative real-time PCR gave the possibility to recognize the increased expression of target-encoding gene (*cyp51*) as another mechanism for exclusive resistance to all DMI fungicides. Elevated expression levels of the target gene were reported to account for DMI resistance in *P. digitatum*, *Blumeriella jaapii*, *Venturia inaequalis*, *M. graminicola*, *Sclerotinia homeocarpa*, *C. beticola*, *Phakopsora pachyrhizi*, *P. triticina*, *Pyrenopeziza brassicae*, *A. fumigatus*, and in aflatoxigenic mutant strains of *A. parasiticus*. In most of the cases the indentified responsible alterations were connected with an increased production of target enzyme resulting in lower activity of DMIs (Table 13.4). Increased *cyp51* mRNA levels have been associated with inserts of transposable elements in the promoter upstream region of *cyp51* in a number of plant and human pathogens. Specifically, a 126 bp or 199 bp element in *P. digitatum* (Hamamoto et al. 2000; Sun et al. 2013), a 2Kb–5Kb retrotransposon in *B. jaapii* (Ma et al. 2006), a 553 bp insertion in *V. inaequalis* (Schnabel and Jones 2001; Pfeufer and Ngugi 2012), a 65 bp repetitive element (*mona*) in *M. fructicola* (Luo and Schnabel 2008), a 1.3 Kb retrotransposon in *B. cinerea* (Kretschmer et al. 2009), and a 1.8 Kb *Aft1* in the human pathogenic fungus *A. fumigatus* (Albarrag et al. 2011) were associated with a respective *cyp51* overexpression.

Table 13.4 Reported overexpression of C14-demethylase in DMI-resistant fungal isolates

Pathogen	Strain origin	Overexpression due to	References
Aspergillus parasiticus	Lab	Unknown	Doukas et al. (2012)
Blumeriella jaapii	Field	Insert upstream	Ma et al. (2006)
Cercospora beticola	Field	Unknown	Nikou et al. (2009) and Bolton et al. (2012)
Monilinia fructicola	Field	*"mona" insert in promoter area*	Luo and Schnabel (2008) and Chen et al. (2012)
Mycosphaerella graminicola	Field	Promoter insert	Cools and Fraaije (2012)
Penicillium digitatum	Field	Promoter insert	Nakaune et al. (1998), Hamamoto et al. (2000), Ghosoph et al. (2007), and Sun et al. (2013)
Phakopsora pachyrhizi	Field/lab	Unknown	Schmitz et al. (2014)
Puccinia triticina	Field	Unknown	Stammler et al. (2009)
Pyrenopeziza brassicae	Field	Insert upstream	Carter et al. (2014)
Sclerotinia homeocarpa	Field	Unknown	Hulvey et al. (2012)
Venturia inaequalis	Field	Promoter insert/or not	Schnabel and Jones (2001) and Pfeufer and Ngugi (2012)
Aspergillus fumigatus	Clinical isolates	Promoter insert	Chamilos and Kontoyiannis (2005) and Mellado et al. (2007)

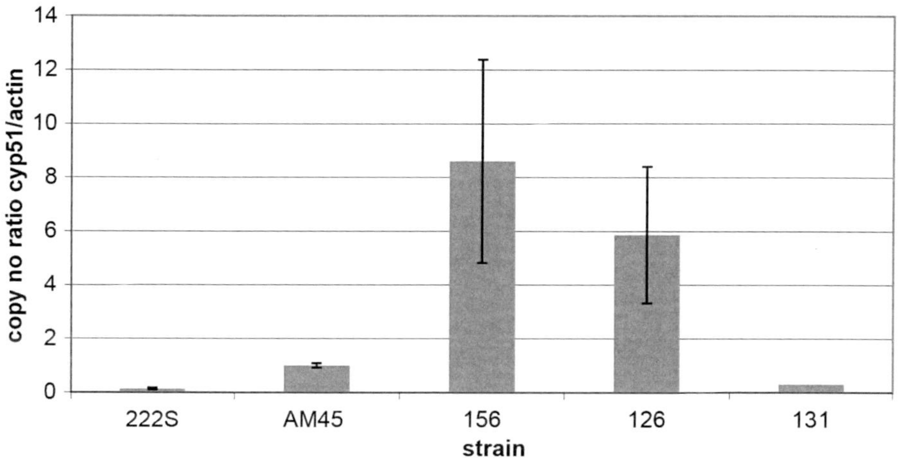

Fig. 13.6 Quantitative PCR analysis of *C. beticola* mRNA expression levels. mRNA copy number of *cyp51* transcript relative to the number of actin transcripts for the DMI-sensitive (222S and 131), moderately DMI-resistant (AM45), and highly DMI-resistant (156 and 126) isolates was calculated (From Nikou et al. 2009)

In contrast with the site modification mechanism described previously, mutants overexpressing the *cyp51* gene are not resistant to specific DMIs but exhibit a wide range of resistance to all DMIs studied. However, the increased gene transcripts compared with the wild type are in most of the cases low (5–12-fold), and the resistance levels are generally lower than those resulting from the site modification mechanism. *C. beticola* seems to be an exception since field isolates with resistance factors 65–115 to epoxiconazole exhibited up to 160–230-fold increased gene transcripts compared with wild-type isolates (Fig. 13.6). No amino acid substitutions were found in highly resistant isolates indicating that the mechanism of target gene overexpression was responsible for the highly resistant phenotype.

Furthermore, the absence of cross-resistance relationship between DMIs and fungicides with different modes of action distinguishes the overexpressing mechanism from that of membrane transporters described before.

13.2.5 Multiple CYP51s

To date, very few studies exist on the function of multiple *cyp51s* while the reason for their existence in some species and their role in fungal biology remain unknown. Until now three paralogues of *cyp51* gene (*cyp51A*, *cyp51B*, *cyp51C*) have been identified in a number of plant and human pathogenic species belonging to the *Aspergillus*, *Fusarium*, *Magnaporthe*, *Penicillium*, and *Rhynchosporium* genera. Soon after the completion of sequencing of the full genome of *F. graminearum*, the existence of three *cyp51* paralogue genes in this species was revealed. Targeted gene disruption showed that only *cyp51A* and *cyp51C* paralogues are involved in DMI sensitivity (Liu et al. 2011). In *A. fumigatus* two paralogues, *cyp51A* and *cyp51B*, with different roles in fungal physiology were identified. *cyp51A* was found to be responsible for the reduced sensitivity to azoles, and *cyp51B* was found to play a role in fungal growth and shape maintenance but not in DMI resistance (Mellado et al. 2006; Garsia-Effron et al. 2008). Both *cyp51A* and *cyp51B* were identified and associated with reduced DMI sensitivity in *P. digitatum* (Sun et al. 2013). Study of transcriptional levels of C14-demethylase gene showed that overexpression of *cyp51A* was strongly associated with DMI resistance in an aflatoxigenic mutant strain of *A. parasiticus* (Doukas et al. 2012). Constitutive overexpression of the *cyp51A* rather than *the cyp51B* gene has also been found to be responsible for a broad-spectrum DMI resistance in *A. flavus* (Krishnan-Natesan et al. 2008), *A. niger* (Mellado et al. 2001), and *M. oryzae* (Yan et al. 2011). This was also the case in *Rhynchosporium commune* where an upregulation of *CYP51A* transcripts in response to disruption of ergosterol biosynthesis by DMI exposure was responsible for resistance (Hawkins et al. 2014). Two paralogues of the *cyp51* gene, attributed as *cyp51A* and *cyp51B*, were also identified in *F. asiaticum* and *R. secalis*, but their specific role as DMI-sensitivity determinant has not yet been elucidated (Yin et al. 2009; Hawkins et al. 2011).

Some *cyp51* paralogues were shown to be functionally redundant but could be induced immediately after ergosterol depletion due to the presence of DMI fungicides (Cools et al. 2013) allowing a fungal species to respond in ergosterol requirements. The presence of multiple *cyp51* paralogues in a fungal species could increase the frequency of resistance mutations leading to target-site modification or *cyp*51 overexpression. Any biological costs from *cyp51* gene mutation for fungicide resistance could be surpassed by the presence of an unchanged enzyme with wild-type activity from the available *cyp51* paralogues. Beyond this, the presence of additional paralogues of the *cyp51* gene could lead to reduced intrinsic DMI sensitivity, as proposed by paralogue knocking out experiments in *A. fumigatus* (Mellado et al. 2005) and *F. graminearum* (Fan et al. 2013). Moreover, the reduction of *R. commune* sensitivity to azoles associated with the reemergence of the *cyp51A* paralogue present only in less sensitive isolates (Hawkins et al. 2014) underlines the significance of this mechanism and is also valuable for the understanding of the evolutionary mechanisms involved in fungicide resistance.

13.3 Conclusions

Overall, taking into account all available data over the decades of successful use of DMI fungicides, resistance to DMIs is a complex phenomenon. Their typical stepwise manner of resistance development was, and partly still may be, attributed to a polygenic control of resistance. Certainly, early studies on DMI resistance utilizing crossing experiments have demonstrated the involvement of up to nine different genes with additive effects on the reduced sensitivity of isolates of various pathogens to DMIs supporting the polygenic control model. The simultaneous involvement of some of the abundant number of available drug transporter genes (belonging to the families of ABC or MFS transporters) could explain the respective phenotypes in terms of resistance levels, fitness penalties, and cross-resistance profiles of those first studies. Since then, the advance of molecular methodology has provided the means for the identification of a number of genes responsible for the observed resistance phenotypes in a large number of plant and human pathogenic species. In most cases, resistance occurred from point mutations in the *cyp51* gene, which is coding the target site of DMIs or in the respective regulatory region of this gene. Wherever target-site modifications are involved, variable levels of resistance, fitness penalties associated with certain amino acid substitutions, and the lack of cross-resistance between different members of the DMI group could be the reason for the delayed resistance development. Resistant isolates of pathogens overexpressing the *cyp51* gene or *cyp51* paralogues (constitutively expressed or reemerged) express their resistance to all DMI members, but resistance levels are usually low and so resistance can be surpassed by higher dose rates that are also consistent with the observed pattern of DMI-resistance development. Making matters more complicated, combinations of mechanisms including target-site modifications, target gene overexpression, utilization of *cyp51* paralogues, and increased efflux can contribute simultaneously to the reduced sensitivity profiles in some cases of resistant isolates, a fact in accordance with the polygenic model by definition. Whatever the type of genetic control and biochemical mechanism involved in DMI resistance, the option to use members from a wide variety of different DMIs and MDR reversal agents to combat resistant isolates makes predictions for the long-term effectiveness of these fungicides still very optimistic. Certainly, it is very important to preserve DMIs in the farmer's arsenal as a valuable asset for future world agricultural production and also for human health.

References

Akins RA (2005) An update on antifungal targets and mechanisms of resistance in *Candida albicans*. Med Mycol 43:285–318

Albarrag AM, Anderson MJ, Howard SJ, Robson GD, Warn PA, Sanglad D, Denning DW (2011) Interrogation of related clinical pan-azole-resistant *Aspergillus fumigatus* strains: G138C, Y431C, and G434C single nucleotide polymorphisms in cyp51A, upregulation of cyp51A, and integration and activation of transposon Atf1 in the cyp51A promoter. Antimicrob Agents Chemother 55:5113–5121

Albertini C, Gredt M, Leroux P (2003) Polymorphism of 14α-demethylase gene (CYP51) in the cereal eyespot fungi *Tapesia acuformis* and *Tapesia yallundae*. Eur J Plant Pathol 109:117–128

Andrade AC, Del Sorbo G, Nistelroy JGM, De Waard MA (2000) The ABC transporter from *Aspergillus nidulans* is involved in resistance to all major classes of fungicides and natural toxic compounds. Microbiol-SGM 146:1987–1997

Baloch RI, Mercer EI, Wiggins T, Baldwin BC (1984) Inhibition of ergosterol biosynthesis in *Saccharomyces cerevisiae* and *Ustilago maydis* by tridemorph, fenpropimorph and fenpropidin. Phytochemistry 23:2219–2226

Becher R, Wirsel SGR (2012) Fungal cytochrome P450 sterol 14α-demethylase (CYP51) and azole resistance in plant and human pathogens. Appl Microbiol Biotechnol 95:825–840

Bolton MD, Birla K, Rivera-Varas V, Rudolf KD, Secor GA (2012) Characterization of CbCyp51 from field isolates of *Cercospora beticola*. Phytopathology 102:298–305

Brent KJ, Hollomon DW (1988) Risk of resistance against sterol biosynthesis inhibitors in plant protection. In: Berg D, Plempel M (eds) Sterol biosynthesis inhibitors. Ellis Horwood Ltd, Chichester, pp 332–346

Brent KJ, Hollomon DW (1998) Fungicide resistance: the assessment of risk. FRAC monograph no. 2. Global Crop Protection Federation, Brussels, pp 1–48

Buchenauer H (1977) Mode of action and selectivity of fungicides which interfere with ergosterol biosynthesis. In: Proceedings of the British crop protection conference. Brighton, UK, pp 699–711

Cañas-Gutierrez GP, Angaria-Velasquez MJ, Restrepo-Florez JM, Rodriguez P, Moreno CX, Arango R (2009) Analysis of the CYP51 gene and encoded protein in propiconazole-resistant isolates of *Mycosphaerella fijiensis*. Pest Manag Sci 65:892–899

Carter HE, Fraaije BA, West JS, Kelly SL, Mehl A, Shaw MW, Cools HJ (2014) Alterations in the predicted regulatory and coding regions of the sterol 14α-demethylase gene (CYP51) confer decreased azole sensitivity in the oilseed rape pathogen *Pyrenopeziza brassicae*. Mol Plant Pathol 15:513–522

Chamilos G, Kontoyiannis DP (2005) Update on antifungal drug resistance mechanisms of *Aspergillus fumigatus*. Drug Resist Updates 8:344–358

Chen FP, Arang R, Zhow T, Liu XL, Liu R, Schnabel G (2012) Baseline sensitivity of *Monilinia fructicola* from China to the DMI fungicide SYP-Z048 and analysis of DMI-resistance. Plant Dis 96:416–422

Chong P, Arango R, Stergiopoulos I (2011) Analysis of azole fungicide resistance in *Mycosphaerella fijiensis*, causal agent of Black Sigatoka. In: Dehne HW, Deinsing HB, Gisi U, Kuck KH, Russel PE, Lyr H (eds) Modern fungicides and antifungal compounds VI: proceedings of the 16th international Reinhardsbrunn symposium on modern fungicides and antifungal compounds, 2011. Deutsche Phytomedizinische Gesellschaft, Braunschweig, pp 217–222

Cools HJ, Fraaije BA (2012) Update on mechanisms of azole resistance in *Mycosphaerella graminicola* and implication for future control. Pest Manag Sci 69:150–155

Cools HJ, Hawkins NJ, Fraaije BA (2013) Constraints on the evolution of azole resistance in plant pathogenic fungi. Plant Pathol 62:36–42

De Waard MA, Van Nistelroy JGM (1982) Antagonistic and synergistic activities of various chemicals on toxicity of fenarimol to *Aspergillus nidulans*. Pestic Sci 13:279–286

De Waard MA, Van Nistelrooy JGM (1984) Differential accumulation of fenarimol by a wild-type isolate and fenarimol-resistant isolates of *Penicillium italicum*. Neth J Plant Pathol 90:143–153

De Waard MA, Andrade AC, Hayashi K, Schoonbeek HJ, Stergiopoulos I, Zwiers LH (2006) Impact of fungal drug transporters on fungicide sensitivity, multidrug resistance and virulence. Pest Manag Sci 62:195–207

Del Sorbo G, Schoonbeek H, De Waard MA (2000) Fungal transporters involved in efflux of natural toxic compounds and fungicides. Fungal Genet Biol 30:1–15

Delye C, Laigret F, Corio-Costet MF (1997) A mutation in the 14α-demethylase gene of *Uncinula necator* that correlates with resistance to a sterol biosynthesis inhibitor. Appl Environ Microbiol 63:2966–2970

Delye C, Bousset L, Corio-Coster MF (1998) PCR cloning and detection of point mutations in the eburicol 14α-demethylase (CYP51) gene from *Erysiphe graminis* f.sp. *hordei*, a recalcitrant fungus. Curr Genet 34:399–403

Demopoulos VP, Ziogas BN (1994) Studies on the mechanism of expression of a major gene mutation for resistance to triadimenol in the filamentous phytopathogenic Ascomycete *Nectria haematococca* var. *cucurbitae*. Pestic Biochem Physiol 50:159–170

Diaz-Guerra TM, Mellado E, Cuensa-Estrella M, Rodriguez-Tudela JI (2003) A point mutation in the 14α-sterol demethylase gene *cyp51a* contributes to itraconazole resistance in *Aspergillus fumigatus*. Antimicrob Agents Chemother 47:1120–1124

Do Nascimento AM, Goldman MH, Goldman GH (2002) Molecular characterization of ABC transporter-encoding genes in *Aspergillus nidulans*. Genet Mol Res 1:337–349

Doukas EG, Markoglou AN, Vontas JG, Ziogas BN (2012) Effect of DMI-resistance mechanisms on cross-resistance patterns, fitness parameters and aflatoxin production in *Aspergillus parasiticus* Speare. Fungal Genet Biol 49:792–801

Fan J, Urban M, Parker JE, Brewer HC, Kelly SL, Hammond-Kosack KE, Fraaije BA, Liu X, Cools HJ (2013) Characterization of the sterol 14 alpha-demethylases of *Fusarium graminearum* identifies a novel genus-specific CYP51 function. New Phytol 198:821–835

Fuchs A, De Waard MA (1982) Resistance to ergosterol biosynthesis inhibitors. I Chemistry and phenomenological aspects. In: Dekker J, Georgopoulos SG (eds) Fungicide resistance in crop protection. Pudoc, Wageningen, pp 71–86

Garsia-Effron G, Dilger A, Alcazar-Fuoli L, Park S, Mellado E, Perlin DS (2008) Rapid detection of triazole antifungal resistance in *Aspergillus fumigatus*. J Clin Microbiol 46:1200–1206

Georgopoulos SG (1987) The development of fungicide resistance. In: Wolfe MS, Caten CE (eds) Populations of plant pathogens: their dynamics and genetics. Blackwell, Oxford, pp 239–251

Ghosoph JM, Schmidr LS, Margosan DA, Smilunick JL (2007) Imazalil resistance linked to a unique insertion sequence in the Pdcyp51 promoter region of *Penicillium digitatum*. Postharvest Biol Technol 44:9–18

Guan J, Kapteyn JC, Kerkenaar A, De Waard MA (1992) Characterisation of energy-dependent efflux of imazalil and fenarimol in isolates of *Penicillium italicum* with a low, medium and high degree of resistance to DMI-fungicides. Neth J Plant Pathol 98:313–324

Hamamoto H, Hasegawa K, Nakaune R, Lee YJ, Makizumi Y, Akutsu K, Hibi T (2000) Tandem repeat of a transcription enhancer upstream of the sterol 14α demethylase gene (*cyp51*) in *Penicillium digitatum*. Appl Environ Microbiol 66:3421–3426

Hawkins NJ, Cools HJ, Shaw MW, Sierotzki H, Fraaije BA (2011) Molecular mechanisms of altered fungicide sensitivity in *Rhynchosporium secalis*. In: Dehne HW, Deising HB, Gisi U, Kuck KH, Russell PE, Lyr H (eds) Modern fungicides and antifungal compounds VI: proceedings of the 16th international Reinhardsbrunn symposium on modern fungicides and antifungal compounds, 2011. Deutsche Phytomedizinische Gesellschaft, Braunschweig, pp 129–134

Hawkins NJ, Cools HJ, Sierotzki H, Shaw MW, Knogge W, Kelly SL, Kelly DE, Fraaije BA (2014) Paralog re-emergence: a novel, historically-contingent mechanism in the evolution of anti-microbial resistance. Mol Biol Evol 31:1793–1802

Hayashi K, Schoonbeek H, Sugiura H, De Waard MA (2001) Multidrug resistance in *Botrytis cinerea* associated with decreased accumulation of the azole fungicide oxpoconazole and increased transcription of the ABC transporter gene BcatrD. Pestic Biochem Physiol 70:168–179

Hayashi K, Schoonbeek H-J, De Waard MA (2002) Bcmfs1, a novel major facilitator superfamily transporter from *Botrytis cinerea*, provides tolerance towards the natural toxic compounds camptothecin and cercosporin and towards fungicides. Appl Environ Microbiol 68:4996–5004

Hayashi K, Schoonbeek H, De Waard MA (2003) Modulators of membrane drug transporters potentiate the activity of the DMI fungicide oxpoconazole against *Botrytis cinerea*. Pest Manag Sci 59:294–302

Hollomon DW (1993) Resistance to azole fungicides in the field. Biochem Soc Trans 21:1047–1051

Hollomon DW, Butters J, Clark J (1984) Genetic control of triadimenol resistance in barley powdery mildew. In: Proccedings 1984 British crop protection conference, vol. 2. Brighton, UK, pp 477–482

Hulvey J, Porko JT Jr, Hyunkyu J, Berg A, Jung C (2012) Overexpression of ShCYP51B and ShatrD in *Sclerotinia homeocarpa* isolates exhibiting practical resistance to a demethylation inhibitor fungicide. Appl Environ Microbiol 78:6674–6682

Kalamarakis AE, Demopoulos VP, Ziogas BN, Georgopoulos SG (1989) A highly mutable major gene for triadimenol resistance in *Nectria haematococca* var. *cucurbitae*. Neth J Plant Pathol 95:109–120

Kalamarakis AE, De Waard MA, Ziogas BN, Georgopoulos SG (1991) Resistance to fenarimol in *Nectria haematococca* var. *cucurbitae*. Pestic Biochem Physiol 40:212–220

Kato T, Tanaka S, Ueda M, Kawase Y (1974) Effects of fungicide S-1358, on general metabolism and lipid biosynthesis in *Monilinia fructigena*. Agric Biol Chem 38:2377–2384

Kerkenaar A (1990) Inhibition of the sterol Δ14-reductase and Δ8-7-isomerase in fungi. Biochem Soc Trans 18:59–61

Koeller W (1992) Antifungal agents with target sites in sterol functions and biosynthesis. In: Koeller W (ed) Target sites of fungicide action. CRC Press, London, pp 119–206

Kretschmer M, Leroch A, Mosbach A, Walker AS, Fillinger S, Memke D (2009) Fungicide-driven evolution and molecular basis of multidrug resistance in field populations of grey mould fungus *Botrytis cinerea*. PLoS Pathog 5:e1000696

Krishnan-Natesan S, Chandrasekar PH, Alangaden GJ, Manavathu EK (2008) Molecular characterisation of cyp51A and cyp51B genes coding for P450 14α-lanosterol demethylases A (CYP51Ap) and B (CYP51Bp) from voriconazole-resistant laboratory isolates of *Aspergillus flavus*. Int J Antimicrob Agents 32:519–524

Kuck KH, Mehl A (2004) Prothioconazole: sensitivity profile and anti-resistance strategy. Pflanzenschutz Nachrichten Bayer 57:225–236

Leroux P, Walker AS (2011) Multiple mechanisms account for resistance to sterol 14α-demethylation inhibitors in field isolates of *Mycosphaerella graminicola*. Pest Manag Sci 67:44–59

Leroux P, Albertini C, Gautier A, Gredt M, Walker A (2007) Mutations in the CYP51 gene correlated with changes in sensitivity to sterol 14α-demethylation inhibitors in field isolates of *Mycosphaerella graminicola*. Pest Manag Sci 63:688–698

Leroux P, Gredt M, Remuson F, Micoud A, Walker AS (2013) Fungicide resistance status in French populations of the wheat eyespot fungi *Oculimacula acuformis* and *Oculimacula yallundae*. Pest Manag Sci 69:15–26

Liu X, Yu F, Schnabel G, Wu J, Wang Z, Ma Z (2011) Paralogous *cyp51* genes in *Fusarium graminearum* mediate differential sensitivity to sterol demethylation inhibitors. Fungal Genet Biol 48:113–123

Loeffler RST, Butters JA, Hollomon DW (1984) The lipid contents of powdery mildews and their sensitivity to fungicides. In: Proc 1984 British crop protection conference. Brighton, UK, pp 911–916

Luo C-X, Schnabel G (2008) The cytochrome P450 lanosterol 14α-demethylase gene is a demethylation inhibitor fungicide resistance determinant in *Monilinia fructicola* field isolates from Georgia. Appl Environ Microbiol 74:359–366

Ma Z, Michailidis TJ (2005) Advances in understanding molecular mechanisms of fungicide resistance and molecular detection of resistant genotypes in phytopathogenic fungi. Crop Prot 24:853–863

Ma Z, Proffer TJ, Jacobs JL, Sundin GW (2006) Overexpression of the 14α-demethylase target gene (cyp51) mediates fungicide resistance in *Blumeriella jaapii*. Appl Environ Microbiol 72:2581–2585

Maan PA, Parmegiani RM, Wei SQ, Mendrick CA, Li X, Loebenberg D, Di Domenico B, Hare RS, Walker SS, McNicholas PM (2003) Mutation in *Aspergillus fumigatus* resulting in reduced susceptibility to posaconazole appears to be restricted to a single amino acid in the cytochrome P450 14α-demethylase. Antimicrob Chemother 47:577–581

Mellado E, Diaz-Guerra TM, Cuenca-Estrella M, Rodriguez-Tudela JL (2001) Identification of two different 14α-sterol demethylase related genes (*cyp51A* and *cyp51B*) in *Aspergillus fumigatus* and other *Aspergillus* species. J Clin Microbiol 39:2431–2438

Mellado E, Garcia-Effron G, Buitrago MJ, Alcazar-Fuoli L, Cuenca-Estrella A, Rodriguez-Tudela JL (2005) Targeted gene disruption of the 14-alpha sterol demethylase (*cyp51A*) in *Aspergillus fumigatus* and its role in azole drug susceptibility. Antimicrob Agents Chemother 49:2536–2538

Mellado E, Alcazar-Fuoli L, Garcia-Effron G, Alastruey-Izquierdo A, Cuenca-Estrella M, Rodriguez-Tudela JL (2006) New resistance mechanisms to azole drugs in *Aspergillus fumigatus* and emergence of antifungal drugs-resistant *A. fumigatus* atypical strains. Med Mycol 44:367–371

Mellado E, Garcia-Effron G, Alcazar-Fuoli L, Melchers WJG, Verweij PE, Cuenca-Estralla M, Rondriguez-Tudela JL (2007) A new *Aspergillus fumigatus* resistance mechanism conferring *in vitro* cross-resistance to azole antifungal involves a combination of *cyp51A* alterations. Antimicrob Agents Chemother 51:1897–1904

Mullins JGL, Parker JE, Cools HJ, Togawa RC, Lukas A, Fraaije BA, Kelly DE, Kelly SL (2011) Molecular modelling of the emergence of azole resistance in *Mycosphaerella graminicola*. PLoS One 6(6):e20973

Nakajima M, Suzuki J, Hosaka T, Hibi T, Akutsu K (2001) Functional analysis of an ATP-binding cassette transporter gene in *Botrytis cinerea* by gene disruption. J Gen Plant Pathol 67:212–214

Nakaune R, Adachi K, Nawata O, Tomiyama M, Akutsu K, Hibi T (1998) A novel ATP-binding cassette transporter involved in multidrug resistance in the phytopathogenic fungus *Penicillium digitatum*. Appl Environ Microbiol 64:3983–3988

Nakaune R, Hamamoto H, Imada J, Akutsu K, Hibi T (2002) A novel ABC transporter gene PMR5, is isolated in multidrug resistance in phytopathogenic fungus *Penicillium digitatum*. Mol Gen Genomics 267:172–185

Nascimento AM, Goldman GH, Park S, Marras SA, Delmas G, Oza V, Lolans K, Dudley MN, Mann PA, Perlin DS (2003) Multiple resistance mechanisms among *Aspergillus fumigatus* mutants with high-level resistance to itraconazole. Antimicrob Agents Chemother 47:1719–1726

Nikou DC, Malandrakis AA, Konstantakaki M, Vontas JG, Markoglou AN, Ziogas BN (2009) Molecular characterization and detection of overexpressed C-14 alpha-demethylase-based DMI resistance in *Cercospora beticola* field isolates. Pestic Biochem Physiol 95:18–27

Parker JE, Warrilow AGS, Cools HJ, Martel CM, Nes WD, Fraaije BA, Lucas JA, Kelly DE, Keelly SL (2011) Mechanism of binding of prothioconazole to *Mycosphaerella graminicola* cyp51 differs from that of other azole antifungals. Appl Environ Microbiol 77:1460–1465

Pfeufer EE, Ngugi HK (2012) Orchard factors associated with resistance and cross resistance to sterol demethylation inhibitor fungicides in population of *Venturia inaequalis* from Pennsylvania. Phytopathology 102:272–282

Pierce AM, Pierce HD Jr, Unrau AM, Oehlschlager AC (1978) Lipid composition and polyene antibiotic resistance of *Candida albicans* mutants. Can J Biochem 56:135–142

Pontzen R, Poppe B, Berg D (1990) Mode of action of sterol biosynthesis inhibitors in obligate parasites. Pestic Sci 30:357–360

Prasad H, Kapoor K (2005) Multidrug resistance in yeast *Candida*. Int Rev Cytol 242:215–248

Ragsdale NN, Sisler HD (1972) Inhibition of ergosterol biosynthesis in *Ustilago maydis* by the fungicide triarimol. Biochem Biophys Res Commun 46:2048–2053

Reimann S, Deising HB (2005) Inhibition of efflux transporter-mediated fungicide resistance in *Pyrenophora tritici-repentis* by a derivative of 4′-hydroxyflavone and enhancement of fungicide activity. Appl Environ Microbiol 71:3269–3275

Sanchez-Torres P, Tuset JJ (2011) Molecular insights into fungicide resistance in sensitive and resistant *Penicillium digitatum* strains infecting citrus. Postharvest Biol Technol 59:159–165

Schmitz HK, Medeiros CA, Craig IR (2014) Sensitivity of *Phakopsora pachyrhizi* towards quinone-outside-inhibitors and demethylation-inhibitors, and corresponding resistance mechanisms. Pest Manag Sci 70:378–388

Schnabel G, Jones AL (2001) The 14α-demethylase (CYP51) gene is overexpressed in *Venturia inaequalis* strains resistance to myclobutanil. Phytopathology 91:102–110

Sisler HD, Walsh RC, Ziogas BN (1983) Ergosterol biosynthesis: a target of fungicidal action. In: Miyamoto J, Kearny PC (eds) UPAC pesticide chemistry human welfare and the environment. Pergamon Press, New York, pp 129–134

Stammler G, Cordero L, Koch A, Semar M, Schlehuber S (2009) Role of the Y134F mutation in cyp51 and overexpression of cyp51 in the sensitivity response of *Puccinia triticina* to epoxiconazole. Crop Prot 28:891–897

Sun X, Xu Q, Ruan R, Zhang T, Zhu C, Li H (2013) PdMLE1, a specific and active transposon acts as a promoter and confers *Penicillium digitatum* with DMI resistance. Environ Microbiol Rep 5:135–142

Tobin MB, Peery RB, Skatrud PL (1997) Genes encoding multiple drug resistance-like proteins in *Aspergillus fumigatus* and *A. flavus*. Gene 200:11–23

Trocha PJ, Jasne SL, Sprinson DB (1977) Yeast mutants blocked in removing the methyl group of lanosterol at C14: separation of sterols by high-pressure liquid chromatography. Biochemistry 16:4721–4726

Vijaya-Palani P, Lalithakumari D (1999) Resistance of *Venturia inaequalis* to the sterol biosynthesis-inhibiting fungicide penconazole [1-(2-(2,4-dichlorophenyl) pentyl)-1H-1,2,4-triazole]. Mycol Res 103:1157–1164

Walsh RC, Sisler HD (1982) A mutant of *Ustilago maydis* deficient in sterol C14-demethylation: characteristics and sensitivity to inhibitors of ergosterol biosynthesis. Pestic Biochem Physiol 18:122–131

Wyand RA, Brown JKM (2005) Sequence variation in the CYP51 gene of *Blumeria graminis* with resistance to sterol demethylase inhibiting fungicides. Fungal Genet Biol 42:726–735

Yan X, Ma W-B, Wang H, Que Y-W, Ma Z-H, Talbot NJ, Wang Z-Y (2011) A sterol 14α-demethylase is required for conidiation and for mediating sensitivity to sterol demethylation inhibitors by the rice blast fungus *Magnaporthe oryzae*. Fungal Genet 48:144–153

Yin Y, Liu X, Li B, Ma Z (2009) Characterization of sterol demethylation inhibitor-resistant isolates of *Fusarium asiaticum* and *F. graminearum* collected from wheat in China. Phytopathology 99:487–497

Ziogas BN, Markoglou AN (2010) Agricultural pharmacology: biochemistry, physiology, mechanism of action and uses of plant protectants, 2nd edn. Athens, pp 1–872

Ziogas BN, Sisler H, Lusby W (1983) Sterol content and other characteristics of pimaricin-resistant mutants of *Aspergillus nidulans*. Pestic Biochem Physiol 20:320–329

Ziogas BN, Oesterhelt G, Masner P, Steel CC, Furter R (1991) Fenpropimorph: a three site inhibitor of ergosterol biosynthesis in *Nectria haematococca* var. *cucurbitae*. Pestic Biochem Physiol 39:74–83

Zwiers LH, Stergiopoulos I, Van Nistelrooy JGM, De Waard MA (2002) ABC transporters and azole susceptibility in laboratory strains of the wheat pathogen *Mycosphaerella graminicola*. Antimicrob Agents Chemother 46:3900–3906

Chapter 14
Sterol Biosynthesis Inhibitors: C-4 Demethylation

Danièle Debieu and Pierre Leroux

Abstract Two available fungicides block sterol C-4 demethylation in ergosterol biosynthesis: the hydroxyanilide fenhexamid and the amino-pyrazolinone fenpyrazamine, introduced in the early 2000s and 2010s, respectively. Both these fungicides specifically target 3-ketoreductase (Erg27), the third enzyme of the C-4 demethylation enzyme complex. Highly effective against *Botrytis cinerea*, causing gray mold, they are also effective against other pathogens, such as *Monilinia* spp., *Sclerotinia* spp., and *Oculimacula* spp. Up to now, most available data on resistance phenomena have been recorded with fenhexamid on gray mold. The mechanism of acquired resistance to fenhexamid is linked to target modification in most resistant strains. Amino-acid changes due to *erg27* gene mutations possibly decrease the affinity of fenhexamid for 3-ketoreductase. Twenty-three individual mutations, sometimes, but rarely, occurring in pairs, have been identified in moderately resistant strains. The principal highly resistant strains display one out of four detected amino-acid substitutions at position 412. In fungal species naturally resistant to fenhexamid, this intrinsic resistance also appears to be related to a low affinity of the hydroxyanilide for the 3-ketoreductase. Gray mold has recently been shown to be caused by a complex of two cryptic species living in sympatry: *B. cinerea*, the predominant species, and *Botrytis pseudocinerea. B. pseudocinerea* is naturally resistant to fenhexamid, due to its cytochrome P450 protein Cyp684 rather than *erg27* polymorphism. The function of Cyp684 is unknown, but it may be involved in fenhexamid detoxification. Moreover, multidrug-resistant strains exhibiting reduced susceptibility to fenhexamid have been detected in *B. cinerea* and *Oculimacula yallundae*.

Keywords *Botrytis cinerea* • *Botrytis pseudocinerea* • Gray mold • Fungicide resistance • Resistance mechanisms • Fenhexamid • Sterol 3-ketoreductase • *erg27* mutation • Cyp684

D. Debieu (✉) • P. Leroux
INRA AgroParisTech, UMR 1290 BIOGER, Avenue Lucien Brétignières, F-78850 Thiverval-Grignon, France
e-mail: debieu@versailles.inra.fr

H. Ishii, D.W. Hollomon (eds.), *Fungicide Resistance in Plant Pathogens*,
DOI 10.1007/978-4-431-55642-8_14

14.1 Introduction

Sterol biosynthesis inhibitors (SBIs) have been used over the last 40 years, in both medicine and agriculture, to control fungal diseases. Ergosterol, the major final sterol in many fungi, is essential for both bulk membrane and metabolic functions. Inhibition of the enzymes involved in ergosterol biosynthesis leads to ergosterol depletion and the accumulation of precursor or abnormal sterols, resulting in fungitoxicity due to alterations of the membrane and metabolic functions (Köller 1992; Debieu et al. 1998, 2001; Akins 2005). The available SBIs can be distributed into four classes on the basis of the enzyme targeted (Fig. 14.1), a membrane-bound enzyme of the endoplasmic reticulum (Leroux et al. 2008). Five of the 15 post-squalene steps in fungal sterol biosynthesis are targeted by these fungicides. Squalene epoxidase (Erg1) is inhibited by allylamines (e.g., terbinafine) and thiocarbamates (e.g., tolnaftate, pyributicarb), which are used mostly in medicine. The 14α-demethylase cytochrome P450 monooxygenase (Erg11 or Cyp51) is the target of commercially important azole drugs (e.g., fluconazole, ketoconazole, posaconazole) and polyvalent DMI (14α-demethylation inhibitor) fungicides (e.g., prochloraz, tebuconazole, epoxiconazole, prothioconazole). The Δ^{14}-reductase (Erg24) and $\Delta^8 \rightarrow \Delta^7$-isomerase (Erg2) are inhibited by tertiary amines (e.g., amorolfine, fenpropimorph, tridemorph, fenpropidin, spiroxamine). These two enzymes may be differently inhibited in different fungi and with different compounds (Debieu et al. 1998, 2000). Amines are mostly used to combat powdery mildews in agriculture. The third enzyme of the sterol C-4 demethylation complex,

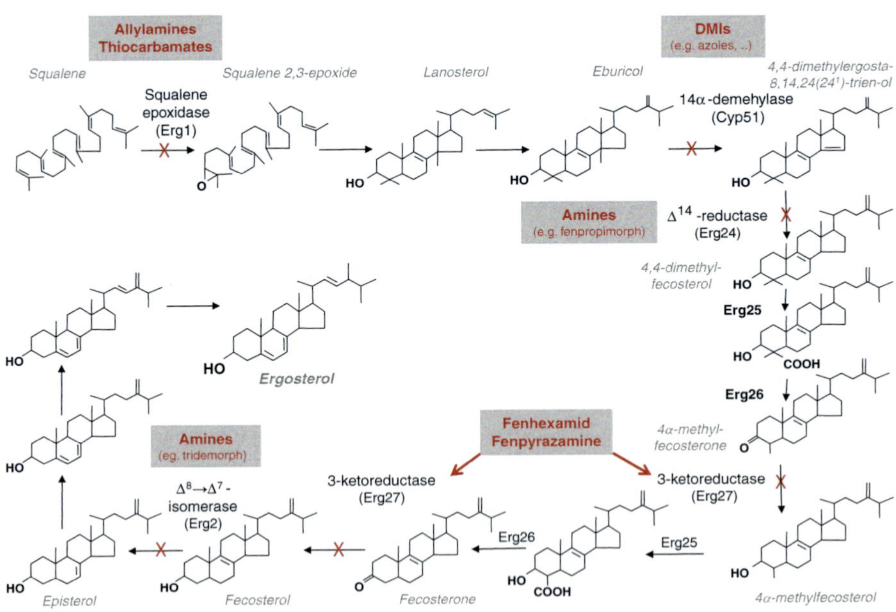

Fig. 14.1 Sterol biosynthesis inhibitors used to control fungal diseases

3-ketoreductase (Erg27), is inhibited by the more recently developed fungicides. This class of fungicides was established in the 2000s, following the elucidation in our laboratory of the mode of action of fenhexamid (Debieu et al. 2001), which was developed by Bayer in 1999 (Krüger et al. 1999). Fenhexamid (1-methyl-cyclohexanecarboxylic acid (2,3-dichloro-4-hydroxy-phenyl)-amide), a hydroxy-anilide derivative, was the only member of this new class of SBIs targeting the sterol 3-ketoreductase until 2012, when Sumitomo Chemical Co. Ltd developed a novel fungicide, fenpyrazamine (S-allyl 5-amino-2,3-dihydro-2-isopropyl-3-oxo-4-(o-tolyl)pyrazole-1-carbothioate), an amino-pyrazolinone derivative targeting the sterol 3-ketoreductase (Tanaka et al. 2012, 2013) (Fig. 14.2). Sterol C-4 demethylation in fungi involves three enzymes (Mercer 1984). A C-4 methyloxidase (Erg25) reaction first converts the 4α-methyl group into an alcohol, then an aldehyde, and, finally, a carboxylic acid (Bard et al. 1996). A C-3 sterol dehydrogenase (Erg26) then removes the 3α-hydrogen, leading to decarboxylation of the 3-ketocarboxylic acid sterol intermediate (Gachotte et al. 1998). Finally, the 3-ketoreductase (Erg27) converts the 3-keto group into a 3β-hydroxyl group (Gachotte et al. 1999). Two consecutive rounds of C-4 demethylation are required to eliminate the two methyl groups on C-4 (Fig. 14.1). In addition to the three enzymes, a protein encoded by *erg28* is thought to facilitate protein-protein interactions within the C-4 demethylation enzyme complex (Gachotte et al. 2001; Mo and Bard 2005). Fenhexamid and fenpyrazamine have a narrow spectrum of activity. They are highly effective against the causal agent of gray mold, *Botrytis cinerea* (Rosslenbroich and Stuebler 2000; Tanaka et al. 2012).

The intensive use of SBIs in medicine and agriculture is leading to the selection of fungal strains with various degrees of resistance to these fungicides. This acquired resistance is often determined by qualitative changes in the target enzymes, possibly resulting in a decrease in the affinity of inhibitors for these proteins (Leroux et al. 2008). Following treatment with azole drugs and DMI fungicides, overexpression of the target *cyp51* gene or of genes encoding membrane transporters increasing the efflux of inhibitors has often been reported (Akins 2005; Cools et al. 2013). This chapter devoted to the sterol 3-ketoreductase inhibitors will review current knowledge about the mechanisms of acquired and natural (intrinsic) resistance to this class of SBIs. Most of the data considered were obtained with fenhexamid.

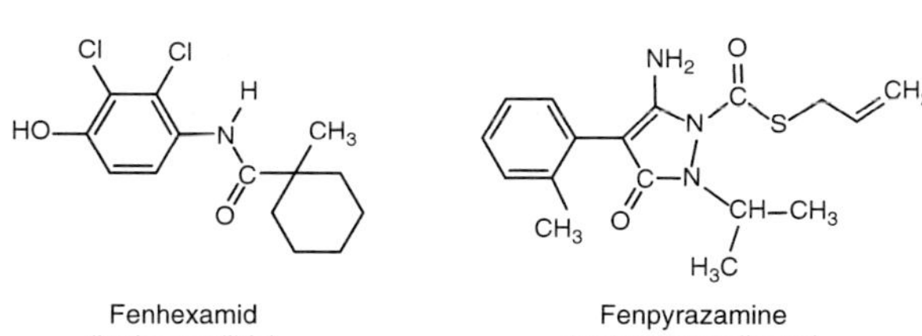

Fig. 14.2 Chemical structures of fenhexamid and fenpyrazamine

14.2 Is the Narrow Spectrum of Fenhexamid and Fenpyrazamine Activity Related to Their Affinity for the 3-Ketoreductase (3-KR)?

The two available fungicides for blocking sterol C-4 demethylation, fenhexamid and fenpyrazamine, are highly toxic to various *Botrytis* species but especially to *B. cinerea* and the related pathogens *Monilinia* spp. and *Sclerotinia* spp. The two species responsible for cereal eyespot, *Oculimacula acuformis* and *O. yallundae,* are also susceptible to these SBIs. By contrast, these fungicides display little or no activity against many other eumycetes, including powdery mildews, rusts, smuts, *Alternaria* spp., *Fusarium* spp., *Mycosphaerella* spp., *Pyrenophora* spp., *Pyricularia* spp., *Rhizoctonia* spp., *Trichoderma* spp., and yeasts (Rosslenbroich 1999; Tanaka et al. 2012; Debieu et al. 2013; Leroux et al. 2013).

In vitro studies on susceptible species at various stages of development have indicated that these fungicides do not affect spore germination; instead, low concentrations of these compounds inhibit germ tube elongation and mycelial growth (Table 14.1). Furthermore, the alterations to germ tube morphology observed are similar to those recorded with other classes of SBIs (Leroux et al. 1999, 2013; Tanaka et al. 2013). The sterol profile in the mycelia of susceptible species (e.g., *B. cinerea* and *Oculimacula* spp.) treated with fenhexamid is characterized by a decrease in ergosterol content and the production of sterones (steryl compounds with a 3-keto rather than a 3β-hydroxyl function) in large quantities (Debieu et al. 2001 and unpublished, Tanaka et al. 2012). Sterones are fungitoxic (Debieu et al. 2001), particularly because the 3β-hydroxyl group is essential for membrane function and cannot therefore be replaced with a 3-keto group (Nes et al. 1993). The three major sterones are 4α-methylfecosterone, fecosterone, and episterone (derived from fecosterone $\Delta^8 \rightarrow \Delta^7$-isomerization), consistent with inhibition of the 3-ketoreductase (Erg27; 3-KR) (Fig. 14.1). The low concentrations of fungicide required to halve the activity of the *B. cinerea* 3-KR (I_{50}) in microsomal assays indicate that fenhexamid and fenpyrazamine are good inhibitors of this enzyme (Debieu et al. 2007, 2013; Tanaka et al. 2013). By contrast, the sterol profile observed in the fenhexamid-treated mycelia of intrinsically resistant fungi (e.g., *Nectria haematococca*, *Pyrenophora teres*) is similar to that of the control, suggesting that the 3-KR enzyme of these species cannot be inhibited by fenhexamid. For fenhexamid, the I_{50} value for *N. haematococca* 3-KR is about 200 times higher than that for *B. cinerea* (Debieu et al. 2013). Thus, intrinsic resistance to this fungicide seems related to a low affinity for the target site. It could therefore be hypothesized that the narrow activity spectra of fenhexamid and, probably, fenpyrazamine reflect their affinity for the 3-KR.

Most fungal NADPH-dependent 3-KR enzymes are characterized by two common features of short-chain dehydrogenase/reductase enzymes: an active catalytic site with the consensus sequence YXXXK and a terminal coenzyme binding site with a characteristic GXXXGXG motif (GANSGI/LG in filamentous fungi but with

Table 14.1 Resistance to the hydroxyanilide fenhexamid in field strains of the causal agents of gray mold

Characteristics	*Botrytis cinerea* Wild-type	HydR2[a]	HydR3⁻	HydR3⁺	MDR[b]	*Botrytis pseudo-cinerea*[c] (HydR1)
In vitro response[d] *to fenhexamid:*						
GT elongation	0.04[e]	S[f]	MR	HR	LR	S
Myc growth	0.016	MR	MR	HR	S	HR
In vitro response[d] *to edifenphos*[g]:						
Myc growth	R	R	R	R	R	S
Strong synergism:						
DMI fungicides	No	Yes	No	No	No	Yes
Resistance basis	–	Detoxification: un-known P450	*Erg27* mutations: reduced affinity of fenhexamid to 3-KR		Efflux pump *BcmfsM2* overproduction	Detoxification: Cyp684(P450)

[a]HydR2, $HydR3^-$ $HydR3^+$ phenotypes resistant to the hydroxyanilide fenhexamid (Leroux et al. 2002a; Fillinger et al. 2008)
[b]MDR: multidrug-resistant strains, MDR2 and MDR3 phenotypes (Kretschmer et al. 2009; Leroux and Walker 2013)
[c]New species, previously referred to as the HydR1 phenotype, naturally resistant to fenhexamid (Walker et al. 2011)
[d]GT elongation: germ-tube elongation; Myc growth: mycelial growth
[e]EC_{50} value expressed in $mg.l^{-1}$ for fenhexamid in wild-type strains
[f]Levels of resistance to fenhexamid estimated as the ratio EC_{50} for the resistant phenotype/EC_{50} for wild-type strains, below 3= S (sensitive); 3–10 = LR (low resistance); 10–100 = MR (moderate resistance) or greater than 100 = HR (high resistance)
[g]EC_{50} values for edifenphos: below 1 $mg.l^{-1}$ = S or higher than 10 $mg.l^{-1}$ = R

N instead of the second G in yeasts) (Oppermann et al. 2003). Moreover, species susceptible to fenhexamid (e.g., *B. cinerea, S sclerotiorum*) have a longer Erg27 sequence than fungal species intrinsically resistant to this fungicide. The two largest extensions are located in the N-terminal part of the molecule, on either side of the NAGI motif, the function of which remains unknown. Finally, the presence of a putative transmembrane domain (TMD) with helical structure at the C-terminal end of the protein is not observed in all fungal species (Fig. 14.3) (Albertini and Leroux 2004; Fillinger et al. 2008; Billard et al. 2011; Debieu et al. 2013). Three-dimensional molecular homology modeling of the fungal 3-KR will provide insight into the differences in susceptibility between species resulting in the narrow spectrum of activity of fenhexamid and fenpyrazamine.

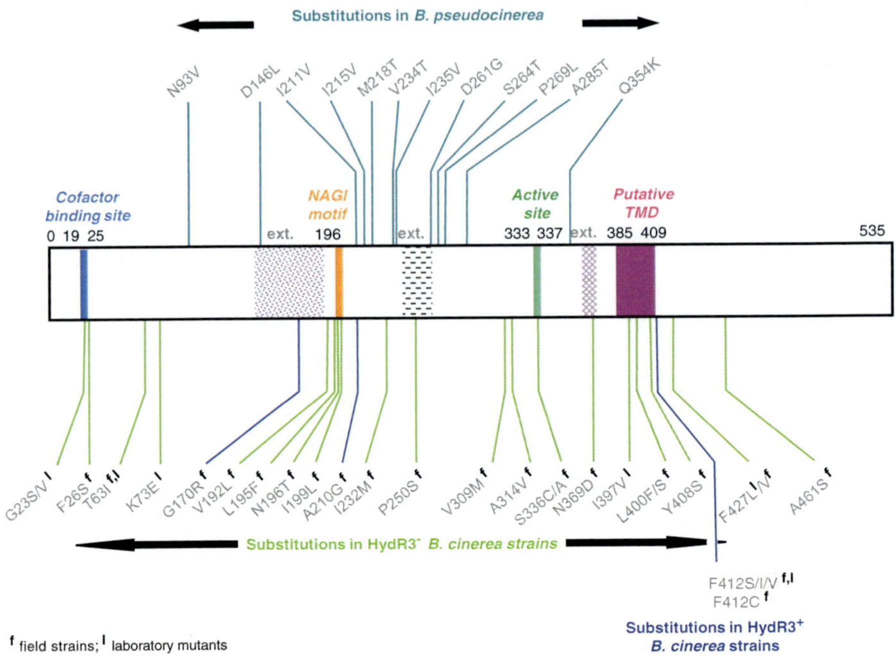

Fig. 14.3 Schematic representation of Erg27 polymorphism in *Botrytis cinerea* and *Botrytis pseudocinerea* (Adapted from Billard et al. 2011). The 12 amino acids that differ between Erg27 *B. pseudocinerea* and *B. cinerea* are indicated above the sequence; the amino-acid substitutions resulting from *erg27* mutations in *B. cinerea* fenhexamid-resistant strains that have been published to date are shown below the sequence. The cofactor (NADPH) binding site, NAGI motif (unknown function), active site, putative transmembrane domain (TMD), and the three main extensions (ext) of the peptide sequence relative to other resistant fungal species (Debieu et al. 2013, supporting information) are indicated in *boxes*

14.3 *Botrytis pseudocinerea* Is a New Species Naturally Resistant to Fenhexamid

Before the release of fenhexamid onto the fungicide market, strains intrinsically resistant to this hydroxyanilide were detected within field populations of *B. cinerea* in Europe (Leroux et al. 1999; Suty et al. 1999). These strains, in which germ tube elongation was sensitive to fenhexamid, but mycelial growth was highly resistant to this fungicide, were referred to as phenotype HydR1 strains (Leroux et al. 2002a, b, 2004). These HydR1 strains with natural resistance to fenhexamid actually correspond to the recently described species *Botrytis pseudocinerea* causing gray mold in sympatry with *B. cinerea* (Walker et al. 2011). These two cryptic species are morphologically indistinguishable, but they display genetic polymorphism for several of the nuclear genes studied, including *erg27* in particular (Albertini and Leroux 2004; Walker et al. 2011). The Erg27 protein sequence of *B. pseudocinerea* differs from that of *B. cinerea* by 12 amino-acid residues (i.e., N93V, D146N, I211V,

I215L, M218T, V234A, I235V, D261G, S264T, P269L, A285T, and Q354K) (Albertini and Leroux 2004). These 12 different amino-acid residues (Fig. 14.3), distributed along the length of the protein, are not located at the NADPH binding site, the NAGI motif, the active site, or the predicted transmembrane domain (TMD) (Albertini and Leroux 2004). In microsomal assays, a higher concentration of fenhexamid was required to inhibit the 3-KR of *B. pseudocinerea* than that of *B. cinerea* (Debieu et al. 2013). Moreover, it has been shown that *erg27* is not overexpressed in *B. pseudocinerea* (Billard 2011), suggesting a decrease of affinity of fenhexamid for *B. pseudocinerea* 3-KR. However, *B. cinerea* transformants harboring the *B. pseudocinerea erg27* gene exhibit a low resistance to fenhexamid, indicating that the *B. pseudocinerea erg27* polymorphism is not the major player in the natural resistance observed (Billard 2011; Billard et al. 2011). Thus, the natural resistance of *B. pseudocinerea*, expressed at mycelial growth stage, appears to be slightly related to the enzyme targeted by fenhexamid, Erg27, a supplemental resistance mechanism that has to be found.

Suty et al. (1999) showed that *B. pseudocinerea* strains metabolized fenhexamid, suggesting a possible resistance mechanism unlinked to the target of the fungicide. Moreover, the strong synergism observed between fenhexamid and DMI fungicides in *B. pseudocinerea* strains but not in *B. cinerea* strains suggests the possible involvement of a cytochrome P450 monooxygenase in fenhexamid detoxification (Leroux et al. 2000, 2002a). This detoxification possibly involved hydroxylation at various positions in the cyclohexyl ring (Leroux et al. 2002b). Among the available *B. cinerea* genome database (Amselem et al. 2011) (http://urgi.versailles.inra.fr/species/Botrytis), 60 putative cytochrome-encoding genes similar to *B. cinerea cyp51* were identified by blastp. Only one of these genes was found to be overexpressed in fenhexamid-treated *B. pseudocinerea* mycelia and not in *B. cinerea* (Billard et al. 2011; Azeddine 2014). According to the nomenclature of P450s, this *Bc1G_02902* gene was named *cyp684*. Its inactivation in *B. pseudocinerea* has been shown to lead to the loss of both natural fenhexamid resistance and synergism between fenhexamid and DMI fungicides (Billard 2011; Billard et al. 2011; Azeddine 2014). Moreover, *B. pseudocinerea* can also be distinguished from *B. cinerea* biologically on the basis of its susceptibility to edifenphos (Table 14.1) (Leroux et al. 2000, 2002a), a phosphorothiolate used against *P. oryzae*. This susceptibility is lost following *cyp684* inactivation (Billard 2011; Billard et al. 2011). Edifenphos is a profungicide, yielding unstable intermediate metabolites (i.e., phenylmercaptan or oxidized derivatives) on P-S cleavage that inhibit fungal phospholipid biosynthesis (Uesugi and Takenaka 1993; Sugiura et al. 1993). These results suggest that Cyp684 plays a major role in detoxifying fenhexamid and activating edifenphos in *B. pseudocinerea*.

In *B. cinerea* and *B. pseudocinerea*, *cyp684* encodes a 500-amino-acid protein with a sequence typical of eukaryotic P450 class II (Werck-Reichhart and Feyereisen 2000). The best protein blast hits are other fungal P450s from species for which complete genome sequences have been published, such as *Aspergillus nidulans*, *Fusarium oxysporum*, and *N. haematococca*. However, the functions of these proteins are unknown. Surprisingly, the *cyp684* gene is absent from *S. sclerotiorum*, which is phylogenetically closer to *B. cinerea* and *B. pseudocinerea* than fungal species listed above. The *Botrytis* spp. Cyp684 sequence comprises three characteristic P450 domains: (1) the heme-binding site including the sequence FGGGSRVCLG (consensus FXXGXRXCXG) with a cysteine residue acting as a fifth ligand for heme iron; (2) the absolutely conserved EXXR motif, which takes the form ETLR in *Botrytis* spp. This motif is located in helix K, close to the heme, and is probably required to stabilize the core structure and (3) the AGSDTT sequence (consensus A/G GX D/E T T/S) of the central part of helix I, involved in proton transfer to heme. Moreover, *Botrytis* spp. Cyp684 contains a membrane-targeting hydrophobic domain at its N-terminal end, mediating anchorage to the endoplasmic reticulum (Werck-Reichhart and Feyereisen 2000; Billard 2011; Azeddine et al. 2012; Azeddine 2014). A comparison of Cyp684 proteins from *B. cinerea* and *B. pseudocinerea* revealed the presence of four polymorphic amino acids, two of which were located in the TMD (V30I and I39T) and the other two being located in the β helix (A100T) or the loop between the F and G helices (for I236M) (Azeddine et al. 2012; Azeddine 2014). This loop is generally thought to be located within the substrate access channel of eukaryotic P450s (Werck-Reichhart and Feyereisen 2000). Moreover, in *B. pseudocinerea cyp684,* a 25 bp deletion in the promoter region and a 24 bp insertion in the 3′UTR may be responsible for the overexpression of *cyp684* in *B. pseudocinerea* (Billard 2011; Azeddine 2014). Further studies are required to assess the impact of *cyp684* mutations and overexpression on the natural resistance of *B. pseudocinerea* and their relative importance for fungicide metabolization.

14.4 Acquired Resistance to Fenhexamid in *Botrytis cinerea*

Up to now, most available data on acquired resistance mechanisms to fenhexamid have been recorded with *B. cinerea*. Two types of fenhexamid resistance phenotype were detected in *B. cinerea* populations, on the basis of the in vitro effects of this hydroxyanilide. These phenotypes were named HydR2 and HydR3 by Leroux et al. (2002a, b). In the small number of HydR2 strains identified to date, resistance is expressed principally at the mycelial growth stage. By contrast, in the more frequent HydR3 strains, resistance is expressed at both the germ tube elongation and mycelial growth stages (Table 14.1).

The first HydR2 strains were collected in the late 1990s, in trials conducted by Bayer on various crops, in Japan and Germany. Their in vitro response to fenhexamid resembles that of *B. pseudocinerea,* resistance but with a lower level of resistance. These HydR2 strains rarely found are not detected in French vineyards (Walker et al. 2013) but possibly, at low frequency, on table grape and strawberry in Southern Italy (De Miccolis Angelini et al. 2014). Recently, Amiri and Peres (2014) detected isolates with reduced sensitivity from strawberry fields in Florida; they published them as HydR2 phenotype, according to their fenhexamid sensitivity evaluation. These isolates are surprisingly representing 25 % of the Florida population. Their frequencies are higher than that of HydR3$^-$ strains but are controlled in fields. More investigations are required to determine both the phenotype and genotype of these isolates for their identification in monitoring of gray mold populations. With the strains isolated from trials conducted by Bayer, strong synergism was also noted between this hydroxyanilide and DMI fungicides (Leroux et al. 2002a, b). The absence of expressed mutations in the *erg27* gene of HydR2 strains (Albertini and Leroux 2004; Billard 2011) suggests that the principal mechanism of resistance may be the detoxification of fenhexamid by a cytochrome P450 monooxygenase, other than Cyp684, that cannot activate edifenphos (Table 14.1) (Billard 2011). Some HydR2 strains produce the red polyketide pigment bikaverin. The production of this pigment is due to the presence of a gene cluster probably acquired from *Fusarium* spp. by horizontal gene transfer. The HydR2 gene locus is genetically linked to this cluster, but none of its six genes seem to be involved in fenhexamid resistance. Moreover, the expression of both the genes of this cluster and the HydR2 genes is upregulated by Bc VEL1, a protein of the VELVET complex that has been reported to be a global regulator of secondary metabolism and differentiation in fungi (Schumacher et al. 2012, 2013).

HydR3 strains display moderate to high levels of resistance to fenhexamid and, possibly, fenpyrazamine, at both the germ tube elongation and mycelial growth stages (Table 14.1) (Fillinger et al. 2008; Tanaka et al. 2012). This phenotype, which has been found in laboratory mutants and, since 2003, in field strains collected from commercial crops, is associated with mutations of the *erg27* gene resulting in changes to the 3-KR protein. In field strains moderately resistant to fenhexamid ($HydR3^-$ strains), the following individual amino-acid changes have been found, sometimes (but rarely) in pairs: F26S, T63I, V192L, L195F, I 199 L, N196T, I232M, P250S, V309M, A314V, S336C/A, N369D, L400F/S, Y408S, F427V, and A461S (Fig. 14.3) (Fillinger et al. 2008; Esterio et al. 2011; Grabke et al. 2013; Amiri and Peres 2014; Saito et al. 2014). In laboratory, mutants with a similar phenotype, T63I substitution has also been found, but others such as G23S/V, K73E, I397V, and F427L substitutions have been observed (De Miccolis Angelini et al. 2012; Saito et al. 2010). Highly resistant strains ($HydR3^+$ strains), with an individual mutation in *erg27*, causing an amino-acid substitution at position 412, have been found in both field strains and laboratory mutants. The most frequent change is the replacement of a phenylalanine with a serine or, more rarely, an isoleucine or valine residue (Fillinger et al. 2008; Saito et al. 2010; Esterio et al. 2011; De Miccolis Angelini et al. 2012). In the Carolinas strawberry fields, in addition to F412S or F412I changes found in resistant strains, Grabke et al. (2013) recorded for the first time the replacement of a phenylalanine with a cysteine residue. Recently, Amiri and Peres (2014) have detected, in addition to strains bearing F412 mutation, strains exhibiting two single mutations, not yet recorded, leading to an amino-acid substitution at position 170 or 210. The resistance level of these strains, found at low frequency, seems to be intermediate between moderate and high according to resistance factor scale for $HydR3^-$ and $HydR3^+$ phenotypes. For these strains, it would be interesting to make in vitro test at both germ tube elongation and mycelial growth stages to confirm $HydR3^+$ phenotype, as initially characterized by Fillinger et al. (2008). The substitutions observed in *B. cinerea* HydR3 strains occur along the entire length of the Erg27 protein sequence and differ from those in *B. pseudocinerea* (Fig. 14.3). They include three substitutions within or close to the NADPH binding site (at residue 23 or 26), two within the NAGI motif (at residue 196 or 199), two within the active site (at residue 336), or eight (at residue 397, 400, 408, or 412) within or close to the putative TMD (Fig. 14.3).

The F412 substitutions, the most frequent ones observed in HydR3 field strains, are predicted to potentially interfere with the helical structure of the TMD and therefore with the fenhexamid and substrate affinity to 3-KR (Billard et al. 2011, 2012a). Microsomal assays have shown that 3KR activity in HydR3$^+$ strains is only one fifth to one half that in wild-type strains, but, surprisingly, these resistant strains have a sterol profile similar to that of wild-type strains (Debieu et al. 2013). Fenhexamid-resistant mutants were generated by site-directed mutagenesis of the HydR3 *erg27* gene in a sensitive recipient strain, to overcome the impact of the different genetic backgrounds of the field strains collected. This work demonstrated that the fenhexamid resistance of HydR3 strains was entirely due to *erg27* mutation (Billard 2011; Billard et al. 2012a; Fillinger et al. 2008). This observation, together with microsomal assays of 3-KR inhibition by fenhexamid, strongly indicates that amino-acid substitutions decrease the affinity of the hydroxyanilide for the 3-KR (by factors of 20 and 100 for HydR3$^-$ and HydR3$^+$ strains, respectively), as indicated by the corresponding I_{50} values (Billard 2011; Debieu et al. 2013, and unpublished). Further studies should make it possible to identify the essential amino acids involved in the binding of the substrate (sterones) and inhibitors, such as fenhexamid or fenpyrazamine, associated with the structural changes induced by the amino-acid substitutions occurring in HydR3 strains.

14.5 Multidrug Resistance

Multidrug resistance, corresponding to the simultaneous resistance of organisms to various unrelated toxic compounds with different modes of action, is caused by an increase in the cellular efflux of these compounds. This phenomenon, which results from the overproduction of plasma membrane transporters, lowers the cytosolic concentrations of the various toxicants. In fungi, the efflux pumps involved are either ATP-binding cassette (ABC) or major facilitator superfamily (MFS) transporters (Kretschmer 2012). Multidrug-resistant (MDR) field strains displaying resistance to fenhexamid have been detected in *B. cinerea* and *O. yallundae* and were initially selected by other fungicides. In both fungi, this phenomenon is expressed principally at the conidial stage (Table 14.1) and also concerns DMIs, tolnaftate, anilinopyrimidines (e.g., cyprodinil, pyrimethanil), and novel succinate dehydrogenase inhibitors (e.g., boscalid, penthiopyrad) (Leroux et al. 1999, 2002a, 2013; Leroux and Walker 2013). In *B. cinerea*, the transporter involved is of the MFS type, and its overproduction results from a combination of the insertion of a retroelement-derived sequence and a deletion in the promoter of *bcmfsM2*. A survey conducted in French and German vineyards highlighted the existence of two different rearrangements that have occurred separately, possibly once in Champagne and then spreading to *B. cinerea* populations from other regions (Kretschmer et al. 2009; Mernke et al. 2011). The *bcmfsM2* deletion in MDR strains results in a phenotype of susceptibility to all the fungicides concerned, including fenhexamid (Kretschmer et al. 2009). Moreover, verapamil, a known modulator of ABC transporters, acts in strong synergism with fungitoxicants in MDR fungi overproducing BcmfsM2, suggesting that this molecule may also inhibit MFS transporters (Leroux and Walker 2013). The transporter involved in multidrug resistance in *O. yallundae* has yet to be identified (Leroux et al. 2013).

14.6 Conclusions

The SBI fungicides blocking C-4 demethylation (i.e., fenhexamid and fenpyrazamine) are characterized by a narrow spectrum of antifungal activity. This is essentially due to major structural differences between fungal 3-ketoreductase (Erg27) enzymes. This enzyme is one of the three enzymes involved in sterol C-4 demethylation and is the target of fenhexamid and fenpyrazamine. Acquired resistance resulting from qualitative modifications of the 3-ketoreductase (HydR3 strains) was recorded in *B. cinerea* after the introduction of this class of SBIs. Resistance management and monitoring are required, to restrict the selection of such strains. For instance, in French vineyards, restrictions have been imposed such that only one treatment of gray mold with 3-ketoreductase inhibitors is permitted per season (Walker et al. 2013). The characterization of HydR3 strains, particularly for the predominant highly resistant strains with substitutions at position 412 of Erg27 ($HydR3^+$), can be achieved with biological methods at the conidial stage, in the presence of high concentrations of fenhexamid (Weber and Hahn 2011; Walker et al. 2013) or with molecular tools. Several PCR assays have been developed for the qualitative or quantitative detection of SNPs (Billard et al. 2012b; Grabke et al. 2013; Amiri and Peres 2014). A slight fitness cost associated with the $HydR3^+$ *erg27* mutation, as demonstrated in isogenic strains, may account for the moderate increase in their frequency in *B. cinerea* populations (Billard et al. 2012a). The mechanism of resistance at work in the less frequent HydR2 strains has yet to be determined, but target modification has been excluded, and it has been suggested that fungicide metabolism is involved. Strains resistant to fenhexamid were detected among the causal agents of gray mold before the release of this fungicide on the market, but they did not lead to practical resistance. In vineyards, these strains (HydR1 phenotype) correspond to the recently identified species *B. pseudocinerea,* living in sympatry with *B. cinerea* and naturally resistant to fenhexamid (Walker et al. 2011). This natural fenhexamid resistance, expressed principally at the mycelial growth stage, probably results from the detoxification of fenhexamid by a cytochrome P450 monooxygenase. Finally, in MDR strains of *B. cinerea,* fenhexamid resistance is associated with the overproduction of a membrane transporter of the MFS type (Kretschmer et al. 2009). A similar phenomenon probably occurred in *O. yallundae,* the causal agent of cereal eyespot (Leroux et al. 2013). As resistance factors are low, no loss of performance in the field due to this mechanism has ever been reported for agricultural fungicides. However, the use of full doses of the individual products and their rotation are recommended. In France, fungicides are rotated between seasons.

References

Akins RA (2005) An update on antifungal targets and mechanisms of resistance in *Candida albicans*. Med Mycol 43:285–318

Albertini C, Leroux P (2004) A *Botrytis cinerea* putative 3-keto reductase gene (ERG27) that is homologous to the mammalian 17 beta-hydroxysteroid dehydrogenase type 7 gene (17 beta-HSD7). Eur J Plant Pathol 110:723–733

Amiri A, Peres NA (2014) Diversity in the *erg27* gene of *Botrytis cinerea* field isolates from strawberry defines different levels of resistance to the hydroxyanilide fenhexamid. Plant Dis 98:1131–1137

Amselem J, Cuomo CA, van Kan JAL et al (2011) Genomic analysis of the necrotrophic fungal pathogens *Sclerotinia sclerotiorum* and *Botrytis cinerea*. PLoS Genet 7:e1002230

Azeddine S (2014) Résistance au fenhexamid dans le complexe d'espèces *Botrytis cinerea/Botrytis pseudocinerea*: Etudes génétiques et moléculaires. Dissertation, University of Paris-sud, Orsay

Azeddine S, Billard A, Solignac P et al (2012) Characterization of the cyp684 gene responsible for fenhexamid resistance in the species *Botrytis pseudocinerea*. In: 11th European conference on fungal genetics, Marburg, Allemagne, 30 March–2 April 2012

Bard M, Bruner DA, Pierson CA et al (1996) Cloning and characterization of ERG25, the *Saccharomyces cerevisiae* gene encoding C-4 sterol methyl oxidase. Proc Natl Acad Sci U S A 93:186–190

Billard A (2011) Fenhexamid: mode d'action et résistance chez le complexe d'espèces *Botrytis* spp., responsable de la pourriture grise de la vigne. Dissertation, University of Paris-sud, Orsay

Billard A, Fillinger S, Leroux P et al (2011) Fenhexamid resistance in the *Botrytis* species complex, responsible for grey mould disease. In: Thajuddin N (ed) Fungicides – beneficial and harmful aspects. InTech, Rijeka (www.intechopen.com), pp 61–78

Billard A, Fillinger S, Leroux P et al (2012a) Strong resistance to the fungicide fenhexamid entails a fitness cost in *Botrytis cinerea*, as shown by comparisons of isogenic strains. Pest Manag Sci 68:684–691

Billard A, Laval V, Fillinger S et al (2012b) The allele-specific probe and primer amplification assay, a new real-time PCR method for fine quantification of single-nucleotide polymorphisms in pooled DNA. Appl Environ Microbiol 78:1063–1068

Cools HJ, Hawkins NJ, Fraaije BA (2013) Constraints on the evolution of azole resistance in plant pathogenic fungi. Plant Pathol 62:36–42

Debieu D, Bach J, Lasseron A et al (1998) Effects of sterol biosynthesis inhibitor fungicides in the phytopathogenic fungus, *Nectria haematococca*: ergosterol depletion versus precursor or abnormal sterol accumulation as the mechanism of fungitoxicity. Pestic Sci 54:157–167

Debieu D, Bach J, Arnold A et al (2000) Inhibition of ergosterol biosynthesis by morpholine, piperidine, and spiroketalamine fungicides in *Microdochium nivale*: effect on sterol composition and sterol delta(8) ->delta(7)-isomerase activity. Pestic Biochem Physiol 67:85–94

Debieu D, Bach J, Hugon M et al (2001) The hydroxyanilide fenhexamid, a new sterol biosynthesis inhibitor fungicide efficient against the plant pathogenic fungus *Botryotinia fuckeliana* (*Botrytis cinerea*). Pest Manag Sci 57:1060–1067

Debieu D, Bach J, Fillinger et al (2007) Fenhexamid: mode of action and resistance in the phytopathogenic fungus *Botrytis cinerea*. In: TERPNET: 8th international meeting: biosynthesis and function of isoprenoids in plants, microorganisms and parasites, Strasbourg, 30 April–4 May 2007

Debieu D, Bach J, Montesinos E et al (2013) Role of sterol 3-ketoreductase sensitivity in susceptibility to the fungicide fenhexamid in *Botrytis cinerea* and other phytopathogenic fungi. Pest Manag Sci 69:642–651

De Miccolis Angelini RM, Pollastro S, Faretra F (2012) Genetics of fungicides resistance in *Botryotinia fuckeliana* (*Botrytis cinerea*). In: Thind TS (ed) Fungicide resistance in crop protection: risk and management. CABI Wallingford, Oxfordshire, pp 237–250

De Miccolis Angelini RM, Rotolo C, Masiello M et al (2014) Occurrence of fungicide resistance in populations of *Botryotinia fuckeliana* (*Botrytis cinerea*) on table grape and strawberry in southern Italy. Pest Manag Sci 70:1785–1796

Esterio M, Ramos C, Walker AS et al (2011) Phenotypic and genetic characterization of Chilean isolates of *Botrytis cinerea* with different levels of sensitivity to fenhexamid. Phytopathol Mediterr 50:414–420

Fillinger S, Leroux P, Auclair C et al (2008) Genetic analysis of fenhexamid-resistant field isolates of the phytopathogenic fungus *Botrytis cinerea*. Antimicrob Agents Chemother 52:3933–3940

Gachotte D, Barbuch R, Gaylor J et al (1998) Characterization of the *Saccharomyces cerevisiae* ERG26 gene encoding the C-3 sterol dehydrogenase (C-4 decarboxylase) involved in sterol biosynthesis. Proc Natl Acad Sci U S A 95:13794–13799

Gachotte D, Sen SE, Eckstein J et al (1999) Characterization of the *Saccharomyces cerevisiae* ERG27 gene encoding the 3-keto reductase involved in C-4 sterol demethylation. Proc Natl Acad Sci U S A 96:12655–12660

Gachotte D, Eckstein J, Barbuch R et al (2001) A novel gene conserved from yeast to humans is involved in sterol biosynthesis. J Lipid Res 42:150–154

Grabke A, Fernandez-Ortuno D, Schnabel G (2013) Fenhexamid resistance in *Botrytis cinerea* from strawberry fields in the Carolinas is associated with four target gene mutations. Plant Dis 97:271–276

Köller W (1992) Antifungal agents with target sites in sterol functions and biosynthesis. In: Köller W (ed) Target sites of fungicide action. CRC Press, Boca Raton, pp 119–206

Kretschmer M (2012) Emergence of multi-drug resistance in fungal pathogens: a potential threat to fungicide performance in agriculture. In: Thind TS (ed) Fungicide resistance in crop protection: risk and management. CABI, Wallingford, Oxfordshire, pp 251–267

Kretschmer M, Leroch M, Mosbach A et al (2009) Fungicide-driven evolution and molecular basis of multidrug resistance in field populations of the grey mould fungus *Botrytis cinerea*. PLoS Pathog OA (US) 5:e1000696

Krüger BW, Etzel W, Goehrt A (1999) Fenhexamid (KBR 2738) – a botryticide from a new chemical class. Pflanzenschutz-Nachrichten Bayer 52:119–126

Leroux P (2004) Chemical control of *Botrytis* and its resistance to chemical fungicides. In: Elad Y et al (eds) *Botrytis*: biology, pathology and control. Kluwer Academic Publishers, Dordrecht, pp 195–222

Leroux P, Walker AS (2013) Activity of fungicides and modulators of membrane drug transporters in field strains of *Botrytis cinerea* displaying multidrug resistance. Eur J Plant Pathol 135:683–695

Leroux P, Chapeland F, Desbrosses D et al (1999) Patterns of cross-resistance to fungicides in *Botryotinia fuckeliana* (*Botrytis cinerea*) isolates from French vineyards. Crop Prot 18:687–697

Leroux P, Chapeland F, Arnold A et al (2000) New cases of negative cross-resistance between fungicides, including sterol biosynthesis inhibitors. J Gen Plant Pathol 66:75–81

Leroux P, Debieu D, Albertini C et al (2002a) The hydroxyanilide botryticide fenhexamid: mode of action and mechanisms of resistance. In: Dehne HW, Gisi U, Kuck KH, Russell PE, Lyr H (eds) Modern fungicides and antifungal compounds III 13th international Reinhardsbrunn symposium. AgroConcept GmbH & Co.KG Verlag, Nordring, Gelsenkirchen, pp 29–40

Leroux P, Fritz R, Debieu D et al (2002b) Mechanisms of resistance to fungicides in field strains of *Botrytis cinerea*. Pest Manag Sci 58:876–888

Leroux P, Bach J, Debieu D et al (2008) Modes of action of sterol biosynthesis inhibitors (SBIs) and resistance phenomena. In: Dehne HW, Gisi U, Kuck KH, Russell PE, Lyr H (eds) Modern fungicides and antifungal compounds V 15th international Reinhardsbrunn symposium. AgroConcept GmbH & Co.KG Verlag, Nordring, Gelsenkirchen, pp 85–92

Leroux P, Gredt M, Remuson F et al (2013) Fungicide resistance status in French populations of the wheat eyespot fungi *Oculimacula acuformis* and *Oculimacula yallundae*. Pest Manag Sci 69:15–26

Mercer EI (1984) The biosynthesis of ergosterol. Pestic Sci 15:133–155

Mernke D, Dahm S, Walker AS et al (2011) Two promoter rearrangements in a drug efflux transporter gene are responsible for the appearance and spread of multidrug resistance phenotype MDR2 in *Botrytis cinerea* isolates in French and German vineyards. Phytopathology 101:1176–1183

Mo C, Bard M (2005) Erg28p is a key protein in the yeast sterol biosynthetic enzyme complex. J Lipid Res 46:1991–1998

Nes WD, Janssen DD, Crumley FG et al (1993) The structural requirements of sterols for membrane function in *Saccharomyces cerevisiae*. Arch Biochem Biophys 300:724–733

Oppermann U, Filling C, Hult M et al (2003) Short-chain dehydrogenases/reductases (SDR): the 2002 update. Chem Biol Interact 143–144:247–253

Rosslenbroich HJ (1999) Efficacy of fenhexamid (KBR 2738) on *Botrytis cinerea* and related fungal pathogens. Pflanzenschutz-Nachrichten Bayer 52:127–144

Rosslenbroich HJ, Stuebler D (2000) *Botrytis cinerea* – history of chemical control and novel fungicides for its management. Crop Prot 19:557–561

Saito S, Furuya S, Takayanagi T et al (2010) Phenotypic analyses of fenhexamid resistant *Botrytis cinerea* mutants. In: Carisse O (ed) Fungicides. InTech, Rijeka (www.intechopen.com), pp 247–260

Saito S, Cadle-Davidson L, Wilcox WF (2014) Selection, fitness, and control of grape isolates of *Botrytis cinerea* variably sensitive to fenhexamid. Plant Dis 98:233–240

Schumacher J, Pradier JM, Simon A et al (2012) Natural variation in the VELVET gene *bcvel1* affects virulence and light-dependent differentiation in *Botrytis cinerea*. PLoS One 7:e47840

Schumacher J, Gautier A, Morgant G et al (2013) A functional bikaverin biosynthesis gene cluster in rare strains of *Botrytis cinerea* is positively controlled by *VELVET*. PLoS One 8:e53729

Sugiura H, Hayashi K, Tanaka T et al (1993) Mutual antagonism between sterol demethylation inhibitors and phosphorothiolate fungicides on *Pyricularia oryzae* and the implication for their mode of action. Pestic Sci 39:193–198

Suty A, Pontzen R, Stenzel K (1999) Fenhexamid – sensitivity of *Botrytis cinerea*: determination of baseline sensitivity and assessment of the resistance risk. Pflanzenschutz-Nachrichten Bayer 52:149–161

Tanaka S, Ishikawa R, Armengaud P et al (2012) General characteristics of fenpyrazamine, a novel fungicidal compound for controlling gray mold. In: AFPP 10th international conference plant diseases, Tours, 3–5 December 2012

Tanaka S, Kiguchi S, Iwahashi F et al (2013) General characteristics of fenpyrazamine, a novel fungicidal compound for controlling gray mold. In: XVI international Botrytis symposium, Bari, 23–28 June 2013

Uesugi Y, Takenaka M (1993) The mechanism of action of phosphorothiolate fungicides. In: Lyr H, Polter C (eds) Proceedings of the 10th international symposium on systemic fungicides and antifungal compounds, Reinhardsbrunn, 3–9 May 1992, pp 159–164

Walker AS, Gautier A, Confais J et al (2011) *Botrytis pseudocinerea*, a new cryptic species causing grey mould in French vineyards in sympatry with *Botrytis cinerea*. Phytopathology 101:1433–1445

Walker AS, Micoud A, Rémuson F et al (2013) French vineyards provide information that opens ways for effective resistance management of *Botrytis cinerea* (grey mould). Pest Manage Sci 69:667–678

Weber RWS, Hahn M (2011) A rapid and simple method for determining fungicide resistance in *Botrytis*. J Plant Dis Prot 118:17–25

Werck-Reichhart D, Feyereisen R (2000) Cytochrome P450: a success story. Genome Biol 1: reviews3003.1-3003.9

Chapter 15
Multidrug Efflux Transporters

Matthias Hahn and Michaela Leroch

Abstract Fungicide resistance is an increasing problem in agriculture and threatens the performance of fungicides to protect plants against fungal diseases. The most common mechanism of resistance in plant pathogenic fungi is target site alteration, which usually confers medium or high levels of resistance against a specific group of fungicides. Resistance based on increased drug efflux mediated by membrane transporters, called MDR (multidrug resistance), has been discovered as a significant alternative resistance mechanism in *Botrytis cinerea* field populations and also in a few other fungi. MDR is caused by mutations leading to overexpression of ABC- or MFS-type multidrug transporters, resulting in low to medium resistance levels against different classes of fungicides. The main MDR type in *B. cinerea*, called MDR1, is of practical relevance because it significantly reduces the sensitivity against the phenylpyrrole fungicide fludioxonil and the anilinopyrimidines. In several countries, *B. cinerea* populations have been observed in recent years that have acquired both MDR1 and target site resistance against all site-specific fungicides registered against *Botrytis*, which make chemical control of grey mould increasingly difficult.

Keywords MDR • ABC transporter • MFS transporter • Drug efflux • Overexpression • Gain-of-function mutations • Multiresistance • *Botrytis cinerea*

15.1 Introduction

Fungicide resistance of plant pathogenic fungi seriously threatens the efficacy of chemical control of plant diseases. It occurs as a consequence of selection pressure by fungicide applications on fungal populations in agricultural environments and is a result of mutations that reduce the sensitivity of the organism against certain drugs. This chapter focuses on fungicide resistance based on increased

M. Hahn (✉) • M. Leroch
Department of Biology, University of Kaiserslautern,
P.O. box 3049, Kaiserslautern, Germany
e-mail: hahn@biologie.uni-kl.de

H. Ishii, D.W. Hollomon (eds.), *Fungicide Resistance in Plant Pathogens*,
DOI 10.1007/978-4-431-55642-8_15

transporter-mediated drug efflux. In order to better understand the agronomical relevance of this type of resistance, other fungicide resistance mechanisms in plant pathogenic fungi will be first briefly discussed.

By far, the most important mechanism is the so-called target site resistance, resulting from changes in the fungicide target protein and subsequent loss or reduction of its affinity to the drug. Many examples of target site resistance exist, which are discussed in detail in other chapters of this volume. The resistance is usually specific for a group of fungicides with the same target protein. For example, the G143A mutation in the gene encoding cytochrome *b* confers high levels of cross resistance in many fungi against all available Qo inhibitors. However, examples of incomplete or even lack of cross resistance of some target resistance mutations exist in other chemical classes, which are explained by differences in the structures of inhibitors and their binding sites in their common target (Sierotzki and Scalliet 2013). Target site resistance only occurs against site-specific fungicides, which dominate the fungicide market since the 1970s. In contrast, fungicides that act as multisite inhibitors, most of which have been developed in the 1960s, are not prone to target site resistance. The probability of target site mutations is usually high provided they don't significantly impair the fitness of the pathogen. A mechanism related to target site resistance is target overexpression. It occurs by mutations that boost the promoter activity of the gene encoding the fungicide target, resulting in higher accumulation of the target protein and reduced fungicide sensitivity. Often, these mutations are insertions of mobile DNA elements (Ma and Michailides 2005). Other mechanisms of resistance, including reduced uptake or increased metabolism of the fungicide, seem to be of low practical importance for the commonly used fungicides.

15.2 General Role of Drug Efflux Transporters in Fungi

Drug efflux transporters are membrane proteins that occur in all living organisms. They mediate the efflux of a variety of toxic substrates from the cell, thus preventing their accumulation to toxic concentrations. In fungi, the role of drug efflux transporters includes the export of self-produced toxins and of any toxic compounds that have entered the cells from the environment. The latter include antifungals produced by bacteria or other microorganisms or plant defence compounds that are particularly relevant for plant pathogens. Finally, drug transporters can also transport synthetic fungicides and therefore play a role in fungicide sensitivity or resistance (Coleman and Mylonakis 2009).

The two major groups of drug transporters in fungi and other eukaryotes are ATP-binding cassette (ABC) transporters and major facilitator superfamily (MFS) transporters. ABC transporters appear to be the dominant drug transporters. They are composed of two halves that can occur in different arrangements, each half containing six transmembrane domains (TMD) and one so-called nucleotide binding domain (NBD). The NBD is located in the cytoplasm and involved in binding and hydrolysis of ATP during the transport process. The tight coupling of ATP cleavage and transport enables ABC transporters to mediate an efficient export of toxic compounds (Lamping et al. 2010). An important property of ABC drug transporters is their low substrate specificity, i.e. their ability to transport a variety of structurally different compounds. For example, for the main drug efflux transporter of yeast, PDR5, several hundred toxic transport substrates have been described (Bauer et al. 1999). Drug efflux transporters are usually only weakly expressed in cells grown under non-stressed conditions. In the presence of drugs, their expression is induced rapidly, and increased efflux occurs within 15–30 min (de Waard et al. 2006). By this means, the cells can quickly adapt to the presence of a multitude of toxic compounds in the environment.

The genomes of filamentous fungi, including the plant pathogens, encode several dozens of ABC transporters, for example, 50 genes are predicted for *Magnaporthe oryzae*, 54 for *Fusarium graminearum* and 42 for *Botrytis cinerea* (Coleman and Mylonakis 2009; Amselem et al. 2011; Ammar et al. 2013). While the role of the large majority of them is still unknown, several ABC transporters from plant pathogenic fungi have been functionally characterized, and some of them will be presented in the following examples.

In the grey mould fungus *Botrytis cinerea*, a major pathogen of many fruits, vegetables and ornamental flowers, *BcatrB* has been shown to transport several antibiotic compounds produced by *Pseudomonas* spp., plant defence compounds such as camalexin and resveratrol and several fungicides (de Waard et al. 2006). *B. cinerea BcatrB* mutants showed increased sensitivity to camalexin and reduced virulence on leaves of *Arabidopsis thaliana*. On the leaf surface, *BcatrB* was induced already in germinated spores by camalexin which was released by the plant in response to the pathogen (Stefanato et al. 2009). In addition, *BcatrB* mutants were slightly more sensitive against the phenylpyrrole fungicides fenpiclonil and fludioxonil, indicating that *BcatrB* and other drug transporters can play a role in the baseline sensitivity of plant pathogenic fungi (Schoonbeek et al. 2001). Thus, due to the wide range of its transported substrates, *BcatrB* plays a significant role in the interaction of *B. cinerea* with other microorganisms, in plant pathogenesis and in fungicide tolerance. As will be shown below, overexpression of *BcatrB* in *B. cinerea* field strains has given rise to the agronomically most important type of multidrug resistance, called MDR1. Another ABC transporter, BcatrD, was found to be an important determinant of sensitivity to azole fungicides. In response to azole treatments, *BcatrD* transcript levels were increased, and *BcatrD* mutants exhibited increased sensitivity to several azoles. Furthermore, laboratory strains that overexpressed *BcatrD* were more tolerant to azoles (de Waard et al. 2006). The natural substrates of BcatrD are unknown. In the rice blast fungus *M. oryzae*, the ABC transporters MgABC1 and MgABC3 have been identified as virulence factors. MgABC1 was found to be induced by several fungicides but did not seem to be required for tolerance to these fungicides but probably to an unknown host defence compound (Urban et al. 1999). Unexpectedly, the substrate of MgABC3 has recently been shown to be an endogenous digoxin-like steroidal glycoside of unknown function for the fungus (Patkar et al. 2012). In the cereal pathogen *F. graminearum*, transcriptome studies revealed 15 genes encoding ABC transporters that are induced by treatment with tebuconazole. The genes for four of these transporters were functionally characterized. Mutants in *FgABC3* and *FgABC4* showed increased sensitivity to triazoles and fenarimol, and mutants in *FgABC1* and *FgABC3* showed reduced virulence on wheat, barley and maize. Based on this and further evidence, it was concluded that FgABC1 might be a transporter for a fungal toxin involved in pathogenesis, and FgABC3 could be involved in the protection against host defence metabolites (Ammar et al. 2013).

Compared to ABC transporters, MFS drug efflux transporters are less well characterized, but they can also play a significant role in drug efflux. They act as proton antiporters that are driven by the proton-motive force across the fungal plasma membrane. In *B. cinerea*, Bcmfs1 was identified as an MFS efflux transporter with a variety of substrates, ranging from the plant toxin camptothecin and the fungal toxin cercosporin to several sterol demethylation inhibitor fungicides (DMIs). A double mutant lacking both *Bcmfs1* and the ABC transporter encoding gene *BcatrD* was more sensitive to DMIs than the *BcatrD* single mutant, indicating that Bcmfs1 contributes to a lower background sensitivity of *B. cinerea* towards these fungicides (Hayashi et al. 2002). In the wheat pathogen *Zymoseptoria tritici* (anamorph of *Mycosphaerella graminicola*), the substrate spectrum of the MFS transporter MgMfs1 was analysed by heterologous expression in a drug-hypersensitive strain of yeast and found to include a broad range of natural toxins and fungicides (Roohparvar et al. 2007). In *Cercospora nicotianae,* the MFS transporter CTB4 is required for efflux of the fungal toxin cercosporin. Mutants in *ctb4* showed drastically reduced production of cercosporin and impaired fungal virulence (Choquer et al. 2007). In addition to CTB4, *C. nicotianae* produces another cercosporin efflux transporter, the ABC transporter ATR1 (Amnuaykanjanasin and Daub 2009).

Fungal drug efflux transporters are often encoded by genes located in secondary metabolite biosynthesis gene clusters, encoding proteins for toxin synthesis, regulation and export. One example is Tri12p, an MFS transporter involved in the efflux of trichothecenes (Kimura et al. 2007; Mernke et al. 2011).

15.3 Efflux Transporter-Mediated Multidrug Resistance (MDR)

As mentioned above, drug efflux transporters can contribute to the baseline sensitivity of fungi towards fungicides. If they were missing, fungi would be much more sensitive against fungicides and other toxins. This is exemplified by the hypersensitive phenotype of the yeast strain *Saccharomyces cerevisiae* AD1-8u- in which eight major ABC drug transporters have been deleted (Decottignies et al. 1998). It therefore seems a paradox that fungicides with excellent antifungal activity could be developed, despite the large arsenal of drug efflux systems. The main reason for this seems to be the regulation of drug transporter genes (Fig.15.1): Most of them are silent or weakly expressed in the absence of drugs and other stresses but rapidly

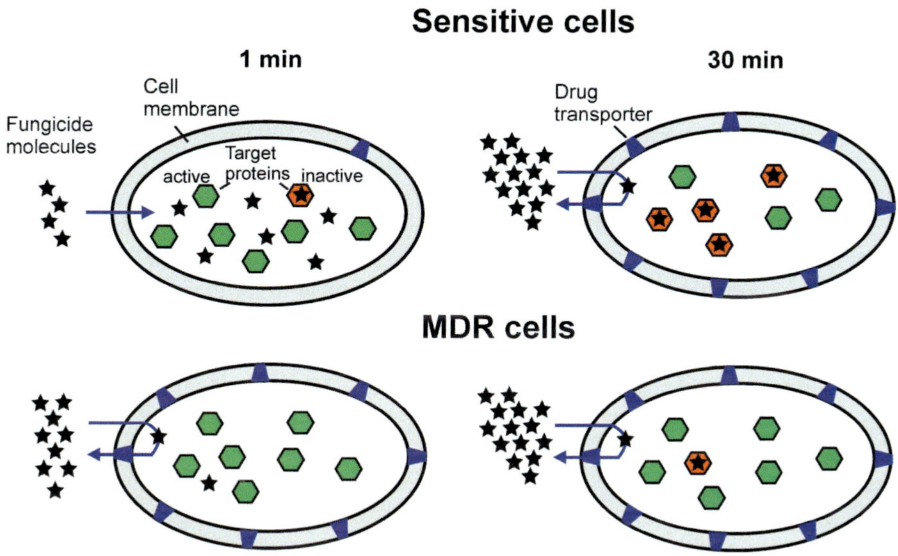

Fig. 15.1 Model for differential, time-dependent uptake of fungicide molecules by sensitive cells with inducible drug transporters and MDR cells with constitutively expressed transporters. The delayed efflux by the sensitive cells results in higher degree of damage, compared to immediate efflux by MDR cells

induced when the cells are exposed to drugs (de Waard et al. 2006). However, although drug efflux can be observed already 10–15 min after the beginning of drug exposure, this lag period seems to be sufficient for many toxins to diffuse into the cells, to bind their target sites and to inhibit fungal metabolism and growth. The lag period can be avoided if the appropriate transporter is constitutively expressed and already 'waiting' for the drugs. In this case, drug uptake is blocked and the cellular drug targets remain active.

In *S. cerevisiae*, the major ABC drug transporter PDR5 is under control of the transcriptional regulator PDR1 and induced by various drugs. Point mutations in PDR1 can result in constitutive activation and PDR5 overexpression, leading to simultaneous resistance to a large variety of drugs, called multidrug resistance (MDR; in yeast called pleiotropic drug resistance (PDR)) phenotype (Balzi et al. 1994). In the human pathogenic yeast *Candida albicans*, expression of CDR1, the orthologue of PDR5, or of the MFS transporter MDR1 was found to be upregulated in clinical isolates showing fluconazole resistance, which demonstrated for the first time a naturally occurring MDR phenotype in a pathogenic fungus (Morschhauser 2010). Similar to the situation in *S. cerevisiae*, *C. albicans* MDR strains carried mutations in the transcription factors Tac1 (regulating CDR1) and Mrr1 (regulating MDR1). These are called gain-of-function mutations because they transform the transcription factors from a drug-inducible into a permanently active, drug-independent state (Morschhauser 2010). As will be shown below, the MDR1 phenotype in *B. cinerea* is caused by similar gain-of-function mutations.

15.4 Preliminary Reports of MDR in Plant Pathogenic Fungi

The first report about field isolates with MDR phenotypes came from *Penicillium digitatum*. Three isolates obtained from infected lemon fruits were resistant against the sterol demethylation inhibitor (DMI) fungicides triflumizole, fenarimol, bitertanol and pyrifenox. The isolates had increased expression levels of the ABC transporter PMR1 compared to DMI sensitive strains. Disruption of PMR1 in one of the MDR isolates resulted in loss of resistance against three of the DMIs, which led to the conclusion that overexpression of PMR1 is responsible for the MDR phenotype (Nakaune et al. 1998). However, in two later studies, this was not confirmed: All three DMI-resistant isolates contained a fivefold amplification of a 126-bp sequence in the promoter of *cyp51* encoding the DMI target, sterol 14α-demethylase, which was correlated with its overexpression. Transformation of a sensitive strain with *CYP51* including its rearranged promoter from a DMI-resistant strain also resulted in DMI resistance (Hamamoto et al. 2000). In contrast, introduction of PMR1 under the control of a strong promoter into a sensitive strain did not lead to DMI resistance (Hamamoto et al. 2001). Therefore, the increased expression of *CYP51* rather than that of *PMR1* was probably responsible for the DMI resistance of these isolates.

Chemical control of *Z. tritici*, causing *Septoria* leaf spot on wheat, mainly relies on the use of azoles (DMIs). The major DMI resistance mechanism is alteration of the target site gene, *cyp51*. Since isolates of *Z. tritici* with the same *cyp51* sequence often have a wide range of sensitivities to certain azoles, it has been proposed that efflux mechanisms could account for this variation. Until now, no clear correlation between expression levels of ABC transporter genes and azole sensitivity has been found (Cools and Fraaije 2013). However, a paper by Leroux and Walker (2011) provided circumstantial evidence for an involvement of drug efflux. The authors classified field isolates of *Z. tritici* according to their levels of resistance. Isolates showing highest resistance levels to DMIs were found to be also resistant to several non-DMI inhibitors. This indicated that, in addition to the *cyp51* mutations, an efflux-based MDR mechanism was active. Further evidence came from experiments using inhibitors (modulators) of ABC transporters. The inhibitors showed a synergistic action with DMIs and some other fungicides only with strains that were phenotypically classified as MDR but not with other strains. The authors concluded that in the *Z. tritici* strains analysed, ABC transporters did not play a major role in DMI resistance but further reduced their fungitoxicity (Leroux and Walker 2011). Increased fungicide efflux, sensitive to various modulators, was demonstrated in two MDR strains of *Z. tritici*, which was correlated with 80-fold overexpression of the *MFS1* transporter gene compared to sensitive strains. Inactivation of *MFS1* abolished the MDR phenotype in one strain, corroborating its requirement for fungicide resistance (Fillinger personal communication).

Oculimacula yallundae causes eyespot on wheat. Resistance monitoring in French wheat fields has revealed the occurrence of a few strains showing evidence for efflux-based multidrug resistance based on their resistance profiles and their sensitivity to modulators (Leroux et al. 2013).

Pyrenophora tritici-repentis is the causal agent of wheat tan spot. Fungal isolates from continuously fungicide-treated wheat fields in Germany showed reduced sensitivity to DMI and QoI fungicides, compared to isolates from untreated fields (Reimann and Deising 2005). To investigate the mechanism of their reduced fungicide sensitivity, sensitive *P. tritici-repentis* isolates were adapted to increased concentrations of DMI and QoI fungicides. The adapted isolates maintained increased tolerance to these fungicides after several transfers on nonselective media. Evidence for increased drug efflux activity of these isolates was obtained by staining of the hyphae with the fluorescent dyes Hoechst 33342 or ethidium bromide, which resulted in reduced nuclear and cytoplasmic staining and increased peripheral staining only of the adapted isolates. Addition of a transporter inhibitor resulted in the disappearance of peripheral fluorescence (Reimann and Deising 2005).

15.5 Discovery and Molecular Basis of MDR Phenotypes in *Botrytis cinerea* Vineyard and Strawberry Populations

In 1994, monitoring of *B. cinerea* field populations in French vineyards for fungicide resistance revealed, in addition to target site resistances, new resistance phenotypes. Compared to strains that showed high levels of resistance to anilinopyrimidines (called AniR1), probably due to target site mutations, these strains (originally called AniR2 and AniR3) were only partially resistant to anilinopyrimidines and showed different patterns of low or intermediate resistance against other fungicides (Chapeland et al. 1999). AniR2 (later called MDR1) strains were partially resistant to two major groups of anti-*Botrytis* fungicides, namely, the anilinopyrimidines (e.g. cyprodinil and pyrimethanil) and the phenylpyrrole fludioxonil. AniR3 (later called MDR2) strains were partially resistant to anilinopyrimidines and three other *Botrytis* fungicides, namely, the hydroxyanilide fenhexamid, the dicarboximide iprodione and the succinate dehydrogenase inhibitor boscalid. A few years later, a combined phenotype of MDR1 and MDR2 was identified, called MDR3, showing resistances of both MDR1 and MDR2 strains. Genetic crosses between MDR strains and sensitive strains revealed that MDR1 and MDR2 were each controlled by a single gene and that MDR3 indeed represents a recombinant of MDR1 and MDR2. The involvement of drug efflux in these strains was confirmed by transport assays using several ^{14}C-labelled fungicides. While the fungicides were transiently taken up by non-MDR strains, indicating the lack of efflux activity in cells immediately after fungicide addition, very little uptake was observed with MDR strains, which was explained by their constitutive activity of efflux transporters (Fig. 15.1; Chapeland et al. 1999; Kretschmer et al. 2009).

To identify the efflux transporters involved in the MDR phenotypes, the expression levels of previously characterized ABC and MFS transporters of *B. cinerea* were analysed. This resulted in the identification of *(Bc)atrB* encoding the ABC transporter AtrB as being constitutively upregulated in MDR1 strains. The transporter responsible for MDR2 was identified by microarray hybridization: In MDR2 strains, *mfsM2* encoding a MFS drug efflux transporter was dramatically upregulated. As described above, MDR strains in *S. cerevisiae* and *C. albicans* carry gain-of-function mutations in transcription factors that activate efflux transporter genes. For *B. cinerea* MDR1 strains, a similar mechanism of *atrB* overexpression was discovered. Using a map-based cloning approach, a gene encoding a zinc-cluster transcription factor called Mrr1 that activates expression of *atrB* was identified. Knock-out of *mrr1* in an MDR1 strain resulted in a mutant that was unable to activate *atrB* and showed similar fungicide sensitivity as an *atrB* mutant, including hypersensitivity to fludioxonil. Significantly, all MDR1 strains tested carried point mutations in the *mrr1* coding region, leading to amino acid exchanges. Transformation of an *mrr1* allele from an MDR1 strain into an *mrr1* mutant resulted in strains with MDR1 phenotype, which confirmed that the point mutations in *mrr1* are responsible for permanent activation of Mrr1 and overexpression of *atrB* (Kretschmer et al. 2009). In MDR2 strains, overexpression of *mfsM2* in MDR2 strains was caused by a different mechanism. All MDR2 and MDR3 strains carried two similar types of rearrangements in the *mfsM2* promoter region, caused by complex insertion-deletion events involving transposable element sequences (Fig. 15.2). Knock-out of *mfsM2* in an MDR2 strain resulted in loss of MDR2 phenotype, and transformation of *B. cinerea* wild type with *mfsM2* under control of a strong promoter led to transformants with MDR2 phenotype (Kretschmer et al. 2009; Mernke et al. 2011).

In vineyards, fungicide sprayings against *B. cinerea* usually occur only once or twice per season. Because fungicides with different modes of action are used in one season, the selection pressure on the fungal population remained moderate and did not lead to excessive frequencies of resistance until now (Leroch et al. 2011; Walker et al. 2013). In contrast to vineyards, strawberry fields are sprayed weekly during

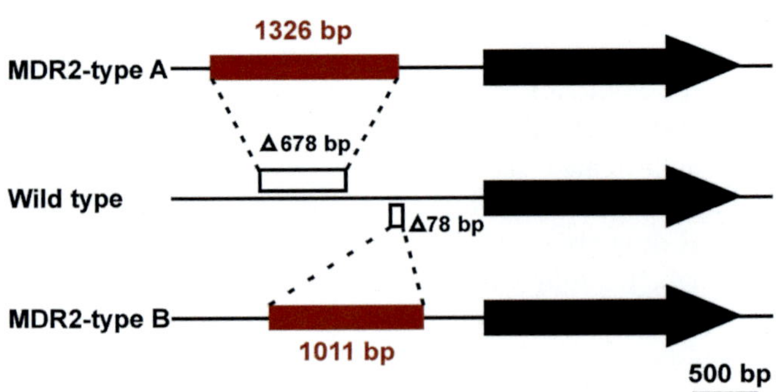

Fig. 15.2 MDR2-related promoter rearrangements in *mfsM2*. Deleted sequences are indicated by *open bars*, insertions derived from transposable element sequences shown in *red*. The dominating type A mutation was observed in MDR2 and MDR3 isolates from German and French vineyards; the type B mutation was found only in a few MDR2 isolates in the Champagne (Modified from Mernke et al. 2011)

flowering time for 4 or more weeks. Repeated sprayings with the same fungicides or fungicide mixtures in one season are common, resulting in strong selection pressure. Consequently, resistance frequencies of grey mould populations in strawberry fields in Germany, Italy and the Southeastern USA have dramatically increased in the last years (Leroch et al. 2013; Amiri et al. 2013; Li et al. 2014; De Miccolis Angelini et al. 2014; Konstantinou et al. 2014). In German strawberry fields, frequencies of MDR1 strains were also high. Closer inspection of these strains revealed that many of them had a 'stronger' phenotype, called MDR1h, i.e. the resistance levels against fludioxonil and cyprodinil were two- to threefold higher than those of MDR1 strains. Furthermore, MDR1h isolates showed higher overexpression levels of *atrB* than MDR1 strains, and they all contained a novel mutation in the *mrr1* gene, namely, a 3-bp deletion resulting in loss of an amino acid (ΔL^{497}) in Mrr1. Surprisingly, the MDR1h strains were found to be genetically different from the common *B. cinerea* strains, as revealed by comparative sequencing of several genes and were therefore referred to as *B. cinerea* group S (Leroch et al. 2013).

In Fig. 15.3, *mrr1* mutations and their effects in MDR1 strains are illustrated. The higher expression levels of *atrB* and increased resistance of MDR1h strains compared to MDR1 strains is correlated with the ΔL^{497} deletion in *mrr1*. We assume that it leads to an even stronger activation of Mrr1 than the various point mutations found in MDR1 strains and that the increased overexpression of *atrB* is responsible for the higher resistance levels. According to the model shown in Fig. 15.2b, in

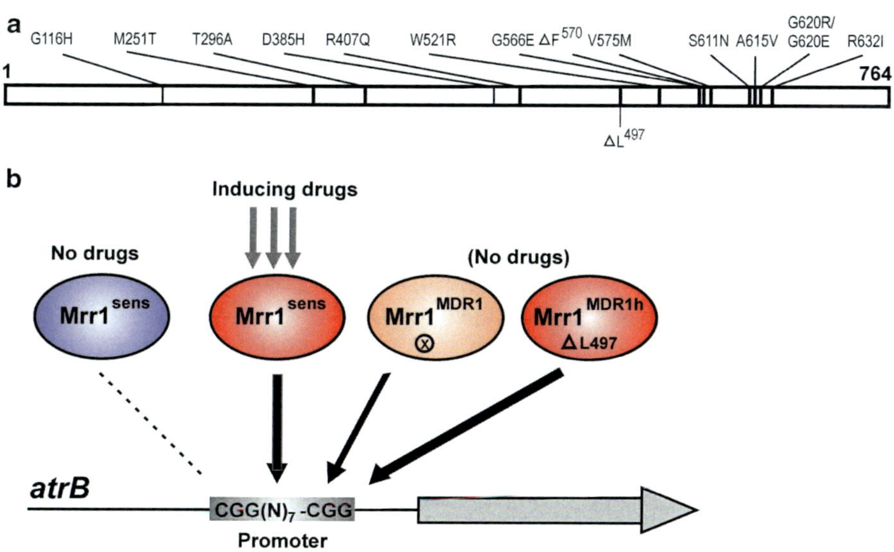

Fig. 15.3 (**a**) MDR1(h)-related mutations in the transcription factor Mrr1. *Letters* and *numbers* indicate amino acids and their positions. MDR1-related mutations are shown above, the MDR1h-related ΔL^{497} mutation below the bar. (**b**) Model for activation of the Mrr1 transcription factor in sensitive, MDR1 and MDR1h strains. The activation state of Mrr1 is indicated by the colour (*blue*, inactive; *orange* and *red*, active)

sensitive but not in MDR1(h) strains, expression of *atrB* requires the presence of inducing drugs (Kretschmer et al. 2009; Leroch et al. 2013). In sensitive cells, activation of Mrr1, which is located in nuclei, might occur by direct interaction of drug molecules (Leroch et al. 2013; Hahn unpublished data), and consequently Mrr1 would act as a nuclear drug receptor (Fig. 15.3). Experimental evidence for this model has been provided in yeast for the transcriptional activator of the PDR5 drug transporter, PDR1, which belongs to the same family of zinc-cluster transcription factors as Mrr1 (Thakur et al. 2008).

15.6 Frequency of Occurrence of *B. cinerea* MDR Field Strains

After the discovery of *B. cinerea* MDR1 and MDR2 strains in 1994, their frequency in French vineyards steadily increased. Several years later, MDR3 strains (carrying MDR1 and MDR2 mutations) appeared. Since 2005, approximately half of the total grey mould population in Champagne vineyards was found to consist of MDR strains (Kretschmer et al. 2009; Walker et al. 2013), and high frequencies of MDR strains were also observed in other vineyards in France and Germany (Leroch et al. 2011). The unusual promoter rearrangement of MDR2 strains indicated that all MDR2 strains with the same mutation in *mfsM2* might have the same origin. This was confirmed by a microsatellite marker-based population analysis which revealed that all French and German MDR2 (and MDR3) strains analysed shared significantly more similarity in the *mfsM2* chromosomal region than non-MDR2 strains. Based on these data, it was suggested that most MDR2 strains were descendants of a single founder cell in a French vineyard and that some of them have migrated eastwards into German vineyards (Mernke et al. 2011). Up to now, MDR2 and MDR3 strains have not been identified anywhere outside of French and German vineyards.

High frequencies of MDR1 and MDR1h strains have been found to occur in German strawberry fields. In 2009–2011, MDR1 strains represented 21–42 % and MDR1h strains 31–61 % of the total population (Leroch et al. 2013). MDR1 phenotypes in field strains can be identified by their partial fludioxonil resistance. This is because *B. cinerea* strains with high levels of fludioxonil resistance due to target site mutations are not observed in the field. The test for cyprodinil sensitivity often fails to provide an answer because many MDR1 strains in strawberry fields are highly resistant to anilinopyrimidines due to target site mutations. Confirmation of MDR phenotypes can be obtained by an assay for resistance to tolnaftate (Kretschmer et al. 2009; Leroux and Walker 2011; Leroux et al. 2012). However, this test has not been used in the publications cited below. Because reduced fludioxonil sensitivity in *B. cinerea* has been always found to be correlated with MDR1, we refer to such isolates as probably being MDR1. In strawberry fields in North and South Carolina, no reduced sensitivity to fludioxonil was observed in 217 *B. cinerea* isolates obtained in 2011, although 47 % of the isolates were resistant to cyprodinil

(Fernandez-Ortuno et al. 2013). In contrast, grey mould populations from strawberry fields in Florida, which showed high frequencies of target site resistance to diverse fungicides, also revealed reduced sensitivity to fludioxonil (MDR1 phenotypes) for 18 % of the isolates (Amiri et al. 2013). In Italy, a significant proportion of *B. cinerea* strains isolated from table grapes (De Miccolis Angelini et al. 2014) showed partial fludioxonil resistance. Most of these isolates were highly resistant to anilinopyrimidines. In Greece, resistance against multiple fungicides, including fludioxonil, has been observed in isolates obtained from seedling pome- and stone-fruit rootstock plants cultivated in greenhouses. These isolates possessed the ΔL^{497} deletion in *mrr1*, and, thus, they were identified as of the MDR1h phenotype (Karaoglanidis personal communication).

Taken together, while MDR2 and MDR3 phenotypes seem to be restricted to vineyards in France and Germany and maybe some adjacent wine-growing regions, MDR1 and MDR1h phenotypes are widely distributed and might further increase in frequencies in the coming years (Table 15.1).

Table 15.1 Properties of MDR types in *B. cinerea*

	MDR1 (AniR2)	MDR1h	MDR2 (AniR3)	MDR3
Fungicides affected (resistance factors)	Cyprodinil (5–10x)	Cyprodinil (13–18x)	Fenhexamid (10x)	Cyprodinil (15x)
	Fludioxonil (5–10x)	Fludioxonil (15–20x)	Cyprodinil (6x)	Fludioxonil (11x)
	Tolnaftate[a] (20x)	Tolnaftate[a] (20x)	Boscalid (6x)	Fenhexamid (10x)
			Iprodione (5x)	Boscalid (13x)
			Cycloheximide[b] (14x)	QoIs[c] (5-16x)
			Tolnaftate[a] (>25x)	Iprodione (6x)
				Cycloheximide[b] (14x)
				Tolnaftate[a] (>25x)
Occurrence: (a) Host plants	Grapes, strawberries, other soft fruits	Strawberries, other soft fruits, pome- and stone-fruit rootstock seedlings	Grapes	Grapes
(b) Countries	France, Germany, USA	Germany, France, Greece, Spain, USA	France, Germany	France, Germany
Drug transporter involved	AtrB	AtrB	MfsM2	AtrB and MfsM2
Mutations	Various point mutations in *mrr1*	3-bp deletion in *mrr1*	Rearrangements in *mfsM2* promoter	As in MDR1 and MDR2

[a]Not used against *B. cinerea*

[b]Antibiotic. Data from Kretschmer et al. (2009), Leroux et al. (2010, 2013), and unpublished results

[c]Azoxystrobin, pyraclostrobin and trifloxystrobin (Kretschmer et al. 2009; Leroux et al. 2010)

15.7 Practical Relevance of MDR in Multiresistant *B. cinerea* Populations

The increasing occurrence of *B. cinerea* MDR strains in many vineyards and strawberry fields raises the question whether MDR strains can compromise fungicide control. Resistance levels of MDR strains against *Botrytis* fungicides are low (resistance factors up to tenfold) or medium (up to 20-fold), whereas medium to high resistance levels (resistance factors up to several hundredfolds) are achieved by target site mutations. Therefore, MDR will not completely abolish fungicide efficacy. So far, no evidence for a decrease of fungicide efficacy due to MDR strains has been reported in French vineyards (Petit et al. 2010; Walker et al. 2013). Nevertheless, experimental verification for this is difficult to achieve, and it could well be that some losses have occurred. Inoculation tests with MDR1 and sensitive *B. cinerea* strains in vineyards indicated that the protective effect of fludioxonil and cyprodinil can be reduced by MDR1 strains (Leroch and Hahn unpublished data).

Compared to the situation in vineyards, the situation in strawberry fields is much more threatening, because of the occurrence of multiresistant (MR) strains with combinations of target site resistance and MDR. In German strawberry fields, *B. cinerea* strains with MDR1h phenotypes (belonging to group S) had significantly higher frequencies of multiple target site resistance than MDR1 or non-MDR strains. In fact, the first 'super-resistant' *B. cinerea* isolate was recovered in 2011 that had a combination of MDR1h and target site resistances against all site-specific fungicides which have been used previously (benzimidazoles, dicarboximides) or which are currently registered (fenhexamid, anilinopyrimidines, QoI and boscalid) in Germany and other countries (Leroch et al. 2013). Super-resistant strains including MDR1h with the ΔL^{497} mutation have recently been identified also in the Southeastern USA and in Greece (Schnabel and Karaoglanidis personal communications) and have increased in frequency in German strawberry fields in the last few years (Hahn unpublished). The practical relevance of MDR1(h) is evident for strains that show multiple target site resistance against QoIs, boscalid, fenhexamid and anilinopyrimidines. By conferring partial resistance to fludioxonil, MDR1(h) partially closes the gap that is left in *B. cinerea* by its inability to develop target site resistance against fludioxonil. Switch®, a mixture of cyprodinil and fludioxonil, is the only fungicide that maintains partial activity against *B. cinerea* MR strains. However, if target site resistance to anilinopyrimidines is combined with MDR1h, the activity of Switch® is severely reduced (Amiri et al. 2013; Hahn and Weber unpublished).

15.8 Fitness of MDR Strains

If the fitness of fungicide resistant strains is markedly reduced, they will not appear at all in the fields or only transiently after fungicide treatments. For *B. cinerea* MDR strains, no evidence for fitness defects was obtained. The steady increase of these strains in Champagne vineyards to high population levels confirms that they efficiently compete with non-MDR strains both during and between the seasons (Kretschmer et al. 2009). Similarly, the increasing populations with MDR1h phenotypes in strawberry fields indicate that they don't suffer significant fitness penalties. Therefore, constitutive overexpression of the drug efflux transporters *mfsM2* and *atrB* does not affect growth and reproduction under field conditions. This is confirmed by the phenotypic characterization of artificial *B. cinerea* MDR strains that have been generated with constructs leading to overexpression of *mfsM2* and *atrB* (Kretschmer et al. 2009; Hahn and Leroch unpublished). Comparative tests for growth, sporulation and other fitness parameters did not reveal differences between these strains and the sensitive parent strain (Hahn and Leroch unpublished). Nevertheless, slight fitness defects of the MDR strains under field conditions cannot be excluded.

15.9 Control of MDR Strains by Inhibitors of Efflux Transporters

Efflux-based MDR is a major problem in medicine and one of the most important resistance mechanisms in cancer cells and in azole-resistant *Candida* spp. (Cannon et al. 2009). The problems with increasing failures of cancer chemotherapy and fungicide treatments have stimulated research aimed at the development of chemicals that inhibit efflux-based drug resistance (Monk and Goffeau 2008). One approach is the use of transporter inhibitors, the so-called MDR modulators, to increase the activity of currently used drugs. Various compounds that are active against P-glycoprotein, the main human ABC-type drug transporter, have been described; however, none of them have been registered for medical treatments yet (Ponte-Sucre 2007). Several modulators have been tested for their activity in plant pathogenic fungi. In *Z. tritici*, increased azole activity was observed in the presence of some of these compounds *in vitro*, but no synergism between the fungicide and the modulators was observed when the experiments were performed on wheat seedlings (Stergiopoulos and de Waard 2002; Roohparvar et al. 2007). For *B. cinerea*, the phenothiazine chlorpromazine and the macrolide tacrolimus showed modulator activity towards oxpoconazole. Significantly, the degree of synergism between the fungicide and the modulator was highest in a strain that overexpressed the ABC transporter AtrD, while it was absent in an *atrD* knock-out mutant (Hayashi et al. 2003). This supports the idea that the modulators directly inhibit the AtrD efflux transporter. In a recent study, synergism between modulators and DMIs was confirmed. No such synergism was observed between modulators and fungicides that are substrates of AtrB, but surprisingly, synergism was observed between modulators and fungicides that are substrates for the MFS transporter MfsM2 (Leroux and Walker 2013). From these results, the authors concluded that the modulator verapamil is able to inhibit the efflux activity of MfsM2.

In summary, the available data demonstrate that modulators can inhibit drug efflux transporters in plant pathogenic fungi to some extent. However, the most important mechanism of fungicide resistance in plant pathogenic fungi is the target site alteration. Practical relevance of MDR in the field has unambiguously been demonstrated so far only for *B. cinerea*, but for the most important MDR type, MDR1(h), which is mediated by AtrB, no modulator has been identified so far for this transporter. Therefore, it appears to be unlikely that modulators will be used for improving fungicide efficacy in agriculture.

15.10 Concluding Remarks

So far, efflux-based resistance mechanisms have been observed only in few plant pathogens. Because of the widespread use of fungicides worldwide, it can be expected that they will be discovered in more fungi in the future. Whether they will become of practical relevance depends on the resistance levels they confer. A potential problem is the fact that MDR can lower the efficacy of new fungicides that have not yet been released, including multisite inhibitors which are not prone to target site resistance.

Acknowledgments We are grateful to our colleagues George Karaoglanidis, Roland Weber and Sabine Fillinger for sharing unpublished data for this manuscript.

References

Amiri A, Heath SM, Peres NA (2013) Phenotypic characterization of multifungicide resistance in *Botrytis cinerea* isolates from strawberry fields in Florida. Plant Dis 97(3):393–401
Ammar A, Tryono R, Döll K et al (2013) Identification of ABC transporter genes of *Fusarium graminearum* with roles in azole tolerance and/or virulence. PLoS One 8(11):e79042
Amnuaykanjanasin A, Daub ME (2009) The ABC transporter ATR1 is necessary for efflux of the toxin cercosporin in the fungus *Cercospora nicotianae*. Fungal Genet Biol 46(2):146–158
Amselem J, Cuomo CA, van Kan JA et al (2011) Genomic analysis of the necrotrophic fungal pathogens *Sclerotinia sclerotiorum* and *Botrytis cinerea*. PLoS Genet 7(8):e1002230
Balzi E, Wang M, Leterme S et al (1994) Pdr5, a novel yeast multidrug resistance conferring transporter controlled by the transcription regulator Pdr1. J Biol Chem 269(3):2206–2214
Bauer BE, Wolfger H, Kuchler K (1999) Inventory and function of yeast ABC proteins: about sex, stress, pleiotropic drug and heavy metal resistance. Biochim Biophys Acta 1461(2):217–236
Cannon RD, Lamping E, Holmes AR et al (2009) Efflux-mediated antifungal drug resistance. Clin Microbiol Rev 22(2):291–298
Chapeland F, Fritz R, Lanen C et al (1999) Inheritance and mechanisms of resistance to anilinopyrimidine fungicides in *Botrytis cinerea* (*Botryotinia fuckeliana*). Pestic Biochem Physiol 64(2):85–100
Choquer M, Fournier E, Kunz C et al (2007) *Botrytis cinerea* virulence factors: new insights into a necrotrophic and polyphageous pathogen. FEMS Microbiol Lett 277(1):1–10
Coleman JJ, Mylonakis E (2009) Efflux in fungi: La pièce de résistance. PLoS Pathog 5(6):e1000486
Cools HJ, Fraaije BA (2013) Update on mechanisms of azole resistance in *Mycosphaerella graminicola* and implications for future control. Pest Manag Sci 69(2):150–155

De Miccolis Angelini RM, Rotolo C, Masiello M et al (2014) Occurrence of fungicide resistance in populations of *Botryotinia fuckeliana* (*Botrytis cinerea*) on table grape and strawberry in southern Italy. Pest Manag Sci 70(12):1785–1796

De Waard MA, Andrade AC, Hayashi K et al (2006) Impact of fungal drug transporters on fungicide sensitivity, multidrug resistance and virulence. Pest Manag Sci 62(3):195–207

Decottignies A, Grant AM, Nichols JW et al (1998) ATPase and multidrug transport activities of the overexpressed yeast ABC protein Yor1p. J Biol Chem 273(20):12612–12622

Fernandez-Ortuno D, Bryson PK, Grabke A et al (2013) Monitoring for resistance in *Botrytis cinerea* from strawberry to seven chemical classes of fungicides in the eastern United States. Phytopathology 103(6):43

Hamamoto H, Hasegawa K, Nakaune R et al (2000) Tandem repeat of a transcriptional enhancer upstream of the sterol 14 alpha-demethylase gene (CYP51) in *Penicillium digitatum*. Appl Environ Microbiol 66(8):3421–3426

Hamamoto H, Nawata O, Hasegawa K et al (2001) The role of the ABC transporter gene PMR1 in demethylation inhibitor resistance in *Penicillium digitatum*. Pestic Biochem Physiol 70(1):19–26

Hayashi K, Schoonbeek HJ, De Waard MA (2002) Bcmfs1, a novel major facilitator superfamily transporter from *Botrytis cinerea*, provides tolerance towards the natural toxic compounds camptothecin and cercosporin and towards fungicides. Appl Environ Microbiol 68(10):4996–5004

Hayashi K, Schoonbeek HJ, De Waard MA (2003) Modulators of membrane drug transporters potentiate the activity of the DMI fungicide oxpoconazole against *Botrytis cinerea*. Pest Manag Sci 59(3):294–302

Kimura M, Tokai T, Takahashi-Ando N et al (2007) Molecular and genetic studies of *Fusarium* trichothecene biosynthesis: pathways, genes, and evolution. Biosci Biotechnol Biochem 71(9):2105–2123

Konstantinou S, Veloukas T, Leroch M et al (2014) Population structure, multiple fungicide resistance profile, and *sdhB* mutation frequency of *Botrytis cinerea* from strawberries and greenhouse-grown tomatoes in Greece. Plant Dis 99(2):240–248

Kretschmer M, Leroch M, Mosbach A et al (2009) Fungicide-driven evolution and molecular basis of multidrug resistance in field populations of the grey mould fungus *Botrytis cinerea*. PLoS Pathog 5(12):e1000696

Lamping E, Baret PV, Holmes AR et al (2010) Fungal PDR transporters: phylogeny, topology, motifs and function. Fungal Genet Biol 47(2):127–142

Leroch M, Kretschmer M, Hahn M (2011) Fungicide resistance phenotypes of *Botrytis cinerea* isolates from commercial vineyards in South West Germany. Phytopathology 159(1):63–65

Leroch M, Plesken C, Weber RW et al (2013) Gray mold populations in German strawberry fields are resistant to multiple fungicides and dominated by a novel clade closely related to *Botrytis cinerea*. Appl Environ Microbiol 79(1):159–167

Leroux P, Walker AS (2011) Multiple mechanisms account for resistance to sterol 14 alpha-demethylation inhibitors in field isolates of *Mycosphaerella graminicola*. Pest Manag Sci 67(1):44–59

Leroux P, Walker AS (2013) Activity of fungicides and modulators of membrane drug transporters in field strains of *Botrytis cinerea* displaying multidrug resistance. Eur J Plant Pathol 135(4):683–693

Leroux P, Gredt M, Leroch M, Walker AS (2010) Exploring mechanisms of resistance to respiratory inhibitors in field strains of *Botrytis cinerea*, the causal agent of gray mold. Appl Environ Microbiol 76(19):6615–6630

Leroux P, Gredt M, Remuson F et al (2013) Fungicide resistance status in French populations of the wheat eyespot fungi *Oculimacula acuformis* and *Oculimacula yallundae*. Pest Manag Sci 69(1):15–26

Li XP, Fernandez-Ortuno D, Grabke A et al (2014) Resistance to fludioxonil in *Botrytis cinerea* isolates from blackberry and strawberry. Phytopathology 104(7):724–732

Ma Z, Michailides TJ (2005) Advances in understanding molecular mechanisms of fungicide resistance and molecular detection of resistant genotypes in phytopathogenic fungi. Crop Prot 24(10):853–863

Mernke D, Dahm S, Walker AS et al (2011) Two promoter rearrangements in a drug efflux transporter gene are responsible for the appearance and spread of multidrug resistance phenotype MDR2 in *Botrytis cinerea* isolates in French and German vineyards. Phytopathology 101(10):1176–1183

Monk BC, Goffeau A (2008) Outwitting multidrug resistance to antifungals. Science 321(5887):367–369

Morschhauser J (2010) Regulation of multidrug resistance in pathogenic fungi. Fungal Genet Biol 47(2):94–106

Nakaune R, Adachi K, Nawata O et al (1998) A novel ATP-binding cassette transporter involved in multidrug resistance in the phytopathogenic fungus *Penicillium digitatum*. Appl Environ Microbiol 64(10):3983–3988

Patkar RN, Xue YK, Shui G et al (2012) Abc3-mediated efflux of an endogenous digoxin-like steroidal glycoside by *Magnaporthe oryzae* is necessary for host invasion during blast disease. PLoS Pathog 8(8):e1002888

Petit AN, Vaillant-Gaveau N, Walker AS et al (2010) Determinants of fenhexamid effectiveness against grey mould on grapevine: respective role of spray timing, fungicide resistance and plant defences. Crop Prot 29(10):1162–1167

Ponte-Sucre A (2007) Availability and applications of ATP-binding cassette (ABC) transporter blockers. Appl Microbiol Biotechnol 76(2):279–286

Reimann S, Deising HB (2005) Inhibition of efflux transporter-mediated fungicide resistance in *Pyrenophora tritici-repentis* by a derivative of 4′-hydroxyflavone and enhancement of fungicide activity. Appl Environ Microbiol 71(6):3269–3275

Roohparvar R, Huser A, Zwiers LH et al (2007) Control of *Mycosphaerella graminicola* on wheat seedlings by medical drugs known to modulate the activity of ATP-binding cassette transporters. Appl Environ Microbiol 73(15):5011–5019

Schoonbeek H, Del Sorbo G, de Waard MA (2001) The ABC transporter BcatrB affects the sensitivity of *Botrytis cinerea* to the phytoalexin resveratrol and the fungicide fenpiclonil. Mol Plant-Microbe Interact 14(4):562–571

Sierotzki H, Scalliet G (2013) A review of current knowledge of resistance aspects for the next-generation succinate dehydrogenase inhibitor fungicides. Phytopathology 103(9):880–887

Stefanato FL, Abou-Mansour E, Buchala A et al (2009) The ABC transporter BcatrB from *Botrytis cinerea* exports camalexin and is a virulence factor on *Arabidopsis thaliana*. Plant J 58(3):499–510

Stergiopoulos I, de Waard MA (2002) Activity of azole fungicides and ABC transporter modulators on *Mycosphaerella graminicola*. J Phytopathol 150(6):313–320

Thakur JK, Arthanari H, Yang FJ et al (2008) A nuclear receptor-like pathway regulating multidrug resistance in fungi. Nature 452(7187):604–609

Urban M, Bhargava T, Hamer JE (1999) An ATP-driven efflux pump is a novel pathogenicity factor in rice blast disease. EMBO J 18(3):512–521

Walker AS, Micoud A, Remuson F et al (2013) French vineyards provide information that opens ways for effective resistance management of *Botrytis cinerea* (grey mould). Pest Manag Sci 69(6):667–678

Part III
Monitoring Resistance

Chapter 16
Monitoring Resistance in Obligate Pathogens by Bioassays Relating to Field Use: Grapevine Powdery and Downy Mildews

Marie-France Corio-Costet

Abstract Monitoring is required to detect and estimate the sensitivity status of pathogen populations to fungicides for the effective control of diseases in a setting of sustainable disease management. Several biological methods are available to detect and to quantify the emergence and evolution of fungicide resistance in obligate parasites in the vineyard such as *Erysiphe necator* and *Plasmopara viticola*. To perform good monitoring, particular attention must be paid to the sampling in the vineyard that will serve to determine baseline and discriminant doses. This chapter first describes the methods used to study fungicide resistance in the field and then compares monitoring results for various fungicides and the role of pathogen diversity. This information will help users to choose the most suitable method and to manage resistance.

Keywords CAA • DMI • *Erysiphe necator* • Fungicide • Grapevine • Monitoring resistance • *Plasmopara viticola* • QoI

Abbreviations

CAA	Carboxylic acid amid
A.i.	Active ingredient
DD	Discriminant dose
DMI	C14-sterol demethylase inhibitor
EC_{50}	Effective concentration of fungicide inhibiting pathogen development at 50 %
MIC	Minimum inhibitory concentration
QiI	Quinone inside inhibitor
QoI	Quinone outside inhibitor

M.-F. Corio-Costet (✉)
INRA, UMR Santé et Agroécologie du vignoble (UMR 1065, SAVE), ISVV, LabexCOTE, Rue Edouard Bourlaux, CS 20032, 33882 Villenave d'Ornon Cédex, France
e-mail: coriocos@bordeaux.inra.fr

H. Ishii, D.W. Hollomon (eds.), *Fungicide Resistance in Plant Pathogens*,
DOI 10.1007/978-4-431-55642-8_16

q-PCR Quantitative polymerase chain reaction
RF Resistance factor

16.1 Introduction

Since their arrival on the European continent, the obligate parasites downy (*Plasmopara viticola*) and powdery mildew (*Erysiphe necator*) have caused much damage in European vineyards. In 1952, only 7 years after the appearance of *E. necator* on the European continent, 50–70 % of the grapevine harvest in French vineyards had been destroyed by powdery mildew (Galet 1977). Similarly, the invasion of *P. viticola* in 1878 led to considerable loss of yield, which in 1882 was total in some French vineyards (Lafon and Bulit 1981). Since then, research has led to the development of phytosanitary protection products such as copper, sulfur, and other multisite fungicides, which are still used but sometimes have limited efficacy depending on the climate and epidemic pressure. Numerous fungicides have been patented during the last 60 years. Today, for example, 60 commercial products have been approved in France to control downy mildew and 31 for powdery mildew, but only 9 different modes of action are available (Baudet and Lejeune 2014). However, while these products have a good bioavailability and are often used in low doses, their specificity in terms of mode of action, which is rather uni-site, drives the pathogen population to adapt. The result is the occurrence of resistance, which is characterized by the appearance individuals that are less sensitive and then by resistant populations that can lead eventually to a loss of efficacy in the field.

The term "resistance monitoring" is used to observe, detect, and monitor changes in the sensitivity of a population of targeted pathogens in the field. This involves continuous surveillance over several years and involves many locations under different epidemic pressures. Well-carried-out monitoring is the cornerstone of good resistance management. Fungicide monitoring is usually performed at the national or regional level, but also by associations such as the FRAC (Fungicide Resistance Action Committee). All partners organize monitoring and pool their data, leading to issuing recommendations for the use of different fungicides (Brent and Hollomon 2007a, b).

How are the initial events of resistance emergence discovered, and how can its dispersal, increase, and persistence be prevented (Fig. 16.1). By identifying the resistance risks of fungicides, all the partners determine sensitivity baselines and discriminant doses for the fungicide, which allows a monitoring strategy to be set up, the epidemic to be monitored nationwide, and the surveillance recommendation to be issued, e.g., via the French national memorandum on mildews.

Furthermore, depending on the mode of resistance acquisition, be it monogenic or multigenic, and according to the characteristics of the pathogen, whether it is haploid or diploid (case of recessive resistance), the follow-up methods can vary. Indeed, in the case of a disruptive resistance which appears very suddenly and the presence of an

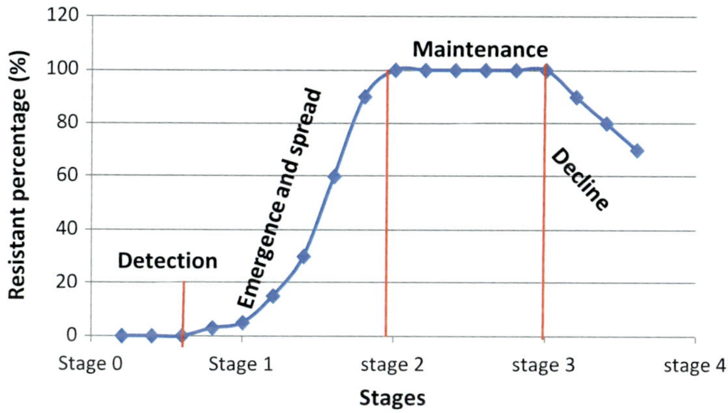

Fig. 16.1 Theoretical evolution of resistance phenomenon with different stages

allele (e.g., allele G143A involved in QoI resistance) confers strong resistance straightaway, a discriminant dose is quite easy to determine (e.g., 100–1000 fold the dose of the MIC for sensitive isolates, Chen et al. 2007; Corio-Costet et al. 2011; Gisi et al. 2002; Grasso et al. 2006). If the gene involved in the resistance is known, it is then quite easy to work out molecular tools (Sirven and Beffa 2003; Baudoin et al. 2008; Dufour et al. 2011). On the contrary, in the event of multigenic or progressive resistance, several years are necessary to observe the appearance of resistance in the field. This is the case for DMI resistance of grapevine powdery mildew (Scheinpflug 1994; Délye and Corio-Costet 1994; Steva and Clerjeau 1990) and the CAA resistance of downy mildew, which is monogenic, but recessive (Gisi et al. 2007; Blum et al. 2010a). In these cases, it is more difficult to determine a discriminant dose, if one wishes to identify the first cases of a decrease of sensitivity and to pay attention to possible false positives, because sympatric populations may sometimes be present (e.g., *E. necator*) (Corio-Costet 2007; Corio-Costet et al. 2003). Often, no naturally occurring resistant isolates are obvious during the first years of viticultural use of these fungicides.

A visible resistance problem is often difficult to predict, and what happens is that monitoring tends to follow the indications for loss of control. Results are then obtained too late to allow any action other than withdrawal of the product. However, caution is needed concerning the use of the term "resistance." More or less resistant isolates can be identified but with low frequency, so this does not systematically lead to a loss of efficacy in the vineyard. Conversely, efficacy losses do not always translate into resistance, because inadequate cultural practices might be involved such as under dosage and poorly, directed, or erroneous treatments. Often, the detected resistant populations reach frequencies superior to 0.01 (1 %) (Fig. 16.1). On the other hand, if monitoring preceded any major decrease in performance and avoidance strategies were either already in operation or being introduced, then the degree of success will vary. Early monitoring from the beginning of the use of a new

Table 16.1 Phases of monitoring and resistance management

Timing	Resistance monitoring activities	Value
Before the beginning of use in the vineyard	Sampling, baseline sensitivity, and distribution of populations	Knowledge of resistance risk and strategy of use
During years of use	Monitoring in different areas (different cultural practices, climates, varieties, etc.)	Practical performance and strategy for use
Case of suspected resistance	Intensification of monitoring and targeted research	Resistance is confirmed or not, study of cross resistance and resistance fitness
Trial of erosion	Plots where fungicide was used at least three times per year	Potential weakening of fungicide and knowledge on resistance selection speed
Subsequently	Monitoring for spread or decline of resistance	Performance and new strategies
In the future used with apparent stability or decline in resistance	Monitoring more spaced out in time and followed by resistance	Value of fungicide in mixture and/or if it is used against other diseases in the same vineyard

Modified from Brent (1994)

molecule makes it easier to track the onset of resistance and its evolution over time. Spot guidance can be obtained and given to growers if the monitoring is rapidly set up, hence the interest of developing molecular methods (e.g., q-PCR), which are easy to develop when the resistance is monogenic and the gene coding for the target is known (Table 16.1).

Below, I describe how to detect resistance during monitoring and what can be done (1) to better detect the emergence of resistance, (2) to develop tools or mapping the risk of resistance, and (3) to assess the dispersal of alleles involved in resistance. Three examples concerning resistance to sterol-C14-demethylase inhibitors (DMIs), quinone outside inhibitors (QoIs), and carboxylic acid amide (CAA) are given to explain the monitoring that can be done in the vineyard with obligate pathogens.

16.2 How Can One Detect Fungicide Resistance in Grapevine Mildews?

Two obligate pathogens, an ascomycete fungus ectoparasite (*E. necator*) and a chromites, oomycete endoparasite (*P. viticola*), are treated very differently in the vineyard: the former with single products and the latter with mixtures of fungicides (2, 3, and nowadays 4).

Firstly, it is essential to determine the variation in sensitivity to a fungicide within the target population. Great care must be taken in extrapolating results of laboratory tests, where one can use isolates or populations that have lost their aggressiveness or doses controlling growth that do not correspond to those used in the field. The latter point requires taking into account the quantity of active ingredient used in the field, and especially the spray volume, because this will have a direct impact on the efficacy of the treatment. This is why field samples should be collected to test the sensitivity levels before the fungicide is used in order to provide baseline data for subsequent comparisons and determine the first discriminant doses.

16.2.1 Sampling

There are two methods of sampling: either extensively in order to create pools and work on a maximum number of geographical sites or by targeting vineyards, where the pressure of disease is usually strong (e.g., vines with flag-shoot symptoms for *E. necator*, environment leading to disease, or vineyards with gobelet pruning, wet areas for *P. viticola*) and subjected to more treatments than other sites where cultural practices restrict mildew spread. The two approaches are complimentary, one giving an overall view of the real status of resistance and the second making it possible to detect and anticipate the spread of resistance.

The quality of the sample plays a big part in monitoring success. Upon arrival in the laboratory, samples with *E. necator* or *P. viticola* are examined with a dissecting microscope to determine the presence of conidia or sporangia. If the leaves or grapes are sporulating, inoculum can be used directly for subculturing or bioassays, or the leaves and grapes are placed in a moist chamber to reinitiate sporulation. However, it is always possible to subculture inoculum even from old leaves infected with E. *necator* and even those harvested in October in Europe when there are cleistothecia (Délye and Corio-Costet 1998; Cartolaro and Steva 1990; Gadoury and Pearson 1991; Evans et al. 1996; Miazzi et al. 1997). For downy mildew, it is possible to recover sporangia after incubation of symptoms in a wet chamber and then to perform a bioassay (Clerjeau et al. 1985; Genet et al. 1997).

Samples are collected during the growing season in different vineyards treated or not treated with fungicide. Usually for a random test, 10–50 leaves exhibiting lesions are collected from each plot. Collecting can be arbitrary at several points in a vineyard. The leaves are cryopreserved in an icebox or packed with other leaves and put in newspaper before rapid postal dispatch. Most samplings are performed in August or September after the final spraying has taken place. However, sampling at the beginning of the season can be very useful for knowing the extent of selection pressure during the previous growing season and for having data about the fitness of resistant isolates.

A sufficient number is needed because single conidia or monosporangial isolates can then be used to build reference collections (Délye et al. 1997b; Erickson and Wilcox 1997; Sierotzki et al. 2005; Toffolatti et al. 2007).

16.2.2 Multiplication and Collection of Isolates

16.2.2.1 Grapevine Powdery Mildew (*E. necator*)

E. necator, which is a strict parasite, requires grapevine leaves of good quality as nutritional support. Susceptible grapevine cultivars are used for experiments (e.g., Cinsaut, Cabernet Sauvignon, Chardonnay, Blauberger, etc.). To perform calibrated bioassays, young cuttings (6–8 weeks) are used. The more susceptible polished leaves are used for inoculum multiplication and bioassays (Debieu et al. 1995; Erickson and Wilcox 1997; Evans et al. 1996; Miazzi et al. 1997; Cartolaro and Steva 1990; Gadoury and Pearson 1991; Colcol et al. 2012). A technique with *E. necator* isolates is to culture them on leaves or on leaf disks (9–20 mm diameter) either surface sterilized for 10 min by immersion in calcium hypochlorite solution (50 g l^{-1}) or disinfected in 10 % bleach with 0.1 % Tween 20 for 2 min, or with 50 % ethanol for 30s, then rinsed in sterile water and blotted dry between paper towels, and kept alive on water agar (20 mg l^{-1} and 30 mg.l^{-1} of benzimidazole or 15 g l^{-1} agar) (Délye et al. 1997b; Erickson and Wilcox 1997). Other authors inoculate *E. necator* conidia under sterile conditions onto the upper surface (adaxial side) of grape leaves in a Plexiglas spore-settling tower (Aslam and Schwarzbach 1980; Reifschneider and Boiteux 1988; Debieu et al. 1995), by blowing 700–1000 conidia per cm^2 of leaf (counting with Malassez cell or Coulter counter) with an air pump connected to a flexible plastic tube terminating in a Pasteur pipette. The inoculated leaves in Petri dishes are removed from the tower and the lids replaced. They are then allowed to grow for 12 days in a growth chamber at 22 °C with 16 h day^{-1} light. Afterward, the inoculum can be used for different tests.

To create an isolate collection, monoconidial isolates can be obtained by picking a single powdery mildew conidium from mildewed samples within a laminar flow hood, after observation with a dissecting microscope, using an eyelash or a camel hair fastened to a holder or a platinum thread. Infections are established on 18 mm diameter leaf disks excised from leaves decontaminated as described above. Between each picking, the eyelash or the thread is disinfected by immersion in ethanol (70%, v/v). Afterward, the inoculated disks or leaves are placed in a growth chamber at 22 °C for 11–14 days (Corio-Costet 2007; Debieu et al. 1995; Erickson and Wilcox 1997).

16.2.2.2 Grapevine Downy Mildew (*P. viticola*)

As for the experiment with *E. necator*, grapevine leaves should be available from cuttings in a greenhouse. Lesion-bearing leaves are cut, moistened, and incubated in a plastic bag overnight at 20–22 °C to promote sporangial growth. Leaves with lesions are usually washed with water to remove any fungicide residue, or freshly produced sporangia may be harvested with a paintbrush and dispersed in water at 4 °C (Bissbort and Schlösser 1991). Sporangial populations are subcultured one or more times without fungicide pressure to obtain sufficient material to carry out biological tests. Sporangia are collected and suspended in sterile water at 4 °C. Inoculation is performed by depositing fifteen 10–20 µl droplets onto the abaxial face of each leaf. After inoculation, leaves in Petri dishes are kept in the dark overnight at 22 °C with a 16 h day^{-1} photoperiod; the next day droplets are sucked up. After 6 or 7 days, freshly produced sporangia are harvested to inoculate the test units (Chen et al. 2007; Andrieu et al. 2001; Sirven and Beffa 2003; Genet and Vincent 1999).

To create a collection, field isolates of *P. viticola* may be collected from grape leaves from different sites in various countries, and single sporangiophore isolates may be produced from this mass by picking sporangia from sporangiophores with a fine needle (Blum et al. 2010a) or by release of sporangia on water agar and picking them and transposed them onto leaf disks (Corio-Costet et al. 2011).

16.2.3 Which Biological Test Can Be Used?

While there are many testing procedures, none seems to affect the conclusions concerning resistance. However, it is better to have standardized methods in order to compare results from different monitoring sessions. Before monitoring, it is generally best to use a range of fungicide concentrations to determine a discriminant dose.

16.2.3.1 Powdery Mildew: *E. necator*

According to the fungicide under study, biologic tests may vary. For example, since DMIs have no effect on spore germination, it is pointless to perform a germination bioassay. On the other hand, a test based on mycelium length or growth bioassay (including sporulation) is better suited for leaves.

16.2.3.1.1 Test of Mycelium Extension

A rapid test of mycelial extension can be performed with conidia being deposited on fungicide-treated or untreated leaf disks and then measuring the length of mycelium 72 h after inoculation. This test is efficient for comparing populations. Leaf disks are taken from each treatment and the germinating conidia removed with cellotape, stained with lactophenol cotton blue, and fixed on glass slides for microscopic observations of conidial germ tube length and morphology. Generally, a minimum of 100 germinating conidia showing a germ tube length of more or less 250 µm are counted (Cartolaro and Steva 1990; Thind et al. 1998). Untreated inoculated leaf disks served as controls.

16.2.3.1.2 Fungicide Sensitivity Assays (Growth and Sporulation)

Many authors have reported assay methods (Cartolaro and Steva 1990; Debieu et al. 1995; Erickson and Wilcox 1997; Délye et al. 1997b). Briefly, leaf disks of a susceptible cultivar are prepared as described in paragraph 2.2.1. Eight disks of eight different leaves are placed in Petri dishes (adaxial surface facing upward). A range of fungicide doses can be applied by spraying disks and incubation for 1 or 24 h before inoculation (Délye and Corio-Costet 1998) or by imbibition on a filter paper impregnated with 3 ml of water (untreated control) or 3 ml of fungicide solution (abaxial surface). After 30 min or 24 h, the disks are turned over and placed in Petri dishes and dried before inoculation (Steva and Clerjeau 1990; Debieu et al. 1995). In another technique, leaf disks are submerged for 1 h with regular agitation in a glass beaker containing fungicide solution and then dried between paper towels (Wong and Wilcox 2002).

After treatment with the various fungicide concentrations, leaves or disks are inoculated homogeneously with conidia at the rate of 600–800 spores by cm^2 with a single isolate or a population using a settling tower, and incubated in a growth chamber at 22 °C (Cartolaro and Steva 1990; Debieu et al. 1995), or are inoculated at a single point with two or three conidial chains (15–30 conidia) from an isolate (Wong and Wilcox 2002). After 12 or 14 days, the percentage of leaf surface infected is determined by scoring between 0 and 10 to indicate the surface area under attack and the degree of sporulation. Alternatively, observation is performed after 7 days of mycelial extension from the point of inoculation using an ocular micrometer and a stereomicroscope. The data are presented graphically by fitting a negative logistic regression curve, and the EC_{50} or EC_{90} and MIC are determined for each isolate or population (Fig. 16.2).

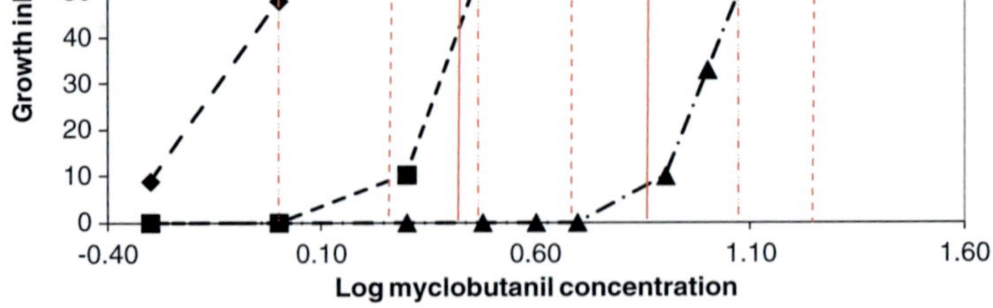

Fig. 16.2 Example of dose response curves to myclobutanil, a DMI fungicide, of three isolates of powdery mildew collected in France in 2007

16.2.3.2 Downy Mildew: *P. viticola*

Often the biological assays are not performed directly on the sample. After subculturing on new leaves in Petri dishes and incubation for 7 days, the new inoculum is used to test the sensitivity of isolates. With fungicide acting directly on the germination or motility of zoospores, it is possible to test a range of concentrations to assess the distribution of sensitivity. Such is the case with fungicide resistance that is recessive (e.g., CAA) or for detecting new forms of resistance and identifying whether they are disruptive or progressive (e.g., QiI, benzamide).

Various methods for testing fungicides can be used:

- One method is based on floating disks (Clerjeau and Simone 1982) in which 8 disks are cut and fungicide solutions dispensed into Petri dishes (20 ml/dish). Leaf disks are placed abaxially on water or fungicide solution. Each disk is inoculated with one droplet (25 μl) of sporangial suspension (20,000–200,000 spores ml^{-1}). Then, the inoculated disks are incubated at 22 °C with 14 h light $day^{-1}$and the droplets dried.
- A second method involves application of fungicides by droplets of 15 μl containing 1 volume of fungicide and 1 volume of sporangia at 200,000 spores ml^{-1} on the abaxial face of the leaf disks. Each Petri dish contains 10 disks from different leaves with 3 droplets (15 μl at 100,000 spores ml^{-1} final concentration). After incubation in the dark at 22 °C for 24 h to promote stomata opening and zoospore encystment, the droplets are sucked up, and scoring is performed 7 days after inoculation with a visual scale. The average of the 30 observations (10 disks × 3 droplets) is used to calculate the relative inhibition and to determine EC_{50} and MIC concentrations (Magnien et al. 2012).
- In a third method, disks are sprayed with fungicide solution (2 ml/Petri dish) (Herzog and Schuepp 1985).
- A fourth method involves inoculated disks by placing one drop (5 μl) of inoculum (40,000 sporangia ml^{-1}) mixed with various concentrations of fungicides on each disk (Reuveni 2003).

Inhibition of sporangia production can be assessed by counting with a hemocytometer or with a Coulter counter.

16.2.3.2.1 Sensitivity Test on Sporangia

To perform monitoring, discriminant doses of a fungicide (e.g., 10 mg to 1,500 mg l^{-1} of famoxadone) may be applied. For sensitive populations, or isolates that do not grow at fungicide concentrations higher than 10 mg l^{-1}, other concentrations are applied (from 0.1 mg to 10 mg l^{-1}). Fungicides are sprayed using a handheld sprayer onto the lower side of 10 grapevine leaf disks (Wong and Wilcox 2000; Chen et al. 2007; Herzog and Schuepp 1985), 1, 12, or 24 h before inoculation depending on the test. The disks are inoculated with 3 droplets of 10 μl with an inoculum of 20,000 or 10,000 sporangia per ml, and 12 h later, droplets are sucked up. It is possible to inoculate with 50,000 sporangia per ml, leaf disks beforehand, placed on water agar in 24-well plates and treated 12 h before inoculation with different concentrations of a QoI (Sierotzki et al. 2005). After 7 days of incubation, the development of mildew on each disk is visually estimated as the proportion of leaf area with sporulation (Genet et al. 1997). For each concentration, the average score is converted to a percentage of inhibition by comparison with untreated disks and the EC_{50} and MIC determined. Different classes can be identified according to their sensitivity to a fungicide (e.g., famoxadone: sensitive isolate MIC < 10 mg l^{-1}) and highly resistant (MIC > 100 mg l^{-1}) (Chen et al. 2007). Depending on the authors, the fungicides used are either the commercial products, and the concentrations are calculated in terms of a.i. (Chen et al. 2007) or technical compounds dissolved in, e.g., 0.1 % of DMSO or acetone (0.2 %) with 0.005 % Tween 20 (Wong and Wilcox 2000; Sierotzki et al. 2005).

It is also possible to examine the effect of fungicide on zoospore release, motility, and integrity and on direct germination of sporangia by incubation of sporangial suspensions at 2.5×10^4 to 5×10^4 sporangia per ml with fungicides at various concentrations in 96-well microtiter plates at 20 °C for 24 h. Sporangia germination, zoospore motility, and zoospore integrity are calculated as a percentage compared to untreated sporangia (Andrieu et al. 2001; Blum et al. 2010a). To examine the development of haustoria, infected leaf pieces are cleared with boiling alcoholic lactophenol for 15 min and are transferred to alcoholic lactophenol containing 0.7 % of aniline blue and boiling again (10 min). Pieces are incubated in the staining solution until cold. Finally, the stained samples are removed and placed in saturated chloral hydrate solution for at least 1 day. Mycelial development is related to the number of haustoria formed per infection site after observations under a microscope (Herzog and Schuepp 1985).

16.2.3.2.2 Test on Oospores

Resistance to QoIs provides a good example. As the genetic basis of the resistant allele G143A is mitochondrial, it is worthwhile observing resistance either on oospores (year N) (Toffolatti et al. 2007) or on the first lesion arising from macroconidia germination (year N + 1) (Chen et al. 2007). Leaves showing mosaic symptoms at the end of the growing season (August–October) are randomly collected from vineyards. Three nylon bags per plot (pore size 100 μm), each containing 50 leaf fragments with oospores, are overwintered in the vineyards (Toffolatti et al. 2007). Germination test of oospores is carried out from January, and they are separated from the leaves by filtration through two nylon filters (100–45 μm) and resuspended in water. Oospore germinability is then assessed on 1 % water agar at 20 °C. Oospores are inoculated onto water agar containing fungicide (e.g., 1 mg l^{-1} azoxystrobin) in the dark for 14 days and observed under a microscope in order to assess the number of germinated oospores in comparison with oospores on untreated water agar. The oospores able to germinate on medium containing fungicide are considered as resistant. The percentage of resistant oospores is calculated as the formula $G_{treated} \times 100/G_{untreated}$, and there is a good correlation between percentages of resistant oospores and the presence of the resistant allele G143A.

16.2.4 Baseline and Discriminant Dose

Before identifying the emergence and the presence of resistance (Fig. 16.1), it is imperative to know the sensitivity levels of pathogen populations. Usually, classes of sensitivity frequencies of natural populations with various doses of fungicides are assessed according to the geographical origin of populations or isolates. The results obtained may differ depending on whether one is dealing with mono-allelic disruptive resistance (e.g., QoI resistance) or pluri-allelic progressive resistance, be it mono or polygenic (e.g., DMI resistance). Commonly, fungicide sensitivity follows a Gaussian curve, but sometimes, there are outlier individuals (case of QoI fungicide) (Fig. 16.3). It is also possible to use isolates of laboratory collections to obtain baseline sensitivity. However, the best baseline is obtained from recently collected field isolates.

To determine the baseline, EC_{50} and MIC for pathogen isolates are estimated, by sensitivity bioassays with a range of fungicide doses, and then a discriminant dose (DD) is established for monitoring (Russell 2007). The DD commonly begins with a dose five- to tenfold higher than the MIC average of the pathogen population to detect the stage of emergence (Fig. 16.2) and even more (e.g., 100-fold for QoI resistance) or at least tenfold higher than EC_{50}. After detection and obtaining EC_{50} and MIC on resistant isolates, it is possible to develop a new DD to monitor the evolution of resistance and its maintenance if the fungicide resistance is progressive. EC_{50} is a good value to estimate resistance and is often better than the MIC or the EC_{90}. However, it should be determined with a number of isolates and/or

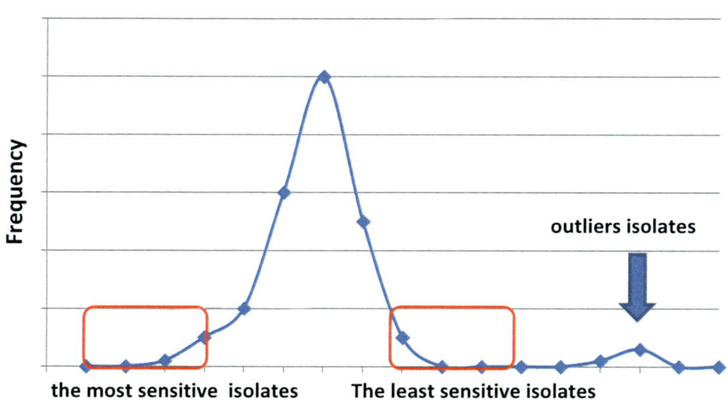

Fig. 16.3 Distribution of sensitivity of pathogen to fungicides (e.g., QoI) with a Gaussian distribution and some isolates that are outliers. This is typical of resistant isolates already naturally present in the field before the use of a fungicide

populations sampled from various vineyards to provide enough variability. During these efficacy tests, it is always important to observe differences in growth rate, sporulation, the aspect of the mycelium, and the pathogenicity of the isolates. These data provide information on the fitness of the resistant isolate collected in the vineyard.

16.2.4.1 EC_{50} and MIC

A range of fungicide concentrations (6–10) is sprayed and allowed to dry before inoculation as described in 2.3. The percentage of leaf tissue infected with powdery mildew is evaluated, and the data converted to EC_{50} values or MIC. Measurements are made for each fungicide concentration, and the mean values for eight replicates of leaf disks per treatment are used to calculate the relative growth and then the efficacy, according to the formula:

Efficacy = 100 × [Cont-Fung/Cont], with **Cont,** which is the average of attack on untreated disks, and **Fung,** that on treated disks. Dose response curves for isolates or populations are generated by plotting the relative growth values against the $\log_{10}$ of the fungicide concentration used, and the $\log_{10}$ effective dose to reduce growth by 50 % (EC_{50}) is calculated from the regression equation generated through the linear portion of the sigmoid curve or graphically (Fig. 16.2). The minimal inhibitory concentration (MIC) and EC_{90} can also be calculated (Wong and Wilcox 2000, 2002; Délye and Corio-Costet 1998; Sombardier et al. 2010).

16.2.4.2 Resistance Factor

A comparison of isolate (or population) sensitivity to fungicide based on the EC_{50} and the MIC of various sensitivities makes it possible to obtain a resistance factor (RF). The resistance factor is calculated for isolates displaying decreased sensitivity to fungicide as follows: $RF = EC_{50R}/EC_{50}S$, where EC_{50R} is the EC_{50} value for isolates or populations studied and the EC_{50S} is the mean of EC_{50} for all reference sensitive isolates or populations. According to the size of this RF and the diversity of the pathogen population, it is possible to estimate the risk of loss of efficacy in the field (Fig. 16.4).

16.3 Monitoring in the Vineyard

16.3.1 Monitoring for Progressive Resistance (DMI Resistance) in E. necator

DMI fungicides inhibit sterol biosynthesis and represent an important fungicide family and include up to five different chemical families (Worthington 1988) and more than 18 molecules. These fungicides have been very successful in view of the low

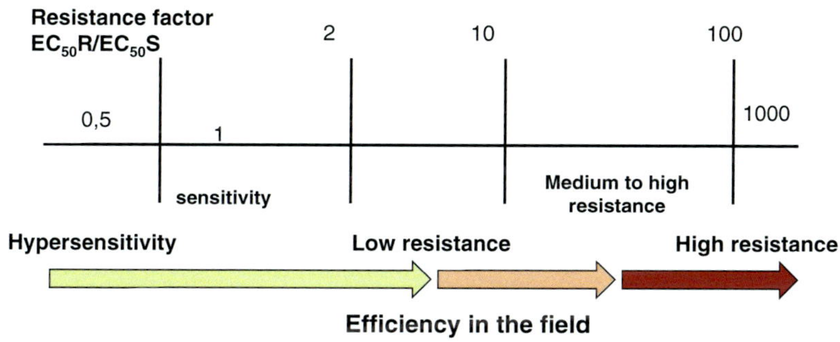

Fig. 16.4 Evolution of resistance factor linked to efficacy of fungicides in the vineyard. *Yellow arrow* = effective treatments, *beige arrow* = effective treatments with sometimes loss of efficiency, *brown* = loss of efficiency in the field

doses used (from 17 to 100 g/ha) and their penetrating and systemic properties. They all inhibit eburicol-C14-demethylase, an enzyme involved in methylene-24-cholesterol biosynthesis, the major sterol of *E. necator* (Délye and Corio-Costet 1994).

Monitoring DMI sensitivity identified the first case of resistance in grape powdery mildew in Portugal in 1984 (Steva et al. 1989), then in France in 1988 (RF varying from 2–50 to triadimenol) (Steva and Clerjeau 1990), and in Italy (RF = 6 for fenarimol) (Aloi et al. 1991) (Table 16.2).

Introduced in 1982 in California, triadimefon was effective until 1986, and resistance was officially described in 1988 (Ogawa et al. 1988). From the 1990s, resistance toward 5 molecules was identified in Austria (Redl and Steinkellner 1996). Myclobutanil resistance was observed in the New York state 6 years after its introduction, and in Australia, the RF varied from 0.6 to 50. From 2000, all vineyards seem to be affected by DMI resistance. Isolates and populations have been identified that exhibited variable RF from 2 to 180 (Table 16.2).

From the sensitivity baselines (EC_{50}, MIC), it is possible to carry out monitoring with discriminant doses of the order of 1–10 mg l^{-1} (a.i.) for each fungicide, showing that all the isolates exhibiting an RF (>5) have lost some of their sensitivity (Fig. 16.4). The gene (*cyp 51*) coding for the target enzyme was sequenced and the presence of one allele involved in the resistance identified (Délye et al. 1997c, d, 1999). A single modification at codon 136 of the *cyp 51* gene replaced a phenylalanine residue with a tyrosine (Y136F, Délye et al. 1997c, d) and which correlated perfectly with RF to triadimenol greater than or equal to 8. Identification of this allele, added to biological assays, helps assess the evolution of resistance in the field and give quantitative results (Dufour et al. 2011) to edit national mapping of resistance status of vineyards or the global evolution of resistance each year (Fig. 16.5). This helps making recommendations to vine-growers so that they can avoid the loss of efficacy of fungicides (Corio-Costet et al. 2014a, b; French national memorandum on mildews 2014). Monitoring of this allele marker by q-PCR (Fig. 16.5) showed an increase of the presence of the resistance marker in *E. necator* populations from 2008 to 2012, with a stability between 2010 and 2012, and even a decrease in

Table 16.2 Detection of DMI resistance in vineyards worldwide

Year[a]	Country	RF	Fungicide	References
1984	Portugal	2–140	Triadimenol	Steva et al. (1989)
1986	USA (California)	0.3–180	Triadimenol, Myclobutanil, Fenarimol, Tebuconazole, Triflumizole	Boubals (1987), Ogawa et al. (1988), Gubler et al. (1996), Erickson and Wilcox (1997), Ypema et al. (1997), Northover and Homeyer (2001), Miller and Gubler (2004), Wong and Wilcox (2002), and Colcol et al. (2012)
1997	USA (New York)			
1987	France	2–140	Triadimenol, Myclobutanil, Fenarimol	Steva and Clerjeau (1990), Debieu et al. (1995), Délye et al. (1997b), and Dufour et al. (2011)
1988	Turkey			Ari and Delen (1988)
1988	Italy		Fenarimol, Triadimefon	Aloi et al. (1991)
1991	Switzerland			Pezet and Bolay (1992)
1995	Austria	1–6	Penconazole, Pyrifenox, Myclobutanil, Triadimenol	Redl and Steinkellner (1996) and Steinkellner and Redl (2001)
1995	Australia	0.6–50	Triadimenol	Savocchia et al. (2004)
1997	India			Thind et al. (1998)
2000	South Africa	1–48	Penconazole, Flusilazole, Triadimenol	Halleen et al. (2000)

[a]Date of first observation of loss of efficacy in the vineyard

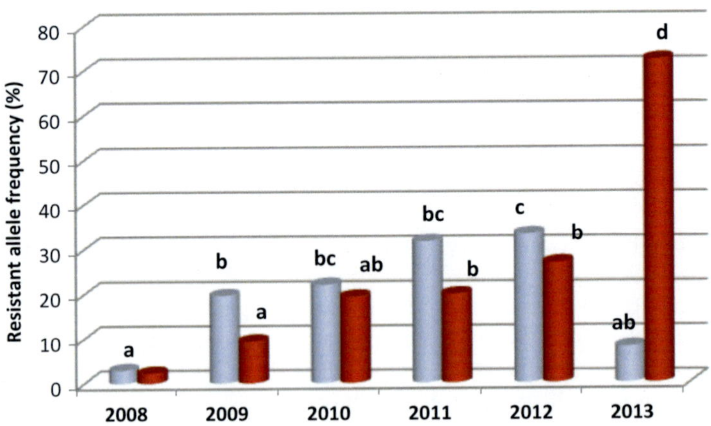

Fig. 16.5 Average regional frequencies of alleles conferring resistance to DMI fungicide (Y136F, in blue or gray) and to QoI fungicide (G143A, in *red* or *black*) in France, in *E. necator* populations between 2008 and 2013. The number of populations analyzed varied from 33 to 78 per year (From Corio-Costet et al. 2014a)

2013, which was partially due to the campaign to limit DMI use in vineyards since 2009, by never using them twice consecutively, and to be careful when treating. However, DMI resistance is polygenic, and besides modifications of the gene (*cyp 51*) coding for the target enzyme (C14-demethylase), there are other mechanisms of resistance (Stergiopoulos et al. 2003; Buchenauer 1995). Monitoring of the Y136F allele makes it possible to anticipate the evolution of the population and, when complemented with a discriminant dose of 5–10 mg L^{-1} of tebuconazole, confirms resistance, and treatment protocols can then be modified.

One problem with progressive polygenic resistance is the discrete onset of the phenomenon, which is closely linked to the mode of reproduction of the pathogen and the selection pressure exerted by the fungicide. This is particularly the case for the genetic group B that survives during the winter either in the dormant buds or by sexual reproduction. Group B contains the majority of isolates on which fungicide selection is applied. On the contrary, genetic group A in France survives essentially as dormant mycelium and undergoes little fungicide pressure (Corio-Costet 2007). Estimating population diversity before and during monitoring can also be as important as having good molecular tools for detection or a good test of discrimination.

16.3.1.1 False Positive: Case of Sympatric Populations of *E. necator* in the Vineyard

It is possible to detect changes in sensitivity and the beginning of resistance (Table 16.3), but interpretation errors due to pathogenic diversity can occur. For example, mildew samples collected in vineyards in the South of France, before flowering, and with young leaves totally mildewed from budbreak (Corio-Costet 2007; Amrani and Corio-Costet 2006; Steva 1991) were sensitive to tebuconazole (mean EC_{50} of 0.12 mg $l^{-1)}$. A second sampling from the same area 6–8 weeks later (after flowering) considered mildew as resistant (mean EC_{50} of 0.73 mg l^{-1}, Table 16.3), and an RF of 6 between the two samplings suggested disruptive resistance. In fact, monitoring was applied to two sympatric populations belonging to two different genetic groups, one more sensitive than the other one (Délye et al. 1997a, b; Corio-Costet et al. 2003). None could really be considered as resistant when compared to the baseline sensitivity of *E. necator* isolates. Resistant isolates collected in the same area had a mean EC_{50} of 5.84 mg l^{-1}. Simply, group A isolates are more sensitive than group B isolates (Corio-Costet et al. 2003; Miazzi and Hajjeh 2011).

16.3.1.2 Cross-Resistance

Resistance to one DMI fungicide can confer positive cross-resistance, i.e., resistance to triadimenol also leads to resistance to other molecules such as myclobutanil, fenarimol, penconazole, or cyproconazole (Erickson and Wilcox 1997; Délye et al. 1997b). In South Africa in 1989, *E. necator* isolates showed reduced sensitivity

Table 16.3 Example of variation in sensitivity of *E. necator* isolates to triadimenol or tebuconazole

	Isolates	Year of sampling	$EC_{50} \pm SE$ (mg l^{-1})	$EC_{50} \pm SE$ average (mg l^{-1})	RF
Triadimenol					
Sensitive isolates	Coc2	1988	0.11 ± 0.02	0.06 ± 0.045	
	Moa2	1989	0.02 ± 0.00		
	Aor3	1990	0.05 ± 0.01		
Resistant isolates	Moa5	1989	3.58 ± 0.45	5.06 ± 1.75	84 (R/S)
	Aor2	1990	4.57 ± 0.22		
	Coc5	1990	7.13 ± 0.22		
	Mr6	1993	3.28 ± 0.37		
	Mr3	1994	6.47 ± 0.19		
Tebuconazole					
Genetic group A*	10 sensitive isolates (SA)	1998		0.12 ± 0.02	1
Genetic group B*	10 sensitive isolates (SB)	1998		0.73 ± 0.33	6 (SB/SA)
Genetic group B	10 resistant isolates (RB)	1998		5.84 ± 0.73	8 (RB/SB) or 48 (RB/SA)

From Steva (1991), Délye (1997), and Corio-Costet et al. (2003)
* = sensitive isolates without resistant allele in position 136 (Y136T), considered as sensitive
RF resistant factor, *SB* sensitive (SB) or resistant (RB) isolates belonging to genetic group B. *SA* sensitive isolates belonging to genetic group A

to triadimenol, flusilazole, and penconazole (Halleen et al. 2000), but the different combinations found between the three fungicides corroborated the polygenic nature of DMI resistance. However, in France, among more than 175 isolates tested with seven DMIs, none was resistant to more than 5 of these fungicides. Recently in Virginia, the comparison between sensitive isolates and isolates from vineyards treated with DMIs showed RF varying from 31 (triflumizole) to 144 (myclobutanil), with RF of 53, 75, and 45, for fenarimol, tebuconazole, and triadimefon, respectively (Colcol et al. 2012).

16.3.2 Monitoring in the Vineyard for Disruptive Resistance: Case of QoI Fungicides

The market launch of successful molecules inhibiting mitochondrial respiration at the end of the 1990s, such as the QoIs (quinone outside inhibitors), highlighted a particular case of resistance, although the genetic support (cytochrome b gene) is mitochondrial. Commonly from 1997, the existence of *P. viticola* populations showing a positive cross-resistance to QoIs (azoxystrobin, pyraclostrobin, kresoxim-methyl, trifloxystrobin, famoxadone, and fenamidone) has been described (Heaney et al. 2000; Jordan et al. 1999; Hollomon et al. 2005; Gisi et al. 2002). Baselines quickly revealed the existence of isolates or populations 100–1000 times less sensitive than sensitive populations of both *E. necator* and *P. viticola* (Heaney et al. 2000; Sierotzki et al. 2005, 2008; Baudoin et al. 2008; Chen et al. 2007; Dufour et al. 2011; Corio-Costet et al. 2008; Corio-Costet 2012; Toffolatti and Vercesi 2012; Sirven et al. 2002). These compounds are recommended for preventive applications. Some are systemic and are capable of circulating via the translaminary route and via the vascularization, which partially explains their strong action on sporulation (Gisi et al. 2002).

16.3.2.1 QoI Fungicide Resistance in *P. viticola*

The QoI fungicides possess strong anti-oomycete activity at low doses (from 50 to 400 g ha^{-1}). They have good persistence and strongly limit differentiation of spores and the release of zoospores (Reuveni 2001; Andrieu et al. 2001). Complete inhibition of the sporulation is obtained with trifloxystrobin at 5 mg l^{-1} applied preventively on young inoculated plants (Reuveni 2001). A dose of famoxadone between 0.01 and 0.1 mg l^{-1} is sufficient to lyse the zoospores of *P. viticola* (Andrieu et al. 2001).

Similar results have also been obtained with azoxystrobin, showing an average EC_{50} of 0.04 mg l^{-1} for 81 isolates, with values ranging from 0.05 to 0.94 mg l^{-1} (Wong and Wilcox 2000). The distribution of sensitivity to famoxadone in 103 European populations of *P. viticola* was described as unimodal with a mean EC_{50} of 0.029 mg l^{-1} and variations ranging from 0.001 to 0.15 mg l^{-1} (Genet and Vincent 1999).

Hence, like other single-site fungicides, the interesting properties of these molecules (wide spectra of efficacy, curative effect) have quickly led to a strong selection process. This in turn has led to the appearance of resistance not only in French vineyards (Magnien et al. 2003, 2012) but also in other vineyards worldwide (Brunelli et al. 2003; Baudoin et al. 2008; Corio-Costet et al. 2008, 2011; Furuya et al. 2010; Heaney et al. 2000; Gullino et al. 2004; Gisi et al. 2000; Sierotzki et al. 2008).

In most cases, resistance is attributed to the presence of a major mutation in the mitochondrial cytochrome b gene, leading to the substitution of a glycine by an alanine at position 143 (G143A), including grapevine downy and powdery mildew (Gisi et al. 2002; Grasso et al. 2006; Chen et al. 2007). Another mutation can add to G143A and involves an alteration of codon 129, where a phenylalanine residue is replaced by a leucine (F129L), even if this modification confers only a low level of resistance (Gisi et al. 2002). As QoI resistance is correlated with the presence of allele G143A in the vineyard, various molecular tests have been developed after identification of discriminant doses in order to identify the resistant isolates or populations. These tests are based on PCR-RFLP (Corio-Costet et al. 2011; Chen et al. 2007; Baudoin et al. 2008; Furuya et al. 2010) or on quantitative PCR and quantification (Sirven et al. 2002; Sirven and Beffa 2003; Collina et al. 2005; Sierotzki et al. 2005, 2008; Toffolatti et al. 2007).

Studies of the gene coding for the target enzyme (*CYT bc1*) have highlighted the existence of two different genetic mitochondrial haplotypes (I and II), demonstrating that at least two independent events led to QoI resistance in European populations (Chen et al. 2007). The same study also showed the existence of several possible subspecies of *P. viticola* in American populations.

Biological assays carried out on bulk samples provide qualitative information if a discriminant dose is used. Another test developed for assessing QoI resistance is based on the germinability of oospores at discriminant doses. A reference dose for resistant field isolates is 10 mg l^{-1} (Heaney et al. 2000; Chen et al. 2007; Toffolatti et al. 2007).

In France, samples are dispatched by a national surveillance network and are accompanied by information index cards clarifying the treatment programs, locality, cultivar, and if losses in efficacy have been observed. Data for the previous years in the vineyards concerned as well as the rate of damage are also noted. Tests are performed either from lesions of *P. viticola* on leaves or grapes or from new sporulations, which are obtained in wet chambers after 24 h of incubation.

In 2005, QoI resistance quickly reached a mean of 29 %, but it varied significantly from one vineyard to another (from 0 to 70 %) (Corio-Costet et al. 2008; Fig. 16.6). In Champagne vineyards, similar results have been obtained with an average of 33 % of resistant lesions and a range from 0.01 to 77 %. A study on diversity performed with microsatellite markers (Delmotte et al. 2006) showed that the genotypic diversity is smaller when the frequency of resistance exceeds 30 % and that this diversity tends to decrease during the season in the resistant plots (Corio-Costet et al. 2008). This decreasing diversity of resistant populations was also demonstrated on plots treated with azoxystrobin (Matasci et al. 2008). In plots not having received any treatment with QoIs, the frequency of the resistance allele G143A in oospores populations was 10 %. Conversely, prevalence reached 90 % in plots treated 5–6 times (Toffolatti et al. 2011).

The persistence of resistance should also be monitored (Fig. 16.1), and various studies show that QoI resistance is rather persistent. All isolates are collected between 2002 and 2007 under fungicide selection pressure or not remained resistant. Furthermore, a study on the fitness of resistant isolates of *P. viticola* during the asexual phase demonstrated that QoI-resistant isolates exhibit good fitness (germination, sporulation, and infection frequency) and that they are competitive (Corio-Costet et al. 2011). Nonetheless, some studies show that after sexual reproduction, a decrease in QoI resistance occurs at the beginning of the season when there is no selection pressure (Toffolatti et al. 2007). In Brazil in 2002 after several years of use of QoI fungicides, the allele (G143A) frequency reached 100 %. By interrupting the application of QoI fungicides for 2 years, the resistance frequency declined to 2 %. On the other hand, 5 consecutive applications led to resurgence of resistance frequency of 100 % during the season, and it decreased again in 2005 (Sierotzki et al. 2008). Indeed, since the mutation is in a mitochondrial gene, this suggests that there is a maternal mitochondrial heritability. According to whether mitochondria "females" carry the resistant allele or not,

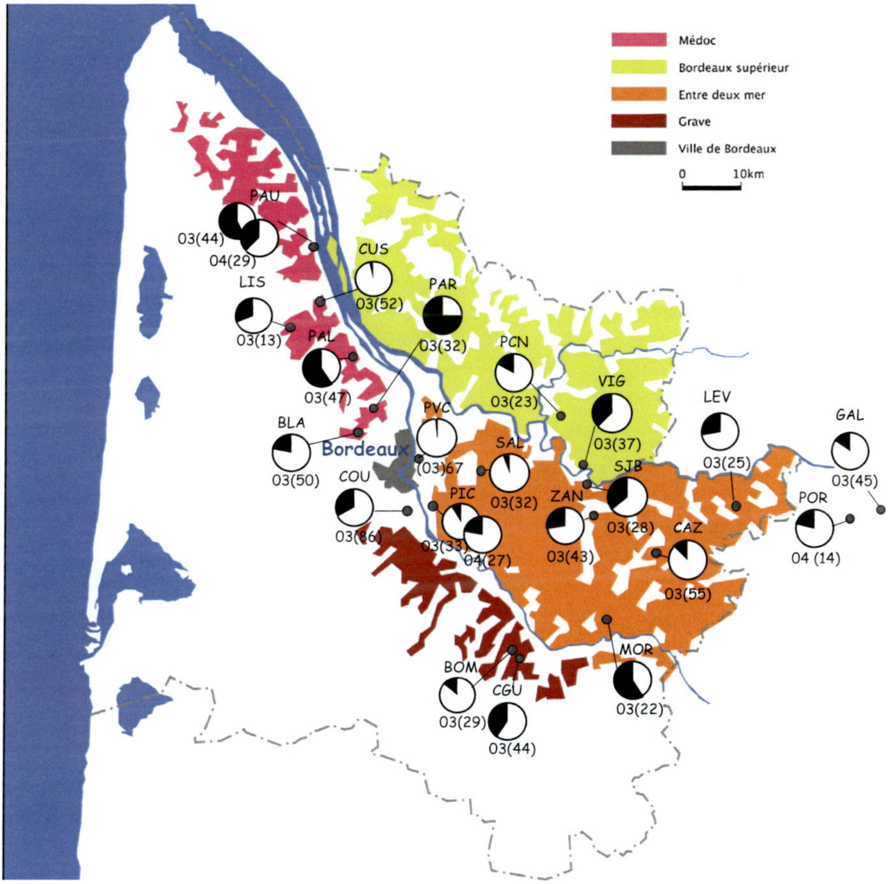

Fig. 16.6 Frequency of QoI fungicide resistance among 22 populations of *P. viticola* in Bordeaux vineyards collected in 2003 and 2004. Within the circle, QoI fungicide-resistant isolates are presented in black and sensitive isolates in white. The number of isolates analyzed is in brackets. Bioassays were performed with a discriminant dose of 50 mg l^{-1} of famoxadone and detection of allele G143A by PCR-RFLP (Chen et al. 2007)

one may expect a "dilution" of resistance phenomena, even its regression after every sexual cycle (Corio-Costet et al. 2011). This also suggests that resistant isolates might be less fertile or that resistant ones (Sierotzki et al. 2008).

16.3.2.2 QoI Fungicide Resistance and *E. necator*

The first QoIs used in the vineyard against *P. viticola*, such as famoxadone or trifloxystrobin, were also partially effective against *E. necator* and favored the selection of QoI resistance in *E. necator*. Methods used are similar to those described in paragraph 2.3, but as QoI fungicides act on respiration and can inhibit germination, a germination bioassay can be performed (Reuveni 2001; Miles et al. 2002).

16.3.2.2.1 Germination Bioassay

The test is performed on microscope slides covered with 3 ml of water agar amended with the commercial formulation of the fungicide at different concentrations (e.g., from 0.0001 to 100 mg l^{-1}). The medium can be also supplemented with 100 mg l^{-1} of salicylhydroxamic acid (SHAM) to block the alternative oxidase (AOX). After 24 h inoculation (conidia transferred with sterilized paintbrush), slides were observed under fluorescent light and 50 conidia observed for germination (Miles et al. 2002). All isolates with germination inhibited by trifloxystrobin at 1.1–100 mg l^{-1} exhibited the resistant allele G143A. A concentration of 0.1 mg l^{-1} inhibited germination of *E. necator* conidia of sensitive isolates completely (Reuveni 2001).

16.3.2.2.2 Discriminant Doses and Monitoring

Wong and Wilcox (2002) described the baseline of *E. necator* sensitivity to azoxystrobin from samples collected in 1999 from untreated vineyards. EC_{50} values ranged from 0.0037 to 0.028 mg l^{-1} with a mean of 0.097 mg l^{-1}. In California, EC_{50} obtained in 2002 varied from 0.00003 to 0.343 mg l^{-1} of trifloxystrobin (Miller and Gubler 2004). A concentration of 0.1 mg l^{-1} inhibited germination of *E. necator* conidia completely (Reuveni 2001). In European vineyards, the first case of resistance was identified in 2007 in Hungary (Bencené pers. comm), and resistant isolates were first detected in French vineyards in 2008 (Dufour et al. 2011).

However, it was not difficult to design allele-specific molecular tools based on the presence or not of the mutation at codon G143A. Therefore, q-PCR tools were soon used to detect and monitor the evolution of resistance in *E. necator* populations in the United States (Baudoin et al. 2008) and in France from 2008 (Dufour et al. 2011). It was noticed that samples exhibiting the QoI-resistant allele frequently also had the DMI-resistant allele. Overall, the mean percentage of QoI-resistant alleles nationwide for 2008 and 2009 reached 5.29 % (Fig. 16.5). However, from 2010, the mean percentage increased to 18 %, then 37 % in 2012, and 72.4 % in 2013. This shows that in spite of recommendations to limit the use of QoI fungicides, both in solo applications and in mixtures, in the French vineyards, they are still used, and the selection continues to be applied on powdery mildew populations. Even moderate, the use of QoI fungicide has rapidly led to resistance selection in *E. necator* in the field in Hungary (Taksonyi et al. 2013).

16.3.3 Monitoring for Recessive Resistance: Case of CAA Resistance in P. viticola

The carboxylic acid amide (CAA) fungicides, including dimethomorph, iprovalicarb, benthiavalicarb, mandipropamid, and valifenalate, are single-site fungicides inhibiting cell wall biosynthesis of oomycetes by inhibition of 1,3-β-glucanase exerting a glucanosyltransferase activity (*PiCesA3* and *PvCesA3* genes) in cellulose biosynthesis in *Phytophthora infestans* and in *P. viticola* (Blum et al. 2010a, b). The mechanism of resistance involving a recessive mutation in the *PvCesA3* is discussed in Chap. 20, but other recessive genes may also be involved. Consequently, resistance of *P. viticola* to CAAs might be reduced if appropriate strategies for commercial fungicides use were implemented in the field (Gisi 2012). Indeed, FRAC and the French national memorandum (2014) recommend the use of CAAs in mixtures with multisite fungicides or other non-cross-resistance groups of fungicides.

At the beginning of CAA use (e.g., dimethomorph), EC_{95} values ranged from 0.25 to 1.15 mg l^{-1} (Bissbort and Schlosser 1991), but this fungicide was used on average only twice per year. When new CAA fungicides entered the market 6–8 years later, EC_{50} values of iprovalicarb showed a wide range of concentrations from <1–30 mg l^{-1}, with a mean of 2 mg l^{-1} (Suty and Stenzel 1999). In 2003, the mean EC_{50} of dimethomorph in the French vineyard was 5.92 mg l^{-1}, i.e., fivefold higher than the EC_{50} found in 1991. In the same isolate collection, iprovalicarb sensitivity showed a mean EC_{50} of 78.7 mg l^{-1} (Corio-Costet 2012), suggesting the appearance of isolates resistant to iprovalicarb and dimethomorph. The EC_{50} for isolates sensitive to CAA fungicides is 5–100 times lower than those of resistant isolates (Table 16.4). Monitoring carried out between 2005 and 2010 on more than 590 French downy mildew populations showed a decrease in the sensitivity of these populations from 74 to 11 % (Fig. 16.7), with a more or less stable intermediate group during the monitoring period and an increase in resistant populations, with a shift from 9 to 32 % (Magnien et al. 2012). However, these results hide a large disparity in resistance behavior according to region and disease pressure. As downy

Table 16.4 CAA sensitivity of *P. viticola* isolates

Fungicides	Mean of EC_{50}[a] in sensitive isolates (mg l^{-1})	Mean of EC_{50} in resistant isolates (mg l^{-1})
Dimethomorph	10.3[b]	>100
Iprovalicarb	6.2	>30
Mandipropamid	0.95	>300
Benthiavalicarb	1.10	>300

[a] Concentration of fungicide inhibiting 50 % *P. viticola* growth
[b]The mean EC_{50} was 0.21 mg l^{-1} in 1989

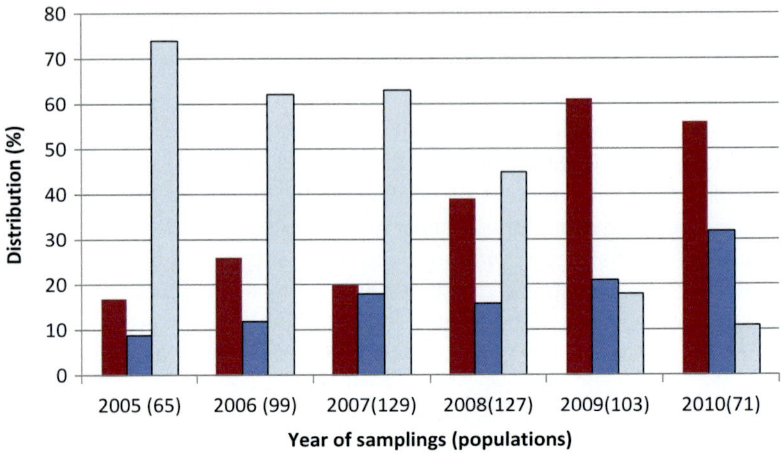

Fig. 16.7 Evolution of CAA resistance in *P. viticola* populations from 2005 to 2010 (From Magnien et al. 2012). *Light blue*, sensitive; *blue*, medium sensitive; *brown*, resistant isolates. Bioassays were performed with discriminant doses of 10 and 100 mg l^{-1} of dimethomorph

mildew populations in French vineyards are now resistant, the efficacy of treatments based on CAA fungicides is directly related to the type and dose of the associated product.

Monitoring can also be used to quantify the number of zoospores encysted in germination with a discriminant dose (e.g., 30 mg l^{-1} for iprovalicarb) and for establishing the resistance level according to the resistant spores (Latorse et al. 2006).

16.4 Conclusion

Data obtained by large-scale sampling on various plots of land in various regions, or in more restricted on trial plots where the application of a fungicide is repeated, provide information about the risk of the emergence of resistance and on the extent to which resistance to a given fungicide may become more widespread. All the data acquired at the regional and national level are used to make recommendations on fungicide use to limit resistance. A national memorandum gives advice on the use of fungicides according to the risk of resistance. Nonetheless, it should be remembered that in the fight against *E. necator* or *P. viticola*, the available formulated products are different and that they are often used solo against powdery mildew or as a mixture against downy mildew.

The diverse monitoring methods provide different data on the sensitivity of isolates or populations, so the most appropriate methods need to be optimized and evaluated for each pathogen-fungicide system (Brent and Hollomon 2007a, b).

Their impact depends on the applicability of the test and the proportion of resistant isolates or populations, together with the characteristics of the targeted genetics, fitness, and biology.

The risk assessment of fungicide resistance depends on the mode of action of fungicide (single site or multisite) and on the genetic support of the target (nuclear or mitochondrial). Generally, the mitochondria genome is more plastic than the nuclear genome, and so the risk of resistance is often greater. To the risk depending on the mode of action of the fungicide should be added the risk connected with the pathogen. Indeed, according to its capacity to reproduce (sexual and/or asexual), the appearance and the spread of resistance will depend on the fungicide selection pressure exerted on the pathogens. Hence, in grapevine powdery mildew which is generally treated with a single product, the evolution of the resistance seems less rapid than that of downy mildew, which is treated with a mixture of fungicide. The FRAC group gives recommendations about the risk of resistance of different pathogens with various fungicides (Brent and Hollomon 2007a).

The dissemination of monitoring results and the communication between the public sector and industrial operators in the field are required, so that the data obtained to date can be interpreted correctly and that some recommendations can be made for the sustainability of the use of fungicides.

These measures will help decrease fungicide selection pressure on obligate pathogens and contribute to limiting the risk of resistance. Spraying should be performed according to the best possible practice (avoiding sub-dosages or untreated areas), and preventive measures should be set up to limit the development of pathogens. The French National Memorandum (2014) recommends the following:

> alternating the use of fungicides having a single-site mode action, with others that have another mode of action. It is important not to alternate different products having the same mode of action. For fungicides for which resistance is widespread such as QoI and CAA, it is advisable to make only one treatment per season.

The 2014 French National Memorandum on Mildew states that the use of QoI fungicides is no longer recommended because of widespread and persistent resistance. Consequently, the efficacy of products containing QoIs is often exclusively related to the type and dose of the associated multisite products used (e.g., copper, mancozeb, cymoxanil, folpet).

Could a new bioassay-based monitoring strategy be of use? At the moment, the QiI (quinone inside inhibitor, cyazofamid) controls the majority of *P. viticola* isolates (94 %) at a dose close to 1 mg l^{-1}. On the other hand, 6 % of isolates from various regions have an MIC higher than 100 mg l^{-1} (Magnien et al. 2012). As for QoIs, there is a Gaussian distribution of sensitivities (Fig. 16.3) with outliers (RF of 100), suggesting that these isolates occurred prior to the use of QiI fungicides, and if used more than once or twice per season, QiI resistance will quickly emerge.

The degree of variation in baseline sensitivity is therefore a good marker of changes in resistance. A wide variation in baseline sensitivity (factor 100 and more) can be interpreted as real resistance provided that the baseline was obtained from field populations of different origin beforehand so as to avoid misinterpretations. A lower evolution of this baseline (RF from 5 to 10) does not necessarily indicate that the resistant isolates are leading to losses of efficiency. Nonetheless, some populations (already tested) which would slide gradually toward higher RF (e.g., DMI resistance or CAA resistance) in time allow the presence of a real risk of progressive or recessive resistance to be detected. Monitoring is also undertaken to investigate complaints from growers of an apparent loss of performance of a fungicide and to guide the selection of future fungicide treatments.

Unfortunately, one of the problems associated with the limitation of fungicide use and improvement of treatment programs is that monitoring is often stopped for financial reasons. Continuing to monitor for several years, even after resistance has been demonstrated, can lead to better understanding of how pathogenic populations adapt to the various pressures of selection exerted in the vineyard. Combining monitoring with the practices of growers (sub-dosages, inadequate adjustment of spraying equipment) can also improve understanding of the onset of resistance and its management. For this reason, monitoring of *E. necator* and *P. viticola* in France is now associated with meta-analyses taking into account of spraying equipment and applied doses. During the last 7 years, monitoring of mildews has shown that it has become more and more frequent to encounter isolates that are resistant to one or two fungicides with different modes of action and even three (DMI, QoI, quinoxyfen) in powdery mildew. In downy mildew, isolates resistant to three and even four different modes of action (e.g., CAA, QoI, mefenoxam, cymoxanil) are now encountered. It should now be made law to monitor the evolution of pathogen populations if they are to be controlled during the next decades, as should the use of other methods of control in association with fungicides (biological, breeding, plant defense stimulation, etc.). Biological tests remain bulwark in monitoring before the development of molecular tools.

Acknowledgments I wish to thank Prof Ray Cooke for reading the manuscript.

References

Aloi M, Gullino ML, Garibaldi A (1991) Reduced sensitivity to fenarimol and triadimefon in field populations of *Uncinula necator*. Pestic Sci 31:114–116

Amrani L, Corio-Costet MF (2006) Single nucleotide polymorphism (SNP) in β-tubulin gene distinguishes two genotypes of *E. necator*: PCR assays from different symptoms in the vineyard. Plant Pathol 55:505–512

Andrieu N, Jaworska G, Genet JL, Bompeix G (2001) Biological mode of action of famoxadone on *Plasmopara viticola* and *Phytophthora infestans*. Crop Prot 20:253–260

Ari M, Delen N (1988) Studies on the fungicide sensitivity of vine mildew (*Uncinula necator* (Schwein.) Burr.) in Aegean region of Turkey. J Turkish Phytopathol 17:19–30

Aslam M, Schwarzbach E (1980) An inoculation technique for quantitative studies of brown rust resistance in barley. Phytopath Z 99:87–91

Baudet A, Lejeune V (2014) Index phytosanitaire ACTA 2014. Acta-Publications, Paris

Baudoin AI, Olaya G, Delmotte F, Colcol JF, Sierotzki H (2008) QoI resistance of *Plasmopara viticola* and *Erysiphe necator* in the Mid-Atlantic United States. Plant Management Network. Plant Health Progr. doi:10.1094/PHP-2008-0211-02

Bissbort S, Schlösser E (1991) Sensitivity of *Plasmopara viticola* to dimethomorph. Med Fac Landbouw Rijksuniv Gent 56:559–567

Blum M, Boehler M, Randall E, Young V, Csukai M, Kraus S, Moulin F, Scalliet G, Avrova AO, Whisson ST, Fonne-Pfister R (2010a) Mandipropamid targets the cellulose synthase-like *PiCesA3* to inhibit cell wall biosynthesis in the oomycete plant pathogen, *Phytophthora infestans*. Mol Plant Pathol 11:227–243

Blum M, Waldner M, Gisi U (2010b) A single point mutation in the novel *PcCesA3* gene confer resistance to the carboxylic acid amide fungicide mandipropamid in *Plasmopara viticola*. Fungal Genet Biol 47:499–510

Boubals D (1987) Dans la lutte contre l'oïdium de la vigne, les Californiens rencontrent des difficultés. Progr Agric Vitic 104:261–262 (in French)

Brent KJ (1994) Monitoring for fungicide resistance. In: Delp CJ (ed) Fungicide resistance in North America, 2nd edn. APS, St. Paul, USA, pp 9–11

Brent KJ, Hollomon DW (2007a) Fungicide resistance in crop pathogens: how can it be managed? FRAC monograph no 1, revised edn. Crop Life International, Brussels

Brent KJ, Hollomon DW (2007b) Fungicide resistance: the assessment of risk. FRAC monograph no 1, revised edn. Crop Life International, Brussels

Brunelli A, Collina M, Guerrini P (2003) La résistance de *Plasmopara viticola* aux fongicides QoI en Italie. In: Proceedings of the 7th international conference on pest and diseases 2003, AFPP Tours, France, p 8 (in French)

Buchenauer H (1995) Resistance of fungi to sterol demethylation inhibitors. In: Lyr H (ed) Modern selective fungicides. Gustav Fischer Verlag, Jena/Berlin, pp 259–279

Cartolaro P, Steva H (1990) Maîtrise de l'oïdium au laboratoire. Phytoma 419:37–44 (in French)

Chen WJ, Delmotte F, Richard-Cervera S, Douence L, Greif C, Corio-Costet MF (2007) At least two origins of fungicide resistance in grapevine downy mildew populations. Appl Environ Microbiol 73:5162–6172

Clerjeau M, Simone J (1982) Apparition en France de souches de mildiou (*Plasmopara viticola*) résistantes aux fongicides de la famille des anilides (métalaxyl, milfuram). Progr Agric Vitic 99:59–61 (in French)

Clerjeau M, Moreau C, Piganeau B (1985) Méthode d'évaluation du taux de souches résistantes aux anilides dans une population de *Plasmopara viticola*: application à la surveillance du vignoble. EPPO Bull 15:423–430 (in French)

Colcol JF, Rallos LE, Baudoin AB (2012) Sensitivity of *Erysiphe necator* to demethylation inhibitor fungicides in Virginia. Plant Dis 96:111–116

Collina M, Landi L, Guerrini P, Branzanti MB, Brunelli A (2005) QoI resistance of *Plasmopara viticola* in Italy: biological and quantitative real-time PCR approaches. In: Dehne HW, Gisi U, Kuck KH, Russell PE, Lyr H (eds) Modern fungicides and antifungal compounds IV. BCPC, Alton, UK, pp 81–88

Corio-Costet MF (2007) Erysiphe necator, Monographie. Tec/Doc Lavoisier, Paris (in French)

Corio-Costet M-F (2012) Fungicide resistance in *Plasmopara viticola* in France and anti-resistance measures. In: Thind TS (ed) Fungicide resistance in crop protection: risk and management. CABI, Oxfordshire, pp 157–171

Corio-Costet MF, Bouscaut J, Delmotte F, Douence L, Richart-Cervera S, Amrani L (2003) Structuration des populations d'oïdium de la vigne et résistance aux fongicides: AFLP et outils moléculaires de détection. In: Annals of the 7th international conference on pest and diseases 2003, AFPP, Tours, France, p 8 (in French)

Corio-Costet M-F, Martinez F, Delmotte F, Douence L, Richart-Cervera S, Chen WJ (2008) Resistance of *Plasmopara viticola* to QoI fungicides: origin and diversity. In: Dehne HW, Gisi U, Kuck KH, Russell PE, Lyr H (eds) Modern fungicides and antifungal compounds V. DPG Selbstverlag, Brauschweig, pp 107–112

Corio-Costet MF, Dufour MC, Cigna J, Abadie P, Chen WJ (2011) Diversity and fitness of *Plasmopara viticola* isolates resistant to QoI fungicides. Eur J Plant Pathol 129:315–329
Corio-Costet MF, Dufour MC, Fontaine S, Micoud A (2014a) Le mildiou et l'oïdium font de la résistance… mais… Union Girondine des vins de Bordeaux, mai, pp 37–41 (in French)
Corio-Costet MF, Fontaine S, Micoud A, Grosman J, Magnien C, Dufour MC (2014b) Assessment of fungicide resistance and pathogen diversity in *Erysiphe necator* using quantitative real-time PCR assays, in France (2008–2012). In: Dehne HW, Deising HB, Fraatje B, Gisi U, Hermann D, Mehl A, Oerke EC, Russell PE, Stammler G, Kuck KH, Lyr H (eds) Modern fungicides and antifungal compounds, vol VI. DPG Publisher, Braunschweig, pp 147–149
Debieu D, Corio-Costet M-F, Steva H, Malosse C, Leroux P (1995) Sterol composition of the wine powdery mildew fungus: *Uncinula necator* sensitive or resistant strains to the sterol biosynthesis inhibitor: triadimenol. Phytochemistry 39:293–300
Delmotte F, Chen WJ, Richard-Cervera S, Greif C, Papura D, Giresse X, Mondor-Genson G, Corio-Costet MF (2006) Microsatellite DNA markers for *Plasmopara viticola*, the causal agent of downy mildew of grapes. Mol Ecol Notes 6:379–381
Délye C (1997) Variabilité de l'agent de l'oïdium de la vigne (*Uncinula necator [schweinitz] Burrill*) et la résistance aux fongicides inhibiteurs de la biosynthèse des stérols. Dissertation, University of Paris VI and XI (in French)
Délye C, Corio-Costet MF (1994) Resistance of grape powdery mildew (*Uncinula necator*) to triadimenol, a sterol biosynthesis inhibitor: biochemical characterisation of sensitive and resistant strains, In: Heaney S, Slawson D, Hollomon DW, Smith M, Russell PE, Parry DW (eds) Fungicide resistance. British crop protection conference 60. PCPC, Surrey, p 87
Délye C, Corio-Costet MF (1998) Origin of primary infections of grape powdery mildew *Uncinula necator*: RAPD analysis discriminates two biotypes. Mycol Res 102:283–288
Délye C, Laigret F, Corio-Costet MF (1997a) RAPD analysis provides insight into the biology and epidemiology of *Uncinula necator*. Phytopathology 87:670–677
Délye C, Laigret F, Corio-Costet MF (1997b) New tools for studying epidemiology and resistance of grape powdery mildew to DMI fungicides. Pestic Sci 60:309–314
Délye C, Laigret F, Corio-Costet MF (1997c) Cloning and sequence analysis of the eburicol 14 alpha-demethylase gene of the obligate biotrophic grape powdery mildew fungus. Gene 195:29–33
Délye C, Laigret F, Corio-Costet MF (1997d) A mutation in the 14alpha-demethylase gene of *Uncinula necator* that correlates with resistance to a sterol biosynthesis inhibitor. Appl Environ Microbiol 63:2966–2970
Délye C, Ronchi V, Laigret F, Corio-Costet MF (1999) Nested allele-specific PCR distinguishes genetic groups in *Uncinula necator*. Appl Environ Microbiol 65:3950–3954
Dufour MC, Fontaine S, Montarry J, Corio-Costet MF (2011) Assessment of fungicide resistance and pathogen diversity in *Erysiphe necator* using quantitative real-time PCR assays. Pest Manag Sci 67:60–69
Erickson EO, Wilcox WF (1997) Distribution of sensitivities to three sterol demethylation inhibitor fungicides among populations of *Uncinula necator* sensitive and resistant to triadimefon. Phytopathology 87:784–791
Evans KJ, Whisson DL, Scott ES (1996) An experimental system for characterizing isolates of *Uncinula necator*. Aust Plant Pathol 100:675–680
FRAC, Fungicide Resistance Action Committee. www.frac.info
French National Memorandum (2014) French National Memorandum on mildews (note technique commune gestion de la résistance 2014, maladies de la vigne, mildiou, oïdium, pourriture grise). http://draaf.pays-de-la-loire.agriculture.gouv.fr/Note-nationale-mildiou (in French)
Furuya S, Mochizuki M, Saito S, Kobayashi H, Takayanagi T, Suzuki S (2010) Monitoring QoI fungicide resistance in *Plasmopara viticola* populations in Japan. Pest Manag Sci 66:1268–1272
Gadoury DM, Pearson RC (1991) Heterothallism and pathogenic specialization in *Uncinula necator*. Phytopathology 81:1287–1293

Galet P (1977) In: Galet P (ed) Les maladies et les parasites de la vigne, vol 1. Paysan du midi, Montpellier, pp 89–222 (in French)

Genet JL, Vincent O (1999) Sensitivity to famoxate of European *Plasmopara viticola* populations. Med Fac Landbouw Rijksuniv Gent 64:559–564

Genet JL, Steva H, Vincent O, Cazenave C (1997) A method for measuring the level of sensitivity of *Plasmopara viticola* populations to cymoxanil. Bulletin OEPP 27:217–225

Gisi U (2012) Resistance to Carboxylic Aid Amide (CAA) fungicides and anti-resistance strategies. In: Thind TS (ed) Fungicide resistance in crop protection: risk and management. CABI, Oxfordshire, pp 96–103

Gisi U, Chin KM, Knapova G, Faerber KR, Mohr U, Parisi S, Sierotzki S, Steinfeld U (2000) Recent developments in elucidating modes of resistance to phenylamide, DMI and strobilurin fungicides. Crop Prot 19:863–872

Gisi U, Sierotzki H, Cook A, McCaffery A (2002) Mechanisms influencing the evolution of resistance to Qo inhibitor fungicides. Pest Manag Sci 58:859–867

Gisi U, Waldner M, Kraus N, Dubuis PH, Sierotzki H (2007) Inheritance of resistance to carboxylic acid amide (CAA) fungicides in *Plasmopara viticola*. Plant Pathol 5:199–208

Grasso V, Palermo S, Sierotzki H, Garibaldi A, Gisi U (2006) Cytochrome b gene structure and consequences for resistance to Qo inhibitor fungicides in plant pathogens. Pest Manag Sci 62:465–472

Gubler WD, Ypema HL, Ouimette DG, Bettiga LJ (1996) Occurrence of resistance in *Uncinula necator* to triadimenol, myclobutanil, and fenarimol in California grapevines. Plant Dis 80:902–909

Gullino ML, Gilardi G, Tinivella F, Garibaldi A (2004) Observations on the behaviour of different populations of *Plasmopara viticola* resistant to QoI fungicides in Italian vineyard. Phytopathol Med 43:341–350

Halleen F, Holz G, Pringle KL (2000) Resistance in *Uncinula necator* to triazole fungicides in South African grapevines. South Afric J Enol Vitic 21:71–80

Heaney SP, Hall AA, Davies SA, Olaya G (2000) Resistance to fungicides in the QoI-STAR cross-resistance group: current perspective, British crop protection conference, BCPC, Farnham, Surrey, pp 755–764

Herzog J, Schuepp H (1985) Haustorial development test to characterize metalaxyl resistance and genetic variability in *Plasmopara viticola*. EPPO Bull 15:431–435

Hollomon DW, Wood PM, Reeve S, Miguez M (2005) Alternative oxidase and its impacts on the activity of Qo and Qi site inhibitors. In: Dehne HW, Gisi U, Kuck KH, Russell PE, Lyr H (eds) Modern fungicides and antifungal compounds IV. BCPC, Alton, UK, pp 55–62

Jordan DB, Livingston RS, Bisaha JJ, Duncan KE, Pember SO, Picollelli MA, Schwartz RS, Sternberg JA, Tang X (1999) Mode of action of famoxadone. Pestic Sci 55:105–108

Lafon R, Bulit J (1981) Downy mildew of the vine. In: Spencer DM (ed) The downy mildews. Academic, London, pp 602–614

Latorse MP, Lagouarde P, Blanc G, Sauzay S (2006) Plan de surveillance de la sensibilité du mildiou de la vigne à l'iprovalicarbe. In: Annals of the 8th international conference on pest and diseases 2003, AFPP, Tours, France, pp 607–611 (in French)

Magnien C, Micoud A, Remuson F (2003) La résistance du mildiou de la vigne aux QoI: resultats des études 2002. In: Annals of the 7th international conference on pest and diseases 2003, AFPP, Tours, France, pp 1–8 (in French)

Magnien C, Remuson F, Le Guellec M, Micoud A, Grosman J (2012) Grapevine downy mildew resistance to fungicides: results of the monitoring directed by the French plant protection organization between 2009 and 2011. In: Annals of the 10th international conference on pests and diseases 2012, AFPP, Tours, France, pp 290–298

Matasci CL, Gobbin D, Schärer HJ, Tamm L, Gessler D (2008) Selection for fungicide resistance throughout a growing season in population of *Plasmopara viticola*. Eur J Plant Pathol 120:79–83

Miazzi M, Hajjeh HR (2011) Differential sensitivity to triadimenol of *E. necator* isolates belonging to different genetic groups. J Plant Physiol 93:729–735
Miazzi M, Natale P, Pollastro S, Faretra F (1997) Handling of the biotrophic pathogen *Uncinula necator* (Schw.) Burr. under laboratory conditions and observations on its mating system. J Plant Pathol 78:71–77
Miles LA, Miles TD, Kirk WW, Schilder AMC (2002) Strobilurin (QoI) resistance in population of *Erysiphe necator* on grapes in Michigan. Plant Dis 96:1621–1628
Miller TC, Gubler WD (2004) Sensitivity of California isolates of *Uncinula necator* to trifloxystrobin and spiroxamine, and update on triadimefon sensitivity. Plant Dis 88:1205–1212
Northover J, Homeyer CA (2001) Detection and management of myclobutanil-resistant grapevine powdery mildew (*Uncinula necator*) in Ontario. Can J Plant Pathol 23:337–345
Ogawa JM, Gubler WD, Manji BT (1988) Effect of sterol biosynthesis inhibitors on disease of stone fruits and grapes in California. In: Berg D, Plempel M (eds) Sterol biosynthesis inhibitors, pharmaceuticals and agrochemicals aspects. Ellis Horwwod, Chichester, pp 262–287
Pezet R, Bolay A (1992) L'oïdium de la vigne: situation actuelle et conséquences pour la lutte. Rev Suisse Vitic Arboric Horticul 24:67–71 (in French)
Redl H, Steinkellner S (1996) Nachweis einer sensitivitäts verminderung von Oidium gegnüber DMI—fungiziden im Österreichschen weinbau. Mitt Klosterneuburg 46:181–188 (in German)
Reifschneider FJB, Boiteux LS (1988) A vacuum-operated settling tower for inoculation of powdery mildew fungi. Phytopathology 78:1463–1465
Reuveni M (2001) Activity of trifloxystrobin against powdery and downy mildew diseases of grapevines. Can J Plant Pathol 23:52–59
Reuveni M (2003) Activity of the new fungicide benthiavalicarb against *Plasmopara viticola* and its efficacy in controlling downy mildew in grapevines. Eur J Plant Pathol 109:243–251
Russell PE (2007) Sensitivity baselines in fungicide resistance research and management, FRAC monograph no 3. Crop Life International, Brussels
Savocchia S, Stummer BE, Wicks TJ, Heeswijck R, Scott ES (2004) Reduced sensitivity of *Uncinula necator* to sterol demethylation inhibiting fungicides in southern Australian vineyards. Aust J Plant Pathol 33:465–473
Scheinpflug H (1994) History of DMI fungicides and monitoring for resistance. In: Delp CJ (ed) Fungicide resistance in North America, 2nd edn. APS, St. Paul, USA, pp 77–78
Sierotzki H, Kraus N, Assemat P, Stanger C, Cleere C, Windass J, Gisi H (2005) Evolution of resistance to QoI fungicides in *Plasmopara viticola* populations in Europe. In: Dehne HW, Gisi U, Kuck KH, Russell PE, Lyr H (eds) Modern fungicides and antifungal compounds IV. BCPC, Alton, UK, pp 73–80
Sierotzki H, Kraus N, Pepin S, Ferandes N, Gisi H (2008) Dynamics of QoI resistance in *Plasmopara viticola*. In: Dehne HW, Gisi U, Kuck KH, Russell PE, Lyr H (eds) Modern fungicides and antifungal compounds V. DPG Selbstverlag, Brauschweig, pp 151–157
Sirven C, Beffa R (2003) Resistance to fenamidone: monitoring by real time quantitative PCR on *Plasmopara viticola*. Pflanz Nachr Bayer 56:23–532
Sirven C, Gonzalez E, Bufflier E, Latorse MP, Beffa R (2002) PCR-based method for detecting mutation allele frequencies for QoI resistance in *Plasmopara viticola*. In: British crop protection conference on pests and diseases 2002, Brighton, pp 823–828
Sombardier A, Dufour MC, Blancard D, Corio-Costet MF (2010) Sensitivity of *Podosphaera aphanis* isolates to DMIs fungicides: distribution and reduced cross-resistance. Pest Manag Sci 66:35–43
Steinkellner S, Redl H (2001) Sensitivity of *Uncinula* necator populations following DMI-fungicide usage in Austrian vineyards. Bodenkult 52:213–219
Stergiopoulos I, Van Nistelrooy JGM, Kema GHJ, De Waard MA (2003) Multiple mechanisms account for variation in base-line sensitivity to azole fungicides in filed isolates of *Mycosphaerella graminicola*. Pest Manag Sci 59:1333–1343
Steva H (1991) Resistance de l'oïdium de la vigne (*Uncinula necator*) aux fongicides inhibiteurs de la biosynthèse des stérols. Dissertation N° 177, University of Bordeaux (in French)

Steva H, Clerjeau M (1990) Cross-resistance to sterol biosynthesis inhibitors fungicides in strains of *Uncinula necator* isolated in France and Portugal. Med Fac Landbouw Rijksuniv Gent 55:983–988

Steva H, Clerjeau M, Da Silva MT (1989) Reduced sensitivity to triadimenol in Portuguese field populations of *Uncinula necator*. ISPP Chem Control Newslett 12:30–31

Suty A, Stenzel K (1999) Iprovalicarb sensitivity of *Phytophthora infestans* and *Plasmopara viticola*: determination of baseline sensitivity and assessment of the risk of resistance. Pflanz Nachr Bayer 52:71–82

Taksonyi P, Kocsis L, Matyas KK, Taller J (2013) The effect of quinone outside inhibitor fungicides on powdery mildew in a grape vineyard in Hungary. Scientia Hort 161:233–238

Thind TS, Clerjeau M, Sokhi SS, Mohan C, Jailloux F (1998) Observations on reduced sensitivity in *Uncinula necator* to triadimefon in India. Ind Phytopathol 51:97–99

Toffolatti SL, Vercesi A (2012) QoI resistance in *Plasmopara viticola* in Italy: evolution and management strategies. In: Thind TS (ed) Fungicide resistance in crop protection: risk and management. CABI, Oxfordshire, pp 157–171

Toffolatti ST, Serrati L, Sierotzki H, Gisi U, Vercesi A (2007) Assessment of QoI resistance in *Plasmopara viticola* oospores. Pest Manag Sci 63:194–201

Toffolatti ST, Prandato M, Serrati L, Sierotzki H, Gisi U, Vercesi A (2011) Evolution of QoI resistance in *Plasmopara viticola* oospores. Eur J Plant Pathol 129:331–338

Wong FP, Wilcox WF (2000) Distribution of baseline sensitivities to azoxystrobin among isolates of *Plasmopara viticola*. Plant Dis 84:275–281

Wong FP, Wilcox WF (2002) Sensitivity to azoxystrobin among isolates of *Uncinula necator*: baseline distribution and relationship to myclobutanil sensitivity. Plant Dis 86:394–404

Worthington PA (1988) Chemistry of sterol biosynthesis inhibitors: piperazines, pyridines, pyrimidines, imidazoles, triazoles, morpholines, piperidines, allylamines. In: Berg D, Plempel M (eds) Sterol biosynthesis inhibitors, pharmaceuticals and agrochemicals aspects. Ellis Horwwod, Chichester, pp 19–55

Ypema HL, Ypema M, Gubler WD (1997) Sensitivity of *Uncinula necator* to benomyl, triadimefon, myclobutanil and fenarimol in California. Plant Dis 81:293–297

Chapter 17
Monitoring Resistance by Bioassay: Relating Results to Field Use Using Culturing Methods

Guido Schnabel, Meng-jun Hu, and Dolores Fernández-Ortuño

Abstract Monitoring for fungicide resistance is becoming more important in light of increased selection for resistance in key agricultural pathogens. *Botrytis cinerea*, the causal agent of gray mold of small fruits, is considered a high-risk fungus and has developed widespread resistance to multiple fungicides in many strawberry production areas managed with site-specific fungicides. Rapid determination of location-specific resistance profiles in strawberry fields allows growers to identify weaknesses in their spray program and tailor spray sequences to farm-specific needs. Growers also appreciate the educational value when participating in monitoring programs. But a monitoring program also benefits the research scientists and allows for the detection of emerging new phenotypes as well as an improved understanding of the molecular basis of resistance and the origin and spread of fungicide-resistant isolates. This information can then be used to design better resistance management strategies. Management of isolates resistant to multiple fungicides is a new challenge that will need to be addressed with an integrated approach involving the implementation of resistance management strategies at the nursery and crop producer levels.

Keywords *Botrytis cinerea* • Gray mold • Grey mould • Multifungicide resistance • Nursery • IPM • Strawberry

G. Schnabel (✉) • M.-j. Hu
Department of Agricultural and Environmental Sciences, Clemson University, Clemson, SC, USA
e-mail: schnabe@clemson.edu

D. Fernández-Ortuño
Dept. de Microbiología, Instituto de Hortofruticultura Subtropical y Mediterránea "La Mayora", Universidad de Málaga-Consejo Superior de Investigaciones Científicas (IHSM-UMA-CSIC), Campus de Teatinos, 29071 Málaga, Spain

H. Ishii, D.W. Hollomon (eds.), *Fungicide Resistance in Plant Pathogens*,
DOI 10.1007/978-4-431-55642-8_17

17.1 Introduction

Monitoring of resistance in fungal pathogens of agriculturally important crops is typically done on a larger, area-wide scale to detect shifts in sensitivity to fungicides or to detect the prevalence of resistant strain in certain regions (Brent 1988). While this has value for scientific purposes and to fine-tune generally applicable antiresistance management strategies, it does not recognize differences in sensitivity to fungicides in local pathogen populations. In this chapter we will illustrate the need for farm-specific resistance monitoring and describe how this program provided individual farmers with critical resistance and disease management information, the clues that were obtained about the origin of resistance (nurseries), and the opportunities that were created for the research laboratory to study the emergence, selection, and molecular mechanisms of resistance.

Strawberries are a major US small fruit crop planted across 59,000 acres (USDA, National Agricultural Statistics Service, http://nass.usda.gov/Quick_Stats/). California and Florida are the top strawberry-producing states, but smaller production also occurs in many other states located on the East Coast. The majority of strawberries produced in eastern states (excluding Florida) are produced by farmers growing 10 acres or less. We estimate that in South Carolina alone there are about 100 family farms growing strawberries commercially and that there are hundreds more between Florida and Pennsylvania. Virtually all commercial strawberries in the Southeast and an increasing number of productions in the Northeast are cultivated using plasticulture (http://www.smallfruits.org/SmallFruitsRegGuide/Guides/2005culturalguidepart1bs1.pdf). Strawberry plasticulture is an annual hill training system in which freshly dug bare-root plants or plugs (transplants started from runner tips in nurseries) are planted in late summer to early fall, depending on the climate.

Gray mold is caused by *Botrytis cinerea* Pers, a disease that dictates the spray program for preharvest and postharvest disease control (Sutton 1998). The fungus can enter the production field with the transplants in the form of latent infections of the leaf or crown tissue, from adjacent plant material such as blackberries or grapes or from spores produced in overwintering plant debris (Sutton 1998). Strawberry flowers are most susceptible to the disease and the gateway for the fungus to infect fruit. Therefore sprays are concentrated on the flowering stage of the crop (Bristow et al. 1986). In the United States, thiram and more frequently captan are used by many growers for control of gray mold, anthracnose, leaf spots, and powdery mildew. However, fungicides with single-site modes of action also called site specific fungicides that often have reduced risk for the environment, worker, and consumer are being pushed into integerated pest management (IPM) programs to control fruit diseases (Adaskaveg et al. 2004). They include benzimidazoles (FRAC 1), dicarboximides (FRAC 2), succinate dehydrogenase inhibitors (SDHIs; FRAC 7), anilinopyrimidines (FRAC 9), quinone outside inhibitors (FRAC 11), phenylpyrroles (FRAC 12), and hydroxyanilides (FRAC 17). They are typically applied at much lower rates compared to captan or thiram but still are generally more effective. The drawback is that these fungicides typically have a single-site mode of action and thus are prone to resistance development (Sierotzki and Scalliet 2013; Fernández-Ortuño et al. 2014). This weakness combined with the tremendous ability to generate genetic diversity in *B. cinerea* populations sets the stage for resistance problems in commercial production.

17.2 Fungicide Resistance in *B. cinerea* from Southeastern Strawberry Fields

Extensive fungicide sensitivity testing of populations from individual farms in North Carolina and South Carolina collected in 2011 demonstrated that resistance to fungicides is an acute problem in the southeastern United States (Fernandez-Ortuño and Schnabel 2012; Li et al. 2014b, Fernández-Ortuño et al. 2012, 2013; Grabke et al. 2013, 2014). Populations of *B. cinerea* from North and South Carolina revealed isolates with resistance to 6 of the 7 chemical classes used for gray mold control nor country captan and thiram and resistance came in 15 different combinations or fungicide resistance phenotypes (Fernández-Ortuño et al. 2014; Li et al. 2014a). Locations tended to have a single dominant, location-specific resistance profile (LSRP) that consisted of resistance to multiple fungicides in fields sprayed weekly with site-specific fungicides. Although the most prevalent profile found in conventional strawberry fields consisted of resistance to FRAC 1 thiophanate-methyl, FRAC 11 pyraclostrobin, and FRAC 7 boscalid, differences in predominant profiles between strawberry fields were identified. Multifungicide resistance had accumulated over time by stepwise acquisition of single resistances (Li et al. 2014a).

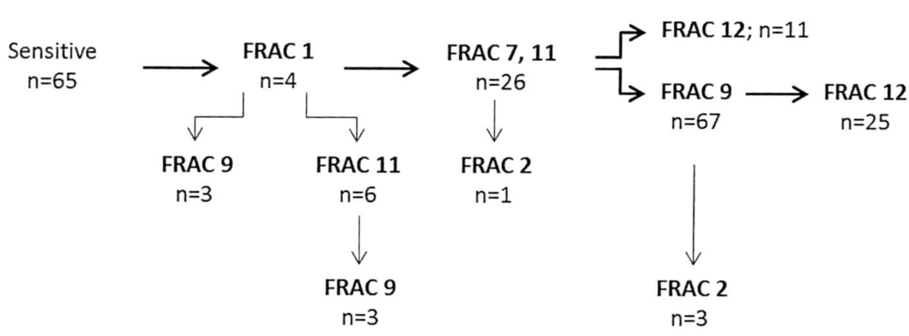

Fig. 17.1 Accumulation (indicated by the *arrows*) of fungicide resistance in *B. cinerea* isolates from commercial strawberry fields of the Carolinas. *Thicker arrows* indicate most likely path of acquistion of resistance to multiple fungicides. FRAC 1 = methyl benzimidazole carbamates, FRAC 11 = quinone outside inhibitors, FRAC 7 = succinate dehydrogenase inhibitors, FRAC 12 = hydroxyanilides, FRAC 9 = anilinopyrimidines, FRAC 2 = dicarboximides, FRAC 12 = phenylpyrroles (low resistance). The number of isolates (*n*) of the accumulated resistance phenotype is indicated

The schematic in Fig. 17.1 shows that all resistant phenotypes likely derived from populations already resistant to FRAC 1 fungicides. For example, isolates resistant to FRAC 12 fungicide were likely to have derived from populations that were already resistant to FRAC 1, 7, and 11 and to FRAC 1, 7, 11, and 9 fungicides (Fig. 17.1).

Multifungicide resistance in *B. cinerea* isolates is also well known in Florida, the largest strawberry production region on the US East Coast. In a study carried out between 2010 and 2012, 392 *B. cinerea* isolates were collected from Florida strawberry fields and evaluated for sensitivity to registered site-specific fungicides. The study documented widespread resistance to FRAC 7 boscalid, FRAC 11 pyraclostrobin, FRAC 12 fenhexamid, FRAC 9 cyprodinil, and FRAC 9 pyrimethanil (Amiri et al. 2013). Isolates resistant to two, three, and four fungicides from different chemical groups were also observed (Amiri et al. 2013). Reports from other states are rare or focus on just one chemical class.

17.3 The Necessity and Development of Location-Specific Resistance Monitoring Programs

General resistance management recommendations such as the rotation and/or mixture of registered chemical classes or the restriction of the number of applications per season delay the selection of fungicide resistance and thus extend the life span of site-specific fungicides (Wade 1988). While these strategies are helpful to delay resistance in situations where the pathogen has not yet acquired resistance, some of them may fail in situations when multifungicide resistance is already observed. Let's use North and South Carolina strawberry fields as an example to illustrate how common strategies may backfire. As described above, resistance to FRAC 1, 7, and 11 fungicides was found to be among the most prevalent resistance profiles. The commonly used premixture of FRAC 7 and 11 fungicides would be expected to provide no protection from gray mold in populations with dual resistance because neither FRAC 7 nor FRAC 11 fungicides are effective. Furthermore, a rotation between FRAC 7/11 mixtures and FRAC 1 fungicides would also be expected to provide very little to no control. In such locations the mixture or rotation of FRAC codes will also be an unnecessary expense for growers and pose unnecessary risk to the environment, worker, and consumer.

Because of widespread multifungicide resistance in the gray mold pathogen, it is critical that general resistance management strategies are supported by the identification of chemical classes with little to no resistance issues. Strawberry fields, even when in close proximity, may have a unique fungicide resistance profile (Fernández-Ortuño et al. 2014) (Fig. 17.2) and therefore LSRPs need to be determined to avoid using ineffective chemical classes in standard rotations or mixtures.

This may at first glance seem like an impossible task given the often large number of individual farms, the number of isolates, and the different chemical classes that would need to be screened with traditional methods such as the deter-

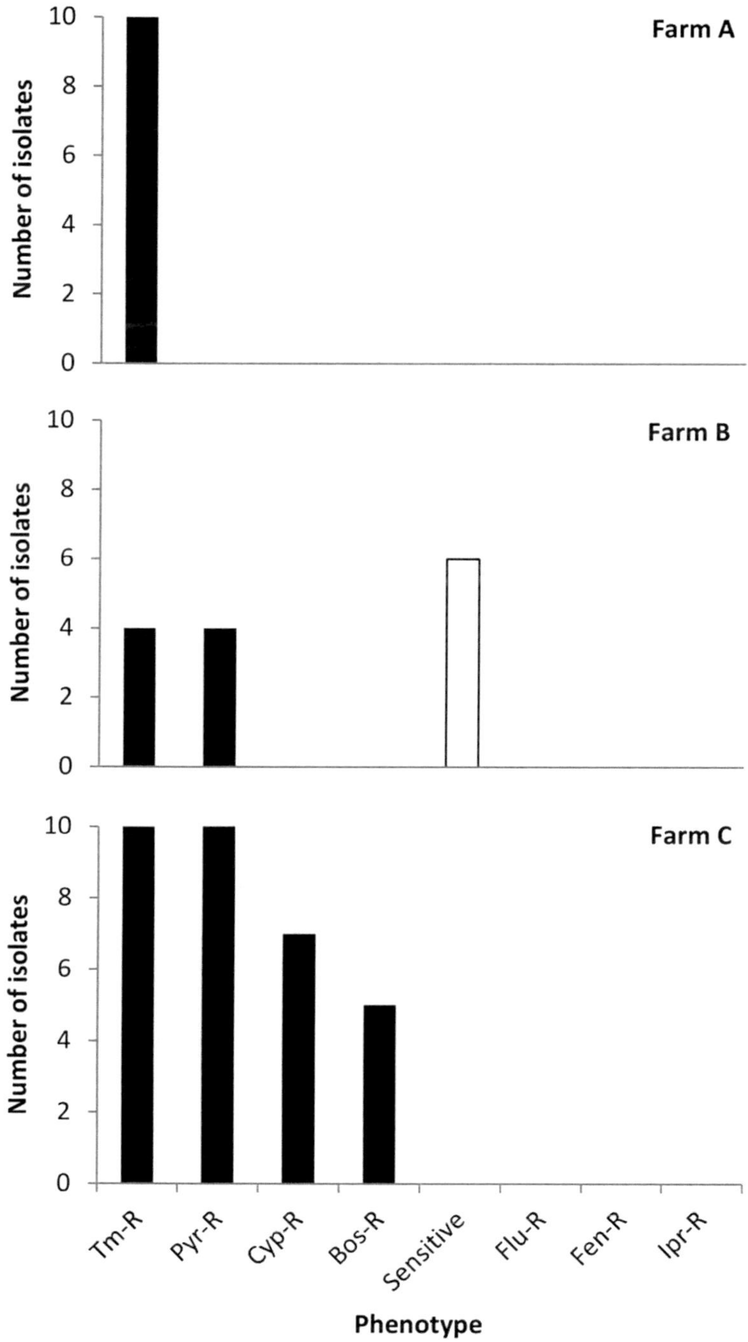

Fig. 17.2 Fungicide resistance profile of *Botrytis cinerea* from strawberry fields of three farms located within 10 km of each other. Ten isolates were investigated for each location and some were resistant to multiple fungicides (farms B and C). Sensitive = sensitive to all fungicides tested Tm = thiophanate-methyl, Pyr = pyraclostrobin, Cyp = cyprodinil, Bos = boscalid, Flu = fludioxonil, Fen = fenhexamid, and Ipr = iprodione

mination of EC_{50} values for a large number of individuals per population (Brent 1988). Molecular methods to detect resistance can be very helpful if the mechanism of resistance is consistent. For example, resistance to FRAC 11 fungicides may be exclusively conferred by G143A in *cyt b* and resistance to FRAC 1 by E198A in beta-tubulin gene. However, molecular methods specific for detecting a single mutation may result in false negatives if other mutations are also involved. For example, resistance to FRAC 7 fungicides can be conferred by different mutations in the *sdh*B gene (Sierotzki and Scalliet 2013) and resistance to FRAC 2 fungicides is conferred by different mutations in the *Bos*1 gene and yet unknown mechanisms of resistance (Grabke et al. 2014). We developed an assay to screen for fungicide sensitivity, first for brown rot of peach (Schnabel et al. 2012) and now for gray mold of strawberry (Fernández-Ortuño et al. 2014), which detects resistance regardless of the mechanism fast and reliably based on mycelium growth on fungicide-amended artificial growth media using a single discriminatory concentration. Although we may not be able to determine the mechanism of resistance, this screening technique will provide information of practical relevance. Advantages of using a single concentration include the time and resource savings compared to EC_{50} determinations. But there are disadvantages. Using a discriminatory concentration may lead to false assumptions. If the discriminatory concentration is too high, results may underestimate the resistance situation. For example, if the concentration is designed to detect qualitative resistance based on point mutations in target genes, isolates with a different mechanism of resistance that confers lower levels of resistance may falsely be rated as sensitive due to lack of growth on the fungicide-amended medium. For this reason the discriminatory concentration selected to detect resistance should be as low as possible to distinguish sensitive from resistant isolates for all practical purposes but high enough to avoid overestimating the resistance situation.

17.4 Selection of Discriminatory Concentrations

The discriminatory concentration used to distinguish a sensitive from a resistant isolate needs to be meaningful. In other words, the level of resistance must be of practical relevance. It is necessary to determine whether isolates with reduced sensitivity are capable of causing disease on fruit or leaves that are treated with label rates of fungicides. Typically, pathogens are collected and isolated from "problem" areas, i.e., areas where disease developed at economically unacceptable levels despite appropriate fungicide applications. Sensitivity testing can be done in vitro using germination and germ tube assays (Weber and Hahn 2011) or mycelium growth assays (Brent 1988). The in vitro sensitivity is traditionally determined by exposing isolates of the pathogen to different concentrations of the fungicide to determine the concentration at which a fungicide inhibits 50 % of the mycelium growth (EC_{50} value) or MIC (minimal inhibitory concentration). Based on in vitro results, isolates with low, medium, and high EC_{50} values (representing the entire range of sensitivities) are selected for in vivo assays. We typically use detached fruit sprayed with label rates of a fungicide and inoculated with the pathogen (Schnabel et al. 2004). It would be expected that isolates with high EC_{50} values (and maybe even those with medium EC_{50} values) are capable of developing disease on fungicide-treated, detached fruit, as well as on the control fruit. A discriminatory concentration can then be inferred based on a comparison between in vitro baseline fungicide sensitivity and in vivo results. The discriminatory concentration may vary from one fungicide to another depending on the chemical class, the pathogen, growth medium used, and incubation conditions.

For *B. cinerea* we found the following fungicides, discriminatory concentrations, and media suitable to distinguish sensitive from resistant isolates (Fernández-Ortuño et al. 2014): fenhexamid at 50 μg/ml, fludioxonil at 0.5 μg/ml, iprodione at 10 μg/ml, thiophanate-methyl at 100 μg/ml, and pyraclostrobin at 10 μg/ml plus the alternative oxidase inhibitor salicylhydroxamic acid (SHAM) at 100 μg/ml in 1 % malt extract agar (MEA); cyprodinil at 4 μg/ml in CzA (CzapeK-Dox agar medium (CzA): 2 g $NaNO_3$, 0.5 g KCl, 0.5 g $MgSO_4.7H_2O$, 0.01 g $FeSO_4$, 1 g K_2HPO_4, 0.01 g $ZnSO_4$, 0.005 g $CuSO_4.5H_2O$, 30 g sucrose, and 15 g agar in 1 l medium); and boscalid at 75 μg/ml in 1 % yeast bacto acetate (YBA) agar. Tests are conducted in wells (15 mm in diameter) of 24-well plates (6×4 wells, 12.5×8.5×2 cm; Thermo Fisher Scientific, Roskilde, Denmark). Inoculated plates are incubated at 22 °C for 4 days, and diametric colony growth is visually assessed in each well: (sensitive, S) for absence of growth, (low resistant, LR) for less than 20 % diametrical growth, (moderately resistant, MR) for less than 50 % but more than 20 % diametrical growth, and (resistant, R) for more than 50 % diametrical growth compared to the 15 mm well diameter. For large-scale resistance monitoring, two sterile 24-well plates named "A" and "B" are used. If held horizontally, 6 wells per row allow for testing six isolates and four fungicides at one discriminatory concentration per plate. Plate "A" contains the control treatment and three fungicide treatments (boscalid, cyprodinil, and fenhexamid, respectively); plate "B" contains four fungicide treatments (fludioxonil, iprodione, pyraclostrobin, and thiophanate-methyl). Four 24-well plates are required (2 plates "A" and 2 plates "B") to test 10 isolates leaving 2 of the 12 columns empty. For simplicity reasons, mycelium growth in compartments is assessed visually. Such visual assessments do not require the use of additional tools such as calipers or rulers and thus facilitate the process of determining the absence or presence of resistance.

17.5 Isolate Collection and Processing

The optimal time for collecting field isolates for testing is in early spring prior to bloom. As mentioned before, the strawberry blossoms are highly susceptible to *Botrytis* infection and knowing what product will best protect the flowers is the key to gray mold control. During winter in South Carolina frost events are common and growers are often protecting their crop using row covers. Still, some frost damage occurs every year resulting in damaged flowers. It is from these dead flowers that we reliably can isolate the *B. cinerea* fungus in spring. We collect about 30 flowers with black torus (likely caused by freeze damage) throughout the field, each from a different plant. After removal of the petals, the torus is dipped in 10 % bleach for 1 min, rinsed with sterile water, and incubated at 100 % humidity at room temperature for 1 day. Then the plant material is kept at 60 to 80 % humidity to encourage sporulation. Typically we have a greater than 50 % recovery rate from flowers. The fungus can also be obtained prior to bloom from latent infections of strawberry leaves. The leaves should be surface sterilized as described above prior to incubation to recover conidia. Some authors recommend to use paraquat at 20 mg ai/L to recover conidia (Sutton and Peng 1992). For fruit samples, 10–12 individual strawberry fruits with small (young) gray mold lesions are obtained from each farm; each fruit came from a different plant with at least five buffer plants between sampled plants. Conidia are collected using individually wrapped sterile cotton swabs (Thermo Fisher Scientific). The cotton tip is rubbed gently against the youngest area of sporulation (periphery of the lesion) of a fruit to capture conidia. The white cotton tip turns from pure white to lightly gray, indicating that sufficient conidia are collected; then, the swab is returned to its wrapper. Conidia from sporulating blossoms and leaves or from the cotton swabs (obtained from fruit) are transferred directly to the center of each well using a sterile wooden toothpick. Inoculated plates are incubated and assessed as described above.

17.6 Interpretation of the Data and Automated Construction of Resistance Management Recommendations

For each location we process 10 isolates. If all isolates from a location test positive, there can be no argument about whether the population is of high risk of resistance. But what if only a fraction of the 10-isolate sample is resistant? Although no data is available specifically for *B. cinerea*, research in other pathosystems has shown that the economic threshold may generally be somewhere between 5 and 20 % resistant isolates in a population. For example, when 5 % of a *Fusarium graminearum* population was resistant to benzimidazole fungicides, the efficacy of carbendazim applications on wheat was 93.2 %. However, efficacy dramatically dropped to 59.7 % when 20 % of the population was resistant (Zhonghua Ma, Zhejiang University, personal communication). In a different pathosystem including populations of azoxystrobin (quinone outside inhibitor)-sensitive and azoxystrobin-resistant downy and powdery mildews, the disease severity on azoxystrobin-treated cucumber leaves climbed to more than 20 % when only 10 % of the pathogen population was resistant to the fungicide (Ishii et al. 2007). We therefore consider a location of high risk of resistance to a fungicide when 2 or more isolates out of 10 (equal or more than 20 %) are resistant. A population is of medium risk of resistance to a particular fungicide if only 1 out of 10 isolates tested resistant.

A web application was developed to accelerate communication transfer and ensure a quick turnaround of results. Web applications are extremely useful due to the ubiquity of web browsers and the ability to update and maintain web applications without distributing and installing software on multiple computers. The ability to update information for a web application also increases the utility of such programs, and in the case of resistance management, new fungicides or resistance conditions on the ground would require immediate changes to the recommendations. A web program gives maximum flexibility in restoring the integrity of the program when changes occur – immediate flexibility. Visual observations from the 24-well plates are entered into the web application together with information about the origin of the isolates and spray history of the field. The web application validates that all required information is entered and requests a final review of the data entered. Then a report is constructed by assembling predrafted resistance management recommendations for each fungicide and corresponding resistance risk category (no risk, low risk, and high risk). The user may save the report in pdf format and instantly share it with the producer by e-mail. A similar web application was designed previously for resistance profiling populations of *Monilinia fructicola*, the causal agent of brown rot of stone fruits (Schnabel et al. 2012). The web application prototype for monitoring resistance in *B. cinerea* is currently located at www.peachdoc.com under Links/strawberry.

17.7 Outcomes

17.7.1 Direct Benefits to Producers

The LSRP provides growers with an assessment of the fungicide resistance situation in a commercial field. Chemical classes are identified that no longer would provide disease protection, while others can then be used to effectively implement general resistance management practices such as rotation or mixtures of different FRAC codes. A poll among participating growers conducted in 2013 indicated improved preharvest and postharvest disease control when following resistance management recommendations based on LSRPs compared to previous years when the resistance profile was unknown. The same poll indicated that producers benefited from the extension program in the form of improved understanding of active ingredients, trade names, FRAC codes, and general resistance management strategies. Often times the resistance profile was reflective of spray habits. For example, growers who sprayed on a calendar-based, weekly regimen and used many different kinds of FRAC codes during the season were much more likely to see a resistance profile reflecting high incidences of multifungicide-resistant isolates. On the other hand, growers with low-input spray histories were more likely to see fewer isolates with resistance (Li et al. 2014a). Growers indicated that knowledge of the resistance risks for each chemical class prevented them from purchasing fungicides that would have not been effective in their fields. Some research and extension specialists working on strawberry but not affiliated with Clemson University complemented this extension program and indicated "how much benefit this service is having to growers! I once said this is the greatest thing to come along since plant tissue sampling, and I meant it!" and "Your [resistance monitoring] work is a really great example of how folks thinking creatively and inclusively can create a program that's win-win for everyone."

17.7.2 Indirect Benefits to the Producers

Three years of region-wide resistance monitoring revealed information that in the long run will help extension specialists improve current disease and resistance management efforts through better understanding of the nature, evolution, and spread of resistance. We obtained a thorough understanding of the molecular basis of resistance to FRAC 1, 2, 7, 11, 12, and 17 fungicides and documented that resistance to multiple fungicides occurs in most production fields. In the eastern United States, resistance in *B. cinerea* to FRAC 1, 7, 11, and 17 fungicides and low and medium resistance to FRAC 2 fungicide are based on target site mutations in the genes encoding β-tubulin, succinate dehydrogenase subunit b, cytochrome b, 3-keto reductase of the ergosterol biosynthesis pathway, and the osmosensing histidine kinase, respectively (Fernández-Ortuño et al. 2012, 2013; Fernández-Ortuño and Schnabel 2012; Grabke et al. 2013, 2014). Resistance to FRAC 12 fungicide is conferred by mutations in transcription factor *mrr1* leading to overexpression of ATP-binding cassette transporter *atrB* and consequently the MDR1 and MDR1h phenotypes (Li et al. 2014b; Fernández-Ortuño et al. 2015). Only the mechanisms conferring high resistance to FRAC 2 fungicide and resistance to FRAC 9 fungicide still remain unknown (Fernández-Ortuño et al. 2013; Grabke et al. 2014).

We also showed that multifungicide resistance has developed via stepwise accumulation of single resistances (Li et al. 2014a). But some findings not discussed in our previous publications are of practical significance. The stepwise accumulation of single resistances started with phenotypes that were resistant to FRAC 1 fungicides. FRAC 1 fungicides such as benomyl were introduced in the late 1960s and had been heavily used by growers until the discovery of widespread resistance. Our data show that the selection for resistance to other, newer fungicides was formed on the backbone of populations already resistant to FRAC 1 fungicides. In other words, at least in the eastern United States, virtually every single isolate resistant to FRAC 2, 7, 9, 11, and/or 12 fungicides is also resistant to FRAC 1 fungicide. This has significant implications for resistance management because with each spray of a FRAC 1 fungicide, phenotypes resistant to other FRAC groups are likely to be selected as well (Fig. 17.3). This may sound strange, and we do not want to imply there to be cross-resistance between FRAC groups. It is simply a consequence of the stepwise accumulation of single resistances in individual isolates of a population

			FRAC code						
			1	11	7	9			
No.	%	Single resistances	Tm	Py	Bo	Cy	Fe	Ip	Fl
65	30.4	0SR	S	S	S	S	S	S	S
4	1.9	1SR	R	S	S	S	S	S	S
3	1.4	2SR	R	S	S	R	S	S	S
6	2.8	2SR	R	R	S	S	S	S/LR	S
3	1.4	3SR	R	R	S	R	S	LR	S
26	12.1	3SR	R	R	R	S	S	S/LR	S
11	5.1	4SR	R	R	R	S	R	S/LR	S
1	0.5	4SR	R	R	R	S	S	R	S
67	31.3	4SR	R	R	R	R	S	S/LR	S
3	1.4	5SR	R	R	R	R	S	R	S
25	11.7	5SR	R	R	R	R	R	S/LR	S

Phenotype is being selected by 4 FRAC codes

Fig. 17.3 A single phenotype can be selected by four different FRAC codes. In this example, the most frequent ($n=67$) *B. cinerea* phenotype with four single resistances (4SR), including thiophanate-methyl (Tm), pyraclostrobin (Py), boscalid (Bo), and cyprodinil (Cy), can be selected by FRAC codes 1, 11, 7, and 9. Data table from Li et al. (2014a). (Fe) fenhexamid, (Ip) iprodione, (Fl) fludioxonil

that had its origin in populations that were resistant to FRAC 1 fungicides. Let's illustrate this phenomenon based on a simplified example. Imagine a population of *B. cinerea* consists of isolates that are resistant to FRAC 1 and FRAC 7 fungicides and of isolates sensitive to both FRAC groups. Each application of a FRAC 1 fungicide will select for the isolates resistant to 1 and 7; thus the application of FRAC 1 fungicides selects indirectly for FRAC 7 resistance. As discussed above, FRAC 1 fungicides are a part of virtually every single resistance phenotype and therefore represent "Selectors" of multifungicide resistance that should for the foreseeable future not be a part of resistance management strategies any longer in the eastern United States.

The resistance monitoring program also provided insights into the origin of fungicide resistance. Growers in South and North Carolina as well as many other eastern states do typically not apply fungicides during the fall and winter months when strawberry plants are nearly dormant. First applications arc usually made in early spring at the onset of bloom. The flowers we used to obtain *B. cinerea* isolates for analysis largely had not been exposed to fungicide applications, and thus there had been no selection pressure for the pathogen population. Still we were isolating multifungicide-resistant isolates from flowers (Fig. 17.4). These pathogens could have derived from strawberry tissue that had not fully been decomposed the previous season or from other hosts such as nearby perennial blackberries or grapes. However, it is also possible that these isolates derived from latent infections in the

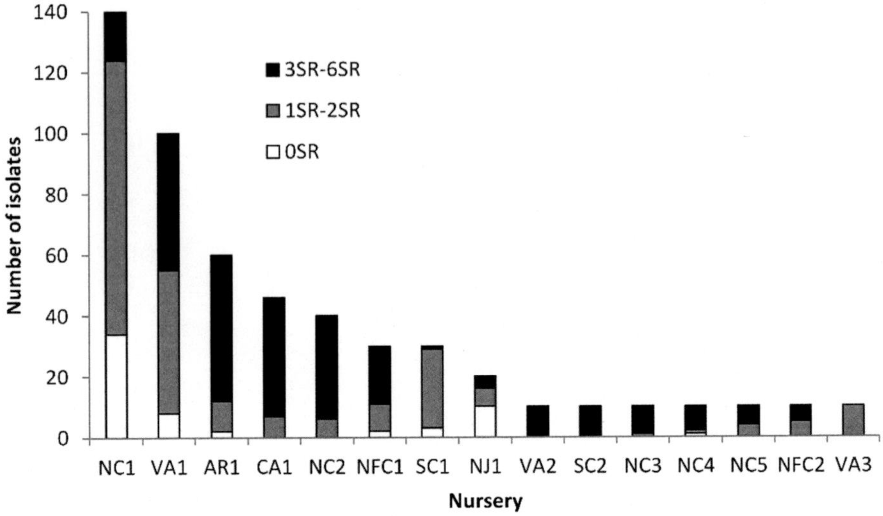

Fig. 17.4 Evidence for nursery-derived, fungicide-resistant *B. cinerea* isolates. The isolates were recovered from strawberry fields located along the US East Coast that had not yet been sprayed with fungicides. Single resistances (SR) ranged from 0 to 6, indicating that some isolates were resistant to up to 6 chemical classes of fungicides

planting stock since many producers do not diversify into other small fruits and because soil preparation for plasticulture leaves the field virtually void of inoculum sources. Very recent studies out of Florida confirm this hypothesis (Oliveira and Peres 2013, 2014). The identification of the nursery as a possible source of resistance provides opportunities for sustained gray mold management with site-specific fungicides. Resistance management in nurseries that produce strawberry tips and plugs for fruit producers must become an integral part of the overall plan to manage resistance.

References

Adaskaveg JE, Förster H, Gubler WD, Teviotdale BL, Thompson DF (2004) Reduced risk fungicides: a new way of managing brown rot and other fungal diseases of stone fruit crops. Calif Agric 59:109–114

Amiri A, Heath SM, Peres NA (2013) Phenotypic characterization of multifungicide resistance in *Botrytis cinerea* isolates from strawberry fields in Florida. Plant Dis 97:393–401

Brent KJ (1988) Monitoring for fungicide resistance. In: Delp CJ (ed) Fungicide resistance in North America. APS Press, St. Paul, pp 9–11

Bristow PR, McNichol RJ, Williamson B (1986) Infection of strawberry flowers by *Botrytis cinerea* and its relevance to grey mould development. Ann Appl Biol 109:545–554

Fernández-Ortuño D, Schnabel G (2012) First report of thiophanate-methyl resistance in *Botrytis cinerea* on strawberry from South Carolina. Plant Dis 96:1700

Fernández-Ortuño D, Chen F, Schnabel G (2012) Resistance to pyraclostrobin and boscalid in *Botrytis cinerea* isolates from strawberry fields in the Carolinas. Plant Dis 96:1198–1203

Fernández-Ortuño D, Chen F, Schnabel G (2013) Resistance to cyprodinil and lack of fludioxonil resistance in *Botrytis cinerea* isolates from strawberry in North and South Carolina. Plant Dis 97:81–85

Fernández-Ortuño D, Grabke A, Bryson PK, Amiri A, Li X, Peres N, Schnabel G (2014) Fungicide resistance profiles in *Botrytis cinerea* from strawberry fields of seven southern U.S. States. Plant Dis 98:825–833

Fernández-Ortuño D, Grabke A, Li XP, Schnabel G (2015) Independent emergence of resistance to seven chemical classes of fungicides in *Botrytis cinerea*. Phytopathology 105:424–432

Grabke A, Fernández-Ortuño D, Schnabel G (2013) Fenhexamid resistance in *Botrytis cinerea* from strawberry fields in the Carolinas is associated with four target gene mutations. Plant Dis 97:271–276

Grabke A, Fernández-Ortuño D, Amiri A, Li X, Peres N, Smith P, Schnabel G (2014) Characterization of iprodione resistance in *Botrytis cinerea* from strawberry and blackberry. Phytopathology 104:396–402

Ishii H, Yano K, Date H, Furuta A, Sagehashi Y, Yamaguchi T, Sugiyama T, Nishimura K, Hasama W (2007) Molecular characterization and diagnosis of QoI resistance in cucumber and eggplant fungal pathogens. Phytopathology 97:1458–1466

Li XP, Fernández-Ortuño D, Chen S, Grabke A, Luo CX, Bridges WC, Schnabel G (2014a) Location-specific fungicide resistance profiles and evidence for stepwise accumulation of resistance in *Botrytis cinerea*. Plant Dis 98:1066–1074

Li XP, Fernández-Ortuño D, Grabke A, Schnabel G (2014b) Resistance to fludioxonil in *Botrytis cinerea* isolates from blackberry and strawberry. Phytopathology 104:724–732

Oliveira MS, Peres NA (2013) The role of nursery plants as a potential source of inoculum for *Botrytis cinerea* and its impact on fungicide sensitivity. Phytopathology:S214-P.

Oliveira MS, Peres NA (2014) Strawberry nursery plants as a source of *Botrytis cinerea* isolates resistant to fungicides. Phytopathology:S208-P

Schnabel G, Bryson PK, Bridges W, Brannen PM (2004) Reduced sensitivity in *Monilinia fructicola* to propiconazole in Georgia and implications for disease management. Plant Dis 88:1000–1004

Schnabel G, Amiri A, Brannen PM (2012) Field kit- and internet-supported fungicide resistance monitoring. In: Thind TS (ed) Fungicide Resistance in Crop Protection: Risk and Management. CAB International, Oxfordshire, pp 116–132

Sierotzki H, Scalliet G (2013) A review of current knowledge of resistance aspects for the next-generation succinate dehydrogenase inhibitor fungicides. Phytopathology 103:880–887

Sutton JC (1998) Botrytis Fruit Rot (Gray Mold) and blossom blight. In: Maas JL (ed) Compendium of strawberry diseases. APS Press, St. Paul, p 28

Sutton JC, Peng G (1992) Biocontrol of *Botrytis cinerea* in strawberry leaves. Phytopathology 83:615–621

Wade M (1988) Strategies for preventing or delaying the onset of resistance to fungicides and for managing resistance occurrences. In: Delp CJ (ed) Fungicide resistance in North America. APS Press, St. Paul, pp 14–15

Weber RWS, Hahn M (2011) A rapid and simple method for determining fungicide resistance in *Botrytis*. J Plant Dis Prot 118:17–25

Chapter 18
Monitoring Resistance Using Molecular Methods

Derek William Hollomon and Hideo Ishii

Abstract Although cultural methods and bioassays remain a cornerstone in the detection and monitoring of fungicide resistance, molecular diagnostics are playing an important part in many aspects of fungicide resistance research. Molecular methods can avoid the need to isolate pathogens and detect resistance alleles at lower frequencies and are less resource-intensive than cultural methods and bioassays. Molecular diagnostics generally identify **s**ingle-**n**ucleotide **p**olymorphisms (SNPs) and so require that point mutations causing resistance are known. The **p**olymerase **c**hain **r**eaction (PCR) is at the core of most methods described here, in which detection is based on hybridization or amplification with allele-specific probes or primers, use of restriction enzymes or sequencing. Results may be presented just as the presence or absence of resistance, but using fluorescent markers, and following hybridization or amplification in real time, allows quantification of the frequency of resistance mutations within pathogen populations. The complexity and equipment requirements for many of these methods prevent the use in small laboratories lacking PCR equipment and expertise. Unlike *Taq* polymerase used in PCR, *Bst* polymerase amplifies DNA without denaturation, so that, using specific primer sequences, SNPs can be detected isothermally using just a water bath and a single temperature. Known as loop-mediated isothermal amplification (LAMP), it offers the potential for accurate and rapid detection of resistance in field samples, even for small diagnostic laboratories. Because the "on-site" methods are still insufficient, molecular diagnostics have not yet proved useful to growers in guiding in-season antiresistance strategies, but further developments in LAMP may overcome this limitation.

D.W. Hollomon (✉)
Pest Management Science, Bristol BS40 8UB, UK
e-mail: orchardhse@ukgateway.net

H. Ishii
Kibi International University, Minami-Awaji, Hyogo 656-0484, Japan

National Institute for Agro-Environmental Sciences,
Kannondai 3-1-3, Tsukuba, Ibaraki 305-8604, Japan
e-mail: hi481204@yahoo.co.jp

H. Ishii, D.W. Hollomon (eds.), *Fungicide Resistance in Plant Pathogens*,
DOI 10.1007/978-4-431-55642-8_18

Keywords Resistance monitoring • Molecular diagnosis • PCR • PCR-RFLP • ASPCR • Multiplex PCR • DNA sequencing • PCR-Luminex • LAMP

18.1 Why Use Molecular Diagnostics to Monitor Resistance?

Cultural methods are widely employed for testing fungicide sensitivity of single- or mass-spore isolates of a pathogen. Bioassays using plants are also essential to establish the field relevance of resistance found by in vitro laboratory tests and for the evaluation of fungicide sensitivity of obligate pathogens in particular. Both methods thus monitor variation in fungicide sensitivity in pathogen populations and are key components of resistance research as well as resistance management. They usually involve simple techniques and allow changes in sensitivity to be followed and resistance monitored regardless of the mechanisms involved. Often very resource intensive so limiting the number of isolates that can be handled, cultural methods and in vitro bioassays often require isolation of pathogens which may germinate poorly, grow slowly and loose pathogenicity in culture. Obligate pathogens not only require *in planta* assays, but maintenance and routine subculturing of isolates on host plants can be a troublesome long-term commitment, although some methods are now available which maintain isolates without subculturing.

Where mechanisms of resistance are known and are caused by a few point mutations, molecular techniques address many of these limitations associated with cultural methods or bioassays. A review of the limitations of these conventional methods and molecular techniques for the detection and monitoring of fungicide resistance was published earlier (Ishii 1995, 2010; Ma and Michailides 2005). More recently, an extensive review has been published by Beckerman (2013). Based around nucleic acid detection techniques to identify single-nucleotide polymorphisms (SNPs), and especially involving the polymerase chain reaction (PCR; Sambrook et al. 1989), molecular diagnostics identify genotypes rather than phenotypes. Central to all molecular diagnostics is the use of small (15–30 base pairs) oligonucleotide probes or primers designed to hybridise specifically to either the wild-type or the mutant gene sequences. DNA is generally, but not necessarily, extracted from infected plant materials, avoiding the need to first isolate the pathogen. Primers and probes are easily synthesised and can be purchased from many companies. Availability of kits for many of the steps involved, readily prepared reagent mixtures including *Taq* DNA polymerase, automated pipetting equipment, PCR machines, and plate readers, means that the large numbers of samples can be handled efficiently. However, the cost of some equipment restricts these techniques to research laboratories, and unlike rapid "in-field" immunoassays which identify pathogens, "in-field" techniques using molecular technologies to monitor resistance are not yet available commercially.

18.2 Detection Using Allele-Specific Probes

The first fungicide resistance mechanism characterised at a molecular level in many pathogens isolated from crops involved a point mutation at codon 198 or 200 in β-tubulin, causing resistance to carbendazim (MBC) and related benzimidazole fungicides such as benomyl and thiophanate-methyl. This provided an opportunity to translate Southern hybridisation diagnostic methods employed in medicine to detect carbendazim resistance (Koenraadt and Jones 1992; Wheeler et al. 1995). PCR-amplified and denatured β-tubulin gene fragments (amplicons) dot-blotted onto nitrocellulose or nylon membranes could be probed with allele-specific oligonucleotides (ASOs) designed to include a different nucleotide in the middle of the probe and which corresponded to changes within codon 198 or 200. Careful control of the stringency (temperature, salt concentration and pH) ensured probes only hybridised if there was a perfect match between ASO and the target gene sequence. Dot-blotting and using radiolabeled probes to confirm hybridisation was cumbersome and hazardous, and so labelling probes instead with biotin followed by ELISA-based detection using alkaline phosphatase-conjugated streptavidin not only avoided use of radioactivity but allowed transfer of the protocol to microtiter plates.

But the large number of pipetting operations needed increased the likelihood of errors, and simpler methods involving fluorescence were available. Most *Taq* DNA polymerases have good endonuclease activity and digest probes only if hybridised to a target sequence. Resistance mutations could be detected by incorporating a fluorescent marker at one probe end and a quencher at the other (TaqMan®Lee et al. 1993) and measuring fluorescence in a fluorimeter. Molecular beacons (Tapp et al. 2000) do not depend on *Taq* DNA polymerase activity, but fluorescent marker and quencher are sufficiently separated only when the beacon hybridises to its target sequence. Use of TaqMan® probes to identify SNPs causing fenhexamid resistance in *Botrytis cinerea* was recently published by Billard et al. (2012). With these methods detection of an amplicon can be achieved in a single reaction without additional pipetting, although controls (no target DNA) are needed to reveal background fluorescence levels. Unfortunately, the degeneracy of the genetic code means that ASOs are species-specific and unlikely to detect resistance even in closely related species.

PCR-Luminex, a novel system developed for high-throughput analysis of SNPs, was successfully introduced to diagnose resistance in the rice blast fungus, *Magnaporthe oryzae*, to inhibitors of scylatone dehydratase involved in melanin biosynthesis (MBI-D) using ASO probes coupled with fluorescent beads (Figs. 18.1 and 18.2; Ishii et al. 2008). The PCR-Luminex system was further proved to be useful for identifying major species causing *Fusarium* head blight on wheat. With this system we can distinguish up to about 100 different SNPs simultaneously in a single reaction; therefore, it might become a powerful tool for big diagnostic laboratories to identify resistance based on multiple mutations easily and rapidly in the future.

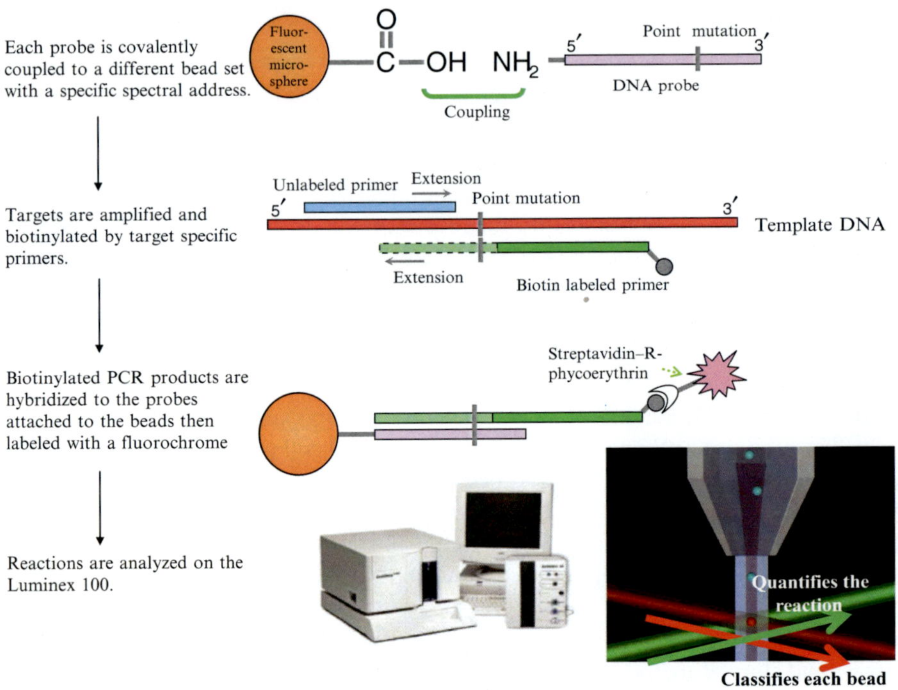

Fig. 18.1 Steps for PCR-Luminex analysis (Reprinted from Ishii et al. (2008))

18.3 Detection Using Restriction Enzymes

18.3.1 PCR-RFLP (Restriction Fragment Length Polymorphism)

Restriction enzymes recognise short specific sequences in double-stranded DNA and split both strands within these specific sites. A single base change can easily be detected when it either destroys an existing recognition sequence or creates a new one. Following PCR amplification of a relevant target gene fragment, incubation with the restriction enzyme, and separation of products by electrophoresis on an agarose gel, changes in the size of the initial amplicon can easily be detected visually by staining the gel with ethidium bromide or GelRed™, dyes that only fluoresce intensively when intercalated within double-stranded DNA. If a mutation-causing resistance creates a restriction site not present in the wild-type, a different fragment pattern will be created confirming resistance. More than 100 different restriction enzymes are readily available, so it is usually possible to purchase an enzyme specific for nucleotide changes involved in resistance. This technique known as PCR-RFLP may occasionally require, depending on the buffer needed for

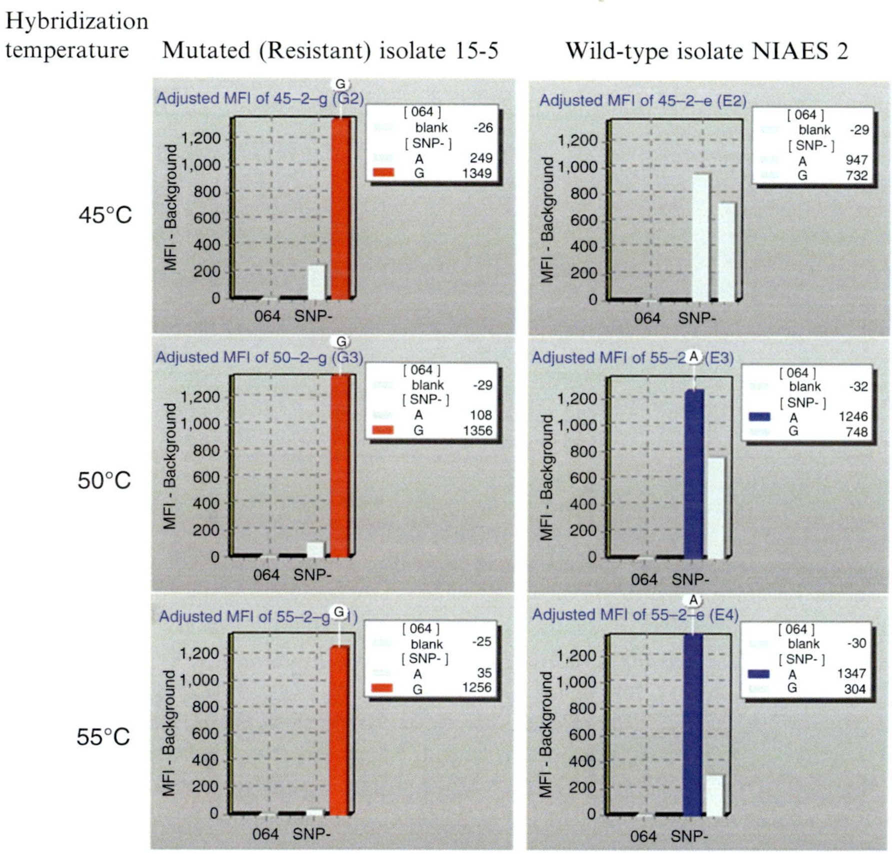

Fig. 18.2 Diagnosis of MBI-D fungicide resistance in *Magnaporthe oryzae* isolates by PCR-Luminex system. Isolate 15–5 is resistant, and NIAES 2 is wild-type sensitive to MBI-D fungicides. Nucleotide changes from GTG in wild-type sensitive to ATG in MBI-D-resistant isolates at amino acid position 75 of scytalone dehydratase gene were clearly identified

restriction enzyme, purification of amplicon prior to enzyme incubation, which means several operations increasing the chance of pipetting errors.

Although RFLP is not quantitative, it does not require sophisticated equipment and consequently is very commonly used in resistance monitoring (Fig. 18.3). Resistance of *Plasmopara viticola* to QoI fungicides, caused by the replacement of glycine by alanine at codon 143 (G143A), is widely monitored in Japan using PCR-RFLP methods developed by Furuya et al. (2009). More recently, the presence of the *PvCesA3* gene allele conferring resistance to the CAA fungicide mandipropamid has also been successfully monitored in *P. viticola* populations by PCR-RFLP (Aoki et al. 2011).

In this way, QoI resistance, mainly caused by a point mutation of the target cytochrome *b* gene, is often identified by PCR-RFLP (Fig. 18.3). The results are reliable

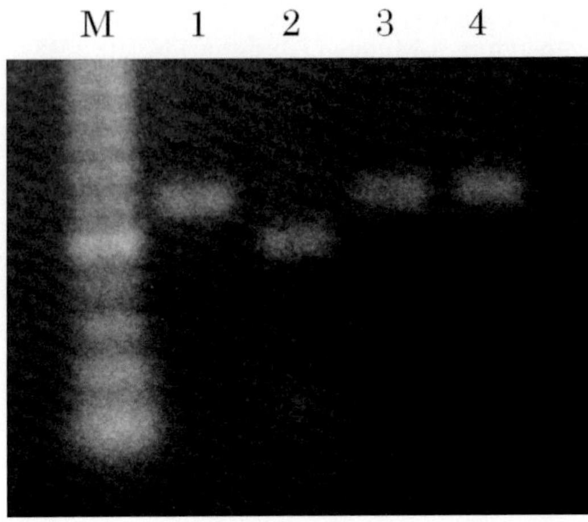

Fig. 18.3 Digestion of the PCR-amplified cytochrome *b* gene fragment from the cucumber powdery mildew fungus *Podosphaera xanthii* by the restriction enzyme *Ita*I. M, 50 bp ladder; *1*, untreated (resistant); *2*, *Ita*I-treated (resistant); *3*, untreated (sensitive); *4*, *Ita*I-treated (sensitive) (Reprinted from Ishii (2012))

when the gene is in the homoplasmic status in the mitochondria (e.g. *P. viticola* and *Mycosphaerella graminicola*). If the proportion of mutated cytochrome *b* gene reached a peak in the mitochondrial genome after receiving heavy fungicide selection pressure, PCR-RFLP would also bring about reliable results. As described before (Ishii 2010; see also Chap. 21), however, heteroplasmy of cytochrome *b* gene has been found in some fungi, such as *B. cinerea* (Ishii et al. 2009) and *Corynespora cassiicola* (Ishii et al. 2007), making this method unreliable. In that case, PCR products from resistant isolates might not be judged to be resistant when little mutated DNA is present. The detection sensitivity of ethidium bromide-stained gels for the mutant allele from QoI-resistant isolates could be at the level of 10 % only (Ishii et al. 2007). In order to better understand the role of heteroplasmy, an allele-specific quantitative PCR method was developed in the apple scab fungus *Venturia inaequalis* (Villani and Cox 2014). Although heteroplasmy of the cytochrome *b* gene was not the primary factor involved in isolates with moderate resistance to trifloxystrobin in this fungus, the relative abundance of A143 allele helps to explain the point of emergence for practical resistance in orchard populations.

18.3.2 PIRA-PCR (Primer-Introduced Restriction Enzyme Analysis PCR)

Resistance of rice blast fungus to carpropamid, an inhibitor of scytalone dehydratase (SDH) in melanin biosynthesis, is caused by a single point mutation in the gene encoding this enzyme resulting in the amino acid substitution of valine at codon 75 with methionine (see Chap. 11). Total DNA can be extracted not only from pure cultured mycelia but also from infected rice leaves using a microwave and target gene fragments are PCR-amplified easily. The fragments from resistant strains are then treated with the restriction enzyme *Sfa*NI producing two bands on a gel specific for resistance mutation. However, this enzyme is fairly expensive so an alternative method named PIRA-PCR was also developed to introduce the restriction site of the more cost-effective enzyme *Xba*I in the sequence of resistant strains (Kaku et al. 2003). This method has been used for a wide range of field monitoring and contributed to resistance management in Japan. Subsequently, PIRA-PCR has also proved useful for the detection of carbendazim resistance in *Fusarium graminearum* (Luo et al. 2009) and boscalid resistance in *B. cinerea* (Veloukas et al. 2011). In the latter fungus, five different mutations (P225F, N230I and H272L/R/Y) in the *sdhB* (B subunit of the fungicide-targeted enzyme succinate dehydrogenase) gene responsible for resistance have been successfully detected.

18.4 Detection Using Allele-Specific PCR (ASPCR) Primers

Specificity can be designed not only into ASO probes, but also into one of the two primers needed for the PCR. Detection of resistance is possible if primers differing at their 3′ ends only in the base pair change linked to resistance are used in separate reactions with a common reverse primer. With the correct stringency conditions, an amplicon will only be generated if there is a perfect match between the allele-specific primer (ASP) and the target DNA. ASPCR was tested to distinguish highly benzimidazole-resistant isolates from moderately resistant or sensitive ones in the very slow growing *Venturia nashicola*, the cause of Asian pear scab (Ishii 2002). In obligate pathogens, ASPCR was also developed to identify DMI-resistant isolates of *Erysiphe necator* (*Uncinula necator*), the grapevine powdery mildew fungus (Délye et al. 1997). More recently, in *Alternaria alternata* on pistachio, ASPCR has been introduced to detect the genotype H277Y in *AaSdhB* gene causing boscalid resistance (Avenot et al. 2008). In this case, the method has been combined with PCR-RFLP using the enzyme *Aci*I designed for the identification of another genotype H277R in *AaSdhB*. One weak point of the study was that some resistant isolates had no mutation in the histidine codon of *AaSdhB*.

Amplicons can be detected by agarose gel electrophoresis or by incorporating a fluorescent marker at the 5′ end of the ASP. By using fluorescent markers with different excitation wavelengths on each ASP, reactions need no longer be carried out in separate tubes or microtiter plate wells, but can be multiplexed in a single reaction.

In boscalid-resistant isolates of *B. cinerea*, two representative mutations (H272R/Y) in the *sdhB* gene have been detected simultaneously in single PCR amplifications using three primers (Yin et al. 2011). Furthermore, in *Alternaria solani* on potato, the five mutations, namely, H278Y and H278R in *AsSdhB*, H134R in *AsSdhC* and H133R and D123E in *AsSdhD* genes, were found to confer boscalid resistance in most isolates, and the former four major mutations were detected by the development of a multiplex PCR system (Mallik et al. 2014).

18.5 Detecting the Frequency of Resistance Alleles

So far we have discussed detection of resistance following analysis of the end product of a PCR. But effective management of resistance in field crops requires information not just on the presence of resistance, but more importantly, the frequency of resistance alleles within a pathogen population. Unfortunately, end-point PCR measurements are not quantitative, since some reagents may have become exhausted so the reaction does not go to completion. Although several different approaches have been tried to address this problem, the method most commonly adopted involves real-time PCR (RT-PCR), including a cyanine dye in the reaction mixture, and measuring fluorescence at the end of each round of PCR replication detects the product when the reaction is linear and no reagents are limiting.

The extent to which a resistance allele is present in the initial DNA template will determine how many PCR cycles are needed before fluorescence exceeds background levels. Differences in the frequency of resistance alleles are reflected in differences in the number of PCR cycles needed for each amplicon to exceed background fluorescence. The more a resistance allele is present in the initial template, the fewer cycles are needed to detect it. RT-PCR requires equipment that combines a PCR machine, a microtiter plate reader incorporating a variable wavelength spectrophotometer and software to manage different protocols and analyse data. Several machines are available from different manufacturers but their cost ensures that RT-PCR is still a research tool and not for small laboratories handling field samples.

In *M. graminicola*, all QoI-resistant isolates carried the G143A mutation in the cytochrome *b* gene. A method integrating spore trapping with RT-PCR assays was developed (Fraaije et al. 2005). This method enabled them to both quantify the number of *M. graminicola* ascospores in air samples and estimate the frequency of resistance allele in ascospore populations. Based on this, extensive monitoring was carried out and revealed that QoI-based treatments rapidly selected isolates carrying resistance within field populations.

RT-PCR assays were also used for quantifying A143 allele in cytochrome *b* gene of *P. viticola* in vineyards (Toffolatti et al. 2007; Toffolatti and Vercesi 2012). The results correlated well with those from the assays of germinated oospores on fungicide-amended agar plates and with the resistance situation in vineyards. QoI resistance was not observed in one vineyard never treated with QoI fungicides. In the vineyard where azoxystrobin had been used in mixture with folpet, selection of QoI-resistant strains was lower, compared with using solely a QoI. In vineyards where QoI treatments had been stopped, a decrease in resistance was generally observed.

Most recently, real-time ASPCR proved to be sensitive and specific for quantitative detection of different SDHI fungicide-resistant genotypes in *Botryotinia fuckeliana* (*B. cinerea*) (De Miccolis Angelini et al. 2014). In this fungus, the different types of *SdhB* gene mutation confer different cross-resistance profiles between boscalid and novel SDHIs such as fluopyram. *SdhB* mutations conferring P225L and P225F substitutions conferred high resistance to boscalid and high or moderate resistance to fluopyram, respectively. Mutants carrying the N230I replacement were moderately resistant to both SDHIs. Substitutions at position H272 responsible for a high level of resistance to boscalid conferred sensitivity (H272R), hypersensitivity (H272Y) or moderate resistance (H272V) to fluopyram. The real-time ASPCR assay confirmed the strict association between resistance profiles and allelic variants of the *SdhB* gene.

18.6 Detection of Resistance by Sequencing

As resistance is caused by DNA changes, it should be possible in many cases to detect resistance using nucleotide sequencing protocols. Unfortunately, the Sanger dideoxynucleotide chain-terminating method is slow and generates sequences far longer than are needed to detect a point mutation, an insertion or deletion causing resistance. Genome sequencing has been transformed by the development of a method (pyrosequencing) based on synthesis rather than on chain termination and that is limited to short (20–50 base pairs) sequences. A primer annealed to the sequence template acts as the starting point for the DNA polymerase. Addition of a complementary base pair to the single-stranded template releases pyrophosphate, which is enzymatically converted to ATP and coupled with the conversion of luciferin to oxyluciferin to generate a light pulse. This bioluminescence is detected by a Charge-Coupled Device (CCD) and is proportional to the amount of a deoxynucleotide (dNTP) incorporated. The light output is measured in real time and recorded on a chart (pyrogram). After enzymatically degrading unincorporated dNTP, the next round of synthesis is initiated by the addition of a different dNTP, either cyclic (GCAT) or programmed where the expected sequence is known. The sequence is generated more quickly than by the Sanger method. A more detailed explanation of pyrosequencing is given by Ahmadian et al. (2006). Similarly with RT-PCR and PCR-Luminex, the current cost of equipment and/or premixed reagents restricts the use of pyrosequencing to large and well-funded laboratories.

A short PCR fragment (50–60 base pairs) which contains the mutational site is amplified using a 5′ biotin-labelled primer so that, after denaturation, a single-stranded DNA template can be captured on streptavidin-coated magnetic beads. Sequencing is initiated when beads are transferred to the premixed reaction mixture. In a microtiter plate format, pyrosequencing can analyse a large number of samples quickly and, depending on the design of the sequencing primer, detect several allele changes within the PCR fragment template. The sensitivity of pyrosequencing at low allele frequencies is not as good as other quantitative diagnostic methods, but careful design of both PCR and sequencing primers can limit this disadvantage. Stammler et al. (2007) monitored for QoI resistance in *M. oryzae* quantitatively using a pyrosequencing method. An example of the use of pyrosequencing, and its limitations, to measure the frequency in field samples of different alleles in the cellulose synthase gene (*CesA3*), which are recessive and cause resistance to CAA fungicides, is described by Sierotzki et al. (2011). The molecular mechanisms affecting sensitivity to MBCs in the oilseed rape pathogen *Pyrenopeziza brassicae* have been identified, and pyrosequencing assays are a powerful tool for quantifying fungicide-resistance alleles, E198A, E198G, F200Y and L240F in β-tubulin gene of pathogen populations (Carter et al. 2013).

18.7 Loop-Mediated Isothermal Amplification (LAMP)

Taq polymerase requires a single-stranded DNA template so PCR protocols involve in each round a denaturation step (usually 94 or 95 °C) followed by lower temperatures for primer annealing and replication. But some DNA polymerases (e.g. *Bst* DNA polymerase) separate DNA strands without the need for denaturation, and DNA is amplified by a process known as strand displacement DNA synthesis. This means that DNA can be replicated using just a water bath or heating block, avoiding the need for a PCR machine. Japanese workers (Notomi et al. 2000; Tomita et al. 2008) exploited this property of *Bst* DNA polymerase and developed an alternative to PCR, which is known as loop-mediated isothermal amplification (LAMP). The process requires the use of two inner primers and two outer primers, which together have six sequences that hybridise to template sequences. Depending on the melting temperature of the primers (T_m), reactions are carried out between 60 and 65 °C and amplify fragments of no more than 300 base pairs. The design of the primers means that the amplified DNA generates stem loops which gives the method its name.

LAMP generates amplicons more rapidly than PCR and reaction times are generally an hour or less. Because primers contain six sequences designed to hybridise to the template, LAMP is more sensitive than PCR methods. Products can be analysed by agarose gel electrophoresis, by turbidity because of the large amounts of magnesium pyrophosphate produced or by fluorescence (Fig. 18.4). Amplification can be followed in real time or as an end product. Because of the large amount of DNA synthesised in LAMP, incorporation of a dye such as PicoGreen, a positive reaction can even be seen visually.

Over the last decade LAMP has been extensively used to identify bacteria, fungi and viruses associated with many different human and plant diseases such as citrus bacterial canker (Rigano et al. 2010), *Erwinia amylovora* (Moradi et al. 2012), *Phytophthora ramorum* (Tomlinson et al. 2007) and *Pythium aphanidermatum* (Fukuta et al. 2013). In the latter case, the pathogen was detected directly from infected tomato roots without specific DNA extraction procedure. With the development of LAMP, a diagnostic kit makes *Fusarium oxysporum* f. sp. *cubense* race 4 detection possible as a simple routine assay in the field (Li et al. 2013). By including

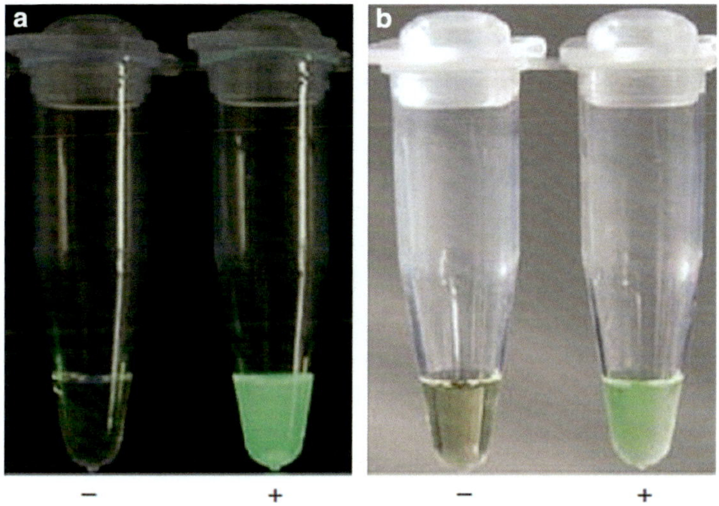

Fig. 18.4 Detection of the loop-mediated isothermal amplification (LAMP) reaction using fluorescent metal indicator. (**a**) Irradiating the tube using a hand-held UV lamp (wavelength: 365 nm) from the *bottom*. (**b**) Under daylight. *Plus sign* denotes reaction (with target DNA); *minus sign* denotes negative reaction (without target DNA) (Reprinted from Tomita et al. (2008))

a mismatch in an appropriate primer, the method can also accurately identify resistance alleles as shown in a recent paper, although involving herbicide and not fungicide resistance (Pan et al. 2015). Because of the isothermal nature of the protocol, it is simpler than PCR and offers some advantages to small diagnostic laboratories lacking thermal cycling equipment needed for PCR. But the LAMP method has very high efficiency of DNA amplification so careful attention needs to be paid to lower the chance of false-positive results caused by undesirable sample contamination. It is therefore essential to include positive and negative controls in every assay (Fukuta and Takahashi 2014).

Recently, Fraaije and colleagues found a new mechanism of DMI resistance in *Zymoseptoria tritici* (*M. graminicola*) isolates, in which a 120 bp insertion in the *CYP51* promoter is linked with 10- to 40-fold *CYP51* overexpression. Other isolates carry promoter variants based on a larger insert of 868 bp but the impact of this insert on the regulation of *CYP51* expression remains unclear. They presented the development and application of LAMP assays for rapid, on the spot detection of *CYP51* promoter inserts carrying CYP51 variants (Fraaije 2014).

A more recent paper (Duan et al. 2014; see also Chap. 19) highlights the potential of LAMP to monitor fungicide resistance rapidly and accurately in field samples without first isolating the pathogen. Primer designs and isothermal conditions were first optimised in the laboratory using well-characterised carbendazim-resistant (F167Y in β_2-tubulin) and carbendazim-sensitive isolates of *F. graminearum* (teleomorph: *Gibberella zeae*). Visual observation of a colour change was caused by the addition of the metal ion indicator hydroxynaphthol blue dye, which interacts with the white magnesium pyrophosphate to give a lighter blue colour. Comparing results with in vitro bioassays indicated an accuracy of at least 95 % in the detection of the F167Y SNP. Subsequently, the authors applied the method directly to DNA extracted from perithecia obtained from 116 different rice stubble samples and were able to show that 41 % of samples were positive, which was consistent with the 44 % resistant samples determined by bioassay. It should be possible to extend this work to evaluate the frequency of the F167Y mutation in field samples by adapting the method to real-time measurement of the colour change.

18.8 What Do Molecular Diagnostics Contribute to Resistance Management?

The stated aim in many papers involving molecular diagnostics is to evaluate a method that can be used by growers to guide in-season treatment decisions in a way that minimises the development of resistance. But the lack of on-site methods means that this aim has not yet been achieved and treatments are still based on predetermined strategies involving mixtures or alternation of products with different modes of action. On-site microarray-based techniques using specific probes which interrogate DNA released by simply boiling samples and are linked through immunoassay may in the future provide sufficiently rapid detection of resistance. But at the moment further development of LAMP offers the best potential for on-site detection, once suitable, portable equipment becomes available. AmplifyRP Acceler8™ (Agdia, Inc., USA) has been developed and commercialised recently for the isothermal DNA amplification of bacterial strains causing citrus greening, also known as huánglóngbìng (HLB). The heat block is lightweight and can easily be transported to the field if necessary. Genie® II (OptiGene Ltd., UK) is a sophisticated portable device that enables sensitive detection of bacteria, viruses and azole-insensitive CYP51-overexpressing strains of *Z. tritici* at a molecular level (Fraaije 2014). This powerful and extremely flexible platform allows isothermal amplification of DNA and RNA to take place in a compact and portable device. Commercial kits using LAMP methods have also been developed for the detection of pathogens such as plum pox virus on stone fruit, citrus greening and tomato yellow leaf curl virus (TYLCV) (Nippon Gene Co., Ltd., Japan).

Fungicide resistance is a phenotype rather than a genotype problem, and although a single target site mutation may be significant, other mechanisms can contribute to resistance. Consequently, bioassays need to partner molecular techniques until it is clear, as in the case of MBC, phenylamide and QoI fungicides, in which particular genotype target-site mutations cause practical resistance.

But molecular diagnostics have had a significant impact on the identification of different resistance mutations within a target site and their evolution within pathogen populations. Indirectly, they can measure the fitness of individual mutations, although resistant isolates are not always clones, in which case other genotype differences will also influence fitness. Molecular methods have also been useful in monitoring the performance of different antiresistance strategies in field trials. Resistance is generally detected first in regions where disease pressures are high, and as a result fungicide use is intensive. Because molecular diagnostics can identify resistance at much lower frequencies than bioassay, in regions where resistance to a particular fungicide has not yet emerged, they offer a way to detect the presence of resistance mutations before they can influence fungicide efficacy. This information is very relevant for registration and extension authorities who may need to make changes to fungicide use strategies to minimise resistance risk.

References

Ahmadian A, Ehn M, Hober S (2006) Pyrosequencing: history, biochemistry and future. Clinica Chem Acta 363:83–94

Aoki Y, Furuya S, Suzuki S (2011) Method for rapid detection of the *PvCesA3* gene allele conferring resistance to mandipropamid, a carboxylic acid amide fungicide, in *Plasmopara viticola* populations. Pest Manag Sci 67:1557–1561

Avenot HF, Sellam A, Karaoglanidis G, Michailides TJ (2008) Characterization of mutations in the iron-sulphur subunit of succinate dehydrogenase correlating with boscalid resistance in *Alternaria alternata* from California pistachio. Phytopathology 98:736–742

Beckerman J (2013) Detection of fungicide resistance. In: Nita M (ed) Fungicides – showcases of integrated plant disease management from around the world. Intech, 281–310. http://dx.doi.org./10.5772 /5591

Billard A, Laval V, Fillinger S, Leroux P, Lachaise H, Beffa R, Debieua D (2012) The allele-specific probe and primer amplification assay, a new real-time PCR method for fine quantification of single-nucleotide polymorphisms in pooled DNA. Appl Environ Microbiol 70:1063–1068

Carter HE, Cools HJ, West JS, Shaw MW, Fraaije BA (2013) Detection and molecular characterisation of *Pyrenopeziza brassicae* isolates resistant to methyl benzimidazole carbamates. Pest Manag Sci 69:1040–1048

De Miccolis Angelini RM, Masiello M, Rotolo C, Pollastro S, Faretra F (2014) Molecular characterisation and detection of resistance to succinate dehydrogenase inhibitor fungicides in *Botryotinia fuckeliana* (*Botrytis cinerea*). Pest Manag Sci 70:1884–1889

Délye C, Laigret F, Corio-Costet MF (1997) A mutation in the 14 alpha-demethylase gene of *Uncinula necator* that correlates with resistance to a sterol biosynthesis inhibitor. Appl Environ Microbiol 63:2966–2970

Duan Y, Zhang X, Ge C, Wang Y, Cao J, Jia X, Wang J, Zhou M. (2014) Development and application of loop-mediated isothermal amplification for detection of the F167Y mutation of carbendazim resistant isolates of *Fusarium graminearum*. Scientific Report 4. doi: 10.1038/srep07094

Fraaije BA (2014) Use of loop-mediated isothermal amplification assays to detect azole insensitive *CYP51*-overexpressing strains of *Zymoseptoria tritici*. Phytopathology 104(Suppl. 3):S3.41 (Abstr)

Fraaije BA, Cools HJ, Fountaine J, Lovell DJ, Motteram J, West JS, Lucas JA (2005) Role of ascospores in further spread of QoI-resistant cytochrome *b* alleles (G143A) in field populations of *Mycosphaerella graminicola*. Phytopathology 95:933–941

Fukuta S, Takahashi R (2014) Development of the detection method for high-temperature-*Pythium* in hydroponic solution by the LAMP assay. Plant Prot 68:303–308 (in Japanese)

Fukuta S, Takahashi R, Kuroyanagi S, Miyake N, Nagai H, Suzuki H, Hashizume F, Tsuji T, Taguchi H, Watanabe H, Kageyama K (2013) Detection of *Pythium aphanidermatum* in tomato using loop-mediated isothermal amplification (LAMP) with species-specific primers. Eur J Plant Pathol 136:689–701

Furuya S, Suzuki S, Kobayashi H, Saito S, Takayanagi T (2009) Rapid method for detecting resistance to a QoI fungicide in *Plasmopara viticola* populations. Pest Manag Sci 65:840–843

Ishii H (1995) Monitoring of fungicide resistance in fungi: biological to biotechnological approaches. In: Singh RP, Singh US (eds) Molecular methods in plant pathology. CRC Press, Boca Raton, pp 493–503

Ishii H (2002) DNA-based approaches for diagnosis of fungicide resistance. In: Clark JM, Yamaguchi I (eds) Agrochemical resistance: extent, mechanism, and detection. American Chemical Society, Washington, DC, pp 242–259

Ishii H (2010) QoI fungicide resistance: current status and the problems associated with DNA-based monitoring. In: Gisi U, Chet I, Gullino ML (eds) Recent developments in management of plant diseases, plant pathology in the 21st century 1. Springer, Dordrecht, pp 37–45

Ishii H (2012) Resistance to QoI and SDHI fungicides in Japan. In: Thind TS (ed) Fungicide resistance in crop protection – risk and management. CAB International, Wallingford, pp 223–234

Ishii H, Yano K, Date H, Furuta A, Sagehashi Y, Yamaguchi T, Sugiyama T, Nishimura K, Hasama W (2007) Molecular characterization and diagnosis of QoI resistance in cucumber and eggplant fungal pathogens. Phytopathology 97:1458–1466

Ishii H, Tanoue J, Oshima M, Chung WH, Nishimura K, Yamaguchi J, Nemoto F, So K, Iwama T, Yoshimatsu H, Shimizu M, Kozawa T (2008) First application of PCR-Luminex system for molecular diagnosis of fungicide resistance and species identification of fungal pathogens. J Gen Plant Pathol 74:409–416

Ishii H, Fountaine J, Chung WH, Kansako M, Nishimura K, Takahashi K, Oshima M (2009) Characterisation of QoI-resistant field isolates of *Botrytis cinerea* from citrus and strawberry. Pest Manag Sci 65:916–922

Kaku K, Takagaki M, Shimizu T, Nagayama K (2003) Diagnosis of dehydratase inhibitors in melanin biosynthesis inhibitor (MBI-D) resistance by primer-introduced restriction enzyme analysis in scytalone dehydratase gene of *Magnaporthe grisea*. Pest Manag Sci 59:843–846

Koenraadt H, Jones AL (1992) The use of allele-specific oligonucleotide probes to characterize resistance to benomyl in field strains of *Venturia inaequalis*. Phytopathology 82:1354–1358

Lee LG, Connell CR, Bloch W (1993) Allelic discrimination by nick translation PCR with fluorogenic probes. Nucl Acid Res 99:1483–1489

Li B, Du J, Lan C, Liu P, Weng Q, Chen Q (2013) Development of a loop-mediated isothermal amplification assay for rapid and sensitive detection of *Fusarium oxysporum* f. sp. *cubense* race 4. Eur J Plant Pathol 135:903–911

Luo QQ, Xu JQ, Hou YP, Chen CJ, Wang JX, Zhou MG (2009) PIRA-PCR for detection of *Fusarium graminearum* genotypes with moderate resistance to carbendazim. Plant Pathol 58:882–887

Ma Z, Michailides TJ (2005) Advances in understanding molecular mechanisms of fungicide resistance and molecular detection of resistant genotypes in phytopathogenic fungi. Crop Prot 24:853–886

Mallik I, Arabiat S, Pasche JS, Bolton MD, Patel JS, Gudmestad NC (2014) Molecular characterization and detection of mutations associated with resistance to succinate dehydrogenase-inhibiting fungicides in *Alternaria solani*. Phytopathology 104:40–49

Moradi A, Nasiri J, Abdollahi H, Almasi M (2012) Development and evaluation of a loop-mediated isothermal amplification assay for detection of *Erwinia amylovora* based on chromosomal DNA. Eur J Plant Pathol 133:609–620

Notomi T, Okayama H, Maoubuchi H, Yonekawa T, Watanabe K, Amino N, Hase T (2000) Loop-mediated isothermal amplification of DNA. Nucl Acid Res 28, E63

Pan L, Li J, Zhang W, Doug L (2015) Detection of I1781L mutation in fenoxaprop-*p*- methyl resistant American sloughgrass (*Beckmannia syzigachne* Stend) based on the loop mediated isothermal amplification method. Pest Manag Sci 71:123–130

Rigano LA, Marano MR, Castagnaro AP, Do Amaral AM, Vojnov AA (2010) Rapid and sensitive detection of citrus bacterial canker by loop-mediated isothermal amplification combined with simple visual evaluation methods. BMC Microbiol 10:176. doi:10.1186/1471-2180-10-176

Sambrook J, Fritsch EF, Maniatis T (1989) Molecular cloning: a laboratory manual (2nd edn). Cold Spring Harbor Laboratory Press, Cold Spring Harbor

Sierotzki H, Blum M, Olaya G, Waldner-Zulauf M, Buttrago C, Wullschleger I, Cohen Y, Gisi U (2011) Sensitivity to CAA fungicides and frequency of mutations in cellulose synthase (*CesA3*) of oomycete pathogen populations. In: Dehne MW, Deising HB, Gisi U, Kuck KH, Russell PW, Lyr H (eds) Modern fungicides and antifungal compounds V1. DPG Selbstverlag, Braunschweig, pp 103–118

Stammler G, Itoh M, Hino I, Watanabe A, Kojima K, Motoyoshi M, Koch A, Haden E (2007) Efficacy of orysastrobin against blast and sheath blight in transplanted rice. J Pestic Sci 32:10–15

Tapp I, MalmbergE RE, Wik M, Syvänen AC (2000) Homogenous scoring of single-nucleotide polymorphisms: comparison of the 5′-nuclease TaqMan™ assay and molecular beacon probes. Biotechniques 4:732–738

Toffolatti SL, Vercesi A (2012) QoI resistance in *Plasmopara viticola* in Italy. Evolution and management strategies. In: Thind TS (ed) Fungicide resistance in crop protection. CAB International, Wallingford, pp 172–183

Toffolatti SL, Serrati L, Sierotzki H, Gisi U, Vercesi A (2007) Assessment of QoI resistance in *Plasmopara viticola* oospores. Pest Manag Sci 63:194–201

Tomita N, Mori Y, Kanda H, Notomi T (2008) Loop-mediated isothermal amplification (LAMP) of gene sequences and simple visual detection of products. Nat Protoc 3:877–882

Tomlinson A, Barker I, Boonham N (2007) Faster, simpler, more-specific methods for improved molecular detection of *Phytophthora ramorum* in the field. Appl Environ Microbiol 73:4040–4047

Veloukas T, Leroch M, Hahn M, Karaoglanidis GS (2011) Detection and molecular characterization of boscalid-resistant *Botrytis cinerea* isolates from strawberry. Plant Dis 95:1302–1307

Villani SM, Cox KD (2014) Heteroplasmy of the cytochrome *b* gene in *Venturia inaequalis* and its involvement in quantitative and practical resistance to trifloxystrobin. Phytopathology 104:945–953

Wheeler IE, Kendall SJ, ButtersJA HDW, Hall L, Hollomon DW (1995) Using allele specific probes to characterise benzimidazole resistance in *Rhynchosporium secalis*. Pestic Sci 43:201–209

Yin YN, Kim YK, Xiao CL (2011) Molecular characterization of boscalid resistance in field isolates of *Botrytis cinerea* from apple. Phytopathology 101:986–995

Part IV
Resistance Management in Major Crops

Chapter 19
Wheat Pathogens in China

Ming-guo Zhou and Xiao-jing Jia

Abstract Wheat production and consumption account for a major share of China's food system (Du and Wang 2004), and *Fusarium* head blight (FHB), rusts, powdery mildew and sharp eyespot are the most important diseases on wheat, causing huge economic losses. They are generally treated by fungicides, except for rusts where resistant cultivars are widely grown. Resistance has developed to one or more fungicides, including the resistance of FHB to benzimidazole fungicides and of wheat powdery mildew to DMI fungicides. Besides resistance management, some fundamental research on fungicide resistance is also discussed.

Keywords *Fusarium* head blight • Wheat powdery mildew • Sharp eyespot • Fungicide resistance • Benzimidazole fungicides • Carbendazim • DMI fungicides • Phenamacril • Jinggangmycin • Resistance management

19.1 Introduction

Wheat has accompanied humans as a bread cereal for many centuries. Its cultivation was known as early as the Neolithic Age, about 7000 years BC, as evidenced by archaeological discoveries in the Middle East and China. To this day, this crop does not lose its importance in China, where it was planted on more than 24.2 million hectares, being the second most important crop. Environmental conditions vary considerably because elevation increases and rainfall decreases from east to west. This variation in environmental conditions has a major influence on the prevalence and incidence of specific wheat diseases across the country, including viral, fungal and nematode diseases, which are a major cause of yield and quality loss in wheat. The head smuts became unimportant after systemic fungicide application to treat cereal seeds in the last three decades in China. Stripe rust, *Fusarium* head blight (FHB), wheat powdery mildew and sharp eyespot are major diseases caused by fungi of the genera *Puccinia*, *Fusarium*, *Blumeria* and *Rhizoctonia*, respectively.

M.-g. Zhou (✉) • X.-j. Jia
College of Plant Protection, Nanjing Agricultural University, Nanjing 210095, China
e-mail: mgzhou@njau.edu.cn; jiaxiaojing@gmail.com

H. Ishii, D.W. Hollomon (eds.), *Fungicide Resistance in Plant Pathogens*,
DOI 10.1007/978-4-431-55642-8_19

FHB is a major economically important wheat disease because high-resistance cultivars are not available, and few fungicides can now be used because benzimidazole resistance has developed in the pathogen population in some areas during the last decade.

19.2 *Fusarium* Head Blight

19.2.1 Occurrence of FHB

In China, the cultivation area for wheat has reached to 24.2 million hectares, and the total wheat production has been more than 1.22×10^8 t (Yearbook 2007), which means that China is the largest wheat producer in the world (Jing and Chang 2003). FHB, also known as scab, is one of the devastating diseases on wheat (Chen et al. 2000; McMullen et al. 1997). Its epidemic occasionally occurs in the northeastern spring wheat region and northern and western winter wheat regions and is frequently damaging in the winter wheat regions of the middle and lower reaches of both the Yangtze and Huaihe rivers of China. FHB is particularly serious in Jiangsu, Anhui and Shanghai Provinces because the climate with high humidity and warm temperature during flowering favoured the disease occurrence (Bai and Shaner 1994; McMullen et al. 1997; Parry et al. 1995). High-yield culture and a moist climate have increased FHB occurrence in recent years. Plenty of inocula are available, and providing at least three continuous rainy and overcast days and the temperature at above 15° Celsius during the period of wheat flowering, FHB epidemics will develop. Losses can be very serious; it was estimated that FHB caused a three billion US dollar loss in the United States alone in the 1990s (Windels 2000) and reduced wheat baking quality (Seitz et al. 1986). Similar harm also occurred in China recently in 2010 and 2012. Based on our investigation, at least 0.73 million tons of grain was lost in Jiangsu Province alone, where 2.3 million hectares of wheat was planted in 2012 (unpublished data).

Scab especially can result not only in a quantitative yield loss but also in qualitative losses due to lower food quality, decreased potential for storage and mycotoxin contamination. In addition to reducing seed vigour and quality, the fungus contaminates grain with toxic metabolites, including nivalenol (NIV), deoxynivalenol (DON) (Proctor et al. 1995; Snijders 1990; Bottalico 1998; Jones and Mirocha 1999) and their acetylated derivatives, which threaten human and livestock health (Bennett and Klich 2003; Fan et al. 2009). *Fusarium* species can also directly infect humans, causing localized necrotic diseases (Chang et al. 2006) and invasive infection, especially in immunocompromised individuals (Dignani and Anaissie 2004).

In wheat, it is characterized by chalky kernels which appear thin and shrunken. Infected kernels also have a white or pinkish fibrous mould in the crease area and sometimes in the germ of the kernel. The initial inoculum source of the pathogen is mainly from diseased plant debris. Although FHB can also be caused by other *Fusarium* spp., *Fusarium graminearum* and *F. asiaticum* (teleomorph: *Gibberella zeae* (Schwein.) Petch) are by far the most important causal organisms. The former is distributed in the north of Huaihe and the latter in the south; in general *F. asiaticum* appears more virulent and produces higher mycotoxin levels (Dai et al. 2013).

19.2.2 FHB Control

A variety of strategies have been used to control FHB and reduce mycotoxin contamination of grain. Based on the difference in host resistance to FHB, a series of Yangmai cultivars, which appeared to have some degree of resistance, have been planted in Jiangsu Province in the last 30 years, though they are not immune resistance cultivars. The cultivars Sumai 3 and Wangshuibai have much higher resistance but are linked with lower yields. Therefore, most FHB control programmes rely on fungicide application particularly in areas where epidemics are frequent. Benzimidazole fungicides, especially carbendazim (MBC), have been used to control FHB uniquely in China since the middle of the 1970s. Carbendazim was used as a spray, first at an early stage in wheat flowering and followed by a second treatment 3–5 days later. To get efficacy of 80 % FHB control, 300–375 g a.i./ha of MBC was used in the early days. With sensitivity to MBC decreasing within the pathogen population, triple dosage of MBC was used for two or three times from flowering to milking stage at present.

19.2.3 Benzimidazole Fungicide Resistance

19.2.3.1 Development of MBC Resistance

Benzimidazole fungicides, which comprise a well-known group of β-tubulin inhibitors (carbendazim, benomyl, thiabendazole, fuberidazole and thiophanate-methyl), have been extensively used to control various plant diseases caused by fungi. Unfortunately, after these have been used for 2 or 3 years, they have generally lost their efficacy because of their selection of resistant genotypes (Bollen and Scholten 1971; Schroeder and Provvidenti 1969). Resistance to benzimidazole fungicides has usually resulted from certain point mutations in the target β-tubulin gene, which generates very high resistance levels. Because of the high risk of rapid development of benzimidazole resistance, the Chinese government has paid attention to monitoring MBC resistance for FHB pathogens and has supported a monitoring programme since 1985. In order to detect resistance at the early stage, it is necessary to test an inordinate number of fungal strains.

Authors and their colleagues have detected in vitro each year MBC resistance in 500–5000 isolates obtained from separate scab heads.

Recently molecular-phylogenetic studies on the species of the genus *Fusarium* and its allies based on comparative analyses of DNA sequences were much progressed (O'Donnell et al. 2004). The first MBC-resistant isolate of *Fusarium asiaticum* (in the early reports considered as *F. graminearum*) was detected in a sample of 423 isolates collected from Haining in Zhejiang Province in 1992, some 20 years after the fungicide was first used. MBC resistance was subsequently found in Jiangsu Province, north, to Haining, in 1994. MBC resistance developed slowly within the FHB pathogen population, and disease control remained effective in these early years after resistance was first detected. However, the frequency of MBC resistance in Zhejiang Province rose from 3.5 % in 1997 to 18.8 % in 1998 because of a serious FHB epidemic. The efficacy of MBC against FHB then declined because there were no effective MBC-resistance management measures available. As a result wheat production in Zhejiang Province almost ceased and now only a few hectares are grown.

MBC-resistance monitoring focused on the FHB epidemic area in Jiangsu and its neighbour provinces Anhui, Shanghai and Shandong after 1998. Resistance detection also was carried out in the other wheat main production area of Henan and Hubei Provinces. MBC resistance spread similarly in Jiangsu, Shanghai and Anhui depending on the different level of epidemics and significantly decreased the efficacy of MBC against FHB. MBC-resistant isolates were also detected at the frequency of 1–5 % in some regions of Shandong, Henan and Hubei Provinces. The dynamics of MBC-resistance development in Jiangsu Province is shown in Fig. 19.1. It is interesting that all of MBC-resistant isolates collected from wheat fields have been identified as *F. asiaticum* and no MBC-resistant isolate has been detected in *F. graminearum*.

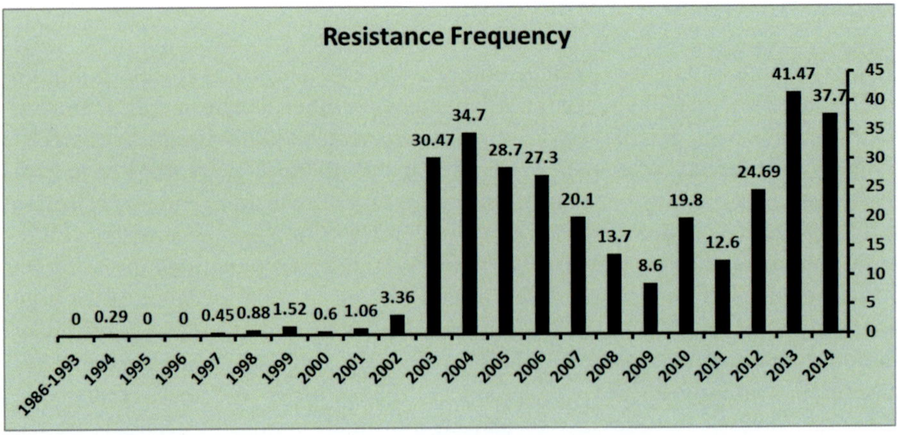

Fig. 19.1 Development of MBC resistance in the FHB pathogen in Jiangsu Province

19.2.3.2 Resistance Characteristics

Frequency of MBC-Resistance Mutation MBC resistance emerged slowly in *F. asiaticum* populations, only after the fungicide had been applied for over two decades in China for FHB control (Yuan and Zhou 2005). But the frequency (10^{-6}–10^{-7}) of MBC-resistance mutations in *F. asiaticum*, generated in the laboratory using UV irradiation, is much lower than that in other fungi such as *Botrytis cinerea* (frequency 10^{-5}) (Gu and Liu 1990; Zhou et al. 1994). This may be one reason why resistance appeared in the field after many years of MBC use.

Sensitivity and Resistance Level Research showed that EC_{50} for MBC inhibition of mycelium growth of wild-type isolates of *F. asiaticum* or *F. graminearum* on the PDA is about 0.5 μg/ml, and MIC (minimal inhibition concentration) value is 1.4 μg/mL MBC. The sensitivity of *Fusarium* to MBC seems less than for *B. cinerea* and most other fungi. More work revealed that the deletion of the β_2-tubulin gene in wild-type or in MBC-resistant isolates showed supersensitivity to MBC. The EC_{50} value of MBC inhibition of mycelium growth of these deletion mutants was about 0.10 μg/mL and the MIC less than 0.5 μg/mL. This sensitivity of β_2-tubulin gene deletion mutants was similar to that of wild-type isolates of *B. cinerea* and *Sclerotinia sclerotiorum*. Increased MBC sensitivity following β_2-tubulin gene deletion implied that the β_2-tubulin gene may play a negative regulation role on the MBC sensitivity in *F. asiaticum*.

The growth of MBC-resistant isolates of *F. asiaticum* appeared a little slower on medium containing MBC (5 μg/mL) than without the fungicide. All resistant isolates with an EC_{50} less than 20 μg/mL are considered as moderately resistant, whilst a few highly resistant isolates could grow on medium containing 100 μg/mL MBC (Chen et al. 2008). Therefore, 5 and 100 μg/mL of MBC treatment could be used to identify moderate and highly resistant isolates in the monitoring programme. Clearly the level of MBC resistance in *F. asiaticum* is generally lower than that of other fungi where resistance is governed by the β_1-tubulin gene.

Pattern of Cross-Resistance All MBC-resistant isolates from field crops had positive cross-resistance with other benzimidazole fungicides, including benomyl, thiabendazole and thiophanate-methyl, but there was no negative cross-resistance between the phenylcarbamate diethofencarb and MBC. This contrasts with the negative cross-resistance of laboratory MBC-resistant mutants. No cross-resistance was found with other fungicides.

Fitness of Resistance Growth rate of mycelium, capacity of sporulation, perithecia formation and virulence were compared between 10 randomly selected field-resistant isolates and 10 wild-type isolates. Results showed that resistant isolates kept their normal mycelium growth, sporulation and perithecia formation, but were more virulent than wild types. Infection of flowering wheat heads by a single conidium or ascospore of either MBC-resistant or wild-type isolates was similar, but the scab severity caused by resistant isolates was remarkably greater than that caused by sensitive isolates (Zhou and Wang 2001).

Deoxynivalenol (DON) was considered a virulence factor of *Fusarium* sp. infecting wheat heads (Miller et al. 2004). In our experience, the pathogen of FHB is a complex of *Fusarium* spp. and variably distributed in different areas. The ability of DON production is significantly stronger in *F. asiaticum* than *F. graminearum.* Zhang et al. (2009a) found more trichothecenes ($P<0.05$) were produced by MBC-resistant isolates than by sensitive wild-type isolates, both in shake culture and in the field. Although the incidence of infected spikelets did not differ between all isolates, the amount of *F. asiaticum* DNA was significantly higher in wheat grain infected with resistant isolates than with wild types in both 2007 and 2008. This evidence implied that the MBC-resistant isolates developed more than wild types within grains. Other studies showed that although MBC and azoxystrobin could control FHB from 40.6 to 53.0 %, they did not lessen DON contamination. Despite some control of FHB, the DON content in MBC-treated grains reached 8.1–9.1 mg/kg and was not significantly different from untreated grains (Table 19.1).

On the other hand, DON content in grain that was treated with phenamacril (experimental code JS399-19) or tebuconazole was only 1.0–1.3 mg/kg (Zhang et al. 2009b). In addition, a point mutation in *F. asiaticum* causing moderate MBC resistance increased DON content by a large margin. In culture resistant isolates produced 2.4-fold more DON than sensitive isolates and 2-fold more in infected wheat grains (Zhang et al. 2009a). Further work has found that the expression of DON biosynthesis gene cluster increased greatly in the F167Y MBC-resistant mutant compared with a β_2-tubulin gene deletion mutant DN83 (Fig. 19.2, unpublished results). These results imply that MBC-resistant mutants may regulate in some way gene function involved in DON biosynthesis.

However, studies on the effect of fungicides against mycotoxigenic fungal species have shown conflicting data. In some cases the use of fungicides effectively inhibited the disease development and mycotoxin production (Matthies and Buchenauer 2000; Pirgozliev et al. 2002), but in other cases it is reported that some fungicides increased the mycotoxigenic ability in some fungal species (Milus and Parsons 1994; D'Mello et al. 1998; Ioos et al. 2005).

Table 19.1 Effect of fungicides on the DON production (Zhang et al. 2009b)

Treatment	Dosage (g a.i./ha)	2006			2007		
		Efficacy* (%)	TKW** (g)	DON mg kg^{-1}	Efficacy (%)	TKW (g)	DON mg kg^{-1}
CK	/	/	32.5^{c}	9.3^{a}	e/	39.9^{c}	7.7^{a}
Carbendazim	375	44.8	38.9^{bc}	9.1^{a}	40.6	43.5^{b}	8.1^{a}
Azoxystrobin	250	51.4	39.8^{b}	9.6^{a}	53.0	44.7^{ab}	7.8^{a}
Tebuconazole	125	74.3	44.7	1.3^{c}	73.5	46.3^{ab}	1.2^{c}
Phenamacril (JS399-19)	375	87.9	45.1^{a}	1.1^{b}	89.5	47.3^{a}	1.0^{b}

*Efficacy was measured by infected mycelia with real-time PCR
**TKW showed the weight of thousand kernels
All the superscripts "a", "b", and "c" following the digits mean whether the difference is significant among the data
All data were analyzed with the analysis of variance and means were separated with the Fisher's protected least significant difference test at P = 0.05 using the statistical software SAS GLM(SAS Institute Inc., Cary, NC)

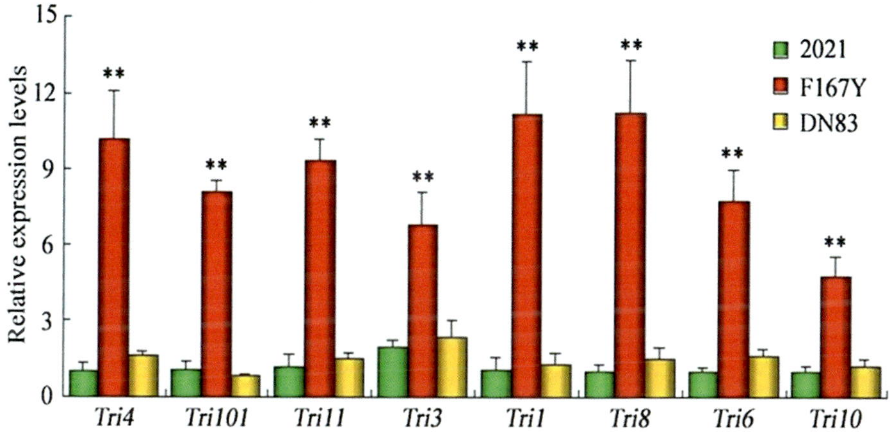

Fig. 19.2 Expression levels of *Tri* genes in DON biosynthesis. 2021 is an original wild-type isolate of MBC resistance mutant F167Y and β_2-tubulin gene deletion mutant DN83

19.2.3.3 Mechanism of MBC Resistance in *F. asiaticum*

Previous data indicated that the mode of action of carbendazim on most filamentous pathogenic fungi, e.g. *Botrytis cinerea*, was inhibition of mitotic division and MBC resistance was governed by one major gene involving mutations in its target β-tubulin gene. In *F. asiaticum*, β-tubulin gene has three fewer introns compared to that in other fungi, although the amino acid sequence homology with different fungi was 95.12–99.30 % similar (Li et al. 2003). But the β-tubulin gene sequence was identical in both MBC-resistant and wild-type isolates of *F. asiaticum*, so we concluded that the FHB pathogen has a different carbendazim resistance mechanism.

Inheritance of MBC resistance in *F. asiaticum* was studied by classical genetics. The segregation of sensitivity of progeny was analysed from the following crosses: MBC-sensitive (S) × S, moderately MBC-resistant (MR) × S, MR × MR, highly MBC-resistant (HR) × S and HR × MR. In crosses between the parents with different sensitivity levels (i.e., MR × S, HR × S, HR × MR), progeny segregated in a 1:1 ratio of the two parental phenotypes. No segregation was found in crosses of S × S and MR × MR. These results indicated that the MR and HR phenotypes in *F. asiaticum* were conferred by different allelic mutations within the same locus. MBC resistancewas not affected by modifying genes or cytoplasmic components (Yuan and Zhou 2005).

By observing the effect of MBC on morphology and mitosis, we observed no differences in conidial germlings and mitosis between wild-type isolates of *F. asiaticum* and *B. cinerea*. In MBC-sensitive *F. asiaticum* isolates, conidia germination was distorted, germ tubes were branched, and elongation was restrained in the presence of carbendazim. Irregular distribution of chromosome masses and abnormal nuclear division were also observed after MBC treatment. The CMI (chromosome mitosis index) also rapidly increased within 60 min and then rapidly dropped. Compared with carbendazim-resistant isolates of *B. cinerea*, resistant isolates of *F. asiaticum* were different in growth and mitosis when treated with carbendazim. Growth rate, morphology of hypha and mitosis of MBC-resistant *B. cinerea* appeared normal, even at high MBC dosages. Whilst hypha of MBC-resistant isolates of *F. asiaticum* grew slowly and produced more branches, abnormal mitosis was not observed following carbendazim treatment. These results indicated that the MBC-resistance mechanism was related to mitosis of *F. asiaticum*, but was different from that at least in *B. cinerea*.

Nucleotide sequence analysis of two tubulin genes involved in mitosis provided circumstantial evidence that MBC resistance and mutations in key codons of the β_2-tubulin gene (FGSG 06611.3,β2tub) were responsible for resistance to MBC in *F. asiaticum* (Chen et al. 2009). Although the β-tubulin gene (FGSG 09530.3, β_1-tubulin gene) is the target of carbendazim in many fungi, it does not relate to MBC resistance in *F. asiaticum* (Qiu et al. 2012). Based on site-directed mutagenesis and transformation experiments, direct evidence was provided that mutations at residue 50 (Tyr to Cys), 167 (Phe to Tyr), 198 (Glu to Lys) and 200 (Phe to Tyr) of the β_2-tubulin gene were responsible for MBC resistance in *F. asiaticum* (Qiu et al. 2011).

Of 2000–3000 resistant isolates obtained in 2010 and 2012, 97.4 % and 99.7 %, respectively, were moderately MBC resistant. Sequences of the β_2-tubulin gene showed that moderate resistance mostly resulted from a point mutation at codon 167 (TTT → TAT, Phe → Tyr), although a few moderately resistant and highly resistant isolates were caused by a point mutation at either 200 (TTC → TAC, Phe → Tyr) or 198 (GAG → CTG, Glu → Leu). It was very interesting that homologous replacement in *F. asiaticum* of the resident β_1-tubulin gene with the resistant E198A mutant β_2-tubulin gene from *B. cinerea* did not cause resistance to carbendazim, even though in *F. asiaticum* the gene mRNA transcript was expressed, and it apparently functioned in mitosis. This result suggested that the MBC-resistance phenotype was not only determined by the target gene mutation but also regulated by other factors in genome.

19.2.3.4 Molecular Diagnosis and Detection of MBC Resistance

PIRA-PCR (primer-introduced restriction analysis PCR) was developed to diagnose isolates of *F. asiaticum* with moderate resistance to MBC. Two primer pairs were designed and synthesized according to the nucleotide sequence of the β_2-tubulin gene from *F. asiaticum*. Fragments of 164 bp were amplified by nested PCR from isolates differing in MBC sensitivity. A *Hin*dIII restriction enzyme recognition site was introduced artificially by inner primers to detect a mutation at codon 167, and *Taa*I (*Tsp*4CI) restriction enzyme was used to detect a mutation at codon 200. The MBC sensitivity of isolates was determined by analysing electrophoresis patterns of the resulting PCR products after simultaneous digestion with both *Hin*dIII and *Taa*I. Results from PIRA-PCR and a conventional method (mycelial growth on agar) were identical, but PIRA-PCR not only recognized MBC-resistance mutations but also required less time (Luo et al. 2009).

A simple but cost-effective method was developed using diagnostic PCR primers that bind only to specific sequence variants, within the 3′-end of the primer overlapping the SNP causing MBC resistance. Primers where 3′ nucleotide covers the SNP are often difficult to optimize for use in quantitative real-time PCR. A deliberate mismatch introduced into the penultimate nucleotide of the primer improved the specific detection of resistance mutations (Chen et al. 2009).

An ARMS-PCR (amplification refractory mutation system-PCR) diagnostic method was developed using four primers to genotype MBC-resistance mutations in *F. asiaticum*. Single nucleotide polymorphisms at codons 167, 198 or 200 were detected using two primer pairs, specifically designed for each codon. Genotypes were identified by the different patterns of amplicons in agarose gels. Tetra-primer ARMS-PCR is not only a useful method for SNP genotyping but also a valuable method to quantify MBC-resistance allele frequencies if combined with cycling probe real-time PCR of bulk *F. asiaticum* DNA samples (Chen et al. 2009).

Based on loop-mediated isothermal amplification (LAMP), a simple, cheap and rapid detection method for the F167Y mutation in *F. asiaticum* isolates was developed recently (Duan et al. 2014). Compared with other conventional molecular detection methods, LAMP is an accurate and easier to perform method (see Chap. 18) and will be useful for future monitoring and management evaluation of MBC resistance in *F. asiaticum*.

19.2.3.5 Loss Caused by MBC-Resistance Epidemic

Seriously FHB epidemics occurred throughout the main wheat-growing regions of China in 2010 and 2012. We randomly investigated the occurrence of FHB and measured its severity in 76 fields, representative of the 2.3 million hectares of wheat grown in Jiangsu Province in 2010 and 2012 (Fig. 19.3).

The average diseased panicle rate in Jiangsu Province in 2010 was 14.71 % and the average disease index was 37.11. According to the statistics announced by the Jiangsu Province Agricultural Commission, the wheat-growing area occupied 2.3 million hectares with 4.8 t production per ha. Therefore we could estimate the yield loss of wheat due to FHB reached 0.44 million tons and the economic loss was at least 0.792 billion RMB in accordance with the minimum purchase price of 1.8 RMB per kilogram authorized by the State Council in 2010.

The average diseased panicle rate in Jiangsu Province in 2012 was 41.86 % and the average disease index was 20.48. This year the estimated yield was 4.9 t per ha, suggesting that the total yield loss of wheat due to FHB reached 0.7317 million tons and the economic loss was at least 1.493 billion RMB in accordance with the minimum authorized purchase price 2.04 RMB per kilogram in 2012.

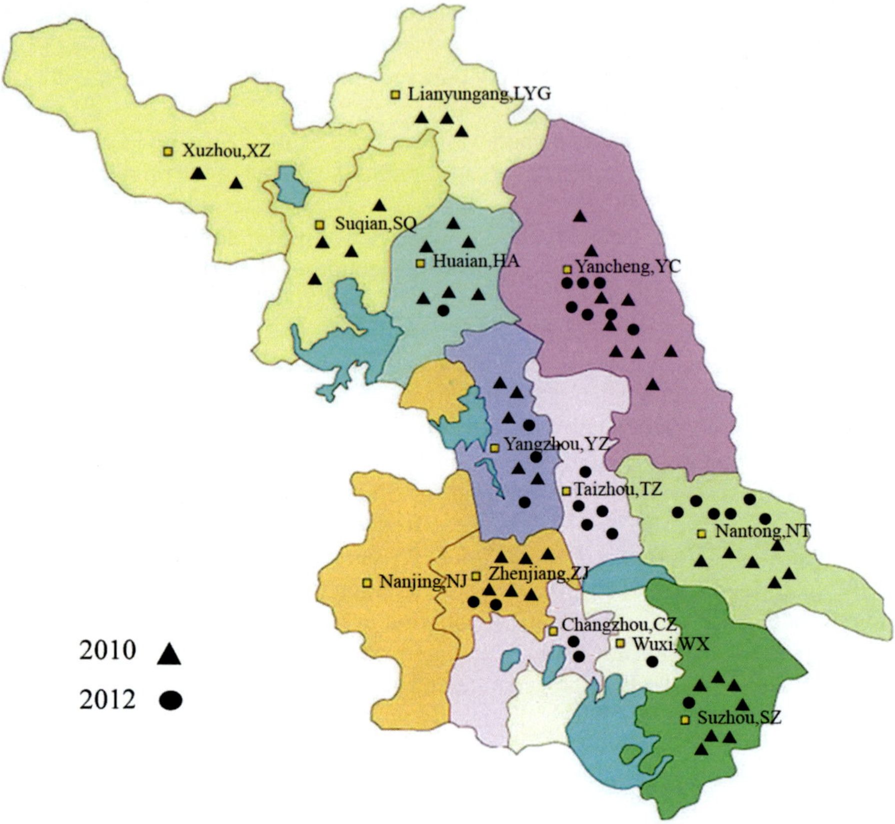

Fig. 19.3 Sample locations in Jiangsu Province

19.2.4 Research on Current Issues

Due to the important influence of FHB on wheat production, more research on current issues is needed. One is on rapid diagnosis technology for MBC resistance, not only to determine the sensitivity of isolates but also the correlation between their genotype and resistance levels. New techniques of fungicide application are needed because the wheat head structure determines how agent droplets diffuse within the ear. Study on formulation, spray time and dosage is pivotal to improve their efficacy. Another is exploration for fungicides with new molecular targets in *F. asiaticum*. Previous reports show that *F. asiaticum* is different from other plant pathogenic fungi in physiological metabolism and gene function. If we could find a sequence of DNA or a noncoding mRNA which triggers expression of a functional gene, it should be feasible to develop fungicides specifically targeting this trigger sequence. Finally the influence of fungicide resistance on DON production would be a valuable research topic.

19.2.5 DMI Fungicides

Studies have indicated that DMI fungicides are efficient in preventing and controlling FHB, not only alleviating the disease and increasing production but also reducing mycotoxin contamination. However, DMI fungicides have been little used to control FHB in China. Tebuconazole has been registered to control FHB since 2010 and is only used on a small scale in China currently. Therefore, no resistance to triazole fungicides has been reported yet. Liu et al. (2011) reported a significant synergism between tebuconazole and triadimefon against the FHB pathogen because their target *cyp*51 genes differ. Some other triazole fungicides are currently being screened to control FHB and combat MBC resistance.

19.2.6 Phenamacril (JS399-19)

Phenamacril is a *Fusarium*-specific fungicide displaying very low activity against other plant pathogenic fungi. Phenamacril has been used to control FHB during the last 5 years, but no resistance has yet been detected in the field. Li et al. (2006) obtained 14 resistant mutants derived from a wild-type *F. asiaticum* isolate through ultraviolet (UV) irradiation and "fungicide training". A frequency of 1.67×10^{-7} for UV mutagenesis was 2.3 times greater than MBC resistance using the same selection conditions. Prolonging the culture time accelerated selection of resistant mutants at sublethal fungicide concentrations. No differences in mycelial growth, perithecia formation and virulence were observed between these laboratory resistant mutants and wild types, but the capacity to sporulate of some mutants decreased remarkably. Genetic studies using nitrate non-utilizing mutants as genetic markers identified that phenamacril resistance was governed by one major gene in *F. asiaticum* (Chen et al. 2007). Recent work has linked phenamacril resistance and virulence with positive regulation by fimbrin gene *FgFim* (Zheng et al. 2014).

Other studies with laboratory mutants (Chen and Zhou 2009) showed that phenamacril was effective in controlling FHB caused by MBC-resistant isolates under field conditions, but was not effective in controlling disease caused by isolates resistant to phenamacril, or those that were resistant to both MBC and phenamacril. Overall phenamacril possessed a high risk of development of resistance in both MBC-resistant and wild-type *F. asiaticum*, and double resistance to both of these fungicides could presumably emerge and lead to major problem since both fungicides are extensively used in China. Consequently, a mixture of phenamacril with triazoles is recommended to preserve the usefulness of this new mode of action.

19.3 Fungicide Resistance in Sharp Eyespot

Sharp eyespot, caused by *Rhizoctonia cerealis*, is an economically important disease on wheat in south China including the provinces of Jiangsu, Shanghai, Anhui, Shandong and Henan. Difenoconazole and tebuconazole are used as seed treatments to control sharp eyespot, and Jinggangmycin, a derivative of validamycin, is applied by spray during the seedling stage. To date there is no credible evidence of fungicide resistance in the pathogen population.

19.4 DMI Fungicide Resistance in Wheat Powdery Mildew

Due to its heavy sporulation, rapid spread short life cycle and susceptible varieties, wheat powdery mildew requires extensive fungicide treatments which increase the selective pressure for the resistance. Since the late 1980s, triadimefon has been extensively used to control wheat powdery mildew in China. At first triadimefon was used at 90–100 g a.i. ha^{-1}, which was enough to keep crops mildew free for 8 weeks. Currently at least 105–210 g a.i. ha^{-1} of triadimefon is required to prevent powdery mildew infection for 2 or 3 weeks. The first report (Yuan et al. 2001) for resistance of powdery mildew to triadimefon revealed resistance factors between 3.22- and 6.74-fold. Yang et al. (2010) tested the sensitivity of 108 isolates, collected from six provinces, to triadimefon by a detached leaf segment method. They found the mean resistance factor of 45.14 and a frequency of triadimefon-resistant isolates of 97 %.

19.5 Epidemiology of Resistance

Human activities also are important factors in the spread of fungicide resistance. Dai et al. (2013) analysed the extension of MBC resistance of *F. asiaticum* and found that resistance did not occur in Zhejiang and Jiangsu Provinces until carbendazim was continuously used for more than 20 years. MBC resistance became detectable in the population in neighbouring Shandong and Henan Provinces in recent years shortly after the intermittent application of carbendazim. Based on the observation that MBC-resistant isolates of *F. asiaticum* were different from the local dominant *F. graminearum* in mycotoxin chemotype, and when resistance was detected, we can speculate that the emergence of carbendazim resistance of FHB pathogen in Shandong and Henan may have spread from southeastern Jiangsu by seed transport and movement of combine harvesters.

19.6 Anti-Resistance Strategies

19.6.1 Boost of Professional Control System for Large-Scale Disease Prevention

Dynamics of the MBC-resistance development in the FHB pathogen population in Jiangsu Province of China over last 30 years has shown that development of resistance dramatically increased when the disease epidemic and the MBC-resistant isolates accumulated above 3 %. Therefore, suppressing disease at lower incidence level is an effective strategy of fungicide resistance management. To minimize the disease occurrence, agronomic ways to reduce wetness and temperature in the plant crown are required. Efficacy of decentralized control cannot be guaranteed to apply fungicides in good time and in the correct way, in severe epidemic years, i.e. professional input into large-scale disease prevention can enhance control and reduce yield loss. Therefore, the Chinese government has encouraged farmers to set up professional cooperation organizations for large-scale disease prevention and subsidized wheat disease control.

19.6.2 Promotion of New Effective Fungicide and Compounds

Minimizing selection pressure of a fungicide on resistance development is another critical important strategy to manage fungicide resistance. In the early stage, China was short of the ability to develop novel fungicides for use in anti-resistance strategies. Mixtures of fungicides with different modes of action are the main strategy to postpone resistance development. The mixture of carbendazim and triadimefon, or of carbendazim, thiram and triadimefon, was applied to control FHB and wheat powdery mildew in China since the 1990s. This strategy played an active role in suppressing development of MBC resistance and DMI resistance in *Fusarium* and powdery mildew populations.

Experience shows that the strategy of screening for novel fungicides is also of critical importance to combat fungicide resistance. As a consequence of widespread MBC resistance in *F. asiaticum*, it is urgent to develop and apply new effective fungicides to overcome both the resistance problem and mycotoxin contamination. Phenamacril (JS399-19) is a novel fungicide developed by a local enterprise, with independent intellectual property rights, that inhibits both disease development and mycotoxin production, but with as yet no monitored resistance (Zhang et al. 2010). Moreover, high efficiency, low residual, low toxicity and yield protection make phenamacril an ideal fungicide to control FHB. Unlike insecticide resistance, in fungicide resistance, cross-resistance generally follows products with the same mode of action. So, as suggested above, phenamacril should be used in a mixture with a partner with a different mode of action. Because mixtures with partners selecting different mechanisms of resistance can provide a sounder scientific strategy, we use mixtures of tebuconazole and thiram to control wheat *Fusarium* head blight and powdery mildew, obtaining 10 % higher crop production and 85 % less DON compared with untreated crops (Zhang et al. 2009a, b, 2010).

19.6.3 Strengthening of Fungicide Application Advice and Guidance

Plant protection departments should strengthen their technical guidance to guarantee effective disease control. Fungicide treatments should be well timed, especially on highly susceptible cultivars in severe disease epidemic years. Fungicides should also be used scientifically, making sure the dosage is optimal. In addition, labels should note the class of mode of action of the fungicide, preventing the same class being used continuously. Therefore, setting up training programme is especially important for Chinese farmers.

References

Bai G, Shaner G (1994) Scab of wheat: prospects for control. Plant Dis 78(8):760–766

Bennett J, Klich M (2003) Mycotoxins. Clin Microbiol Rev 16:497–516

Bollen GJ, Scholten G (1971) Acquired resistance to benomyl and some other systemic fungicides in a strain of *Botrytis cinerea* in cyclamen. Neth J Plant Pathol 77(3):83–90

Bottalico A (1998) Fusarium diseases of cereals: species complex and related mycotoxin profiles, in Europe. J Plant Pathol 80:85–103

Chang DC, Grant GB, O'Donnell K, Wannemuehler KA, Noble-Wang J, Rao CY, Jacobson LM, Crowell CS, Sneed RS, Lewis FM (2006) Multistate outbreak of *Fusarium keratitis* associated with use of a contact lens solution. JAMA 296(8):953–963

Chen CJ, Yu JJ, Bi CW, Zhang YN, Xu JQ, Wang JX, Zhou MG (2009) Mutations in a β-tubulin confer resistance of *Gibberella zeae* to benzimidazole fungicides. Phytopathology 99(12):1403–1411

Chen C, Bi C, Yu J, Wang J, Li H, Luo Q, Zhou M (2008) MBC-resistance and its molecular mechanism in *Gibberella zeae*. J Plant Pathol 90:S2

Chen L, Bai G, Desjardins A (2000) Recent advances in wheat head scab research in China. In: Raupp WJ, Ma Z, Chen P, Liu D (eds) Proceedings of international symposium wheat improvement for scab resistance. Nanjing Agricultural University, Jiangsu, pp 258–273 (in Chinese)

Chen Y, Chen CJ, Wang JX, Jin LH, Zhou MG (2007) Genetic study on JS399-19 resistance in hyphal fusion of *Fusarium graminearum* by using nitrate nonutilizing mutants as genetic markers. J Genet Genom 34(5):469–476

Chen Y, Zhou MG (2009) Characterization of *Fusarium graminearum* isolates resistant to both carbendazim and a new fungicide JS399-19. Phytopathology 99(4):441–446

Dai DK, Jia XJ, Wu DX, Hou YP, Wang JX, Chen CJ, Zhou MG (2013) Analysis of diffusion path of carbendazim-resistance population of Fusarium head blight-based on Fusarium species, mycotoxin chemotype and resistance timing. Chine J Pestic Sci 15(3):279–285 (in Chinese with English Abstr)

D'Mello JF, Macdonald AM, Postel D, Dijksma WT, Dujardin A, Placinta CM (1998) Pesticide use and mycotoxin production in *Fusarium* and *Aspergillus* phytopathogens. Eur J Plant Pathol 104(8):741–751

Dignani M, Anaissie E (2004) Human fusariosis. Clin Microbiol Infect 10(s1):67–75

Du W, Wang HH (2004) Price behavior in China's wheat futures market. China Econ Rev 15(2):215–229

Duan YB, Zhang XK, Ge CY, Wang Y, Cao JH, Jia XJ, Wang JX, Zhou MG (2014) Development and application of loop-mediated isothermal amplification for detection of the F167Y mutation of carbendazim-resistant isolates in *Fusarium graminearum*. Scientific Reports, 4:7094

Fan P, Zhang Y, Zhou M, Chen C, Wang J (2009) Incidence of trichothecenes in wheat-based foods from China. Int J Environ Anal Chem 89(4):269–276

Gu BG, Liu JF (1990) Research on MBC resistance of Fusarium head blight-II. Resistance monitoring and induction. J Nanjing Agric Univ 13(1):57–61 (in Chinese with English Abstr)

Ioos R, Belhadj A, Menez M, Faure A (2005) The effects of fungicides on *Fusarium* spp. and *Microdochium nivale* and their associated trichothecene mycotoxins in French naturally-infected cereal grains. Crop Prot 24(10):894–902

Jing RL, Chang XP (2003) Genetic diversity in wheat (*T. aestivum*) germplasm resources with drought resistance. Acta Botanica Boreali-occidentalia Sinica 23(3):410 (in Chinese with English Abstr)

Jones R, Mirocha C (1999) Quality parameters in small grains from Minnesota affected by Fusarium head blight. Plant Dis 83(6):506–511

Li HK, Chen CJ, Wang JX, Zhou MG (2006) Study on baseline-sensitivity of *Fusarium graminearum* to JS399-19 and assessment of the risk of resistance *in vitro*. Acta Phytopathol Sinica 36(3):273–278 (in Chinese with English Abstr)

Li HX, Lu YJ, Wang JX, Zhou MG (2003) Cloning of β-tubulin gene from *Gibberella zeae* and analysis its relationship with carbendazim-resistance. Acta Microbiologica Sinica 43(4):424–429 (in Chinese with English Abstr)

Liu X, Yu FW, Schnabel G, Wu JB, Wang ZY, Ma ZH (2011) Paralogous *cyp51* genes in *Fusarium graminearum* mediate differential sensitivity to sterol demethylation inhibitors. Fungal Genet Biol 48:113–123

Luo Q, Xu J, Hou Y, Chen C, Wang J, Zhou M (2009) PIRA-PCR for detection of *Fusarium graminearum* genotypes with moderate resistance to carbendazim. Plant Pathol 58(5):882–887

Matthies A, Buchenauer H (2000) Effect of tebuconazole (Folicur®) and prochloraz (Sportak®) treatments on Fusarium head scab development, yield and deoxynivalenol (DON) content in grains of wheat following artificial inoculation with *Fusarium culmorum*. Z PflKrankh PflSchutz 107(1):33–52

McMullen M, Jones R, Gallenberg D (1997) Scab of wheat and barley: a re-emerging disease of devastating impact. Plant Dis 81(12):1340–1348

Miller SS, Chabot DMP, Ouellet T et al (2004) Use of a *Fusarium graminearum* strain transformed with green fluorescent protein to study infection in wheat (*Triticum aestivum*). Can J Plant Pathol 26:453–463

Milus E, Parsons C (1994) Evaluation of foliar fungicides for controlling Fusarium head blight of wheat. Plant Dis 78(7):697–699

O'Donnell K, Ward TJ, Geiser DM, Kistler HC, Aoki T (2004) Genealogical concordance between the mating type locus and seven other nuclear genes supports formal recognition of nine phylogenetically distinct species within the *Fusarium graminearum* clade. Fungal Genet Biol 41:600–623

Parry D, Jenkinson P, McLeod L (1995) Fusarium ear blight (scab) in small grain cereals – a review. Plant Pathol 44(2):207–238

Pirgozliev S, Edwards S, Hare M, Jenkinson P (2002) Effect of dose rate of azoxystrobin and metconazole on the development of Fusarium head blight and the accumulation of deoxynivalenol (DON) in wheat grain. Eur J Plant Pathol 108(5):469–478

Proctor RH, Hohn TM, McCormick SP (1995) Reduced virulence of *Gibberella zeae* caused by disruption of a trichothecene toxin biosynthetic gene. Mol Plant Microbe Interact 8(4):593–601
Qiu J, Xu J, Yu J, Bi C, Chen C, Zhou M (2011) Localisation of the benzimidazole fungicide binding site of *Gibberella zeae* β2-tubulin studied by site-directed mutagenesis. Pest Manag Sci 67(2):191–198
Qiu J, Huang T, Xu J, Bi C, Chen C, Zhou M (2012) β-Tubulins in *Gibberella zeae*: their characterization and contribution to MBC resistance. Pest Manag Sci 68(8):1191–1198
Schroeder W, Provvidenti R (1969) Resistance to benomyl in powdery mildew of cucurbits. Plant Dis Rep 53:271–275
Seitz L, Eustace W, Mohr H, Shogren M, Yamazaki W (1986) Cleaning, milling, and baking tests with hard red winter wheat containing deoxynivalenol. Cereal Chem (USA) 63(2):146–150
Snijders C (1990) Fusarium head blight and mycotoxin contamination of wheat, a review. Neth J Plant Pathol 96(4):187–198
Windels CE (2000) Economic and social impacts of Fusarium head blight: changing farms and rural communities in the northern Great Plains. Phytopathology 90:17–21
Yang P, Zhang R, Duan X-Y, Zhou YL (2010) Sensitivity of *Blumeria graminis* f. sp. *tritici* isolates to triadimefon and quinoxyfen in China in 2008 and its cross-resistance between triadimefon and quinoxyfen. Acta Phytopathol Sinica 40(4):404–410
Yearbook CS (2007) National Bureau of statistics of China. China Statistical Yearbook (in Chinese)
Yuan SK, Ma YM, Zhou MG (2001) Germination of the cleistothecium of *Blumeria graminis* on wheat and its establishment of sensitive base-line to triadimefon. Plant Prot 27(1):1–3
Yuan S, Zhou M (2005) A major gene for resistance to MBC, in field isolates of *Gibberella zeae*. Can J Plant Pathol 27(1):58–63
Zhang YJ, Yu JJ, Zhang YN, Zhang X, Cheng CJ, Wang JX, Hollomon DW, Fan PS, Zhou MG (2009a) Effect of MBC resistance on trichothecene production and aggressiveness of *Fusarium graminearum*. Mol Plant Microbe Interact 22(9):1143–1150
Zhang Y, Fan P, Zhang X, Chen C, Zhou M (2009b) Quantification of *Fusarium graminearum* in harvested grain by real-time polymerase chain reaction to assess efficacies of fungicides on Fusarium head blight, deoxynivalenol contamination, and yield of winter wheat. Phytopathology 99(1):95–100
Zhang YJ, Zhang X, Chen CJ, Zhou MG, Wang HC (2010) Effects of fungicides JS399-19, azoxystrobin, tebuconazole, and MBC on the physiological and biochemical indices and grain yield of winter wheat. Pestic Biochem Physiol 98(2):151–157
Zheng ZT, Gao T, Zhang Y, Hou YP, Wang JX, Zhou MG (2014) FgFim, a key protein regulating resistance to the fungicide JS399-19, asexual and sexual development, stress responses and virulence in *Fusarium graminearum*. Mol Plant Pathol 15(5):488–499
Zhou MG, Ye ZY, Liu JF (1994) Progress of fungicide resistance research. J Nanjing Agric Univ 17(3):33–41 (in Chinese with English Abstr)
Zhou MG, Wang JX (2001) Study on sensitivity base-line of *Fusarium graminearum* to carbendazim and biological characters of MBC-resistant strains. Acta Phytopathol Sinica 31(4):365–370 (in Chinese with English Abstr)

Chapter 20
Origin of Fungicide-Resistant Barley Powdery Mildew in Western Australia: Lessons to Be Learned

Madeline Ann Tucker, Francisco Lopez-Ruiz, Kithsiri Jayasena, and Richard Peter Oliver

Abstract The risk of fungicide resistance is greatest with pathogens with short dormant periods, with both sexual and asexual reproduction cycles, with large population sizes and when fungicides of a single mode of action (MOA) are repeatedly used. Most of the barley growing area in Western Australia (WA) has been seeded with powdery mildew (*Blumeria graminis* f. sp. *hordei* (*Bgh*)) susceptible cultivars for the last 10–15 years. Fungicides from the triazole group dominate the market and are used repeatedly as both seed and foliar treatments. Field failures have been observed leading to losses estimated at AU$100 m annually since 2007. Reduced efficacy has often been found to result from alterations in the gene encoding triazole target 14α-sterol demethylase (*CYP51* syn. *ERG11*). Clear associations were found between accumulations of *CYP51* mutations and reductions in triazole sensitivity. The combination of susceptible cultivars, conducive environmental conditions and repeated use of a single MOA has led with disappointing predictability to perhaps the most costly fungicide resistance epidemic in history.

Keywords *Blumeria graminis* f. sp. *hordei* • Barley powdery mildew • *Mlo* • Triazole • Fungicide resistance • *CYP51*

20.1 Introduction

Powdery mildew is a common fungal disease of many crops (Eichmann and Hückelhoven 2008). In grasses, the disease caused by *Blumeria* (syn. *Erysiphe*) *graminis* is divided into formae speciales (f. sp.) each colonising individual genera

M.A. Tucker (✉) • F. Lopez-Ruiz • R.P. Oliver
Centre for Crop and Disease Management, Department of Environment and Agriculture, School of Science, Curtin University, Bentley 6102, WA, Australia
e-mail: madeline.tucker@curtin.edu.au

K. Jayasena
Department of Agriculture and Food Western Australia, Albany 6330, WA, Australia

H. Ishii, D.W. Hollomon (eds.), *Fungicide Resistance in Plant Pathogens*,
DOI 10.1007/978-4-431-55642-8_20

of the grass family. *Blumeria graminis* f. sp. *hordei* (*Bgh*) is the causal agent of powdery mildew of barley, *Hordeum vulgare* (Pedersen et al. 2002). Barley powdery mildew (BPM) is one of the most damaging diseases of barley throughout the world, with the potential to cause considerable yield losses (Jørgensen 1992; Tacconi et al. 2006). *Bgh* can infect all green tissue of the host plant and is characterised by white fluffy pustules. In acute disease outbreaks, these pustules tend to coalesce, so that a severely infected barley crop may appear yellow at a distance.

Germplasm exploited in European spring cultivars provides complete protection against BPM infection. The MLO protein negatively regulates plant defences against powdery mildew, so cultivars carrying *mlo* alleles are resistant against all pathotypes of *Bgh*. The use of *mlo* in spring cultivars has remained a success story in agriculture for nearly 40 years (Kim et al. 2002). However, there are reported downsides to the use of *mlo* (Bjørnstad and Aastveit 1990). Cultivars with some *mlo* alleles are reported to have enhanced susceptibility to other fungal pathogens (Kumar et al. 2001), and it is claimed that *mlo* cultivars exhibit a yield penalty even in the absence of any infection (McGrann et al. 2014). It is for these reasons that breeders have been reluctant to include *mlo* germplasm in Australian barley breeding programmes (Dreiseitl and Platz 2012).

Breeders in WA have historically focussed on malt quality, and many cultivars released over the last few decades have been susceptible to BPM. Therefore, growers in regions prone to infection have relied heavily on the application of fungicides for disease control. Triazole fungicides are the most widely used class of antifungal agents for the control of fungal diseases of humans, animals (Chamilos and Kontoyiannis 2005) and plants (Poole and Arnaudin 2014). This fungicide class dominates those registered for barley diseases in WA, being contained in all registered foliar formulations until 2013 (Table 20.1). Triazole resistance is now widely reported in both clinical and agricultural sectors including powdery mildew populations from other barley growing regions of the world (Wyand and Brown 2005). The use of formulations containing only one class of fungicide coupled with the popularity of BPM susceptible cultivars was a perfect recipe for fungicide resistance. Growers in WA are now faced with an epidemic of highly virulent (Tucker et al. 2013) and fungicide-resistant BPM (Tucker et al. 2014). Here, we discuss the aetiology which has led to such an epidemic and discuss future prospects for the billion dollar barley industry in Australia.

20.2 Historical Cultivar Use in Western Australia

Australian barley growers have a worldwide reputation for producing a reliable supply of high-quality malting barley. Annually, WA produces on average 2.5 million tonnes of barley grain (ABARES 2014) of which the majority is exported for malting, shochu production, human consumption and animal feed (Paynter et al. 2012). The Wheat Belt is the common name applied to the arable area of WA and typically has 4 Mha of wheat, 1.5 Mha of barley and 1.5 Mha of other crops (mainly canola

Table 20.1 The year of first registration of the foliar active compounds permitted for barley powdery mildew control in Western Australia

Active ingredient/s[a]	Year first registered
Tebuconazole	1995
Flutriafol	1997
Propiconazole	1998
Triadimefon	1998
Azoxystrobin + cyproconazole	2004
Propiconazole + cyproconazole	2004
Epoxiconazole	2005
Epoxiconazole + **pyraclostrobin**	2008
Tebuconazole + flutriafol	2009
Tebuconazole + prothioconazole	2009
Azoxystrobin + tebuconazole	2012
Spiroxamine	2013

[a]Active ingredients other than triazoles are indicated in bold type (APVMA 2014)

and lupins). Barley production covers much of the Wheat Belt and powdery mildew infection has proven to be a massive threat to production in the recent years (GRDC 2012). Field surveys and crop reports have mapped the pathogen's spread over the cereal cropping area with extremely high levels of disease pressure particularly in the cooler, high rainfall areas along the south coast (Fig. 20.1).

There are a range of barley cultivars available to growers, each with different characteristics of yield potential, agronomic features, disease resistance profiles and grain quality and are marketed as malt, feed or food grades depending on their intended use postharvest. Growers decide which cultivar to grow depending on a number of factors including (1) the end price paid for grain (2) and the likelihood that a particular cultivar will meet the required grade (3) likely yield potential and (4) agronomic and disease resistance characteristics. Deliveries that meet malt grade specifications receive a premium price which normally outweighs the lower yield achievable compared to feed or food cultivars. Factors such as weather and disease pressure, which vary season to season, can result in downgrading of malt cultivars.

Malt grade deliveries of cultivars Baudin, Hamelin, Gairdner and Stirling attract about AU$37/t more than food or feed grades (DailyGrain 2014). The market leader for export malt, Baudin, is highly susceptible to powdery mildew (Paynter et al. 2012) as it possesses only the *Ml-a8* resistance gene which has now been completely defeated by *Bgh* in WA (Tucker et al. 2013). The three other cultivars of high popularity harbour the same resistance gene and are thus susceptible (Table 20.2).

On a typical farm of 4,500 ha where 67 % is cropped (Rowe 2012), growers will earn an extra AU$300,000 per year if their deliveries achieve malt grade (based on the average yield of 2.91 t/ha (NVT 2014)). The mildew susceptible cultivars have occupied on average 95 % of the total barley growing area for most of the last decade (Fig. 20.2 (CBH 2014)). This has gifted the pathogen with perfect conditions for proliferation and hence adaptation, key contributing factors to the current BPM epidemic.

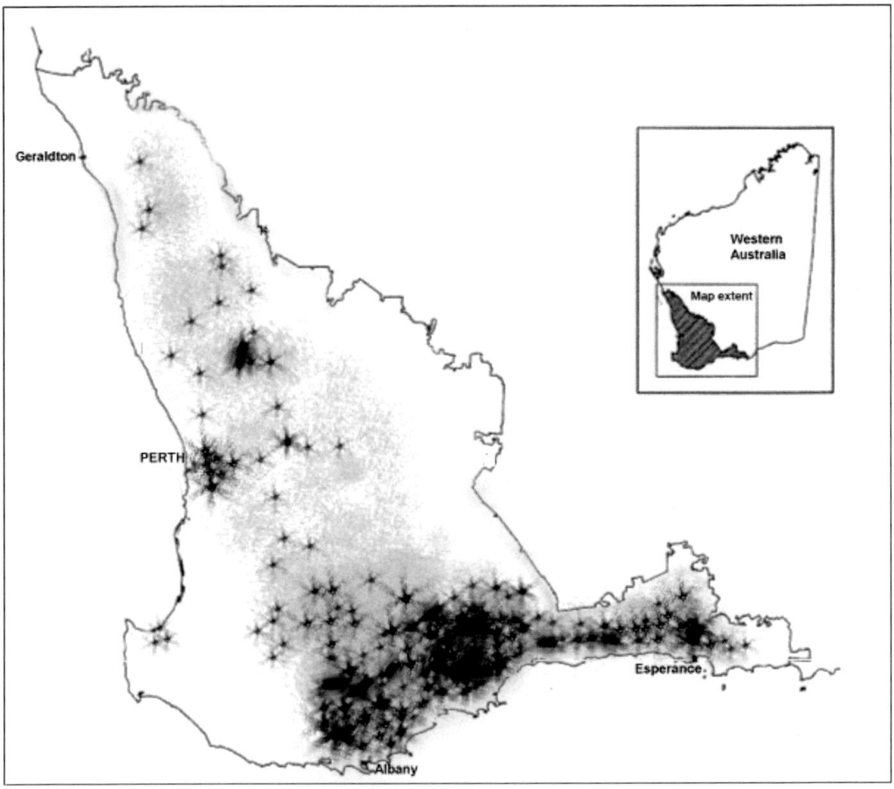

Fig. 20.1 Map of the Western Australian Wheat Belt. *Grey shading* indicates areas that barley is grown with mildew densities indicated in *black*

Table 20.2 Popular cultivars grown in the Western Australia with resistance genes and powdery mildew susceptibility indicated

Cultivar	Resistance genes[a]	Relevant virulence present in WA *Bgh*[b]	Powdery mildew susceptibility[c]
Baudin	*Ml-a8*	Yes	VS
Vlamingh	*Ml-a8*	Yes	S
Gairdner	*Ml-g*	Yes	S
Hamelin	*Ml-g*	Yes	S
Stirling	None	–	S
Bass	*Ml-a8*	Yes	MS
Hindmarsh	*Ml-a8, Ml-La*	Yes, Yes	MRMS
Buloke	*Ml-a7, Ml-La*	Yes, Yes	MR

[a]As postulated by Dreiseitl and Platz (2012)
[b]Results published by Tucker et al. (2013)
[c]Mildew susceptibility ratings by the Department of Food and Agriculture Western Australia

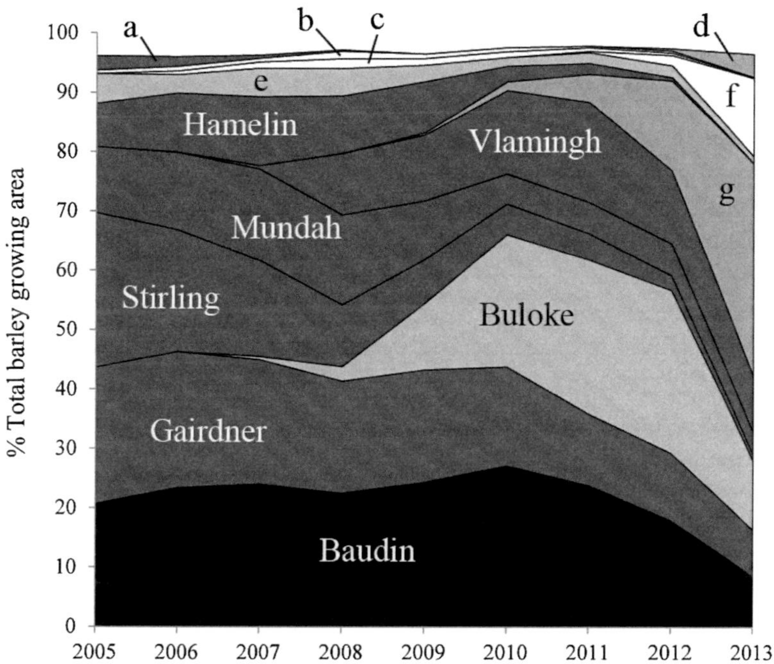

Fig. 20.2 Percent of Western Australian growing area seeded with barley cultivars from 2005 to 2013. Colour blocking indicates resistance profile to barley powdery mildew and is as follows: very susceptible, *black*; susceptible, *dark grey*; moderately resistant, *light grey*; and resistant, *white*. *Letters* indicate cultivars (*a*) Schooner, (*b*) Barque, (*c*) Dash, (*d*) Bass, (*e*) Yagan, (*f*) Scope and (*g*) Hindmarsh (Data obtained from CBH grower estimates 2005–2013)

Recently, three *mlo* cultivars have been examined in WA. These cultivars, Westminster, Henley and Granger, which were bred using European spring barley germplasm, consistently performed better than the market leader Baudin in regard to yield (NVT 2014). All three were resistant to BPM infection but have so far not yet obtained malt certification. The effect of *mlo* on other pathogens and yields in WA conditions clearly needs further study. New high-yielding cultivars which have increased resistance to BPM infection are currently undergoing malt accreditation (Barley Australia 2014). If accepted, we can expect that they will replace Baudin and the other susceptible cultivars as long as they possess a full suite of desirable characteristics.

20.3 Fungicide Use in Western Australia

20.3.1 Triazole Use and the Outbreak of Resistance

Fungicide use in Australian broad acre cropping has risen sharply over the last decade. According to a 10 year review published by Murray and Brennan (2010), nearly 95 % of the land sown to barley in mildew-prone areas of WA received some form of fungicide treatment, with greater than 35 % having multiple applications each season. The registration and use of formulations is tightly controlled by Australian government bodies, and product labels specify the crop, specific diseases and the permitted minimum and maximum doses. Although there are numerous different MOAs available (FRAC 2014), prior to 2013, all registered formulations for barley powdery mildew in WA contained a triazole (Table 20.1; Fig. 20.3). Triazole fungicides have a specific single-site MOA, targeting the product of the *CYP51* gene (sterol 14α-demethylase), thereby disrupting the synthesis of Ergosta-5,24(24^{1})-diene-3β-ol, the main sterol component of cell membranes in powdery mildews (Loeffler et al. 1992). Resistance to this class is now well documented in both clinical and agricultural fungal species where it can be acquired by (1) alterations in the *CYP51* gene which disturbs binding of the triazole at the target site (Mullins et al. 2011), (2) *CYP51* overexpression thereby increasing levels of sterol 14α-demethylase (Ma et al. 2006) and (3) an increased efflux of triazoles mediated through the presence of membrane-bound transported proteins (Da Silva Ferreira

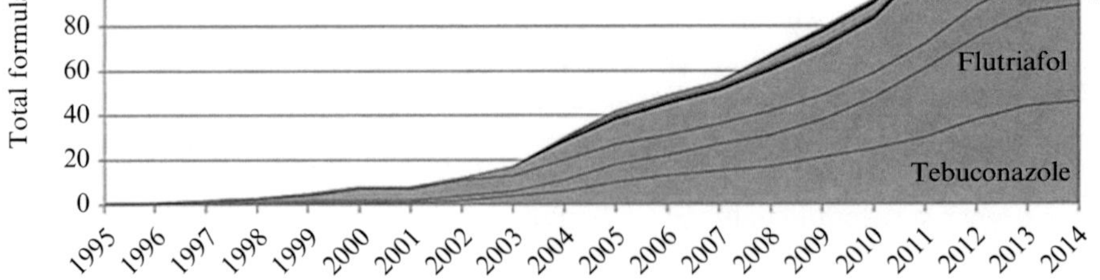

Fig. 20.3 Number of registered formulations for use on barley powdery mildew from 1995 to 2014 (APVMA 2014). Single-triazole formulations are shaded grey. Formulations mixing two triazoles are indicated with *thick black lines*, and formulations containing fungicides with a different MOA to triazoles are unfilled. *Letters* indicate formulations containing (*a*) spiroxamine, (*b*) azoxystrobin and tebuconazole, (*c*) flutriafol and tebuconazole, (*d*) epoxiconazole and pyraclostrobin, (*e*) epoxiconazole, (*f*) cyproconazole and propiconazole, (*g*) azoxystrobin and cyproconazole and (*h*) triadimefon

et al. 2004). As yet mutations in *Bgh CYP51* have been reported only in the UK isolates of *Bgh* that were less sensitive to the triazole fungicide triadimenol (Wyand and Brown 2005).

For almost a decade, growers in WA have been reporting a decline in the efficacy of the widely used triazole tebuconazole (GRDC 2012). First registered in 1995 (APVMA 2014) (Table 20.1), tebuconazole provided protective, eradicative and curative action against many barley diseases common in WA including rusts (*Puccinia* spp.) and powdery mildew (MacBean 2012). Few new products were registered for BPM control from this time until 2004 when the expiry of patents saw the market flood with formulations from international manufacturers, supplying chemicals at a fraction of the previous price (Fig. 20.3) (Paton 2014). The first strobilurin (QoI) fungicide, azoxystrobin, was introduced in 2004 as a mixture with a triazole, cyproconazole (APVMA 2014). Additional formulations containing QoIs were incorporated in subsequent years but always mixed in patent formulations containing various triazole partners.

Local reports of triazole field failures have been circulating since 2005. Despite this, new formulations that were devoid of triazole fungicides were not registered until 7 years later. The 2013 cropping seasons saw the first spiroketal-amine product registered in a formulation containing spiroxamine (Table 20.1, Fig. 20.3) (APVMA 2014).

Did the widespread and prolonged use of triazoles contribute to the build-up of resistant BPM isolates in WA? To answer this, we began a study in 2009 to examine whether resistance to triazoles had evolved in the *Bgh* population in WA.

20.3.2 Genetics of Triazole Resistance

Several studies have dissected the relationship between mutational changes in the triazole target gene, *CYP51*, with triazole efficacy failures in the field. To date, two mutational changes in *CYP51* have been reported in BPM isolates from the UK. A tyrosine to phenylalanine substitution at amino acid 136 (Y136F) and a lysine to glutamine change at amino acid 147 (K147Q) were found in isolates that had reduced sensitivity to the triazole triadimenol (Wyand and Brown 2005). To determine whether mutational events were causing loss of efficacy in WA fields, a pilot study was initiated in 2009, whereby isolates of *Bgh* were phenotyped in response to tebuconazole and their *CYP51* sequence was obtained. In 2009, the Y136F mutation was identified in WA isolates, although no correlation could be drawn between the presence of the mutation and the failure of tebuconazole in the field as wild-type *Bgh* isolates were never found (Tucker et al. 2014).

CYP51 sequences were obtained from additional isolates in subsequent seasons, and further mutations were identified. The Y136F mutation was found to be ubiquitous throughout Australia. WA isolates harboured additional modifications of which a change from serine to threonine at amino acid 509 (S509T) was the most significant (Tucker et al. 2014). Isolates harbouring the S509T mutation were consistently less sensitive to all formulations tested (Fig. 20.4). The most commonly used triazole, tebuconazole, was most badly affected by the mutation, and this presumably explains its loss of field efficacy. The prevalence of the S509T mutation increased from nonexistent in isolates collected in 2009 to almost universal in WA in 2011 (Fig. 20.5). A homology search revealed that the S509T mutation is aligned with the S524T change in *Zymoseptoria tritici*, previously characterised as conveying reductions in triazole sensitivity (Cools and Fraaije 2013). We conclude that tebuconazole has selected for the Y136F/S509T mutant which is capable of proliferation even in the presence of field rates of tebuconazole.

20.3.3 Strategies for Combatting Triazole Resistance of **Bgh**

Genetic changes occur randomly in the field. They only become a problem when a resistant mutant form proliferates to become dominant in the population (van den Bosch et al. 2014). It is therefore quite feasible that the *Bgh CYP51* mutations Y136F and S509T were present in the WA population long before they were first detected in 2009 and 2010, respectively. Microsatellite studies have determined that these mutations occurred numerous times and in numerous sites around the state (Tucker unpublished). Hence, we conclude that the increase in frequency has been caused by successive selection for Y136F/S509T mutants.

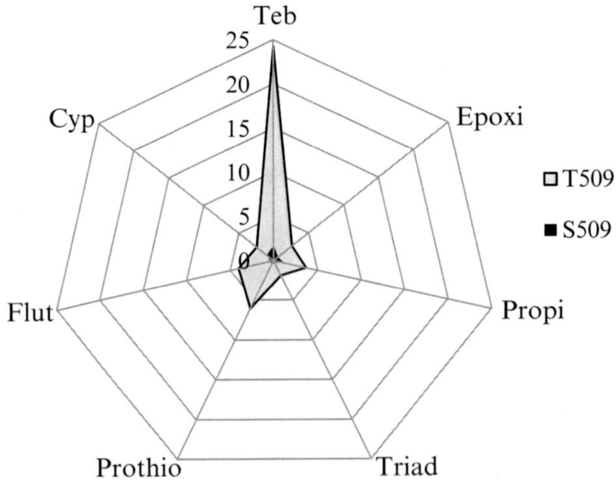

Fig. 20.4 Average EC_{50} ($\mu g\ mL^{-1}$) of a collection of Australian *Bgh* isolates that are either S509 (*black area*) or T509 (*grey area*) in *CYP51*. Each axis represents a triazole EC_{50}. *Teb* tebuconazole, *Epoxi* epoxiconazole, *Propi* propiconazole, *Triad* triadimefon, *Prothio* prothioconazole, *Flut* flutriafol or *Cyp* cyproconazole. Outermost ring corresponds to 25 $\mu g\ mL^{-1}$ with each preceding ring representing a 5 $\mu g\ mL^{-1}$ reduction as indicated

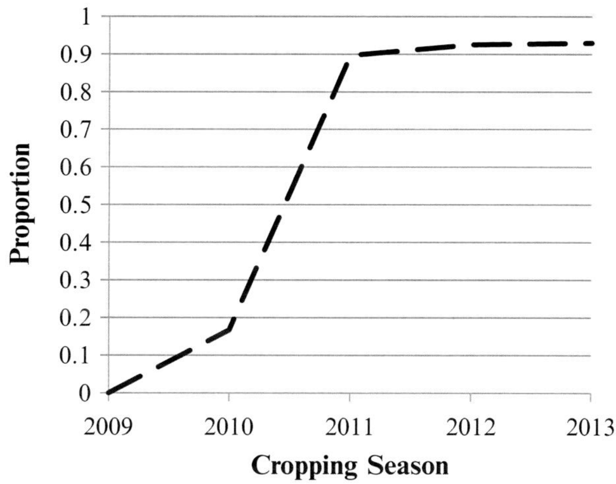

Fig. 20.5 The proportion of *Bgh* isolates with mutant T509 *CYP51* alleles collected in Australia over a 5 year period

Until recently, triazoles dominated the market for foliar fungicides in WA. Products that also contain a QoI were introduced in 2004 and 2008 (Table 20.1). A formulation containing spiroxamine, registered in 2013, is the only product available that does not contain a triazole. In contrast, in the UK, 50 compounds having 15 different MOAs are registered for use on spring barley (FRAG 2014). There are clearly many fungicides that are not currently accessible to WA growers. Furthermore, five of the fungicides registered in the UK are multisite inhibitors, which have proven to have a low risk of resistance development. There is an immediate need to fast track the registration of chemicals with new MOAs that have already been thoroughly assessed for use on many barley diseases.

The primary strategy recommended to manage resistance to any fungicide class is to mix or alternate with another fungicide with a different mode of action (Brent and Hollomon 2007). The at-risk fungicide, which usually provides great disease control, can be protected against resistance breakdown by the inclusion of an active with a different MOA, either in a tank mix or alternating spray applications. The partner compound could ideally be a multisite inhibitor, presumed to have a low resistance risk or an unrelated single-site fungicide. The concept behind this arises from the expectation that mutants that have evolved resistance to one fungicide MOA would still be sensitive to a different MOA (van den Bosch et al. 2014). Thus, there is an absolute requirement for registrations of numerous MOAs that can be rotationally applied in attempt to suppress resistance build-up.

The widespread use of highly susceptible cultivars and the repeated use of a single mode of action fungicide were a perfect recipe for an epidemic of fungicide resistance. For a period of 9 years, there have been reports on field failure of tebuconazole in controlling BPM outbreaks (GRDC 2012). During successive seasons, BPM infections have been highly damaging and widespread resulting in reductions of yield from 2.91 t/ha to 1.76 t/ha and downgrading of grain to feed quality (NVT 2014). Annually, an average of 1.3 m ha (ABARES 2014) is planted in WA and receival ports indicate that 55 % of barley grain taken was susceptible to BPM infection (CBH 2014). Given these figures, we estimate that the epidemic of highly virulent and tebuconazole resistant *Bgh* has caused at least AU$100 M annually in losses in WA alone. This may well be the most damaging and costly fungicide resistance outbreak ever recorded.

20.4 Conclusions and Future Prospects

In the case of BPM, the factors that have been shown to contribute to fungicide resistance had been previously well recognised (Wyand and Brown 2005). BPM resistance to triazole fungicides has been documented since1981 (FRAC 2013; Fletcher and Wolfe 1981). Tebuconazole was registered for *Bgh* control in WA, 14 years later. Strategies for resistance management, such as combining or alternating triazoles with a different MOA, decreasing dose (van den Bosch et al. 2011) or spraying before complete crop infection, could have been introduced.

Unfortunately, for a number of diseases, including BPM, the rate of loss of effective fungicides threatens to exceed the rate of introduction (van den Bosch et al. 2014), and this is very much the present situation in Australia. Maintaining an array of effective fungicides with many different MOAs is critical. Many fungicides with different MOAs are registered overseas, and these could be made available to Australian growers. This would provide more options for resistance management. As well as new fungicides, an ongoing programme monitoring in shifts in resistance in the pathogen population would be needed to prolong the effective life of any newly introduced formulation.

References

ABARES (2014) Australian crop report, vol 169. Australian Bureau of Agricultural and Resource Economics and Sciences, Canberra

APVMA (2014) Public chemical registration information systems search. Australian Government. Accessed 20 June 2014

BarleyAustralia (2014) Barley Malt varieties under evaluation. http://www.barleyaustralia.com.au/varieties-under-malt-evaluation. Accessed 1 July 2014

Bjørnstad Å, Aastveit K (1990) Pleiotropic effects on the *mlo* mildew resistance gene in barley in different genetical backgrounds. Euphytica 46(3):217–226. doi:10.1007/bf00027221

Brent K, Hollomon DW (2007) Fungicide resistance: the assessment of risk, vol 2, 2nd edn. Fungicide Resistance Action Committee, Brussels

CBH (2014) CBH Barley grower estimates. Perth

Chamilos G, Kontoyiannis DP (2005) Update on antifungal drug resistance mechanisms of *Aspergillus fumigatus*. Drug Resist Updat 8(6):344–358

Cools HJ, Fraaije BA (2013) Update on mechanisms of azole resistance in *Mycosphaerella graminicola* and implications for future control. Pest Manag Sci 69(2):150–155. doi:10.1002/ps.3348

Da Silva Ferreira ME, Luiz Capellaro J, Dos Reis ME, Malavazi I, Perlin D, Park S, Anderson JB, Colombo AL, Arthington-Skaggs BA, Goldman MHS, Goldman GH (2004) In vitro evolution of itraconazole resistance in *Aspergillus fumigatus* involves multiple mechanisms of resistance. Antimicrob Agents Chemother 48(11):4405–4413

DailyGrain (2014) Historical barley prices. Perth

Dreiseitl A, Platz G (2012) Powdery mildew resistance genes in barley varieties grown in Australia. Crop Pasture Sci 63(10):997–1006

Eichmann R, Hückelhoven R (2008) Accommodation of powdery mildew fungi in intact plant cells. J Plant Physiol 165(1):5–18

Fletcher JS, Wolfe MS (1981) Insensitivity of *Erysiphe graminis* f. sp. *hordei* to triadimefon, triadimenol and other fungicides. In: Proceedings of the Brighton crop protection conference. Pests and Diseases, Brighton, pp 633–640

FRAC (2013) FRAC list of plant pathogenic organisms resistant to disease control agents

FRAC (2014) FRAC code list 2014: fungicides sorted by mode of action

FRAG (2014) Fungicides approved for winter barley. Accessed 11 July 2014

GRDC (2012) Barley powdery mildew fact sheet: control strategies for powdery mildew. CORETEXT, Western Region

Jørgensen JH (1992) Discovery, characterization and exploitation of *Mlo* powdery mildew resistance in barley. Euphytica 63(1–2):141–152

Kim MC, Panstruga R, Elliott C, Muller J, Devoto A, Yoon HW, Park HC, Cho MJ, Schulze-Lefert P (2002) Calmodulin interacts with *MLO* protein to regulate defence against mildew in barley. Nature 416(6879):447–451

Kumar J, Hückelhoven R, Beckhove U, Nagarajan S, Kogel K-H (2001) A compromised *Mlo* pathway affects the response of Barley to the Necrotrophic Fungus *Bipolaris sorokiniana* (Teleomorph: *Cochliobolus sativus*) and its toxins. Phytopathology 91(2):127–133. doi:10.1094/phyto.2001.91.2.127

Loeffler RST, Butters JA, Hollomon DW (1992) The sterol composition of powdery mildews. Phytochemistry 31(5):1561–1563. doi:http://dx.doi.org/10.1016/0031-9422(92)83106-9

Ma Z, Proffer TJ, Jacobs JL, Sundin GW (2006) Overexpression of the 14α-demethylase target gene (*CYP51*) mediates fungicide resistance in *Blumeriella jaapii*. Appl Environ Microbiol 72(4):2581–2585

MacBean C (ed) (2012) The pesticide manual: a world compendium, vol 16. British Crop Production Council, Hampshire

McGrann GRD, Stavrinides A, Russell J, Corbitt MM, Booth A, Chartrain L, Thomas WTB, Brown JKM (2014) A trade off between *mlo* resistance to powdery mildew and increased susceptibility of barley to a newly important disease, Ramularia leaf spot. J Exp Bot. doi:10.1093/jxb/ert452

Mullins JGL, Parker JE, Cools HJ, Togawa RC, Lucas JA, Fraaije BA, Kelly DE, Kelly SL (2011) Molecular modelling of the emergence of azole resistance in *Mycosphaerella graminicola*. PLoS One 6(6):e20973. doi:10.1371/journal.pone.0020973

Murray GM, Brennan JP (2010) Estimating disease losses to the Australian barley industry. Australas Plant Pathol 39(1):85–96

NVT (2014) Long term yield report – Barley. Australian Crop Accreditation System. Grains Research and Development Corporation

Paton S (2014) Fungicide use in practice: insights from industry. Perth

Paynter B, Hills A, Dhammu H (2012) Barley variety guide for WA 2013, Bulletin 4836. Government of Western Australia, South Perth. Government of Western Australia, South Perth, Australia

Pedersen C, Rasmussen SW, Giese H (2002) A genetic map of *Blumeria graminis* based on functional genes, Avirulence genes, and molecular markers. Fungal Genet Biol 35:235–246

Poole NF, Arnaudin ME (2014) The role of fungicides for effective disease management in cereal crops. Can J Plant Pathol 36(1):1–11

Rowe P (2012) Western Australia's farm debt trends: are we at the tipping point? In: Australian Agricultural and Resource Economics Society. AgEcon, Fremantle/Perth

Tacconi G, Baldassarre V, Collins NC, Bulgarelli D, Stanca AM, Vale G (2006) Haplotype characterization and markers at the barley *Mlo* powdery mildew resistance locus as tools for marker-assisted selection. Genome 49:864–872

Tucker MA, Jayasena K, Ellwood SR, Oliver RP (2013) Pathotype variation of barley powdery mildew in Western Australia. Australas Plant Pathol 42(5):617–623

Tucker MA, Lopez-Ruiz F, Cools HJ, Mullins JGL, Oliver RP (2014) Accumulation of mutations in *Blumeria graminis* f. sp. *hordei CYP51* confers positive and negative cross-resistance to triazole fungicides. Under Review

van den Bosch F, Paveley N, Shaw M, Hobbelen P, Oliver R (2011) The dose rate debate: does the risk of fungicide resistance increase or decrease with dose? Plant Pathol 60(4):597–606. doi:10.1111/j.1365-3059.2011.02439.x

van den Bosch F, Oliver R, van den Berg F, Paveley N (2014) Governing principles can guide fungicide-resistance management tactics. Ann Rev Phytopathol 52 (1):null. doi:10.1146/annurev-phyto-102313-050158

Wyand RA, Brown JKM (2005) Sequence variation in the CYP51 gene of *Blumeria graminis* associated with resistance to sterol demethylase inhibiting fungicides. Fungal Genet Biol 42:726–735

Chapter 21
Rice Pathogens in Japan

Hideo Ishii

Abstract *Magnaporthe oryzae* causes blast, the most serious disease on rice. The fungus is genetically diverse and hence easily develops resistance to fungicides. In Japan, it started from the resistance to kasugamycin and has continued up until MBI-D and QoI fungicide resistance experienced most recently. The history of fungicide resistance occurred on rice blast disease and research and countermeasure taken in the country are summarized in this chapter. Some related information from overseas is also introduced.

Keywords Benzimidazole fungicides • Fungicide resistance • *Gibberella fujikuroi* • *Magnaporthe oryzae* • MBI-D fungicides • QoI fungicides • Rice Bakanae • Rice blast

21.1 Introduction

Rice is one of the most important crops worldwide. Blast, caused by the fungus *Magnaporthe oryzae*, is the most serious disease on rice. This disease is distributed in about 85 countries (Kato 2001). In Japan, applications of chemical fungicides are common to control rice diseases although alternative methods such as biofungicides, hot water seed treatment, and blast-resistant multiline rice cultivars are employed. Chemical control of rice diseases, blast disease in particular, has been reviewed recently (Hirooka and Ishii 2013). Fungicide resistance in rice was also reviewed before (Ishii 2011; Uesugi 1982).

Four decades have passed since fungicide resistance first occurred on rice in Japan that was the resistance of blast fungus *M. oryzae* to kasugamycin. Thereafter, resistance issue was not always very serious on rice although some other cases were reported as reviewed by Uesugi (1982). However, the new problem of resistance to

H. Ishii (✉)
Kibi International University, Minami-Awaji, Hyogo 656-0484, Japan

National Institute for Agro-Environmental Sciences,
Kannondai 3-1-3, Tsukuba, Ibaraki 305-8604, Japan
e-mail: hi481204@yahoo.co.jp

H. Ishii, D.W. Hollomon (eds.), *Fungicide Resistance in Plant Pathogens*,
DOI 10.1007/978-4-431-55642-8_21

MBI-D fungicides caused the decrease of control efficacy, and there was a concern whether QoI resistance might occur in the field populations of rice blast fungus in the future. In this paper, the history of fungicide resistance in rice and current topics related are reviewed.

21.2 Rice Blast Disease

21.2.1 Kasugamycin (Antibiotic)

In the early 1970s, kasugamycin was the most common fungicide for the control of this disease and sprayed frequently (Miura et al. 1975). However, control efficacy against this disease was lost in northern part of Japan in 1971 (Miura 1984).

Resistant isolates of *M. oryzae* exhibited reduced sensitivity to kasugamycin on rice-straw decoction agar medium. Control efficacy of kasugamycin was extremely low in the tests when resistant isolates were inoculated. However, resistant strains declined after withdrawal of this fungicide and control efficacy against blast was recovered gradually (Fukaya and Kobayashi 1982; Miura 1984).

21.2.2 Organophosphorus Fungicides

The fungicide IBP has been used since 1965 and resistance of *M. oryzae* to this fungicide was found in 1976 when the decrease of efficacy was observed (Katagiri et al. 1980; Yaoita et al. 1978). Two levels of IBP resistance were reported but most of the resistant isolates showed a moderate level of resistance. IBP-resistant isolates exhibited cross resistance to another organophosphorus fungicide EDDP and an organosulfur fungicide isoprothiolane (Katagiri and Uesugi 1977). The frequency of IBP-resistant strains also decreased when the selection pressure with the fungicide was removed (Iijima and Terasawa 1987).

21.2.3 MBI-D Fungicides

M. oryzae requires melanized appressoria for host penetration. MBI-D fungicides containing carpropamid, diclocymet, and fenoxanil (Fig. 21.1) inhibit scytalone dehydratase in fungal melanin biosynthesis. Nursery box treatment with MBI-D fungicides, carpropamid in particular, became common in many rice-growing areas as this fungicide exhibited long-lasting control efficacy against blast disease. The treatment was labor cost effective and greatly contributed to diminishing fungicide applications in paddy fields and lowering the pesticide input to the environment.

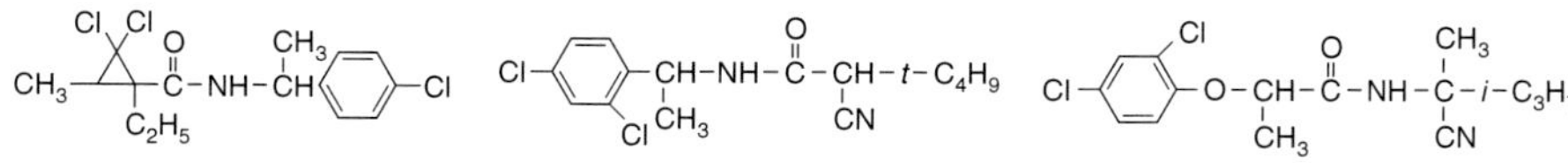

Fig. 21.1 MBI-D fungicides which inhibit scytalone dehydratase in fungal melanin biosynthesis

In 2001, however, the efficacy of carpropamid against leaf blast decreased suddenly in some areas of the southern part of Japan. Results from studies indicated that resistant strains appeared and played a significant role in the decrease of fungicide efficacy (Yamaguchi et al. 2002; Sawada et al. 2004). As of 2013, resistant strains have been detected in 36 out of 47 prefectures within the country although the impact of resistance greatly differed depending on the areas.

Results from monitoring tests suggested that resistant strains seemed to be less fit to the environment as their populations decreased in the absence of the selection pressure by MBI-D fungicides (Suzuki et al 2010; Yasunaga 2007). In a model experiment, resistant isolates showed lower competitive ability than sensitive isolates as the proportion of resistant ones decreased when their mixtures with sensitive isolates were inoculated repeatedly under no fungicide treatment (Kimura 2006).

Despite that, production of carpropamid was stopped and the use of diclocymet and fenoxanil in the same cross-resistance group has been largely reduced (Fig. 21.2). This is a typical example showing how seriously resistance development in major pathogens influences the market share of particular chemical control agents. In this case, MBI-D fungicides were replaced by disease resistance inducers such as probenazole, the inducer of systemic acquired resistance (SAR) commercially introduced first in the world, and two other products tiadinil and isotianil (Fig. 21.3). Probenazole has been widely used for 40 years as a major blasticide with no sign of field resistance development in *M. oryzae*. More details of MBI-D fungicide resistance of rice blast fungus are described in Chap. 11 of this book.

MBI-R fungicides, another class of melanin biosynthesis inhibitors, contain tricyclazole, pyroquilon, and phthalide and the primary target of these fungicides is 1,3,8-trihydroxynaphthalene reductase (Motoyama and Yamaguchi 2003). Laboratory mutants resistant to tricyclazole were obtained in rice blast fungus (Zhang et al. 2006); however, those mutants have not been isolated from the field where decreased efficacy of tricyclazole was reported in China (Zhang et al. 2009). No field isolates resistant to MBI-R fungicides have been found in Japan so far although nearly 30 years have passed since three fungicides in this class were registered and used for rice blast control (Eizuka et al. 2001). In Italy, tricyclazole-based fungicide was registered at the end of the 1990s and has been widely used for rice blast management since then. However, reduced sensitivity to tricyclazole has not been observed in the populations of *M. oryzae* collected from rice fields repeatedly treated with this fungicide over a 12-year period (Kunova et al. 2014).

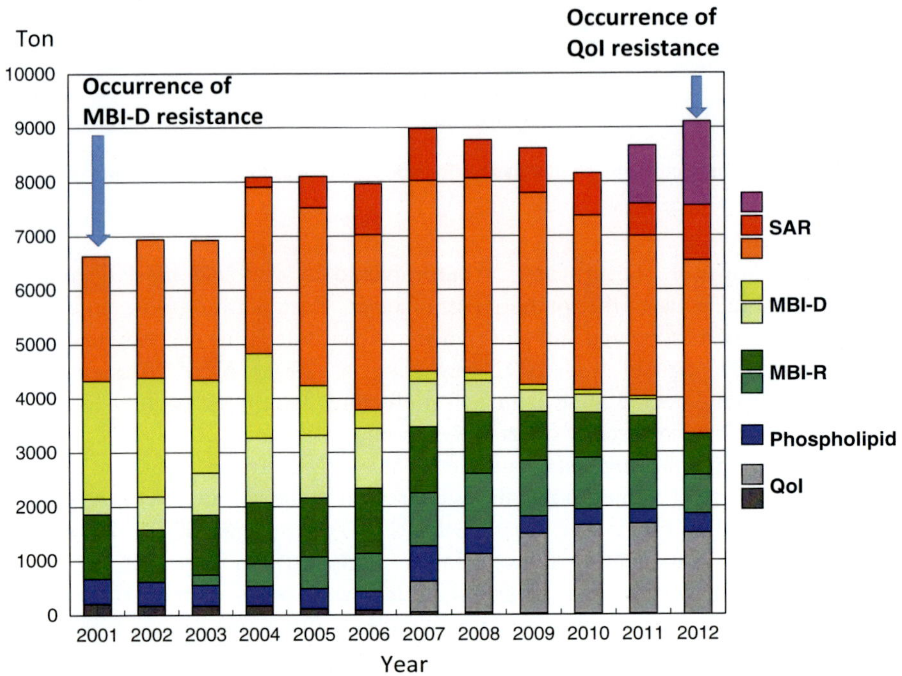

Fig. 21.2 Market value of fungicides used for rice blast in Japan

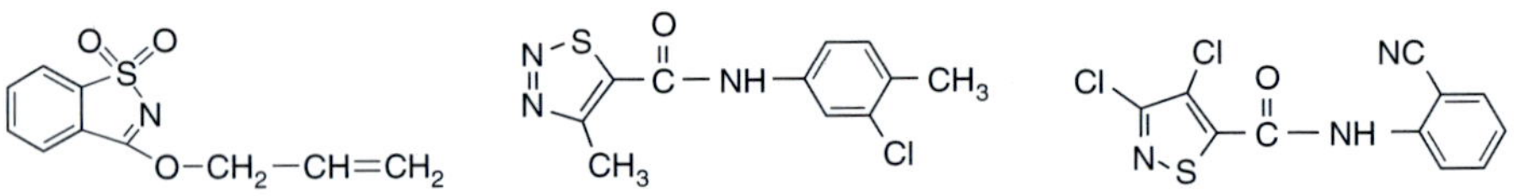

Fig. 21.3 Resistance inducers against rice diseases

21.2.4 *QoI Fungicides*

21.2.4.1 Background of Resistance Development

Occurrence and subsequent widespread of MBI-D resistance in rice blast fungus resulted in the reduction of the use of this class of fungicides rapidly. Alternatively, QoI fungicides became popular particularly when orysastrobin came into the market for nursery box treatment (Hirooka and Ishii 2013). Two other fungicides azoxystrobin and metominostrobin (Fig. 21.4) in the same cross resistance group had already been marketed for applications to paddy field. Nursery box treatment of rice with granule formulations of orysastrobin exhibited long-lasting control efficacy not only against blast but also against sheath blight diseases,

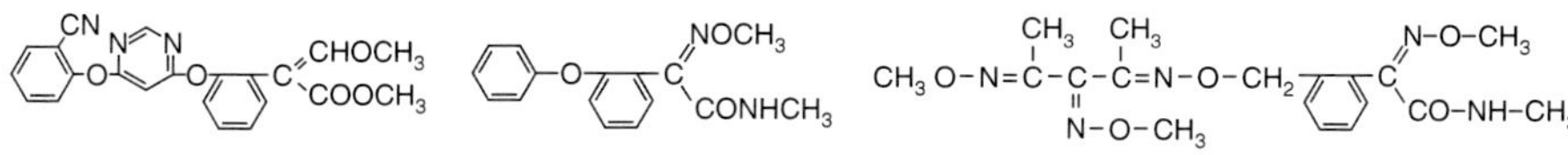

Azoxystrobin Metominostrobin Orysastrobin

Fig. 21.4 QoI fungicides registered for rice disease control

caused by *Rhizoctonia solani*, contributing to the reduction of fungicide applications (Stammler et al. 2007).

It was well known that QoI fungicides possessed high risk for resistance development in target pathogens. In fact, resistance had occurred in many diseases on various crops other than rice within Japan (Ishii 2012). Laboratory mutants of blast fungus resistant to azoxystrobin were obtained on culture media amended with azoxystrobin and SHAM (salicylhydroxamic acid), an inhibitor of AOX (alternative oxidase) in electron transfer pathway of fungal mitochondria (Avila-Adame and Köller 2003). In those spontaneous mutants, point mutations such as G143A (substitution of glycine with alanine at position 143) were found in cytochrome *b* gene which encodes the fungicide target protein.

However, as experienced with MBI-R fungicides mentioned above, it was not certain yet whether QoI fungicide-resistant strains could appear in the field even if resistant mutants were produced in the laboratory. Subsequently, azoxystrobin-resistant strains were reported in *Pyricularia grisea* (*M. oryzae*), closely related with rice blast fungus, from perennial ryegrass grown in the USA where severe outbreaks of gray leaf spot were observed despite the treatment with this fungicide (Vincelli and Dixon 2002). Molecular characterization of resistance was conducted and two mutations, G143A and F129L (substitution of phenylalanine with leucine at position 129) of cytochrome *b* gene, were found in resistant isolates. The former mutation was involved in higher level of resistance to azoxystrobin and trifloxystrobin than the latter one (Kim et al. 2003).

21.2.4.2 Monitoring for Field Resistance on Rice

Monitoring for QoI fungicide sensitivity of rice blast fungus was started in Japan. Araki et al. (2005) first established baseline sensitivity to metominostrobin and then compared it with field isolates collected from 2001 to 2003 when no QoI-resistant isolates were found and QoI sensitivity occurred throughout Japan. In this study, neither G143A nor F129L mutations were observed in cytochrome *b* gene from isolates examined by PCR-RFLP analysis. In 2004 and 2005, Stammler et al. (2007) also monitored for resistance quantitatively using pyrosequencing method but both mutations, G143A and F129L, were not detected. Both of these groups concluded that no QoI-resistant isolates and full QoI-sensitive situation were found throughout Japan. Meanwhile, it was briefly reported that benomyl and azoxystrobin activity against rice blast decreased in field experiments conducted in Louisiana, the USA, over time suggesting the occurrence of resistance (Groth and Rush 2006).

Monitoring studies were also carried out in our laboratory using isolates collected from various regions in 2004 and 2005 but resistant isolates were never found in mycelial growth tests (Wei et al. 2009). Growth of all isolates was inhibited on PDA medium supplemented with 1 mg L^{-1} (a.i.) azoxystrobin plus 1 mM *n*-propyl gallate, an inhibitor of AOX known to be more specific than SHAM.

21.2.4.3 Guideline for Fungicide Use

When orysastrobin was marketed on rice, the Japan Fungicide Resistance Action Committee (J FRAC, http://www.jfrac.com/) made a guideline indicating how to use orysastrobin and other QoI fungicides which had already been in the market. Shortly after that, the Research Committee on Fungicide Resistance (http://www.taiseikin.jp/), the Phytopathological Society of Japan, also proposed a guideline on this subject (So and Yamaguchi 2008). It was proposed to use QoIs only once per year on rice if necessary. In the latter guideline, furthermore, QoIs were recommended to be used in alternation with other unrelated fungicides such as MBI-R fungicides or resistance inducers every 2–3 years when QoIs were employed in nursery box treatment. As rice blast fungus is disseminated not only by wind but also by seeds, it was not recommendable to use QoIs in a paddy field where commercial seeds were produced. The same strategies were also proposed for MBI-D fungicides, if they were still effective.

21.2.4.4 Occurrence of Resistance

Nakamura et al. (2008) showed that three isolates of blast fungus sampled in 2007 were less sensitive to azoxystrobin in inoculation to young rice plants. In the summer 2012 subsequently, heavy outbreak of leaf blast disease has been reported from various regions in the western part of Japan after nursery box was regularly treated with orysastrobin for some years. Results from experiments conducted urgently confirmed that QoI-resistant strains were distributed in a large populations of the fungus (Ishii and Fuji 2013; Miyagawa et al. 2013). Resistant strains have been reported officially from three prefectures, Yamaguchi, Shimane, and Ehime, in 2012. As of October 2014, the presence of resistant strains has been proved in 16 prefectures. Cross resistance among three QoI fungicides, orysastrobin, azoxystrobin, and metominostrobin, was confirmed by fungus inoculation tests (Miyagawa and Fuji 2013). Resistant isolates were clearly distinguished from sensitive ones on PDA medium supplemented with 1 mg L^{-1} (a. i.) azoxystrobin plus 1 mM *n*-propyl gallate (Fig. 21.5).

Monitoring for QoI fungicide sensitivity has been continued, but resistant strains have not been detected from some prefectures such as Nagasaki, and Hiroshima yet. In these areas, the authorities related give rice growers a caution on the use of QoI fungicides. Some prefectures have started adopting the guideline from the Research Committee on Fungicide Resistance described above.

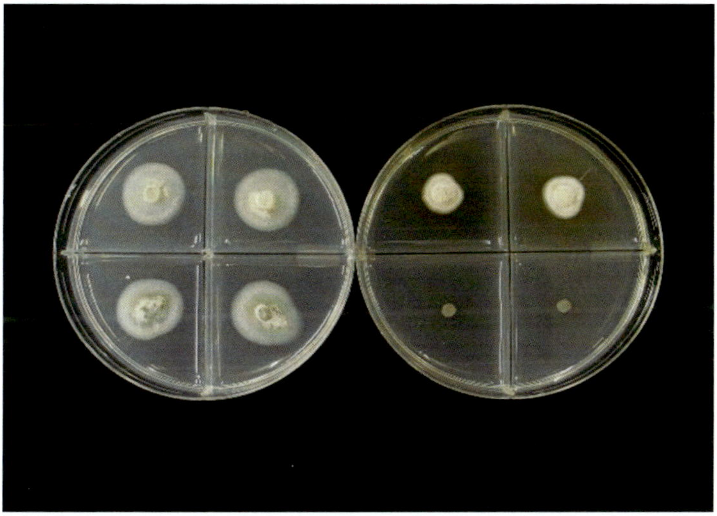

Fig. 21.5 Mycelial growth test for the detection of QoI-resistant isolates of rice blast fungus. *Upper*, resistant isolates; *lower*, sensitive isolates. *Left*, No fungicide treatment; *right*, 1 mg L^{-1} (a. i.) azoxystrobin plus 1 mM *n*-propyl gallate

21.2.4.5 Mechanism of Resistance

It is well known that QoI resistance is mostly caused by a single point mutation of fungicide-targeted cytochrome *b* gene in pathogens (Ishii 2012). In highly resistant isolates of the fungus *Pyricularia grisea* grown in the USA where there were severe outbreaks of gray leaf spot on perennial ryegrass, nucleotide sequences at the position 143 of this gene were converted from GGT to GCT resulting in the amino acid substitution of glycine (G) by alanine (A) (Kim et al. 2003). Another mutation, the change of phenylalanine (F) to leucine (L), was found at the position 129 in moderately resistant isolates.

When the sequence of cytochrome *b* gene was examined, QoI-resistant isolates of rice blast fungus carried the same mutation of G143A (exchange of GGT to GCT at position 143) as expected, but F129L mutation, found in the gray leaf spot fungus in the USA previously, was not recognized (Miyagawa et al. 2013). The partial nucleotide sequences analyzed in resistant and sensitive isolates are shown in Fig. 21.6 (Ishii 2014).

21.2.4.6 Molecular Diagnosis of Resistance

Using the single point mutation specifically found in cytochrome *b* gene of QoI-resistant isolates, a couple of methods have been developed to diagnose resistance. PCR-RFLP analysis is most widely used at present (Miyagawa and Fuji 2013). Fragments of cytochrome *b* gene are PCR-amplified from cultures of isolates with

Sensitive isolate:

5'...GGTTTCCTA..................TTATGAGGTGCTACA.........3'

Resistant isolate:

5'...GGTTTCCTA..................TTATGAGCTGCTACA.........3'

Fig. 21.6 A single point mutation at the cytochrome *b* gene found in QoI-resistant isolates of *Magnaporthe oryzae*. Nucleotide sequences corresponding to position 143 of cytochrome *b* gene are underlined. GGT at sensitive isolates are converted to GCT in resistant isolates resulting in the substitution of glycine by alanine

two primers designed by BASF and the products treated with a restriction enzyme *Fnu*4HI are loaded on an agarose gel. After digestion with this enzyme, the products from resistant isolates show two bands on a gel, whereas those from sensitive isolates remain a single band as they do not possess the restriction site (GCNGC) of *Fnu*4HI at the position 143 of cytochrome *b* gene.

The method of PCR-RFLP can also be applied for genomic DNA extracted from diseased rice leaves using microwave and pathogen-contaminated grains (Wei et al. 2009). Other methods such as ASPCR (allele-specific PCR) with a primer which recognizes the G143A mutation specifically and PCR-Luminex system suitable for high-throughput diagnosis were further developed (Wei et al. 2009). PCR-Luminex was originally introduced for rapidly identifying MBI-D fungicide resistance of rice blast fungus or identifying fungal species causing head blight of wheat (Ishii et al. 2008). In the future, the development of a more simple method like LAMP (loop-mediated isothermal amplification) may be necessary for molecular resistance monitoring on site.

Heteroplasmic status has often been found in cytochrome *b* gene of QoI-resistant fungal isolates (Ishii 2011). It is not known yet whether resistant isolates of rice blast fungus found in Japan carry heteroplasmic cytochrome *b* gene or not. If so, however, molecular methods for identifying QoI resistance would encounter some difficulties as experienced previously (Ishii 2010). Involvement of heteroplasmic cytochrome *b* gene with stability of resistance is described in details in Chap. 3 of this book.

21.2.4.7 Countermeasure with Resistance

In the areas where QoI-resistant strains were found to be distributed widely, the use of QoI fungicides has been stopped. However, in some cases, resistant strains have also been detected in a low frequency from the areas in which QoI fungicides were used in paddy fields only with foliar applications but not as a nursery box treatment and even from the areas with no use history of QoI fungicides. Therefore, it will be an important subject in the future to assess the resistance risk when these fungicides are used for foliar applications.

It is definitely important to save QoI fungicides within one application per year as well as to avoid their yearly successive applications for a nursery box treatment. There have been no reports on resistance to MBI-R fungicides and resistance inducers so far. Alternative use of QoI fungicides with these fungicides will be highly recommended to delay QoI resistance development.

As rice blast disease is disseminated not only by air but also by seeds, it is also quite important how we control seed contamination effectively. In general, it is recommended to growers not to use seeds harvested and stored by themselves and to alternatively purchase seeds disinfected by other sectors previously. However, it is not very rare to see those commercial seeds contain fungal strains resistant to fungicides. Such cases have been claimed on MBI-D as well as QoI fungicide resistance.

Due to this, the guideline recommends that QoI fungicides shouldn't be used in a seed-producing paddy field and its surroundings. In some regions actually, QoI fungicides are never applied through the course of rice seed production. Management such as removal of diseased plant debris, rice straws, and hulls from related facilities is also effective to sanitize the environment resulting in the decrease of infection source.

21.3 Report of QoI and SDHI Fungicide Resistance in Other Diseases Overseas

Occurrence of sheath blight disease, caused by *Rhizoctonia solani*, is increasing recently. Although it hasn't been found in Japan yet, resistance of this pathogen to azoxystrobin has been reported in the USA (Olaya et al. 2012). In addition, control failure using azoxystrobin has been mentioned in *Bipolaris* leaf spot disease on turf grass and reduced fungicide sensitivity of *B. spicifera* isolates was briefly reported (Tomaso-Peterson 2012). Although brown spot disease rarely occurs on rice these days in Japan, this disease is caused by *Cochliobolus miyabeanus* close to *B. spicifera*.

Most recently, QoI resistance has been reported in wheat blast pathogen *M. oryzae* in Brazil (Castroagudin et al. 2015). Control of this disease relied mainly on QoIs and these fungicides were used over the last 15 years. As a result, isolates carrying high frequency of the G143A mutation in cytochrome *b* gene associated with high QoI resistance have been sampled from both wheat and other poaceous host species of *M. oryzae* adjacent to wheat fields. Castroagudin et al. (2015) mentioned that these species may be an important reservoir for the pathogen that could contribute QoI-resistance inoculum during the early stage of a wheat blast epidemic. Furthermore, in *M. grisea*, anastomosis (i.e., hyphal fusion) has been proposed as a possible mechanism of resistance through transmission of the G143A mutation (Avila-Adame 2014).

New generation of succinate dehydrogenase-inhibiting (SDHI) fungicides has been recently developed very actively worldwide. Penflufen, one of them, has been registered in early 2014 for the control of sheath blight disease on rice in Japan.

SDHI fungicides also carry moderate to high risk for resistance development. In fact, isolates of *R. solani* resistant to the preexisting SDHI fungicide thifluzamide have been obtained in the laboratory (Mu et al. 2014). Furthermore, resistant isolates have been detected from the field as well and they reduced the efficacy of thifluzamide in fungus inoculation tests (Liu unpublished).

21.4 Fungicide Resistance in Bakanae Disease

21.4.1 Benzimidazole Fungicide Resistance

The benzimidazole fungicide benomyl, used in a mixture with thiram as a seed disinfectant, effectively controlled Bakanae disease caused by *Gibberella fujikuroi* (*Fusarium moniliforme*). However, isolates of this pathogen resistant to benomyl were detected in 1980 (Ogawa and Suwa 1981), and heavy occurrence of this disease due to resistance was reported in 1984. In 1987, resistant strains were widely distributed in 37 prefectures in Japan (Yoshino 1988).

Recently, Suga et al. (2013) divided the isolates of *F. fujikuroi* complex into two groups: G strains, gibberellic acid producing, thiophanate-methyl resistant, and highly ipconazole sensitive, and F strains, not gibberellic acid producing, thiophanate-methyl sensitive, and less ipconazole sensitive. Molecular mechanism has been studied and resistance was found to result from mutations in β_2*tub* gene [GAG (Glu) → GTG (Val) at codon 198 and TTC (Phe) → TAC (Tyr) at codon 200] but not in β_1*tub* gene (Chen et al. 2014). This finding could explain the following reports from early studies on fungicide sensitivity and binding to target proteins: (1) increased sensitivity (negative cross-resistance) to the *N*-phenylformamidoxime compound *N*-(3,5-dichloro-4-propynyloxyphenyl)-N′-methoxyformamidine (DCPF) was associated with a high level of carbendazim resistance in *Botrytis cinerea* but not with a moderate resistance level and (2) increased sensitivity to DCPF was not observed in carbendazim-resistant isolates of *G. fujikuroi* (Ishii and Takeda 1989).

21.4.2 DMI Fungicide Resistance

21.4.2.1 Reduced Sensitivity in Japan

G. fujikuroi isolates less sensitive to the DMI fungicide triflumizole were detected. MIC values of this fungicide for mycelial growth on PDA were 1,000 mg L^{-1} or more, but the EC_{50} values were less than 1.3 mg L^{-1}, only slightly different from sensitive isolates (Hamamura et al. 1989). Less triflumizole-sensitive isolates reduced pathogenicity remarkably against rice seeds and flowers than sensitive isolates, and such a difference in pathogenicity coincided with their lower production of gibberellic acids. Sensitivity of this fungus was also tested for pefurazoate, and less triflumizole-sensitive isolates were also less sensitive to this DMI fungicide, but pefurazoate still showed high efficacy against these isolates in artificial inoculation tests (Wada et al. 1990).

Similarly, isolates with MIC values lower than 0.78 mg L^{-1} for ipconazole were pathogenic but all the isolates with MIC values higher than or equal to 1.56 mg L^{-1} were not pathogenic to rice seedlings. Low pathogenicity or lack of pathogenicity of the isolates less sensitive to ipconazole may contribute to the stable efficacy of this fungicide (Tateishi and Chida 2000). Subsequently, *G. fujikuroi* species complex was classified into two groups based on MIC of ipconazole: (1) between 0.10 and 0.78 mg L^{-1} and gibberellic acid-producing *F. fujikuroi* and (2) 0.78 to 6.25 mg L^{-1} and no gibberellic acid-producing species such as *F. proliferatum* (Tateishi et al. 2011). Most recently, *G. fujikuroi* isolates resistant to benomyl and less sensitive to DMIs were predominant, and these isolates were pathogenic to rice causing a disease on fungicide-treated seeds (Kudo et al. 2014).

21.4.2.2 Resistance Development in Korea

In Korea, the incidence of Bakanae disease has increased rapidly. Resistance of the pathogen to prochloraz and hexaconazole was detected using the agar dilution method, but there was no evidence of cross-resistance between these two DMI fungicides (Jeong et al. 2009). Subsequently, cross-resistance to prochloraz and tebuconazole was found in some isolates (Lee et al. 2010). Surprisingly, it was reported that degradation of prochloraz might account for the reduced sensitivity to this fungicide (Kim et al. 2010). Relationship of *in vitro* resistance with disease control in the field has not been reported yet in details.

21.5 Recent Topics on Rice Disease Control

There is a strong demand from consumers to reduce pesticide applications. The use of biofungicides and/or hot water treatment has been introduced for seed disinfection of rice, but these treatments tend to increase the occurrence of Bakanae disease. Therefore, the development of alternative methods is still required to control major diseases.

'Koshihikari', the most popular and abundantly grown rice cultivar in Japan, was crossed with a resistant cultivar and progenies were further crossed with 'Koshihikari' five to six times. In 2005, seeds of four blast-resistant multi-lines thus bred were mixed together and cultivated at ca. 80 % of paddy fields in Niigata Prefecture, the most important rice-growing region in Japan. As a result, the occurrence of leaf and panicle blast dramatically decreased and blasticide applications were reduced to one fourth as compared before (Ishizaki 2010).

A mixture of the QoI fungicide orysastrobin and the best-selling disease resistance inducer probenazole (plus an insecticide) was developed and commercialized locally. These strategies are expected to play roles in reducing the risk of fungicide resistance.

References

Araki Y, Sugihara M, Sawada H, Fujimoto H, Masuko M (2005) Monitoring of the sensitivity of *Magnaporthe grisea* to metominostrobin 2001–2003: no emergence of resistant strains and no mutations at codon 143 or 129 of the cytochrome *b* gene. J Pestic Sci 30:203–208

Avila-Adame C (2014) Transmission of the G143A QoI-resistance point mutation through anastomosis in *Magnaporthe grisea*. Pest Manag Sci 70:1918–1923

Avila-Adame C, Köller W (2003) Characterization of spontaneous mutants of *Magnaporthe grisea* expressing stable resistance to the Qo-inhibiting fungicide azoxystrobin. Curr Genet 42:332–338

Castroagudin VL, Ceresini PC, de Oliveira SC, Reges JTA, Maciel JLN, Bonato ALV, Dorigan AF, McDonald BA (2015) Resistance to QoI fungicides is widespread in Brazilian populations of the wheat blast pathogen *Magnaporthe oryzae*. Phytopathology 105:284–294

Chen Z, Gao T, Liang S, Liu K, Zhou M, Chen C (2014) Molecular mechanism of resistance of *Fusarium fujikuroi* to benzimidazole fungicides. FEMS Microbiol Lett 357:77–84

Eizuka T, Sato T, Chida T, Yamaguchi I (2001) Simple method for monitoring the sensitivity of *Pyricularia oryzae* to fthalide. J Pestic Sci 26:385–389

Fukaya T, Kobayashi J (1982) Occurrence of resistant strains of *Pyricularia oryzae* Cavara against kasugamycin in Akita Prefecture and declining of the strains after breaking of chemical application. Ann Rep Soc Plant Protect North Jpn 33:25–28 (in Japanese)

Groth DE, Rush MC (2006) Possible development of resistance by the rice blast fungus to fungicides. Phytopathology 96(6s):S43 (Abstr)

Hamamura H, Kawahara M, Shimoda S (1989) Some characteristics of *Gibberella fujikuroi* (*Fusarium moniliforme*) isolates less-sensitive to triflumizole. Ann Phytopathol Soc Jpn 55:275–280

Hirooka T, Ishii H (2013) Chemical control of plant diseases. J Gen Plant Pathol 79:390–401

Iijima A, Terasawa M (1987) Epidemical studies on the drug-resistant strains of rice blast fungus, *Pyricularia oryzae* Cavara. Bull Nagano Agric Exp Stn Jpn 44:39–94 (in Japanese with English summary)

Ishii H (2010) QoI fungicide resistance: current status and the problems associated with DNA-based monitoring. In: Gisi U, Chet I, Gullino ML (eds) Recent developments in management of plant diseases, Plant pathology in the 21st century 1. Springer, Dordrecht, pp 37–45

Ishii H (2011) Fungicide resistance in rice. In: Dehne HW, Deising HB, Gisi U, Kuck KH, Russell PE, Lyr H (eds) Modern fungicides and antifungal compounds. DPG Selbstverlag, Braunschweig, pp 35–40

Ishii H (2012) Resistance to QoI and SDHI fungicides in Japan. In: Thind TS (ed) Fungicide resistance in crop protection – risk and management. CAB International, Wallingford, pp 223–234

Ishii H (2014) Situation of QoI fungicide resistance in rice blast fungus and countermeasures. Plant Prot 68:274–279 (in Japanese)

Ishii T, Fuji M (2013) Occurrence of *Pyricularia oryzae* with reduced sensitivity to QoI inhibitor in Fukuoka Prefecture. Jpn J Phytopathol 79:197 (Abstr in Japanese)

Ishii H, Takeda H (1989) Differential binding of a *N*-phenylformamidoxime compound in cell-free extracts of benzimidazole-resistant and -sensitive isolates of *Venturia nashicola*, *Botrytis cinerea* and *Gibberella fujikuroi*. Neth J Plant Pathol 95(Suppl 1):99–108

Ishii H, Tanoue J, Oshima M, Chung WH, Nishimura K, Yamaguchi J, Nemoto F, So K, Iwama T, Yoshimatsu H, Shimizu M, Kozawa T (2008) First application of PCR-Luminex system for molecular diagnosis of fungicide resistance and species identification of fungal pathogens. J Gen Plant Pathol 74:409–416

Ishizaki K (2010) Breeding and practical application of 'Koshihikari Niigata BL series' in Niigata Prefecture. Breed Res 12(Suppl 1):6–7

Jeong JY, Lee ZA, Lee YH, Myung IS, Ra DS, Lee SW (2009) Resistance of *Fusarium moniliforme* causing rice Bakanae disease to demethylation inhibitor fungicides. Abstr 1st Japan-Korea Joint Symp:134 (Abstr)

Katagiri M, Uesugi Y (1977) Cross-resistance between isoprothiolane and organophosphorus fungicides. Ann Phytopath Soc Jpn 43:360 (Abstr in Japanese)

Katagiri M, Uesugi Y, Umehara Y (1980) Development of resistance to organophosphorus fungicides in *Pyricularia oryzae* in the field. J Pestic Sci 5:417–421

Kato H (2001) Rice blast disease. Pestic Outlook 12:23–25

Kim YS, Dixon EW, Vincelli P, Farman ML (2003) Field resistance to strobilurin (QoI) fungicides in *Pyricularia grisea* caused by mutations in the mitochondrial cytochrome b gene. Phytopathology 93:891–900

Kim SH, Park MR, Kim YC, Lee SW, Choi BR, Lee SW, Kim IS (2010) Degradation of prochloraz by Bakanae disease pathogen *Fusarium fujikuroi* with differing sensitivity: a possible explanation for resistance mechanism. J Korean Soc Appl Biol Chem 53:433–439

Kimura N (2006) Comparison of biological properties of resistant field isolates of *Magnaporthe grisea* with susceptible isolates to melanin biosynthesis inhibitors targeting scytalone dehydratase (MBI-Ds). Abstr 16th Symp Res Com Fungic Resist. Phytopathol Soc Jpn:41–50 (in Japanese with English abstr)

Kudo G, Koh YJ, Toda T, Furuya H, Fuji S (2014) Fungicides sensitivity of *Fusarium fujikuroi* and their virulence. Jpn J Phytopathol 80:64 (Abstr in Japanese)

Kunova A, Pizzatti C, Bonaldi M, Cortesi P (2014) Sensitivity of nonexposed and exposed populations of *Magnaporthe oryzae* from rice to tricyclazole and azoxystrobin. Plant Dis 98:512–518

Lee YH, Kim S, Choi HW, Lee MJ, Ra DS, Kim IS, Park JW, Lee SW (2010) Fungicide resistance of *Fusarium fujikuroi* isolates isolated in Korea. Korean J Pestic Sci 14:427–432 (in Korean with English abstr)

Miura H (1984) Ecological study on the kasugamycin resistant rice blast fungus. Spec Bull Yamagata Agric Exp Stn 14:1–44 (in Japanese with English summary)

Miura H, Ito H, Takahashi S (1975) Occurrence of resistant strains of *Pyricularia oryzae* to kasugamycin as a cause of the diminished fungicide activity to rice blast. Ann Phytopathol Soc Jpn 41:415–417 (in Japanese with English abstr)

Miyagawa N, Fuji M (2013) Occurrence of QoI fungicide-resistant strains of *Magnaporthe oryzae* on rice and fungicidal effective. Abstr 23rd Symp Res Com Fungic Resist. Phytopathol Soc Jpn:25–36 (in Japanese with English abstr)

Miyagawa N, Fuji M, Kawabata Y (2013) Occurrence of orysastrobin-resistant isolates of rice blast fungus. Jpn J Phytopathol 79:197 (Abstr in Japanese)

Motoyama T, Yamaguchi I (2003) Fungicides, melanin biosynthesis inhibitors. In: Plimmer JR, Gammon DW, Ragsdale NN (eds) Encyclopedia of agrochemicals, vol 2. Wiley, Hoboken, pp 584–592

Mu W, Li B, Chen C, Liu X, Hao J (2014) Molecular mechanisms of thifluzamide resistance in *Rhizoctonia solani*. Phytopathology 104 (Suppl 1):S1.4 (Abstr)

Nakamura N, Fukuda H, Uchida K, Sou K, Takeda T (2008) Monitoring and methods for testing the sensitivity to melanin biosynthesis inhibitor (MBI-D, R) and QoI inhibitor (strobilurin) of *Magnaporthe grisea* isolates. J Plant Pathol 90(Suppl 2):S2.144

Ogawa K, Suwa M (1981) Benomyl sensitivity of rice Bakanae fungus distributed in Iwate Prefecture in 1980. Ann Rep Soc Plant Prot North Jpn 32:160 (in Japanese)

Olaya G (2012) Detection of resistance to QoI fungicides in *Rhizoctonia solani* isolates from rice. Phytopathology 102(7s):S4.88 (Abstr)

Sawada H, Sugihara M, Takagaki M, Nagayama K (2004) Monitoring and characterization of *Magnaporthe grisea* isolates with decreased sensitivity to scytalone dehydratase inhibitors. Pest Manag Sci 60:777–785

So K, Yamaguchi J (2008) Management of MBI-D and QoI fungicide resistance on rice blast caused by *Magnaporthe grisea* (in Japanese with English summary). In: Abstr 18th Symp Res Com Fungic Resist. Phytopathol Soc Jpn:70–80

Stammler G, Itoh M, Hino I, Watanabe A, Kojima K, Motoyoshi M, Koch A, Haden E (2007) Efficacy of orysastrobin against blast and sheath blight in transplanted rice. J Pestic Sci 32:10–15

Suga H, Hunasaka M, Fukasawa E, Arai M, Shimizu M, Kageyama K, Hyakumachi M (2013) Discovery of nonpathogenic lineage in *Fusarium fujikuroi*. Jpn J Phytopathol 79:227 (Abstr in Japanese)

Suzuki F, Yamaguchi J, Koba A, Nakajima T, Arai M (2010) Changes in fungicide resistance frequency and population structure of *Pyricularia oryzae* after discontinuance of MBI-D fungicides. Plant Dis 94:329–334

Tateishi H, Chida T (2000) Sensitivity of *Fusarium moniliforme* isolates to ipconazole. J Gen Plant Pathol 66:353–359

Tateishi H, Miyake T, Saishoji T (2011) Productivity of gibberellins and sensitivity to ipconazole in *Gibberella fujikuroi* species complex isolated from rice plants. Jpn J Phytopathol 77:207–208 (Abstr in Japanese)

Tomaso-Peterson M (2012) Possible alternative mechanisms of azoxystrobin resistance in *Bipolaris* spp. Phytopathology 102 (7s): S4.120 (Abstr)

Uesugi Y (1982) Case study 3: *Pyricularia oryzae* of rice. In: Dekker J, Georgopoulos SG (eds) Fungicide resistance in crop protection. Pudoc, Wageningen, pp 207–218

Vincelli P, Dixon E (2002) Resistance to QoI (strobilurin-like) fungicides in isolates of *Pyricularia grisea* from perennial ryegrass. Plant Dis 86:235–240

Wada T, Kuzuma S, Takenaka M (1990) Sensitivity of *Fusarium moniliforme* isolates to pefurazoate. Ann Phytopath Soc Jpn 56:449–456

Wei CZ, Katoh H, Nishimura K, Ishii H (2009) Site-directed mutagenesis of the cytochrome *b* gene and development of diagnostic methods for identifying QoI resistance of rice blast fungus. Pest Manag Sci 65:1344–1351

Yamaguchi J, Kuchiki F, Hirayae K, So K (2002) Decreased effect of carpropamid for rice blast control in the west north area of Saga Prefecture in 2001. Jpn J Phytopathol 68:261 (Abstr in Japanese)

Yaoita T, Go N, Aoyagi K, Sakurai H (1978) Frequency distribution of sensitivity in rice blast fungus to an organophosphorus fungicide in Niigata Prefecture. Ann Phytopath Soc Jpn 44: 401–402 (Abstr in Japanese)

Yasunaga T (2007) Fluctuation of MBI. D-resistant rice blast fungus and reuse of fungicides. kongetsu-no-Nogyo 61(4):88–92 (Japanese)

Yoshino R (1988) Current status of Bakanae disease occurrence and its control. Plant Prot 42:321–325 (in Japanese)

Zhang CQ, Zhu GN, Ma ZH, Zhou MG (2006) Isolation, characterization and preliminary genetic analysis of laboratory tricyclazole-resistant mutants of the rice blast fungus, *Magnaporthe grisea*. J Phytopathol 154:392–397

Zhang CQ, Huang X, Wang JX, Zhou MG (2009) Resistance development in rice blast disease caused by *Magnaporthe grisea* to tricyclazole. Pestic Biochem Physiol 94:43–47

Chapter 22
Potato Pathogens in Northern and Western Europe

Huub T.A.M. Schepers and Louise R. Cooke

Abstract The climatic conditions in Northern and Western (NW) Europe, which are very suitable for the cultivation of potatoes, are also very favourable for a range of diseases of which late blight, caused by *Phytophthora infestans,* is by far the most important with the potential to cause total crop loss. It is controlled by foliar fungicides, with up to 15–20 sprays being used per season. A wide range of fungicides is approved for the control of late blight including nonsystemics (e.g. mancozeb, fluazinam), numerous translaminar compounds (e.g. cymoxanil, dimethomorph, mandipropamid) but few fully systemic fungicides (viz. the phenylamides including metalaxyl-M, and propamocarb hydrochloride). Resistance to metalaxyl appeared in 1980, within a few years of its introduction, and led to control failures in Ireland and the Netherlands. Subsequently, anti-resistance management strategies were developed which allowed continuing use of phenylamides only in mixtures with nonsystemic multisite-inhibiting fungicides and with a limited number of applications. Resistance to other fungicides used to control late blight or early blight (caused by *Alternaria* spp.) has not developed or has not led to major reductions in control, perhaps in part as a consequence of effective resistance management. The potato is also susceptible to many tuber pathogens, which can cause rots and blemishes and are frequently transmitted via the seed tuber. Relatively few fungicides are approved for tuber or soil application to control tuber diseases. Resistance to thiabendazole, first used on potatoes in the 1970s, developed in several pathogens and resulted in loss of control where it was used alone. Resistance management has focused on avoiding repeated use of thiabendazole during multiple generations of the potato crop and use in mixtures with imazalil. In Europe, resistance has not developed to other fungicides used to control tuber diseases, although pathogen strains resistant to fludioxonil have been reported elsewhere.

H.T.A.M. Schepers (✉)
Applied Plant Research, Wageningen University Research,
PO Box 430, 8200 AK Lelystad, The Netherlands
e-mail: Huub.Schepers@wur.nl

L.R. Cooke
Sustainable Agri-Food Sciences Division, Agri-Food & Biosciences Institute,
Newforge Lane, Belfast BT9 5PX, UK

H. Ishii, D.W. Hollomon (eds.), *Fungicide Resistance in Plant Pathogens*,
DOI 10.1007/978-4-431-55642-8_22

Keywords Late blight • *Phytophthora infestans* • Phenylamides • Metalaxyl • Early blight • *Alternaria* • Dry rot • *Fusarium* • *Rhizoctonia* • *Phoma* • *Colletotrichum* • Thiabendazole

22.1 Introduction

In Northern and Western Europe, potatoes are grown on almost 1 million hectares (Table 22.1). Besides seed potatoes this also includes potatoes for the fresh market and for processing. The climatic conditions in NW Europe are usually very favourable for the development of a range of diseases of which late blight, caused by the oomycete pathogen *Phytophthora infestans*, is by far the most important. The annual losses caused by late blight in the EU are estimated at more than € 1,000,000,000 (Haverkort et al. 2008). Fungicides are an important tool to control late blight also because most of the widely grown and most popular commercial potato varieties are susceptible. The sometimes intensive use of fungicides to control foliar and tuber diseases in potato poses a risk for the development of resistance in the different potato pathogens. In this chapter the occurrence of resistance of the most important potato pathogens in NW Europe and the resistance management strategies developed will be reviewed.

Table 22.1 Potato crop area, yield and production in NW Europe in 2012 (http://faostat.fao.org)

Country	Area (ha)	Yield t/ha	Production (tonnes × 10^3)
Austria	21,800	30.5	665
Belgium	64,500	45.4	2,929
Denmark	39,500	42.1	1,664
Estonia	7,623	18.2	138
Finland	20,700	23.6	489
France	154,229	41.1	6,340
Germany	238,300	44.7	10,665
Iceland	505	19.2	9
Ireland	9,000	25.7	232
Latvia	28,200	19.1	539
Lithuania	31,800	18.0	572
Luxembourg	600	34.3	20
The Netherlands	149,770	45.2	6,765
Norway	126,58	26.3	333
Switzerland	11,012	41.0	451
The UK	149,000	30.6	4,553
Sweden	24,720	32.6	805
Total	951,443	-	37,176

22.2 *Phytophthora infestans*

Late blight caused by the oomycete *P. infestans* is the most damaging potato pathogen in NW Europe. Depending on the climatic conditions and intensity of the potato cultivation, potatoes are sprayed up to 15–20 times per growing season to protect the crop against late blight (Cooke et al. 2011). Copper compounds began to be used to control late blight in the 1890s. Subsequently during the twentieth century, these were largely superseded (except in organic production) by other broad-spectrum contact fungicides such as dithiocarbamates and organotins (the latter are no longer approved as fungicides in Europe).

Phenylamides: the fully systemic phenylamides introduced in the late 1970s were very effective, but soon after their introduction, *P. infestans* developed resistance to these fungicides. Resistance to phenylamides in *P. infestans* was first reported in the Netherlands and Ireland (Davidse et al. 1981; Dowley and O'Sullivan 1981) and subsequently in many NW European countries (César et al. 2012; Cooke 1981; Cooke et al. 2012; Corbière et al. 2010; Davidse 1985; Davidse et al. 1989; Dowley and O'Sullivan 1985; Dowley 1986; Gisi et al. 2011; Hannukkala 2014; Holmes 1984; Kildea et al. 2010; Lehtinen et al. 2007; Nassar et al. 2014; Schmitt and Kleinhenz 2012). The extensive literature on phenylamide resistance was reviewed by Gisi and Cohen (1996), Cooke and Little (2006) and Hermann and Gisi (2012). Since 2004, the widespread occurrence of the aggressive and invasive lineage of *P. infestans* known as 13_A2 or Blue 13 has been associated with an increased incidence of phenylamide resistance since this genotype is invariably resistant to metalaxyl (Cooke et al. 2012). Since the spread of the 13_A2 lineage is probably mainly driven by its ability to outcompete other genotypes, rather than by selection by phenylamides, this increase in phenylamide resistance is not amenable to anti-resistance strategies.

Propamocarb hydrochloride: in a sensitivity monitoring programme of European and US isolates carried out in 1995–1997, no evidence of resistance was found (Bardsley et al. 1998). Some isolates sporulated on leaf discs floating on 500 mg/l propamocarb hydrochloride, but these sporangia were not infective. Only three of the 1726 isolates collected in 1997–2000 from early late blight attacks in Finland were found to sporulate on leaf discs floating on 1000 mg/l propamocarb hydrochloride (Lehtinen et al. 2007). As the infectivity of the sporangia formed was not tested, it remains unclear whether isolates able to sporulate on 1000 mg/l propamocarb hydrochloride were truly resistant to the fungicide. In Norway, Nærstad et al. (2014) found that isolates recovered from fields treated with propamocarb hydrochloride prior to sampling had a higher frequency of reduced sensitivity to propamocarb hydrochloride than isolates from fields without propamocarb hydrochloride treatment. Whether this reduced sensitivity is related to reduced performance of propamocarb hydrochloride in the field is unclear.

Zoxamide: isolates collected in 1997–2000 from zoxamide-treated and non-zoxamide-treated trial plots in Europe varied in sensitivity and were considered to reflect the natural variation in the pathogen population (Cooke et al. 2002). Isolates collected from trials in a range of European countries between 2003 and 2012 again varied in sensitivity. There was no evidence of reduced sensitivity to zoxamide in 12 years of its use to control potato late blight in Europe (Cooke and Edmonds 2014).

CAA fungicides (dimethomorph, benthiavalicarb, mandipropamid): a small proportion of field isolates of *P. infestans* were found to be resistant to dimethomorph in Russia; however, resistant isolates that were induced through repeated treatment of potato plots appeared to have reduced fitness (Dereviagina et al. 1999). Isolates collected in 2001–2005 in mandipropamid-treated and non-mandipropamid-treated fields were all sensitive to mandipropamid. Also an enforced selection pressure did not produce isolates resistant to mandipropamid (Cohen et al. 2007). Seventy isolates collected in Ireland in 2008, 2009 and 2011 were all found to be sensitive to mandipropamid (Kildea et al. 2014). Sensitivity monitoring studies over several years revealed that in populations of *P. infestans*, all isolates were fully sensitive to CAA fungicides (www.frac.info).

QiI fungicides (cyazofamid, amisulbrom): *P. infestans* isolates with a reduced sensitivity to cyazofamid were detected after UV mutagenesis. However these isolates showed a significantly reduced fitness (Ziogas et al. 2006).

Fluazinam: In surveys in Europe no resistant isolates of *P. infestans* against fluazinam were detected up to 2010 (Cooke et al. 1998; Kessel et al. 2011; Räder and Gisi 2010; Schulte 2011). Zoospores in particular are very sensitive to low concentrations of fluazinam. The intensive use in the Netherlands of fluazinam in the past two decades may have exerted sufficient selection pressure against the formation of zoospores. Thus, over the years the balance between direct and indirect germination may have shifted towards direct germination. Kessel et al. (2009) investigated this hypothesis using isolates that originated before and after the introduction of fluazinam in the Netherlands. A shift from indirect to direct sporangial germination was not observed. In 2006–2007 in field trials in Denmark, a significantly lower efficacy of fluazinam to control late blight was observed (Nielsen 2014). In 2011, *P. infestans* isolates were detected in the Netherlands with a new genotype (Green 33, 33_A2). A survey in 2011 showed that 20 % of the isolates belonged to the Green 33 genotype. Reduced efficacy of fluazinam was demonstrated in some of these isolates in zoospore motility tests and field trials carried out in 2011 and 2012 with a high disease pressure (Schepers et al. 2013).

Cymoxanil: Power et al. (1995) studied the in vitro sensitivity to cymoxanil and chlorothalonil of 50 isolates of *P. infestans* collected in 1992–1993 from six European countries plus Mexico and the USA and found that the variation in sensitivity to both these fungicides was low. Subsequently, Hamlen and Power (1998) reported that a collection of approximately 200 isolates of *P. infestans* collected between 1980 and 1996 from ten European countries, Mexico and the USA were all sensitive to cymoxanil and showed no decrease in sensitivity after more than 16 years of use of this fungicide.

Copper, mancozeb, chlorothalonil, fluopicolide, ametoctradin, fenamidone, famoxadone: resistance in *P. infestans* populations is not known (www.frac.info).

22.3 *Alternaria solani*

Early blight on potato is a worldwide problem and is, at least in Europe, considered to be caused by the plant pathogenic fungus *Alternaria solani*. The pathogenic status of *Alternaria alternata* is currently under debate. In recent years early blight has become more prominent leading to increased interest and the formation of an early blight subgroup within the EuroBlight network (www.euroblight.net). Current control of early blight is based on the broad-spectrum contact fungicide mancozeb and the QoI fungicides. Recently also the triazole (DMI) fungicide difenoconazole has been approved in some NW European countries for the control of early blight in potatoes. Due to the single-site nature of QoI fungicides, resistance of various pathogens towards the fungicide has been reported. So far, early blight has been controlled without problems by QoIs in Europe. Although QoI fungicides have been reported to show a reduced efficacy against species of *Alternaria* in some parts of the USA, no problems have been reported from Europe. Reduced sensitivity to QoI fungicides in *Alternaria* spp. is caused by three different nucleotide substitutions in the amino acid at position 129 (referred to as F129L) in the gene encoding cytochrome b. In surveys in Sweden in 2009–2012, none of the *A. solani* isolates possessed the F129L mutation. Isolates from a field with low efficacy of QoI fungicides possessed an unknown sequence in cytochrome b which diverged considerably from the wild type. These strains may be of another genotype more resembling the American strains of *A. solani* (Edin and Andersson 2014). The majority of the *A. alternata* isolates possessed the G143A mutation normally associated with high-level resistance to QoI fungicides. Further investigations will determine whether the occurrence of G143A has consequences for the resistance to QoIs and for the control of disease since the pathogenicity of *A. alternata* to potatoes may be low (Blixt and Andersson 2010; Edin and Andersson 2014). Two hundred and three *A. solani* field isolates collected from 81 locations in Germany between 2005 and 2011 were screened for the presence of the F129L mutation; it was present in 74 isolates. All isolates possessing the F129L mutation had reduced sensitivity to azoxystrobin and, to a lesser extent, to pyraclostrobin. Data suggest an accumulation of F129L isolates in the German *A. solani* population over the years 2009–2011. It is assumed that the application of QoIs has selected for the occurrence of F129L mutations, which may contribute to loss of fungicide efficacy (Leiminger et al. 2014). In 2013 less sensitive isolates of *A. solani* bearing the F129L mutation were found in samples from Belgium, Germany and the Netherlands. All samples tested from France were sensitive (www.frac.com). Resistance factors are significantly lower in comparison with the G143A mutation, and field performance of products was good. In 2013 resistant isolates of *A. alternata* bearing the G143A mutation were found in samples from Austria, Belgium, Germany, France, the UK and the Netherlands (www.frac.com). The role of *A. alternata* in the disease complex remains under discussion.

22.4 Management of Fungicide Resistance in Late and Early Blight

The risk for development of resistance is a combination of the inherent pathogen risk, the inherent fungicide risk and the agronomic risk. The Fungicide Resistance Action Committee (FRAC) rates the inherent risk for the development of resistance of *P. infestans* and *A. solani* to fungicides as medium. The inherent risk of the phenylamide and QoI fungicides is rated as high; the inherent risk of the multisite inhibitors is rated as low (www.frac.info). The agronomic risk for the control of both late and early blight is estimated as high since both pathogens are controlled by numerous sprays per growing season. When a risk for development of resistance is present, a resistance management strategy has to be used. Within the FRAC, working groups discuss the monitoring results regarding the most important fungicide groups. The working groups on QoI and CAA fungicides have formulated use recommendations to manage fungicide resistance in late and early blight. For the other fungicides the FRAC recommends that resistance management is necessary but does not provide recommendations. In NW Europe two regional Fungicide Resistance Action Groups (FRAGs) are active, namely in the UK and in the Netherlands. These regional groups combine the expertise of the industry with the independent sector to provide up-to-date information and advice on fungicide resistance. They translate the information that the FRAC provides into recommendations for their local conditions. The UK group has formulated a resistance management guideline for late blight which has been adapted and translated by the Dutch group (Anonymous 2009).

To reduce the risk of resistance developing in a pathogen population, it is essential to put in place an anti-resistance management strategy at the outset. Managing resistance once it has occurred may not be effective. The first principle of any anti-resistance management strategy should be to reduce the risk from disease by attention to good agronomic practice.

- One of the most effective methods of reducing the risk from late blight is to grow cultivars with as high a disease resistance rating as possible. However, this is difficult to achieve when customers demand a specific cultivar, which may be highly susceptible to late blight.
- Where possible, avoid growing large areas of highly susceptible cultivars, particularly in locations where there is a history of high risk from late blight. Not only are crops at greater risk from becoming infected early in the season but also, once they have become infected, they serve as an inoculum source for neighbouring crops.
- Dumps are an important source of early inoculum. Destroy all dumps of discarded potatoes. Sheeting with heavy-gauge black polythene can prevent haulm growth, or young haulm can be killed by spraying with an approved desiccant. It is important to check the dumps throughout the season for regrowth.
- Control volunteers/groundkeepers. Although they tend to become infected later in the season and are, therefore, less likely to contribute to the early epidemics, they can still provide inoculum and infect crops as they approach harvest.

- Source good-quality certified seed. Discard blight-affected seed tubers. Only about one in 200 blighted tubers produces infected stems. However, 1 % seed infection would produce about two primary infectors per hectare. Under warm moist conditions spores from these primary infectors will spread throughout the crop.
- Make a timely start to spray programmes, when there is a warning of risk. Use forecast schemes and/or local knowledge regarding infection pressure to time applications more accurately.
- Adjust intervals according to risk (weather conditions/crop growth) and characteristics of the fungicides.
- Do not apply fungicides when the disease is well established in the crop, i.e. do not 'chase' the epidemic with fungicide, but consider burning off. This will not only help protect the crop from infection of the tubers but reduce late blight inoculum for neighbouring crops.
- Use mixed formulations of active ingredients with different modes of action or from a different fungicide family, or target specific products in blocks to appropriate growth stages.

The management strategies for the specific fungicides are summarised in Table 22.2. These strategies are a result of the overall resistance risk, the monitoring results and the recommendations on the label of the product. The recommendations on the label can be different from country to country mainly because the agronomic risk in which the intensity of fungicide applications and the disease pressure of the pathogens are included will not be identical in all regions. Except for resistance of *P. infestans* to phenylamides, late blight and early blight have not developed field resistance to fungicides in NW Europe. Although this can never be proven, the resistance management strategies will have played an important role in this.

22.5 Potato Tuber Diseases

The range of fungicides approved for application to potato tubers or to soil at planting for the control of rots and blemishing diseases is relatively limited compared with those approved for the control of late blight (Table 22.3). This reflects the smaller market (particularly as typically only a single annual application of product is made to control tuber diseases compared with the 10–15 annual applications to crops for late blight control) and the costs of obtaining approval for fungicides in Europe. In addition, options for postharvest treatments to give direct control of disease in the treated tubers (as opposed to treatments applied to seed to control disease in progeny tubers) are limited. This is because for ware tubers destined for human consumption, only imazalil alone and thiabendazole alone (but not imazalil plus thiabendazole mixtures) are approved, and treatments approved for postharvest use on seed only have a more restricted market and require segregation of ware and seed tubers.

Table 22.2 The active ingredients, FRAC codes and groups, diseases controlled, resistance risk, resistance reported in NW Europe and the use of the most important active ingredients for late blight and early blight control in NW Europe

Active ingredients	FRAC fungicide code and group	Diseases controlled	Resistance risk	Resistance reported in NW Europe	Use
Ametoctradin	45 QoSI	Late blight	Medium–high	No	No cross-resistance with QoIs. Use restricted to 5 applications
Azoxystrobin, pyraclostrobin, famoxadone, fenamidone	11 QoI	Early blight	High	Yes	Where QoI fungicide products applied solo do not exceed 33 % of the total number of sprays or a maximum of 4. Where mixtures used do not exceed 50 % of the total number of sprays or a maximum of 6 QoI fungicide applications, whichever is lower
		Late blight	High	No	
Cyazofamid, amisulbrom	21 QiI	Late blight	Medium–high	No	No more than three consecutive sprays recommended and should not form more than 50 % of the intended programme
Copper	M1 inorganic	Late blight	Low	No	Throughout the season also in some countries in organic potato
Chlorothalonil	M5 chloronitriles	Early blight	Low	No	Throughout the season. A good partner for at risk active substances
		Late blight	Low	No	
Cymoxanil	27 cyanoacetamideoxime	Late blight	Low–medium	No	Short persistence used on own. Use with a suitable partner
Difenoconazole	3 DMI	Early blight	Medium	No	Use restricted to 3 sprays per season
Dimethomorph	40 CAA	Late blight	Low–medium	No	Apply preferably in a preventive manner. Apply a maximum of 50 % of the total number of intended applications. Alternation with fungicides having other modes of action is recommended
Benthiavalicarb					
Mandipropamid					
Valifenalate					

Fluazinam	29	Late blight	Low	33_A2 genotypes are less effectively controlled	When 33_A2 is present it is recommended to restrict the number of sprays to 4 and in high pressure mix with fungicide with a different mode of action
Fluopicolide	43 benzamides	Late blight	Low	No	Formulated as a mixture with propamocarb. Maximum number of sprays is 4 at full dose
Metalaxyl-M, benalaxyl	4 phenylamides	Late blight	High	Yes	Only available in formulation with a partner of a different group. Best used preventively and in a restricted number of sprays (2–3). The dominant genotype 13_A2 is often associated with resistance
					Mainly resistant to phenylamides. Check with manufacturers for advice on recommended numbers of sprays per crop
Mancozeb	M3 dithiocarbamates	Late blight	Low	No	Can be used throughout the season. A good partner for at risk active substances. Can be used alone
Maneb		Early blight		No	
Propamocarb hydrochloride	28 carbamates	Late blight	Low–medium	No	Best used during period of rapid haulm growth. Use with a suitable partner
Zoxamide	22 benzamides	Late blight	Low–medium	No	Can be used throughout the season. Formulated in a mixture with mancozeb

Source: FRAC, FRAG-UK, FRAG-NL

Table 22.3 Fungicides used for control of potato tuber pathogens

Active ingredients	FRAC fungicide code and group	Diseases controlled or reduced	Pathogens	Application	Resistance reported in potato tuber pathogen(s)
Azoxystrobin, fluoxastrobin	11 QoI	Black dot	*Colletotrichum coccodes*	In-furrow or incorporated at planting	No
		Black scurf/stem canker	*Rhizoctonia solani*		
Fludioxonil	12 phenylpyrrole	Black dot	*Colletotrichum coccodes*	Seed in store or at planting	*Helminthosporium solani* (Tsror, unpublished)
		Black scurf/stem canker	*Rhizoctonia solani*		
		Silver scurf	*Helminthosporium solani*		
Flutolanil	7 SDHI	Black scurf/stem canker	*Rhizoctonia solani*	Seed in store or at planting	No
				Dust applied at planting	
Imazalil	3 DMI	Dry rot	*Fusarium* spp.	Seed in store	No
		Gangrene	*Phoma* spp.	Postharvest to seed and ware	
		Silver scurf	*Helminthosporium solani*		
		Skin spot	*Polyscytalum pustulans*		
Pencycuron	20 phenylurea	Black scurf/stem canker	*Rhizoctonia solani*	Dust applied at planting	No
Thiabendazole	1 MBC	Dry rot	*Fusarium* spp.	Seed in store	*Fusarium* spp.
		Gangrene	*Phoma* spp.	Postharvest to seed and ware	*Helminthosporium solani*
		Silver scurf	*Helminthosporium solani*		*Polyscytalum pustulans*
		Skin spot	*Polyscytalum pustulans*		
Tolclofos-methyl	14 aromatic hydrocarbon	Black scurf/stem canker	*Rhizoctonia solani*	Dust applied at planting	Variation in sensitivity in some *Rhizoctonia solani* AG groups, but not AG 3 (the main potato pathogen)
				Seed in store or at planting	

22.5.1 Dry Rot and Gangrene

Dry rot, caused by a range of soilborne and seed-borne *Fusarium* spp., is one of the most important fungal storage rots of potato tubers in Europe. In the UK and Ireland, the species most commonly isolated from tubers with dry rot is *F. coeruleum* (formerly *F. solani* var. *coeruleum*), followed by *F. culmorum* and *F. avenaceum*, with *F. sambucinum* (formerly *F. sulphureum*) the least common (Choiseul 1996; Choiseul et al. 2007; Peters et al. 2008a). This contrasts with North America where *F. sambucinum* is the most important species (e.g. Gachango et al. 2012). In contaminated tuber samples collected in 1969–1970 in Germany, *F. coeruleum* was isolated in 33–45 % of dry rot tubers, *F. sulphureum* in 15–20 %, *F. avenaceum* in 3–11 %, *F. culmorum* in 6 % and *F. oxysporum* in 2 % (Langerfeld 1978). In 84 samples with dry rot symptoms in seed tubers from the Netherlands, *F. sulphureum* was isolated in 52 % of the samples, *F. culmorum* in 19 % and *F. coeruleum* in 6 % (Dorenbos 1999). Gangrene was also a major storage disease of potatoes in Europe in the 1960s and 1970s, but its importance subsequently declined. It is a seed-borne rot caused by *Phoma foveata* (formerly *P. exigua* var. *foveata*); sometimes the less aggressive species *P. exigua* (formerly *P. exigua* var. *exigua*) may also be associated with gangrene-type rots. Both *Fusarium* and *Phoma* spp. enter tubers through damage points. Dry rot is favoured by temperatures of 7 °C or above and, as tubers increase in susceptibility later in storage (January to April), grading tubers in the spring encourages this disease. In contrast, gangrene is encouraged by damage at or after harvest and by storing, handling and moving tubers in cold conditions.

In the 1970s and 1980s, the application of benzimidazole (MBC) fungicides notably thiabendazole and/or the sterol demethylation inhibitor (DMI) imazalil to tubers going into store was shown to be effective in reducing dry rot and/or gangrene (Hide et al. 1969; Copeland and Logan 1975; Lashin and Henriksen 1977; Hide and Cayley 1980, 1985), but preplanting seed treatment did not consistently reduce either disease in progeny tubers (Copeland and Logan 1975; Logan et al. 1975; Carnegie et al. 1998). The application of thiabendazole was widely adopted in Europe, but during the 1980s, thiabendazole-resistant isolates of *F. sambucinum* were identified in France, Germany and the Netherlands (Langerfeld 1986, 1987; Meijers 1986; Tivoli et al. 1986), and thiabendazole-resistant *F. coeruleum* was reported in Germany (Langerfeld 1990). In the UK in 1992, resistance to thiabendazole was detected in the majority (68 %) of isolates of *F. sambucinum*, in one isolate of *F. culmorum* but not in *F. coeruleum* or *F. avenaceum* (Hide et al. 1992). In further studies in Ireland (1994–1996), Scotland (1997–2000), England and Scotland (2000–2002) and Northern Ireland (2002–2006), between 60 and 70 % of isolates of *F. sambucinum* proved resistant to thiabendazole, whereas resistance was seldom detected in *F. avenaceum* and *F. culmorum*, and no resistant isolates of the most common species *F. coeruleum* were reported (Choiseul 1996; Choiseul et al. 2007; Peters et al. 2008a; Cooke unpublished).

Thus it may be concluded that *F. sambucinum* readily develops resistance to thiabendazole and that where this species is a significant cause of dry rot, treating tubers with thiabendazole rapidly selects for *F. sambucinum* populations in which the majority of strains are resistant. Thus in North America where *F. sambucinum* is the most common cause of dry rot, resistance to thiabendazole is frequent (Gachango et al. 2012). However, thiabendazole resistance appears much more sporadic in other *Fusarium* species associated with potato dry rot, e.g. *F. avenaceum, F. culmorum* and *F. coeruleum*, and this lower incidence of resistance in these species is also supported by studies in the USA (Hanson et al. 1996). There has apparently been only a single report of thiabendazole-resistant *F. coeruleum* from Europe (Langerfeld 1990), although in the USA, thiabendazole-resistant strains of *F. solani* (probably *F. coeruleum*) were reported by Hanson et al. (1996) and Ocamb et al. (2007), but not by Gachango et al. 2012. In contrast to the situation with dry rot, there have been no reports of resistance of *P. foveata* or *P. exigua* to thiabendazole; 33 *P. foveata* and nine *P. exigua* isolates from thiabendazole-treated seed from 1976 to 1977 from Northern Ireland, 170 isolates of *P. foveata* from 1991 to 1992 from Great Britain and 20 isolates of *P. foveata* and *P. exigua* from 2002 to 2008 from Northern Ireland all proved to be thiabendazole sensitive (Logan et al. 1978; Carnegie et al. 1994b; Cooke unpublished).

The selection of thiabendazole-resistant *Fusarium* spp. (and other tuber pathogens; see below) encouraged commercial development for potato tuber disease control of fungicides with different modes of action, notably the DMI fungicide imazalil, and of products containing a combination of thiabendazole and imazalil. Hide and Cayley (1980) had shown imazalil to be effective against *Fusarium* and *Phoma* spp. and Carnegie et al.(1998) demonstrated that fungicides including imazalil and the phenylpyrrole fenpiclonil were effective in reducing dry rot and gangrene when applied to tubers in store and that a mixture of thiabendazole and imazalil was more effective than imazalil alone. Hide et al. (1992) showed that all UK *Fusarium* isolates were sensitive to imazalil. Choiseul et al. (2007) found that all Scottish *Fusarium* isolates from 1997 to 2000 were sensitive to imazalil, except for *F. avenaceum* where some isolates exhibited reduced sensitivity. Similarly, Peters et al. (2008a) reported that all isolates of *F. coeruleum, F. culmorum* and *F. sambucinum* collected from Great Britain in 2000–2002 were sensitive to imazalil, but some isolates of *F. avenaceum* were less sensitive than those of the other species. It has been suggested that increased and widespread use of imazalil seed treatments may favour *F. avenaceum* particularly in Scotland. To date, there is no evidence of selection of potato tuber pathogenic *Fusarium* or *Phoma* strains with reduced sensitivity to imazalil, although imazalil resistance has been reported in storage diseases of other crops caused by *Penicillium* spp. (e.g. citrus green mould caused by *Penicillium digitatum;* Eckert et al. 1994). However, it appears that *F. avenaceum* may be inherently less sensitive to imazalil than other *Fusarium* spp.

22.5.2 Silver Scurf and Skin Spot

Silver scurf, caused by *Helminthosporium solani*, is a surface-blemishing disease that increased in importance during the 1990s (see review by Errampalli et al. 2001) and continues to be a major problem as it reduces the quality of washed fresh-packed ware potatoes. It is mainly seed-borne but can also spread in soil. Infection generally occurs before lifting, but under high humidity, it develops and spreads in store via airborne spores produced on the tuber surface. If infection is severe, the skin may flake off and tubers become dehydrated and shrivelled. Skin spot, caused by *Polyscytalum pustulans*, is a surface-blemishing disease which is both seed- and soilborne. It can affect prepack and processing product value but is a particular problem in seed because it can infect and kill eyes, reducing emergence. Eye killing may be exacerbated if severe silver scurf is present. *P. pustulans* also causes lesions on underground stems, stolons and roots. Skin spot is favoured by wet conditions at harvest, dirty tubers and cool, damp storage. Symptoms take several months to develop and are usually not visible until January or February.

Fungicidal control of silver scurf and skin spot, as for the other fungal storage diseases, relies heavily on thiabendazole and imazalil. In the 1970s, research in the UK showed that thiabendazole applied postharvest prevented the development of silver scurf and skin spot in store and applying thiabendazole as a seed tuber treatment reduced infection of underground stems by *P. pustulans* and skin spot and silver scurf on the harvested tubers (Jellis and Taylor 1977; Cayley et al. 1979; Hide et al. 1980). Because of the risk of thiabendazole resistance, the use of alternative fungicides and fungicide mixtures to control silver scurf and skin spot was evaluated from the 1980s. Hide and Cayley (1981) and Cayley et al. (1981, 1983) found that treatment of seed tubers during storage with either thiabendazole or imazalil decreased silver scurf and skin spot in progeny tubers and that further treatment postharvest improved control. Oxley et al. (1990) showed that imazalil alone or imazalil plus thiabendazole applied to seed tubers reduced silver scurf in progeny, and Tsror and Peretz-Alon (2002) found that imazalil as a combined pre- and post-storage treatment was very effective in controlling silver scurf.

The treatment of potato tubers with thiabendazole alone was, nonetheless, widely adopted in the UK, but in the mid-1980s, severe silver scurf was observed on some UK stocks that had been treated with thiabendazole, and Hide et al. (1988) showed that thiabendazole-resistant strains of *H. solani* and *P. pustulans* were present on tubers grown in England and Scotland. Subsequently, Carnegie and Cameron (1992) reported that thiabendazole-resistant strains of *P. pustulans* were widespread in potato stocks in Scotland; they also found them on a stock from Northern Ireland and detected thiabendazole-resistant spores of *P. pustulans* in the air in potato stores. In Northern Ireland in the 1990s, the majority of isolates of *H. solani* and *P. pustulans* were found to be thiabendazole-resistant (Cooke LR unpublished). Thiabendazole-resistant *H. solani* was also reported in Sweden (Bång 1993) and is believed to be present in many commercial seed tuber stores in Europe as well as North America (Errampalli et al. 2001).

The selection of thiabendazole-resistant strains of *H. solani* was shown to be rapid by Hide and Hall (1993) and Burgess et al. (1994): in both cases after a single application of thiabendazole to seed tubers, over 50 % of isolates of *H. solani* recovered from progeny tubers were thiabendazole-resistant. In studies on the selection of thiabendazole resistance and its impact on control, Hall and Hide (1992) showed that, where seed was infected with thiabendazole-resistant strains of *H. solani* and *P. pustulans,* applying imazalil or thiabendazole plus imazalil to seed tubers decreased silver scurf and skin spot on stored progeny tubers, but thiabendazole alone applied to seed tubers or to progeny tubers after harvest did not reduce the severity of either disease. When a formulated mixture of imazalil and thiabendazole was applied to seed tubers initially infected with thiabendazole-sensitive strains of *H. solani* (Hide and Hall 1993) or *H. solani* and *P. pustulans* (Carnegie et al. 1994a, b; 1998) every year for 3–4 years, this consistently reduced silver scurf and skin spot in progeny tubers and resulted in a lower rate of selection of thiabendazole-resistant pathogen strains compared with the use of thiabendazole alone. Carnegie et al. (1994a, b) concluded that the strategy of controlling disease in seed stocks by applying thiabendazole alone annually to tubers was likely to produce populations of thiabendazole-resistant *H. solani* and *P. pustulans* within 1–3 years and that this would result in a subsequent failure to control silver scurf and skin spot if thiabendazole was applied to ware tubers. In contrast to the situation with thiabendazole and despite the reliance for many years on imazalil and mixtures of thiabendazole and imazalil for the control of tuber diseases including silver scurf and skin spot, there are no reports of reduced sensitivity to imazalil in *H. solani* or *P. pustulans* in Europe.

Other fungicides including iprodione (Oxley et al. 1990; James and Higginbotham 1994), prochloraz (Drummond et al. 1990) and the phenylpyrroles fenpiclonil (Kirk et al. 1993) and fludioxonil (Leadbitter et al. 1994) have been shown to be effective against silver scurf and skin spot, but of these only fludioxonil is now approved for the control of tuber-blemishing diseases in Europe. Fludioxonil is effective against several potato tuber pathogens including *H. solani* and *P. pustulans* (Leadbitter et al. 1994). Tsror and Peretz-Alon (2004) showed that fludioxonil applied to seed tubers in store controlled silver scurf in the subsequent crop. Fludioxonil has since been approved in several European countries as a seed tuber treatment to control silver scurf and black scurf in progeny tubers. In North America, since the late 1990s fludioxonil has been used as a seed piece treatment to control diseases including dry rot (e.g. Gachango et al. 2012), a disease for which it is not approved in Europe. However, since 2008 fludioxonil-resistant isolates of *Fusarium* spp., including *F. sambucinum*, *F. coeruleum* and *F. oxysporum*, have been reported from Canada and the USA (Peters et al. 2008b; Gachango et al. 2011), although these have not been clearly linked to reduced disease control (Gachango et al. 2011). Nonetheless, there appear to have been no published reports of resistance to fludioxonil in *H. solani,* although in Israel isolates of *H. solani* able to grow on fludioxonil concentrations of up to 100 mg/l in vitro have been obtained from potato tubers (Tsror personal communication).

22.5.3 Black Dot

Colletotrichum coccodes, which is both seed- and soilborne, causes a dark brown-grey blemish over the tuber surface similar in appearance to silver scurf but with more irregularly shaped lesions with less well-defined edges on which the eponymous microsclerotia form. In potato-growing regions with hot summers, infection of stems by *C. coccodes* may result in early dying with symptoms similar to *Verticillium* wilt, but this is not normally a problem in NW Europe, where the main impact of this disease is on the quality of washed ware tubers. The incidence of black dot has apparently increased over the past 20 years: in part this may be because it is not affected by the fungicides effective against *H. solani*, and where silver scurf is controlled, levels of black dot increase if soilborne inoculum is present.

The QoI fungicides azoxystrobin and fluoxastrobin reduce black dot. Approvals differ between European countries: in the UK azoxystrobin is the only fungicide with a label claim for reduction in black dot, but in the Netherlands azoxystrobin and fluoxastrobin (which are approved for control of *Rhizoctonia*) both mention reduction of black dot as a side effect. In the UK, Brierley et al. (2015) showed that in-furrow application of azoxystrobin consistently reduced black dot on progeny tubers but stated that its use can only be justified where inoculum is present. There have been no reports of resistance to azoxystrobin in *C. coccodes*.

22.5.4 Other Potato Tuber Diseases

Several other fungi and oomycetes cause potato tuber rots including *Geotrichum candidum* (formerly *Oospora lactis*), cause of rubbery rot; *Phytophthora erythroseptica,* cause of pink rot; and *Pythium* spp., which cause watery wound rot or leak. However, because these diseases are relatively minor and sporadic in occurrence in NW Europe, it is not economic to register fungicides for their control, and consequently none are approved. In North America, products containing the phenylamides metalaxyl and metalaxyl-M have been used for the control of pink rot, and this has selected for phenylamide-resistant strains of *P. erythroseptica* (e.g. Lambert and Salas 1994). The use of metalaxyl (applied co-formulated with mancozeb) for pink rot control was evaluated in Ireland (O'Sullivan and Dowley 1998), but there was no significant reduction in rotting. In 2005, seven isolates of *P. erythroseptica* obtained from tubers of three cultivars with pink rot submitted to the Department of Agriculture and Rural Development in Northern Ireland were tested for metalaxyl resistance as it was considered that they might have been exposed to selection by the use of phenylamide formulations for control of late blight: all proved sensitive (Cooke unpublished).

22.5.5 Black Scurf and Stem Canker

The seed- and soilborne fungus *Rhizoctonia solani* causes black scurf when its sclerotia develop on tuber surfaces and is associated with tuber malformations, which detract from the crop value. Black scurf is also undesirable in seed stocks, since

R. solani attacks developing sprouts, stolons and underground stems, causing brown cankers which may result in leaf rolling, wilting, formation of aerial tubers, stem killing, stolon nipping and reduction of crop uniformity. *R. solani* comprises 13 different anastomosis groups (AGs) which are pathogenic to different plant species and differ in their fungicide sensitivity. Isolates of *R. solani* that infect potatoes are predominantly AG 3; this was confirmed in studies in the UK (Woodhall et al. 2007), Finland (Lehtonen et al. 2008), France (Campion et al. 2003; Fiers et al. 2011) and Poland (Woodhall et al. 2013); see also the review by Tsror (2010). Ritchie (2006) showed that UK potato AG 3 isolates of *R. solani* were significantly more pathogenic than AG 2-1 and AG 5 isolates.

Fungicides for the control of *R. solani* are either dusted or sprayed onto the seed potato tuber before or at planting (if seed-borne inoculum is the major problem) or applied in-furrow at planting if soilborne inoculum is important (Little and Cooke 1988). Most of the fungicides approved for this purpose are specific to *R. solani* and have little effect on other potato tuber pathogens. In the early 1980s, tolclofos-methyl was developed and shown to control potato stem canker and black scurf (e.g. Harris et al. 1988). This was soon followed by pencycuron, which also proved very effective against stem canker and black scurf (Adam and Malcom 1988; Little and Cooke 1988). Iprodione was used to control *R. solani* on potatoes during the 1990s and 2000s (James and Higginbotham 1994) but is no longer approved for application to potatoes in NW Europe. In the early 2000s, flutolanil was introduced for preplanting application to seed tubers for the control of stem canker and black scurf; trials in Northern Ireland confirmed its effectiveness (Little and Cooke 2005), while azoxystrobin was developed for in-furrow application at planting to reduce soilborne infection by *R. solani* and *C. coccodes.* Wale et al. (2004) reported that azoxystrobin was effective against both seed- and soilborne inoculum of *R. solani* in Scotland; reduction in stem and stolon canker was variable, and where there was a high soil inoculum pressure, control was limited, but black scurf was generally well controlled. Hilton et al. (2004) found that azoxystrobin reduced stem canker, but not stolon canker.

Van Bruggen and Arneson (1984) reported resistance to tolclofos-methyl in *R. solani*, but this was obtained by in vitro exposure of the pathogen, and there have been no subsequent reports of resistance in vivo to this fungicide. Kataria et al. (1991a, 1991b) showed variation in sensitivity to fungicides between *Rhizoctonia* spp., between *R. solani* AGs and between isolates within AGs. These authors found that tolclofos-methyl was very active in vitro against all *Rhizoctonia* spp. and isolates but that there was variation in their response to pencycuron, although pencycuron was very active against most AG 3 isolates. Campion et al. (2003) tested the in vitro sensitivities to flutolanil, iprodione and pencycuron of French isolates of *R. solani* from potato and found that all were highly sensitive to flutolanil, but while AG 3 and AG 2-1 isolates were highly sensitive to pencycuron, AG 5 isolates were only moderately sensitive to this fungicide. Similarly Woodhall and Jenkinson (2002) studied 14 *R. solani* isolates obtained from potato stems and tubers mainly from England and Scotland and reported that those from AG 2-1 and AG 3 were all sensitive to pencycuron in vitro, whereas an AG 5 isolate and an AG 8 isolate were insensitive. These studies highlight the importance of selecting the appropriate fungicide for the control of *Rhizoctonia* spp. on different hosts and of confirming the host from which isolates originate when testing for fungicide sensitivity. Crucially, there have been no reports of resistance in *R. solani* AG 3 to any fungicides used for the control of this pathogen on potatoes in Europe. Although there was a recent report of resistance to azoxystrobin in *R. solani* AG 3 isolates from Tunisia (Djebali et al. 2014), this was based on in vitro testing and was not reflected in reduced efficacy in the field; the apparent insensitivity in vitro may have been due to induction of alternative oxidase in agar culture.

22.6 Management of Fungicide Resistance in Potato Tuber Pathogens

Of crucial importance to manage fungicide resistance in tuber pathogens is an appreciation of the clonal nature of the crop and of the sources of pathogen inoculum. For diseases that are tuber-borne, if fungicides with the same mode of action are applied to potato tubers at each generation, then this produces cycles of repeated selection on the same pathogen population. Where the fungicide belongs to a group with a high risk of resistance, e.g. thiabendazole, this can rapidly lead to a build-up of a resistant pathogen population. Thus it is imperative that fungicide treatment histories of a stock throughout the production cycle are made available to growers so that effective anti-resistance strategies can be adopted at all stages of both seed and ware production (Anonymous 2010), thus allowing repeated use of the same fungicide from year to year throughout a multiplication programme to be minimised.

To date, among potato tuber pathogens in Northern Europe, fungicide resistance has been a significant problem only in the case of thiabendazole resistance in *F. sambucinum, H. solani* and *P. pustulans*. This probably reflects the rather restricted range of fungicides approved for the control of potato tuber diseases and perhaps their limited use, rather than any inherent low risk. Of these pathogens, *H. solani* is the most important, since *F. sambucinum* is not the major cause of dry rot in Northern Europe and skin spot is a relatively minor disease. It is unclear why thiabendazole resistance is infrequent in other *Fusarium* spp., particularly *F. coeruleum*.

Adopting an integrated control strategy helps to reduce the risk of resistance (Hide 1994), and for the potato crop, a key element of this is the use of healthy seed, since tuber transmission of pathogens increases disease and results in repeated cycles of selection. Soilborne inoculum is generally less exposed to intensive selection. Second, the use of long rotations and selection of appropriate sites to minimise build-up of inoculum in soil are important. Other measures such as varietal choice, manipulation of planting and lifting dates, dry curing, adoption of good store hygiene and control of temperature and humidity in store all help to reduce disease and thus the amount of inoculum exposed to fungicidal selection.

In Europe, as previously noted, very few fungicides are approved for application to potato tubers, particularly ware tubers destined for human consumption, notably thiabendazole alone or imazalil alone. In line with EU pesticide registration requirements to put in place measures to manage fungicide resistance, the occurrence of benzimidazole-resistant strains of silver scurf, skin spot and dry rot is reported on labels of products containing thiabendazole (e.g. Storite Excel in the UK and Tecto 500 SC in the Netherlands). Labels include resistance management guidance such as 'To reduce the chance of such strains increasing benzimidazole based products should not be used more than once in the cropping cycle' and 'Thiabendazole should only be used on ware potatoes where there is a likely risk of disease during the storage period and in combination with good storage hygiene and maintenance' (Storite Excel UK product label) and 'Do not treat seed lots in which decreased efficacy of thiabendazole has been observed; only treat seed lots with destination ware production or export' (Tecto 500 SC Netherlands product label).

In the UK, the Fungicide Resistance Action Group, FRAG-UK, has developed recommendations to manage fungicide resistance in potato tuber diseases (Anonymous 2010). FRAG-UK is one of four Resistance Action Groups, which are informal, UK-based groups including experts from the Crop Protection Association member companies, other representatives from the agrochemical industry, a range of independent organisations, including public sector research institutes, and the Chemicals Regulation Directorate (CRD). The groups are independent of the CRD and work to produce guidance on pesticide resistance issues. FRAG-UK's guidelines are intended to maintain the effectiveness of thiabendazole for the treatment of ware crops, avoiding build-up of thiabendazole-resistant *H. solani* by advocating the use of imazalil on seed crops intended for ware prepack production so that the use of thiabendazole alone on seed crops is avoided. The use of mixtures of fungicides with different modes of action, notably thiabendazole plus imazalil, contributes to slowing the build-up of resistance, as evidenced above, and FRAG-UK advises that this mixture may be used on seed of processing varieties. In general it is advised that fungicides should only be used routinely if one or more of the following apply:

- The variety grown is highly susceptible to a specific disease.
- There is a history of persistent disease on the farm.
- Significant levels of the disease were present on mother tubers.
- The presence of a disease would have a major effect on marketability.
- Treatment constitutes part of a contract.

These FRAG-UK recommendations have not been updated since fludioxonil was approved for application to potatoes, and this will need to be considered in the future. Although the Netherlands and Ireland have Fungicide Resistance Action Groups, these are not actively involved in resistance management of potato tuber pathogens.

There are no specific recommendations for resistance management in *R. solani* and *C. coccodes*. Although in some countries azoxystrobin and fluoxastrobin may be applied in-furrow at planting to control stem canker, black scurf and black dot, and there are limitations on the number of applications of other QoI fungicides (pyraclostrobin, famoxadone and fenamidone) which may be made to potato foliage for control of late blight and early blight to reduce the risk of selection of QoI-resistant *P. infestans* and *A. solani*, soil-applied azoxystrobin and fluoxastrobin are not taken into account when considering the total number of permitted applications of foliar QoI fungicides (Anonymous 2010). Brierley et al. (2015), however, stated that application of azoxystrobin and fluoxastrobin to control black dot can only be justified where inoculum is present and suggested that a soil test should be used to make informed decisions on the use of this fungicide: in practice this would assist in fungicide resistance management by reducing inappropriate use of azoxystrobin and fluoxastrobin.

References

Adam NM, Malcom AJ (1988) Control of *Rhizoctonia solani* in potatoes with pencycuron. In: Proceedings of Brighton Crop Protection Conference – Pests Disease 1988, pp 959–964

Anonymous (2009) Potato late blight: guidelines for managing fungicide resistance. FRAG-UK, 8 pp

Anonymous (2010) Fungicide resistance management in potatoes: tuber diseases, November 2010. FRAG-UK. http://www.pesticides.gov.uk/Resources/CRD/Migrated-Resources/Documents/F/FRAG_UK_potato_tuber_diseases_November_2010.pdf. Accessed 3 Sept 2014

Bång U (1993) Resistance to thiabendazole in Swedish isolates of *Helminthosporium solani*. Potato Res 36:76

Bardsley RA, Shattock RC, Day JP (1998) Studies comparing the sensitivity of European and USA isolates of *Phytophthora infestans* to propamocarb hydrochloride. In: Proceedings of Brighton Crop Protection Conference – Pests Disease 1988, pp 523–528

Blixt E, Andersson B (2010) Occurrence of *Alternaria solani* in Sweden and its sensitivity to strobilurins. PPO Special Rep nr 14:161–163

Brierley JL, Hilton AJ, Wale SJ, Peters JC, Gladders P, Bradshaw NJ, Ritchie F, MacKenzie K, Lees AK (2015) Factors affecting the development and control of black dot on potato tubers. Plant Pathol 64:167–177

Burgess PJ, Forbes FJ, Oxley SJP, Bowen SA (1994) Resistance of *Helminthosporium solani* to thiabendazole in relation to different strategies of fungicide use during seed potato production. 1994 BCPC Monograph No. 60: Fungicide Resistance, pp 197–206

Campion C, Chatot C, Perraton B, Andrivon D (2003) Anastomosis groups, pathogenicity and sensitivity to fungicides of *Rhizoctonia solani* isolates collected on potato crops in France. Eur J Plant Pathol 109:983–992

Carnegie SF, Cameron AM (1992) Resistance to thiabendazole in isolates of *Polyscytalum pustulans* (skin spot) and *Fusarium solani* var. *coeruleum* (dry rot) in Scotland. Plant Pathol 41:606–610

Carnegie SF, Cameron AM, Hide GA, Hall SM (1994a) The occurrence of thiabendazole-resistant isolates of *Polyscytalum pustulans* and *Helminthosporium solani* on seed potato tubers in relation to fungicide treatment and storage. Plant Pathol 43:961–971

Carnegie SF, Hide GA, Black SM, Cameron AM, Haddon PJ (1994b) Improving tuber health of seed and ware crops 1. Tuber health of Scottish seed potato crops. Potato Marketing Board Project 807/127 Report, 38 pp

Carnegie SF, Cameron AM, Lindsay DA, Sharp E, Nevison IM (1998) The effect of treating seed potato tubers with benzimidazole, imidazole and phenylpyrrole fungicides on the control of rot and skin blemish diseases. Ann Appl Biol 133:343–363

Cayley GR, Hide GA, Lord KA, Austin DJ, Davies AR (1979) Control of potato storage diseases with formulations of thiabendazole. Potato Res 22:177–190

Cayley GR, Hide GA, Read PJ, Dunne Y (1983) Treatment of potato seed and ware tubers with imazalil and thiabendazole for control of silver scurf and other storage diseases. Potato Res 26:163–173

Cayley GR, Hide GA, Tillotson Y (1981) The determination of imazalil on potatoes and its use in controlling potato storage diseases. Pestic Sci 12:103–109

César V, Labbe V, Laguesse L, Rolot J-L (2012) Phenotypic characteristics of Belgian populations of *Phytophthora infestans* (2005–2010). PPO Special Rep nr 15:213–214

Choiseul J, Allen A, Carnegie SF (2007) Fungi causing dry tuber rots of seed potatoes in storage in Scotland. Potato Res 49:241–253

Choiseul JW (1996) The characterisation, pathogenicity and control of *Fusarium* spp. that cause dry rot of potato. PhD thesis, National University of Ireland, 223 pp [unpublished]

Cohen Y, Rubin E, Hadad T, Gottlieb D, Sierotzki H, Gisi U (2007) Sensitivity of *Phytophthora infestans* to mandipropamid and the effect of enforced selection pressure in the field. Plant Pathol 56:836–842

Cooke LR (1981) Resistance to metalaxyl in *Phytophthora infestans* in Northern Ireland. In: Proceedings of Brighton Crop Protection Conference – Pests Disease, pp 641–649

Cooke LR, Carlisle DJ, Wilson DG, McCall RD (2002) Sensitivity of European isolates of *Phytophthora infestans* to zoxamide (RH7281). PPO Special Rep nr 8:39–40

Cooke LR, Edmonds J (2014) Zoxamide sensitivity in European *Phytophthora infestans* isolates, 2003–2012. PPO Special Rep nr 16:285–293

Cooke LR, Kildea S, Mehenni-Ciz J, Quinn L, Little G, Hutton F, Perez FM, Deahl KL, Griffin D (2012) Ongoing changes in the Irish potato late blight population. PPO Special Rep nr 15:75–80

Cooke LR, Little G (2006) Twenty-five years of phenylamide resistance management in potato late blight in Northern Ireland. Aspect Appl Biol 78:73–82

Cooke LR, Little G, Wilson DG (1998) Sensitivity of *Phytophthora infestans* to fluazinam and its use in potato blight control in Northern Ireland. In: Proceedings of Brighton Crop Protection Conference – Pests Disease 1998, pp 517–522

Cooke LR, Schepers HTAM, Hermansen A, Bain RA, Bradshaw NJ, Shaw DS, Evenhuis A, Kessel GJT, Wander JGN, Andersson B, Hansen JG, Hannukkala A, Nærstad R, Nielsen BJ (2011) Epidemiology and integrated control of potato late blight in Europe. Potato Res 54:183–222

Copeland RB, Logan C (1975) Control of tuber diseases, especially gangrene, with benomyl, thiabendazole and other fungicides. Potato Res 18:179–188

Corbière R, Magalon H, Boulard F, Andrivon D (2010) Study of invasive French populations (2006–2008) of *Phytophthora infestans*, the Oomycete causing potato late blight. PPO Special Rep nr 14:289–290

Davidse LC (1985) Resistance to acylalanines in *Phytophthora infestans* in the Netherlands. EPPO Bull 15:403–409

Davidse LC, Henken J, Van Dalen A, Jespers ABK, Mantel BC (1989) Nine years of practical experience with phenylamide resistance in *Phytophthora infestans* in the Netherlands. Neth J Plant Pathol 95(Suppl 1):197–213

Davidse LC, Looijen D, Turkensteen LJ, Van der Wal D (1981) Occurrence of metalaxyl-resistant strains of *Phytophthora infestans* in Dutch potato fields. Neth J Plant Pathol 87:65–68

Dereviagina MK, Elansky SN, Diakov YT (1999) Resistance of *Phytophthora infestans* to the dimethomorph fungicide. Mikologiya Fitopatol 33:208–213

Djebali N, Elkahoui S, Taamalli W, Hessini K, Tarhouni B, Mrabet M (2014) Tunisian *Rhizoctonia solani* AG3 strains affect potato shoot macronutrients content, infect faba bean plants and show in vitro resistance to azoxystrobin. Austr Plant Pathol 43:347–358

Dorenbos J (1999) Droogrotonderzoek NAK. Emmeloord, Report NAK (in Dutch)

Dowley LJ (1986) Factors affecting the survival of metalaxyl-resistant strains of *Phytophthora infestans* (Mont.) de Bary in Ireland. Potato Res 30:473–475

Dowley LJ, O'Sullivan E (1981) Metalaxyl-resistant strains of *Phytophthora infestans* (Mont.) de Bary in Ireland. Potato Res 24:417–421

Dowley LJ, O'Sullivan E (1985) Monitoring metalaxyl resistance in populations of *Phytophthora infestans*. Potato Res 28:531–534

Drummond DJ, Bush MN, Wale SJ, Lang RL, Oxley S (1990) Development of seed tuber treatments for the control of stem and skin diseases of potatoes. Proc Crop Prot North Br : 343–389

Eckert JW, Sievert JR, Ratnayake M (1994) Reduction of imazalil effectiveness against citrus green mold in California packinghouses by resistant biotypes of *Penicillium digitatum*. Plant Dis 78:971–974

Edin E, Andersson B (2014) The early blight situation in Sweden-species abundance and strobilurin sensitivity. PPO Special Rep nr 16:83–84

Errampalli D, Saunders JM, Holley JD (2001) Emergence of silver scurf (*Helminthosporium solani*) as an economically important disease of potato. Plant Pathol 50:141–153

Fiers M, Edel-Hermann V, Heraud C, Gautheron N, Chatot C, Le Hingrat Y, Bouchek-Mechiche K, Steinberg C (2011) Genetic diversity of *Rhizoctonia solani* associated with potato tubers in France. Mycologia 103:1230–1244

Gachango E, Hanson LE, Rojas A, Hao JJ, Kirk WW (2012) *Fusarium* spp. causing dry rot of seed potato tubers in Michigan and their sensitivity to fungicides. Plant Dis 96:1767–1774

Gachango E, Kirk W, Hanson L, Rojas A, Tumbalam P, Shetty K (2011) First report of in vitro fludioxonil-resistant isolates of *Fusarium* spp. causing potato dry rot in Michigan. Plant Dis 95:228

Gisi U, Cohen Y (1996) Resistance to phenylamide fungicides: a case study with *Phytophthora infestans* involving mating type and race structure. Ann Rev Phytpathol 34:549–572

Gisi U, Walder F, Resheat-Eini Z, Edel D, Sierotzki H (2011) Changes of genotype, sensitivity and aggressiveness in *Phytophthora infestans* isolates collected in European countries in 1997, 2006 and 2007. J Phytopathol 159:223–232

Hall SM, Hide GA (1992) Fungicide treatment of seed tubers infected with thiabendazole-resistant *Helminthosporium solani* and *Polyscytalum pustulans* for controlling silver scurf and skin spot on stored progeny tubers. Potato Res 35:143–147

Hamlen RA, Power RJ (1998) Distribution of sensitivity responses to cymoxanil within global populations of *Phytophthora infestans*. Pestic Sci 53:101–103

Hannukkala AO (2014) Changes in epidemiology and population structure of *P. infestans* in Finland 1847–2011. PPO Special Rep nr 16:153–158

Hanson LE, Schwager SJ, Loria R (1996) Sensitivity to thiabendazole in *Fusarium* species associated with dry rot of potato. Phytopathology 86:378–384

Harris RI, Greig RJ, Atkinson RJ (1988) Potato tuber disease control by seed treatment with tolclofos-methyl/prochloraz manganese chloride mixtures. In: Proceedings of Brighton Crop Protection Conference – Pests Disease 1988, pp 901–906

Haverkort AJ, Boonekamp PM, Hutten R, Jacobsen E, Lotz LAP, Kessel GJT, Visser RGF, van der Vossen EAG (2008) Societal costs of late blight in potato and prospects of durable resistance through cisgenic modification. Potato Res 51:47–57

Hermann D, Gisi U (2012) Fungicide resistance in Oomycetes with special reference to *Phytophthora infestans* and phenylamides. In: Thind TS (ed) Fungicide resistance in crop protection: risk and management, pp 133–140. CABI

Hide GA (1994) Fungus diseases of potato tubers and strategies for their control. Brighton Crop Prot Conf – Pests Dis 1994, pp 359–366

Hide GA, Cayley GR (1980) Tests of fungicides for controlling gangrene (*Phoma exigua* var. *foveata*) and dry rot (*Fusarium solani* var. *coeruleum* and *F. sulphureum*) on potatoes during storage. Potato Res 23:395–403

Hide GA, Cayley GR (1981) Control of potato diseases during growth and in store by fungicide seed treatment. Tests of Agrochemicals and Cultivars 2 (a supplement to Ann appl Biol 97), pp 24–25

Hide GA, Cayley GR (1985) Effects of delaying fungicide treatment of wounded potato tubers on the incidence of *Fusarium* dry rot in store. Ann Appl Biol 107:429–438

Hide GA, Cayley GR, Read PJ, Fraser JH (1980) Treatment of seed and ware potato tubers with thiabendazole for control of storage diseases. Ann Appl Biol 96:119–131

Hide GA, Hall SM (1993) Development of resistance to thiabendazole in *Helminthosporium solani* (silver scurf) as a result of potato seed tuber treatment. Plant Pathol 42:707–714

Hide GA, Hall SM, Boorer KJ (1988) Resistance to thiabendazole in isolates of *Helminthosporium solani*, the cause of silver scurf disease of potatoes. Plant Pathol 37:377–380

Hide GA, Hirst JM, Griffith RL (1969) Control of potato tuber diseases with systemic fungicides. In: Proceedings of 5th British insect fungicide conference, pp 301–314

Hide GA, Read PJ, Hall SM (1992) Resistance to thiabendazole in *Fusarium* species isolated from potato tubers affected by dry rot. Plant Pathol 41:745–748

Hilton AJ, Griffin-Walker RD, Matheson SJ, Blackwood JM, Wale SJ (2004) Epidemiology and control of soil-borne *Rhizoctonia solani*, the cause of stem and stolon canker of potato. Proc Crop Prot North Br :255–260

Holmes SJ (1984) The incidence of metalaxyl-insensitivity in late blight of potatoes in the West of Scotland. In: Proceedings of Brighton Crop Protection Conference – Pests and Diseases, pp 108–113

James DA, Higginbotham S (1994). The use of iprodione and imazalil for disease control of seed potato tubers. 1994 BCPC Monograph No. 77: Seed treatment: Progress and Prospects, pp 257–262

Jellis GJ, Taylor GS (1977) Control of silver scurf (*Helminthosporium solani*) disease of potato with benomyl and thiabendazole. Ann Appl Biol 86:59–67

Kataria HR, Hugelshofer U, Gisi U (1991a) Sensitivity of *Rhizoctonia* species to different fungicides. Plant Pathol 40:203–211

Kataria HR, Verma PR, Gisi U (1991b) Variability in the sensitivity of *Rhizoctonia solani* anastomosis groups to fungicides. J Phytopathol 133:121–133

Kessel G, Veloso S, Förch M, Latorse M-P (2009) New Phytophthora populations: a shift from indirect to direct sporangial germination? PPO Special Rep nr 13:171–176

Kessel GJT, Förch MG, van den Bosch T, van Bekkum PJ, van der Zande JC, Evenhuis A (2011) Gevoeligheid van twee cohorten *P. infestans*-isolaten voor Shirlan (in Dutch). http://edepot.wur.nl/164664

Kildea S, Cooke LR, Quinn L, Little G, Armstrong C, Hutton F, Dowley LJ, Griffin D (2010) Changes within the Irish potato late blight population. PPO Special Rep nr 14:147–150

Kildea S, Mehenni-Ciz J, Griffin D, Cooke LR (2014) Sensitivity of Irish *Phytophthora infestans* to the CAA fungicide mandipropamid. PPO Special Rep nr 16:221–222

Kirk WW, Leadbeater AJ, Burgess PJ, Wale SJ, Oxley SJP (1993) Control of potato stem and skin diseases with fenpiclonil seed treatments in Northern Britain. In: Proceedings of Crop Protection in Northern Britain 1993, pp 313–318

Lambert DH, Salas B (1994) Metalaxyl insensitivity of *Phytophthora erythroseptica* isolates causing pink rot of potato in Maine. Plant Dis 78:1010

Langerfeld E (1978) *Fusarium coeruleum* (Lib.) Sacc. Als Ursache von Lagerfäulen an Kartoffelknollen. Mitteilungen aus der Biologischen Bundesanstalt für Land- und Forstwirtschaft (in German), Berlin-Dahlem, p 184

Langerfeld E (1986) Thiabendazole resistance in *Fusarium sulphureum*. Nachrichtenblatt des Deutschen Pfianzenschutzdienstes 38:165–168

Langerfeld E (1987) Thiabendazol-Resistenz bei *Fusarium sulphureum* (in German). Potato Res 30:357

Langerfeld E (1990) Thiabendazole resistance in *Fusarium coeruleum*. Nachrichtenblatt des Deutschen Pfianzenschutzdienstes 42:79

Lashin SM, Henriksen JB (1977) Control of gangrene and Fusarium dry rot on potato tubers with thiabendazole. Særtryk af Tidsskrift for Planteavl 81:310–314

Leadbitter NJ, Nyfeler R, Elmsheuser H (1994) The phenylpyrroles: the history of their development at Ciba. 1994 BCPC Monograph No. 77: Seed treatment: Progress and Prospects, pp 129–134

Lehtinen A, Hannukkala A, Rantanen T, Jauhiainen L (2007) Phenotypic and genetic variation in Finnish potato late blight populations, 1997–2000. Plant Pathol 56:480–491

Lehtonen MJ, Ahvenniemi P, Wilson PS, German-Kinnari, M Valkonen JPT (2008) Biological diversity of *Rhizoctonia solani* (AG-3) in a Northern potato-cultivation environment in Finland. Plant Pathol 57:141:151

Leiminger JH, Adolf B, Hausladen H (2014) Occurrence of the F129L mutation in *Alternaria solani* populations in Germany in response to QoI application, and its effect on sensitivity. Plant Pathol 63:640–650

Little G, Cooke LR (1988) The effect of fungicides applied to seed potato tubers on *Rhizoctonia solani* stem canker and black scurf. Tests of Agrochemicals and Cultivars 9. Ann Appl Biol 112 (Suppl):34–35

Little G, Cooke LR (2005) Evaluation of fungicide treatments against potato black scurf and silver scurf. Tests Agrochem Cultiv 26:6–7

Logan C, Copeland RB, Little G (1975) Potato gangrene control by ultra low volume sprays of thiabendazole. Ann Appl Biol 80:199–204

Logan C, Copeland RB, Little WG (1978) The efficacy of selected chemicals applied as mists for control of potato gangrene during storage. Rec Agric Res (Dept Agric North Ireland) 26:17–20

Meijers CP (1986) What can we do this year on the storage diseases of seed potatoes? Aardappelwereld July, 13–15

Nærstad R, Sharma SS, Elameen A, Hermansen A, Brurberg MB (2014) Potato late blight forecasting and initial inoculum sources in Norway. PPO Special Rep nr 16:41–49

Nassar H, Aav A, Hansen M, Tähtjärv T, Runno-Paurson E (2014) Distribution of mating types and resistance to metalaxyl of *Phytophthora infestans* in southern Estonia. PPO Special Rep nr 16:235–236

Nielsen B (2014) Efficacy of fluazinam for control of potato late blight (*Phytophthora infestans*) in Danish field trials. PPO Special Rep nr 16:113–116

O'Sullivan E, Dowley LJ (1998) Epidemiology and control of pink rot of potatoes. Teagasc Report, 5 pages

Ocamb CM, Hamm PB, Johnson DA (2007) Benzimidazole resistance of *Fusarium* species recovered from potatoes with dry rot from storages located in the Columbia basin of Oregon and Washington. Am J Potato Res 84:169–177

Oxley SJP, Lang RW, Wale S (1990) Control of silver scurf and skin spot by fungicide seed treatment. In: Proceedings of Crop Protection in Northern Britain 1990, pp 351–357

Peters JC, Lees AK, Cullen DW, Sullivan L, Stroud GP, Cunnington AC (2008a) Characterization of *Fusarium* spp. responsible for causing dry rot of potato in Great Britain. Plant Pathol 57:262–271

Peters RD, Platt HW, Drake KA, Coffin RH, Moorehead S, Clark MM, Al-Mughrabi KI, Howard RJ (2008b) First report of fludioxonil-resistant isolates of *Fusarium* spp. causing potato seed-piece decay. Plant Dis 92:172

Power RJ, Hamlen RA, Morehart AL (1995) Variation in sensitivity of *Phytophthora infestans* field isolates to cymoxanil, chlorothalonil and metalaxyl. In: Dowley LJ, Bannon E, Cooke LR, Keane T, O'Sullivan E (eds) Phytophthora 150. Boole Press, Dublin, pp 154–159

Räder T, Gisi U (2010) *Phytophthora infestans*: Veränderungen von Genotypen, Sensitivität und Aggressivität in Europa in den letzten 10 Jahren – Konsequenzen für die Bekämpfung (in German). Kartoffelbau 61:116–120

Ritchie F (2006) Aspects of the biology, epidemiology and control of *Rhizoctonia solani* (Kühn) on potato. PhD thesis, University of Glasgow, 285 pp

Schepers HTAM, Kessel GJT, Evenhuis A, Förch MG, Van den Bosch GBM, Topper C, Strypstein C, Meier-Runge F, Sierotzki H, Scalliet G (2013) Reduced efficacy of fluazinam against some Green 33-isolates of *Phytophthora infestans* in Europe. Abstracts 17th International Reinhardsbrunn Symposium, Friedrichroda, p 102

Schmitt J, Kleinhenz B (2012) Distribution of mating types and resistance to metalaxyl of *Phytophthora infestans* in Germany. PPO Special Rep nr 15:215–222

Schulte M (2011) Fungizidresistenz gegen Phytophthora – Wie sind die Zusammenhänge? (in German) Kartoffelbau 62:28–31

Tivoli B, Deltour A, Molet D, Bedin P, Jouan B (1986) Isolation of thiabendazole-resistant strains of *Fusarium roseum* var. *sambucinum* from potato tubers. Agronomie 6:219–224

Tsror (Lahkim) L, Peretz-Alon I (2002) Reduction of silver scurf on potatoes by pre- and post-storage treatment of seed tubers with imazalil. Am J Potato Res 79:33–37

Tsror (Lahkim) L, Peretz-Alon I (2004) Control of silver scurf on potato by dusting or spraying seed tubers with fungicides before planting. Am J Potato Res 81:291–294

Tsror L (2010) Biology, epidemiology and management of *Rhizoctonia solani* on potato. J Phytopathol 158:649–658

Van Bruggen AHC, Arneson PA (1984) Resistance in *Rhizoctonia solani* to tolclofos-methyl. Neth J Plant Pathol 90:95–106

Wale SJ, Hilton A, Tait M, Mitchell T (2004) The control of diseases of potatoes caused by soil-borne *Rhizoctonia solani* using soil application of azoxystrobin. In: Proceeding of Crop Protection in Northern Britain 2004, pp 291–296

Woodhall JW, Lees AK, Edwards SG, Jenkinson P (2007) Characterization of *Rhizoctonia solani* from potato in Great Britain. Plant Pathol 56:286–295

Woodhall JW, Lutomirska B, Peters JC, Wharton PS (2013) *Rhizoctonia solani* anastomosis group 3 is predominant in potato tubers in Poland. Plant Dis 97:1245

Woodhall JW, Jenkinson P (2002) Differences between *Rhizoctonia solani* isolates from potato crops. In: Proceedings of BCPC Conference – Pests and Disease 2002, pp 757–760

Ziogas BN, Markoglou AN, Theodosiou DI, Anagnostou A, Boutopoulou S (2006) A high multi-drug resistance to chemically unrelated oomycete fungicides in *Phytophthora infestans*. Eur J Plant Pathol 115:283–292

Chapter 23
Sugar Beet Diseases: *Cercospora* Leaf Spot

Mohamed F.R. Khan

Abstract Sugar beet (*Beta vulgaris* L.) was first commercially processed in 1802 in Cunern, now Germany, and since that time has become a provider of 25 % of the world's sucrose requirement. *Cercospora beticola* causes *Cercospora* leaf spot which is one of the most damaging foliar fungal pathogens of sugar beet especially in warm and humid growing areas. Crop rotation, incorporation of infected plant debris, use of partially resistant varieties, and timely application of fungicides are combined to manage *C. beticola*. Fungicides are critical for disease control in areas where the pathogen is endemic. Most classes of fungicides have been used to control *C. beticola*; however, prolonged use and sometimes overuse have resulted in the pathogen developing resistance and rendering the fungicide ineffective in the field. In Minnesota and North Dakota, the United States, the use of different modes of action in a rotation program and/or fungicide mixtures comprising different modes of action in rotation always with different chemistries has contributed to successful control of the disease over the past 15 years, a reduction in the number of fungicide applications, and savings in fungicide use of $14 million annually compared to the last epidemic in 1998.

Keywords *Cercospora beticola* • Fungicide resistance • Sugar beet

23.1 Introduction

Sugar beet (*Beta vulgaris* L.) was first grown at Cunern, Silesia, in 1801 and several other European countries including France, Austria, Russia, and Denmark starting around 1811. In North America, the United States was successful at growing and processing sugar beet at Alvarado, California, in 1870, and growers in Quebec, Canada, started sugar beet production in 1881. In South America, sugar beet was produced in Uruguay until the mid-1990s and continues today in Chile.

M.F.R. Khan (✉)
Plant Pathology Department, North Dakota State University, Fargo, ND, USA 58108

University of Minnesota, St. Paul, MN, USA
e-mail: Mohamed.khan@ndsu.edu

H. Ishii, D.W. Hollomon (eds.), *Fungicide Resistance in Plant Pathogens*,
DOI 10.1007/978-4-431-55642-8_23

Sugar beet was also successfully produced in Asian countries including Japan, Turkey, and China and in the African countries of Morocco, Algeria, Tunisia, and Egypt (Francis 2006).

In the early history of sugar beet production in the United States, sugar companies such as the American Beet Sugar Company provided the sugar beet seeds to be used by growers. In areas such as the Red River Valley of North Dakota and Minnesota, varieties such as American 2 Hybrid B which was moderately resistant to the fungus *C. beticola* were widely cultivated but not as high yielding as varieties developed in Europe. In 1973, growers purchased American Crystal Sugar Company and changed their seed policy to allow the use of the higher yielding European varieties which were also more susceptible to *C. beticola* and required fungicide use to optimize yields (Windels et al. 1998).

Cercospora beticola was first described by Saccardo in 1886, and in a relatively short time, the disease was reported as a serious problem in Europe and the United States. *Cercospora*-resistant materials were developed in Italy and the United States in the early 1900s (Munerati, Skuderna), and these sources of resistance are still being used today. *Cercospora beticola* Sacc. causes *Cercospora* leaf spot (CLS) of sugar beets in production areas worldwide when grown in areas with warm and humid conditions (Ruppel 1986; Kerr and Weiss 1990; Holtschulte 2000). About 36 % of sugar beet production area worldwide has moderate to high incidences of *Cercospora* leaf spot (Holtschulte 2000). Symptoms typically appear after row closure initially on older leaves close to the ground. *Cercospora* leaf spots appear circular, about 3–5 mm in diameter, with light- to dark-tan centers, and brown to reddish-purple borders. Lesions occur on leaf blades, veins, and petioles where they may be elongated. Lesions may appear on the upper part of exposed roots (Giannopolitis 1978). In humid conditions, pseudostromata comprising of conidia on conidiophores form in the center of lesions. As the disease progresses, lesions coalesce and kill entire leaves which remain attached to the plant. Infected plants produce new leaves which may also become infected in favorable conditions. Loss of mature leaves and growth of new leaves result in significant reduction in root yield and recoverable sucrose (Shane and Teng 1992; Khan and Smith 2005) and increase concentration of impurities such as amino nitrogen, potassium, sodium, and betaine (Carrathures and Oldfield 1961; Smith and Martin 1978; Rossi et al. 2000) which increases processing costs (Smith and Ruppel 1973). In the absence of control measures, losses in recoverable sucrose as high as 54 % have been reported in inoculated trials (Khan and Smith 2005; Khan et al. 2008; Shane and Teng 1992). Smith and Ruppel (1973) reported that *Cercospora* leaf spot may predispose sugar beet roots to rot in storage piles which can cause economic damage especially in areas such as North Dakota and Minnesota where beets are stored for 8–9 months before processing.

Cercospora leaf spot is managed in an integrated system which includes planting disease-tolerant varieties, reducing inoculum density by crop rotation with nonhost crops, incorporation of infected crop debris, cultural practices such as planting in fields at least 100 m away from fields with sugar beet the preceding year, and fungicide applications (Miller et al. 1994; Khan and Smith 2005; Khan et al. 2008).

It is difficult to combine the multiple genes responsible for conferring resistance to *C. beticola* with acceptable agronomic characters that will lead to high yield in recoverable sucrose (Smith and Gaskill 1970; Smith and Campbell 1996). Consequently, commercial sugar beet varieties typically have moderate levels of *Cercospora* resistance and require fungicide applications to optimize yields in areas where the disease is endemic (Miller et al. 1994).

23.2 Fungicides

The pattern of fungicide use in sugar beet was similar to that for other agricultural crops. Copper salts were used for CLS control in the 1920s, and the use of 4-4-50 Bordeaux (Hull 1960) was reported to prevent epidemics. The more effective organic tin products were produced in the 1950s. Benzimidazole fungicides (benomyl and thiophanate-methyl) were evaluated in the 1960s, and their effectiveness at low use rates and their systemicity facilitated postinfection application and allowed extended application intervals. Consequently, they were widely used multiple times resulting in the rapid development of resistant *C. beticola* populations which rendered the product ineffective in the field as described later.

In the United States, *Cercospora* leaf spot is a major problem in Minnesota, North Dakota, Michigan, and specific areas of Montana and other Midwest sugar-producing states including Wyoming, Colorado, and Nebraska. North Dakota and Minnesota have three sugar cooperatives – American Crystal Sugar Company, Minn-Dak Farmers Cooperative, and Southern Minnesota Beet Sugar Cooperative – which together produce 57 % of the US sugar beet crop (USDA-ERS 2014). Growers at these cooperatives have controlled *Cercospora* leaf spot with fungicides since the late 1970s (Table 23.1), and their field experiences are rather similar to the situation described in Greece by Georgopoulos (1982).

In North Dakota and Minnesota, growers initially used copper (metallic, hydroxide, sulfate, ammonium carbonate) products, triphenyltin hydroxide products, sulfur, ethylenebisdithiocarbamates (mancozeb) products, and benzimidazoles (benomyl and thiabendazole) for controlling *Cercospora* leaf spot (1977 Sugarbeet Production Guide). The advent of the benzimidazole class of chemistry provided effective control of *Cercospora* leaf spot when first used in 1979. However, field failures were reported soon after and *C. beticola* resistant to benzimidazoles was confirmed in 1981 (Bugbee 1981). The management of sugar cooperatives recommended the use of protectant fungicides in areas with known benzimidazole resistance problems and the alternation of protectants with systemics in areas with no reported problems (Cattanach et al. 1981).

Table 23. 1 List of fungicides and when first used for controlling *Cercospora beticola* on sugar beet in North Dakota and Minnesota

Fungicide	Year of registration or first use
Cupric hydroxide (Kocide 101)	1911 (first reported use in 1983)
Benomyl (Benlate)	1979
Thiabendazole (Mertect)	1979
Thiophanate-methyl (Topsin M)	1979
Triphenyltin hydroxide (Duter)	1979
Triphenyltin hydroxide (Agsco TN)	1979
Triphenyltin hydroxide (Triple Tin)	1979
Triphenyltin hydroxide (Super Tin)	1979
Manganese ethylene-1,2-bisdithiocarbamate (Maneb)	1979
Dithiocarbamates (Mancozeb)	1979
Azoxystrobin (Quadris)	1998
Triphenyltin hydroxide (Agri Tin)	1999
Tetraconazole (Eminent)	1999–2004 (US Environmental Protection Agency granted emergency exemption); full registration in 2005
Kocide (copper hydroxide)	1985
Trifloxystrobin (Gem)	2002
Pyraclostrobin (Headline)	2002
Fenbuconazole (Enable)	2006
Propiconazole (Tilt)	2006
Difenoconazole (Inspire)	2008
Difenoconazole + propiconazole (Inspire XT)	2009
Flutriafol (Topguard)	2010
Pyraclostrobin + fluxapyroxad (Priaxor)	2012

23.3 Resistance to Benzimidazoles

The Fungicide Resistance Action Committee (FRAC) considered benzimidazoles (FRAC 1) as a high-risk group of fungicides. Benzimidazoles affect β-tubulin assembly during mitosis (FRAC 2014). The benzimidazole fungicides were used for a few years in Minnesota after which field failures were reported (Windels et al. 1998), followed by the confirmation of *C. beticola* benzimidazole-resistant strains (Bugbee 1981). Although recommendations were made in 1981 advising growers not to use benzimidazole fungicides for the control of *Cercospora* leaf spot (Cattanach et al. 1981), the continued usage of this class of chemistry by growers quickly resulted in widespread resistance of *C. beticola* to benzimidazoles in North Dakota and Minnesota (Percich and Nickelson 1985). The ineffectiveness of benzimidazole fungicides at controlling *C. beticola* resulted in growers rapidly shifting to the use of only protectant fungicides. An annual growers' survey showed that in 1982, 66 % of the sugar beet acreage was treated with a benzimidazole fungicide, and 83 % of the acreage received a protectant fungicide. By 1986, benzimidazole fungicides were not used (0 %), and 98 % of the acreage received a protectant fungicide with triphenyltin hydroxide (91 %) being most widely used, followed by the EBDC products (7 %) (Dexter et al. 1986).

In Europe, Georgopoulos and Dovas (1973) had already reported *C. beticola* resistance to benzimidazoles by 1973. In the United States, Ruppel and Scott (1974) had reported benomyl-resistant strains of *C. beticola*.

23.4 Resistance to Triphenyltin Hydroxide

Triphenyltin hydroxide (TPTH, FRAC 30) is classified as low to medium risk based on the ability of fungi to develop resistance to this group. Triphenyltin hydroxide inhibits adenosine triphosphate (ATP) synthase in the mitochondria (FRAC 2014). After field failures of benzimidazole fungicides, growers relied on the use of protectants, particularly triphenyltin hydroxide, to control *Cercospora* leaf spot. In 1995, conditions were favorable for disease development resulting in an average of 3.36 fungicide applications per field, with triphenyltin hydroxide applied on average at 2.60 times to each field and an EBDC (ethylenebisdithiocarbamate) fungicide applied on average at 0.51 times. In 1996, triphenyltin hydroxide was used alone 1.95 times on each field and as a mixture with a benzimidazole or an EBDC fungicide 0.18 times on fields when each field received an average of 2.64 fungicide applications. Other protectants including copper and sulfur products were also used, but on less than 5 % of treated fields (Dexter et al. 1996). The overreliance on TPTH eventually resulted in *C. beticola*-resistant isolates first reported in 1994 (Bugbee 1995; Bugbee et al. 1995). By 1998, high inoculum pressure and favorable environmental conditions resulted in a leaf spot epidemic. Growers used on average over all counties 3.74 fungicide applications per acre with some growers in Chippewa and Renville using 11 applications of mainly triphenyltin hydroxide and EBDC fungicides (Dexter and Luecke 1999). In 1998, 65 % of the *C. beticola* isolates tested were resistant to triphenyltin hydroxide at 1.0 ppm, making the fungicide ineffective (Weiland and Smith 1998). It was estimated that growers in North Dakota and Minnesota lost over $100 million due to yield and increased processing costs as a result of the *Cercospora* epidemic (Khan and Smith 2005).

23.5 Resistance to EBDC

Ethylenebisdithiocarbamates (EBDCs; FRAC M3) have a multisite mode of action and are considered a low-risk group of fungicides (FRAC 2014). Since protectants were mainly used for CLS control and *C. beticola* isolates resistant to triphenyltin hydroxide were confirmed, testing was also done to determine the sensitivity of *C. beticola* to EBDC. Weiland and Smith (1998) reported that *C. beticola* had developed resistance to mancozeb at both 5 and 10 ppm, and at Minn-Dak Farmers Cooperative, 64 % of the samples tested were resistant to mancozeb at 5 ppm.

23.6 Resistance to Triazoles

Triazoles (FRAC 3) affect sterol biosynthesis by inhibiting the C14-demethylase enzyme and are considered as a medium-risk group of fungicides (FRAC 2014). Widespread resistance of *C. beticola* to triphenyltin hydroxide, mancozeb, and benzimidazoles resulted in the US Environmental Protection Agency granting a special emergency exemption for the use of azoxystrobin (Quadris) (a QoI fungicide) in 1998 and the use of tetraconazole (Eminent, Sipcam) in 1999 on sugar beet. In field trials, tetraconazole consistently resulted in effective control of *Cercospora* leaf spot and became the fungicide most used in a rotation program with triphenyltin hydroxide. In 2002 and 2003, trifloxystrobin and pyraclostrobin became available, and the latter was widely used and replaced much of the usage of triphenyltin hydroxide.

In the annual collection of *Cercospora* leaf spot samples and testing for fungicide sensitivity, isolates were found with high EC_{50} values to tetraconazole and were determined to be resistant (Secor et al. 2010). Resistance of *C. beticola* to tetraconazole was as a result of overexpression of the cytochrome P450 sterol 14 alpha-demethylase (Cyp51) gene (Bolton et al. 2012). It should be noted that although tetraconazole-resistant *C. beticola* isolates were collected from growers' fields, there have been no reports of field failures in controlling *Cercospora* leaf spot. This may be a result of low populations of resistant isolates in the fields. In addition, management at sugar companies advise growers on using higher resistant varieties and avoiding the use of triazoles in townships/areas identified with *C. beticola* tetraconazole-resistant isolates.

23.7 Resistance to QoI Fungicides

The quinone outside inhibitors (QoIs) (FRAC 11) affect mitochondrial respiration by inhibiting cytochrome bc1 (complex III). QoIs have a specific target site and are considered as a high-risk group of fungicides (FRAC 2014). Azoxystrobin, a QoI fungicide, became available in 1998 but was not widely used for the control of *Cercospora* leaf spot. In 2002, trifloxystrobin became available and was used only on 9 % of the planted sugar beet acreage (Luecke and Dexter 2003). In 2003, growers rapidly adopted pyraclostrobin and used it on 85 % of planted acreage (Luecke and Dexter 2003), and growers have continued the use of this product which replaced much of the use of triphenyltin hydroxide. Although the QoIs pyraclostrobin and to a lesser extent trifloxystrobin have been used for CLS control in North Dakota and Minnesota since 2002, only a few *C. beticola* isolates were found with high EC_{50} values, and growers (>86 %) have reported consistently good to excellent control with fungicides (Carlson et al. 2014).

However, *C. beticola* isolates from sugar beet fields in Michigan where pyraclostrobin failed to control *Cercospora* leaf spot were determined to be resistant to pyraclostrobin (Bolton et al. 2013). *C. beticola* isolates resistant to pyraclostrobin were also reported from sugar beet fields in Ontario, Canada, which produces sugar beet that is processed in Michigan (Trueman et al. 2013). Kirk et al. (2012) also reported *C. beticola* isolates collected from Michigan and Nebraska to be resistant to strobilurins (QoIs). Isolates collected from sugar beet from several provinces in Italy were also found to be resistant to strobilurins (Cioni et al. 2013). *C. beticola* isolates resistant to pyraclostrobin were all characterized by a mutation in cytochrome b as a result of an amino acid exchange from glycine to alanine at position 143 (G143A) (Bolton et al. 2013).

23.8 Management Strategies

Growers in North Dakota and Minnesota are provided with research-based recommendations from extension specialists at North Dakota State University and the University of Minnesota. Specialists annually determine the efficacy of individual fungicides or mixtures and fungicide rotation programs which best control *C. beticola*. The sensitivity of *C. beticola* to the fungicides used is also determined annually from leaf spot samples collected from all factory districts in North Dakota and Minnesota (Secor et al. 2010). Recommendations are made based on the effectiveness of fungicide and fungicide programs and sensitivity of the fungus to fungicides possessing different modes of action (Khan 2014). Growers also use crop rotation of 3–4 years, varieties with partial resistance since all varieties must have a certain level of resistance to the pathogen, and incorporation of infected crop residues which occurs during tillage operations along with timely fungicide applications to manage the disease.

No field failures have been reported after using triphenyltin hydroxide, triazoles, or QoI fungicides for controlling *C. beticola* in North Dakota and Minnesota since 1998. It is possible that the reduction of the use of some fungicides and the use of fungicides with different modes of action have contributed to preventing the fungus from developing high populations of a resistant isolate. For example, about 65 % of *C. beticola* isolates had developed resistance to triphenyltin hydroxide in 1998, but by 2004, after using a QoI and a triazole and significantly reducing or not using TPTH in many areas, the pathogen reverted to high sensitivity to TPTH (Secor et al. 2010). Sugar beet growers in North Dakota and Minnesota have reduced the number of fungicide applications for controlling CLS by about 42 % and have saved over $14 million annually in fungicide use and applications compared to the last epidemic in 1998 (Carlson et al. 2014).

References

Bolton MD, Birla K, Rivera-Varas V, Rudolph K, Secor G (2012) Characterization of CbCyp51 from field isolates of *Cercospora beticola*. Phytopathology 102:298–305

Bolton MD, Rivera-Varas V, Secor G (2013) Identification of the G143A mutation associated with QoI resistance in *Cercospora beticola* field isolates from Michigan, United States. Pest Manag Sci 69:35–39

Bugbee WM (1981) Sugar beet disease research – 1981. SBREB 12:155

Bugbee WM (1995) *Cercospora beticola* tolerant to triphenyltin hydroxide. J Sugar Beet Res 32:167–174

Bugbee WM, Nielsen G, Sundsbak J (1995) A survey for the prevalence and distribution of *Cercospora beticola* tolerant to triphenyltin hydroxide and resistant to thiophanate methyl in Minnesota and North Dakota 1995. SBREB 26:176–178

Carlson AL, Khan MFR, Boetel MA (2014) Survey of fungicide use in sugar beet in Minnesota and eastern North Dakota in 2013. SBREB. http://www.sbreb.org/research/plant/plant13/plant13.htm. Accessed 21 Sept 2014

Carrathures A, Oldfield JFT (1961) Methods for the assessment of beet quality. Int Sugar J 63:72–74

Cattanach A, Dexter A, Bissonnette H, Bugbee WM, Lamey A (1981) Cercospora leaf spot management meeting and resultant recommendations. SBREB 12:165–169

Cattanach A, Dexter A, Bissonnette H, Bugbee W, Lamey A (2000) Cercospora leaf spot management meeting and resultant recommendations. SBREB 12:165–169

Cioni F, Collina M, Maines G, Khan MFR, Secor GA, Rivera VV (2013) A new integrated pest management (IPM) model for Cercospora leaf spot of sugar beets in the Po Valley. Sugar Technol. doi:10.1007/s12355-013-0260-7

Dexter AG, Luecke JL (1999) Survey of fungicide use in sugar beet in eastern North Dakota and Minnesota in 1998. SBREB 29:229–233

Dexter AG, Peters TJ, Cattanach AW (1986) Survey of fungicides use in sugar beets-1986. SBREB 17:111–112

Dexter AG, Luecke JL, Windels CE, Lamey HA, Cattanach AW (1996) Survey of fungicide use in sugar beet in eastern North Dakota and Minnesota 1996. SBREB 27:219–224

FRAC (2014) Fungicide Resistance Action Committee code list 2014: fungicides sorted by mode of action (including FRAC code numbering). http://www.frac.info/. Accessed 23 Oct 2014

Francis SA (2006) Development of sugar beet. In: Draycott AP (ed) Sugar beet. Blackwell Publishing Ltd, Oxford

Georgopoulos SG (1982) Recommended methods for the detection and measurement of resistance of agricultural pests to pesticides. FAO Plant Prot Bull 30:36–71

Georgopoulos SG, Dovas C (1973) A serious outbreak of strains of *Cercospora beticola* resistant to benzimidazole fungicides in Northern Greece. Plant Dis Rep 57:321–324

Giannopolitis CN (1978) Occurrence of strains of *Cercospora beticola* resistant to triphenyltin fungicides in Greece. Plant Dis Rep 62:205–208

Holtschulte B (2000) *Cercospora beticola*—worldwide distribution and incidence. In: Asher MJC, Holtschulte B, Molard R, Rosso F, Steinruecken G, Beckers R (eds) Advances in Sugar Beet Research: *Cercospora beticola* Sacc. Biology, agronomic influence and control measures in sugar beet, vol 2. International Institute for Beet Research, Brussels, pp 5–16

Hull R (1960) Sugar beet diseases (2 edn.). Bulletin No. 142, Ministry of Agriculture, Fisheries and Food, London

Kerr ED, Weiss A (1990) Fungicide efficacy and yield responses to fungicide treatments based on predictions of Cercospora leaf spot of sugar beet. J Sugar Beet Res 27:58–71

Khan MFR (2014) Developing fungicide recommendations for Cercospora leaf spot control in the USA. In: Dehne HW, Deising HB, Fraaije B, Gisi U, Hermann D, Mehl A, Oerke EC, Russell PE, Stammler G, Kuck KH, Lyr H (eds) Modern fungicides and antifungal compounds VII. Deutsche Phytomedizinische Gesellschaft, Braunschweig, pp 245–250

Khan MFR, Smith LR (2005) Evaluation fungicides for controlling Cercospora leaf spot on sugar beet. Crop Prot 24:79–86

Khan J, del Rio LE, Nelson R, Rivera-Varas V, Secor GA, Khan MFR (2008) Survival, dispersal and primary infection site for *Cercospora beticola* in sugar beet. Plant Dis 92:741–745

Kirk WW, Hanson LE, Franc GD, Stump WL, Gachango E, Clark G, Stewart J (2012) First report of strobilurin resistance in *Cercospora beticola* in sugar beet (*Beta vulgaris*) in Michigan and Nebraska, USA. New Dis Rep 26:3

Luecke JL, Dexter AG (2003) Survey of fungicide use in sugar beet in eastern North Dakota and Minnesota 2003. SBREB 34:223–228

Miller SS, Rekoske M, Quinn A (1994) Genetic resistance, fungicide protection and variety approval politics for controlling yield losses from Cercospora leaf spot infection. J Sugar Beet Res 31:7–12

Percich JA, Nickelson LJ (1985) Field evaluation of various fungicides to control Cercospora leaf spot. SBREB 16:179–190

Rossi V, Meriggi P, Biancardi E, Rosso F (2000) Effect of Cercospora leaf spot on sugar beet growth, yield and quality. In: Asher MJC, Holtschulte B, Molard R, Rosso F, Steinruecken G, Beckers R (eds) Advances in sugar beet research: *Cercospora beticola* Sacc. Biology, agronomic influence and control measures in sugar beet, vol 2. International Institute for Beet Research, Brussels, pp 49–76

Ruppel EG (1986) Cercospora leaf spot. In: Whitney ED, Duffus JE (eds) Compendium of beet diseases and insects. American Phytopathological Society, St. Paul, pp 8–9

Ruppel EG, Scott PR (1974) Strains of *Cercospora beticola* resistant to benomyl in U.S.A. Plant Dis Rep 58:434–436

Secor G, Rivera V, Khan M, Gudmestad N (2010) Monitoring fungicide sensitivity of *Cercospora beticola* of sugar beet for disease management decisions. Plant Dis 94:1272–1282

Shane WW, Teng PS (1992) Impact of Cercospora leaf spot on root weight, sugar yield and purity of *Beta vulgaris*. Plant Dis 76:812–820

Smith GA, Campbell LG (1996) Association between resistance to Cercospora and yield in commercial sugar beet. Plant Breed 115:28–32

Smith GA, Gaskill JO (1970) Inheritance of resistance to Cercospora leaf spot in sugarbeet. J Am Soc Sugar Beet Technol 16:172–180

Smith GA, Martin SS (1978) Differential response of sugar beet cultivars to Cercospora leaf spot disease. Crop Sci 18:39–42

Smith GA, Ruppel EG (1973) Association of Cercospora leaf spot, gross sugar, percentage sucrose and root weight in sugar beet. Can J Plant Sci 53:695–696

Trueman CL, Hanson LE, Rosenzweig N, Jiang QW, Kirk WW (2013) First report of QoI insensitive *Cercospora beticola* on sugar beet in Ontario, Canada. Plant Dis 97:1255

USDA-ERS (2014) Sugar and sweeteners yearbook tables. http://www.ers.usda.gov/data-products/sugar-and-sweeteners-yearbook-tables.aspx#25440. Accessed 4 Sept 2014

Weiland JJ, Smith GA (1998) A survey for the prevelance and distribution of *Cercospora beticola* tolerant to triphenyltin hydroxide and mancozeb and resistant to thiophanate methyl in 1998. SBREB. http://www.sbreb.org/Research/plant/plant98/plant98.htm. Accessed 21 Oct 2014

Windels CE, Lamey HA, Hilde D, Widner J, Knudsen T (1998) A Cercospora leaf spot model for sugar beet: In practice by an industry. Plant Dis 82:716–726

Chapter 24
Oilseed Rape Pathogens in France

Annette Penaud and Anne-Sophie Walker

Abstract In France, a dozen of diseases can affect rapeseed, but *Sclerotinia* stem rot is the major disease for which chemical control remains the main way to prevent severe attacks. Five fungicide modes of action are effective against *Sclerotinia sclerotiorum* and some solo or mixed products are registered in France. The history and current status of resistance to these different groups in French oilseed rape crops are presented. Strategies for preventing and managing fungicide resistance are mainly based on the alternation of products from different classes of biochemical modes of action.

Keywords *Sclerotinia sclerotiorum* • *Sclerotinia* stem rot • Oilseed rape • Fungicide resistance • Resistance monitoring • Resistance management

24.1 Introduction

Oilseed rape is the major oilseed crop cultivated in France with 1.5 million ha per year. The increased production of oilseed rape has also increased the risk of diseases. Nowadays, a dozen of fungal diseases are known to infect French oilseed rape crops, but two of them – blackleg caused by *Leptosphaeria maculans* and *Sclerotinia* stem rot (SSR) caused by *Sclerotinia sclerotiorum* – can cause severe damage and important yield losses.

In France, blackleg is mainly controlled by using resistant cultivars, and resistance groups are alternated in order to prevent an overcoming of single resistance genes (Gladders et al. 2006). Chemical control is only performed with DMI fungicides on some susceptible varieties according to the epidemics provided by the monitoring of the release of *L. maculans* ascospores and a forecasting model so that any shift in fungicide sensitivity has never been observed in practice.

A. Penaud (✉)
TERRES INOVIA, Avenue Lucien Brétignières, F-78850 Thiverval-Grignon, France
e-mail: a.penaud@terresinovia.fr

A.-S. Walker
INRA AgroParisTech, UMR 1290 BIOGER, Avenue Lucien Brétignières, F-78850 Thiverval-Grignon, France

H. Ishii, D.W. Hollomon (eds.), *Fungicide Resistance in Plant Pathogens*,
DOI 10.1007/978-4-431-55642-8_24

After two severe epidemics in 1971 and 1979 in the middle of France, SSR has become the second most damaging disease on oilseed rape. Therefore, in the 1980s, its control had been developed. It was usually associated with light leaf spot (*Pyrenopeziza brassicae*) and then white leaf spot (*Mycosphaerella capsellae*) control involving up to three applications. Afterward due to the cropping of more tolerant cultivars to light leaf spot, the chemical control was focused on *Sclerotinia* stem rot. It was based on one or two applications of benzimidazoles used alone or associated with dicarboximides or DMIs. As these fungicides provided more preventive effects than curative effects, the applications have become systematic even if a severe epidemic occurs once or twice per decade (Penaud et al. 2013a). In the 2000s, biological control using *Coniothyrium minitans* has provided a new tool for controlling SSR (Penaud and Michi 2009). Moreover tools for disease management such as risk indicators or forecasting models have been developed to optimize the timing of fungicide application and to reduce fungicide use.

The history and current status of resistance to the different groups of fungicides used in French oilseed rape crops are presented in this chapter. Most data were produced from a network involving the French plant protection service DQSPV, INRA as scientific support, CETIOM as the technical institute for oilseed crops, and recently most chemical companies. Resistance management options, based on the reduction of fungicide applications per season and the alternation of solo products or the mixture of different modes of action, are discussed.

24.2 Anti-microtubule Agents

Carbendazim and its precursor thiophanate-methyl belong to the benzimidazole group and were introduced in the late 1960s in European crops. Their French registration on oilseed rape was achieved in the 1980s (Fig. 24.1). The emergence of resistance to carbendazim in oilseed rape was first reported in 1994 in Burgundy in two fields in which reduced efficacy was suspected (Souliac and Leroux 1995). Four years later in 1998, two new resistant strains were detected. In 1999, the presence of resistant strains was detected in 20 % of the sampled plots with a proportion of resistant isolates to carbendazim ranging from 10 to 100 %. The high level of resistance was associated with a reduction in the efficiency of the fungicidal control (Kaczmar et al. 2000). Since 2000, *Sclerotinia* monitoring was yearly carried out. Depending on disease pressure, between 150 and 340 field locations were sampled for ten sclerotia per site. The sclerotia were tested for carbendazim resistance using two discriminatory doses (Penaud et al. 2003, 2013b). Most resistant strains showed normal growth on the two rates of carbendazim and furthermore exhibited an increased sensitivity to the phenylcarbamate diethofencarb, suggesting a mutation similar to that of the BenR1 phenotype of *Botrytis cinerea* (Leroux et al. 2002). Few weakly carbendazim-resistant strains were insensitive to diethofencarb, suggesting another mutation such as in *B. cinerea* BenR2 phenotype. Benzimidazole fungicides prevent microtubule assembly by binding to β-tubulin (Davidse and Ishii

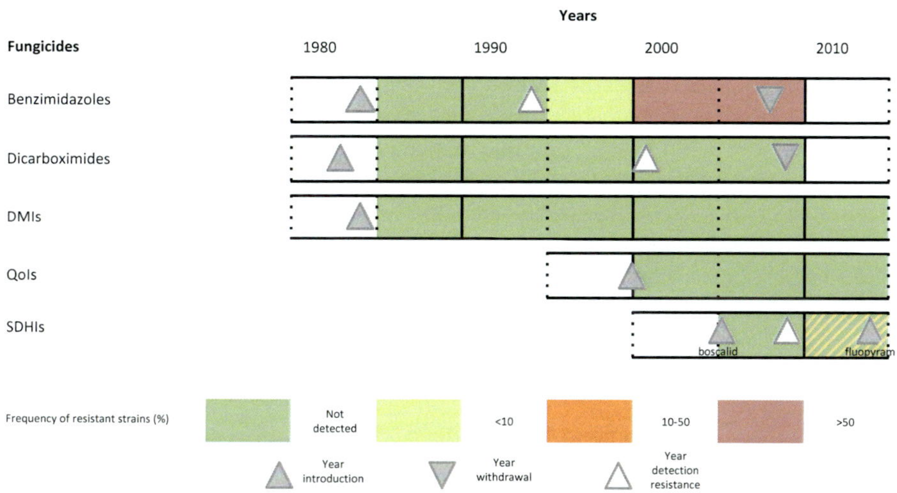

Fig. 24.1 Evolution of resistance in *S. sclerotiorum* populations in winter oilseed crops in France (Data from monitoring conducted by ONPV, CETIOM, ANSES, INRA, and chemical companies)

1995). In *B. cinerea*, resistance is conferred by allelic changes in the gene encoding β-tubulin leading to the change G198A in BenR1 strains and F200T in BenR2 strains (Leroux et al. 2002). Similarly, the G198A change in *S. sclerotiorum* has also been reported in China, and PCR methods have been developed to detect both mutations (Li et al. 2002; Yin et al. 2010).

Over the country, the resistance to carbendazim occurred mainly in northeastern and central regions where 60 to 75 % of monitored fields have produced resistant isolates to carbendazim. In the Lorraine region, up to 90 % of field samples collected in 2005 were resistant. Due to a general wide spread of carbendazim resistance and a future withdrawal of this fungicide, there was no more interest for monitoring *Sclerotinia* resistance to benzimidazole fungicides after the mid-2000s (Penaud et al. 2013b).

An analysis of cultural practices suggested that the development of carbendazim resistance was associated with short rotations of oilseed rape (every 2 or 3 years) and at least five sprays of benzimidazole fungicide during the last 10 years (Penaud et al. 2001). Repeated applications of benzimidazole fungicides have also promoted the emergence of resistant strains of *S. sclerotiorum* in oilseed rape in China or Canadian prairies (Pan et al 1997; Gossen and Rimmer 2001; Wang et al 2014a).

The occurrence of *S. sclerotiorum* resistant to carbendazim led us to develop resistance management strategies, and several recommendations are made to farmers: (1) give up the systematic treatment at the beginning of flowering and prefer the optimum timing to control the disease according to the climate conditions, (2) when protection is needed, apply one single spray at the appearance of the first pods, and (3) choose an efficient fungicide among different fungicide groups without cross-resistance to carbendazim. Until 2007, three groups of fungicides were registered for *Sclerotinia* use in France: (1) dicarboximide fungicides (iprodione, procymidone, or vinclozolin) which have high intrinsic activity, (2) DMIs (tebuconazole, metconazole) which are of lower intrinsic activity than the previous but interesting against other rapeseed diseases such as light leaf spot or powdery mildew, and (3) a strobilurin (azoxystrobin). Thus, using effective fungicides only if necessary and alternating them could make it possible to manage practical resistance and maintain an effective chemical control. At the same time, studies of biological control and of decision support system are investigated for sustainable crop protection (Penaud and Michi 2009; Penaud et al. 2013a).

24.3 Dicarboximides

Dicarboximides (i.e., iprodione, procymidone, vinclozolin) were registered on oilseed rape in the early 1980s firstly for controlling black spot on pods (*Alternaria brassicae*) and afterward for *Sclerotinia* control.

Although the dicarboximide fungicides were the first authorized for *Sclerotinia* control, the high price of these products has restricted their use, with farmers preferring cheaper solutions mainly based on carbendazim alone or in mixture. However, their use has increased as soon as the resistance of *S. sclerotiorum* to carbendazim became widespread across the country. Dicarboximide fungicides used solo and in mixture represented 50 % of fungicide treatments applied to prevent attacks of *Sclerotinia* in the mid-2000s (Fig. 24.2).

Despite the increase in use, there was no practical loss of efficacy in the field. For 6 years (2001–2006) of monitoring for dicarboximide fungicides, *S. sclerotiorum* isolates resistant to iprodione were only detected in 9 field locations (Fig. 24.3).

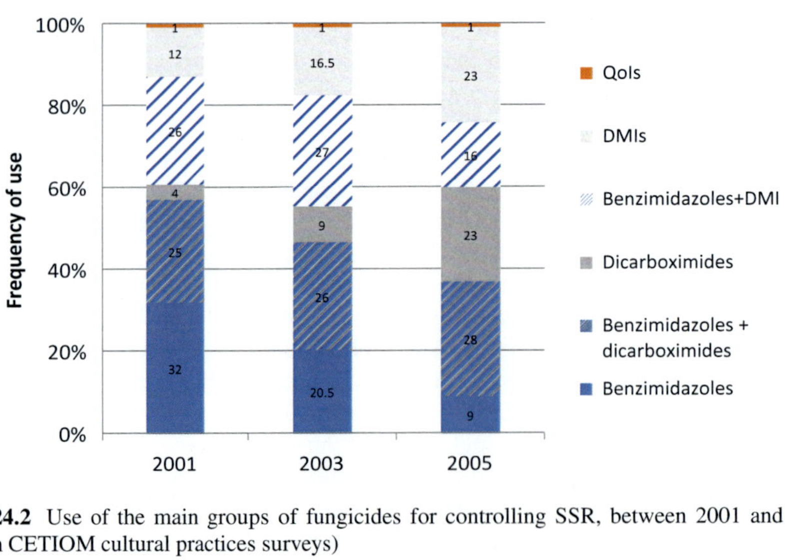

Fig. 24.2 Use of the main groups of fungicides for controlling SSR, between 2001 and 2005 (From CETIOM cultural practices surveys)

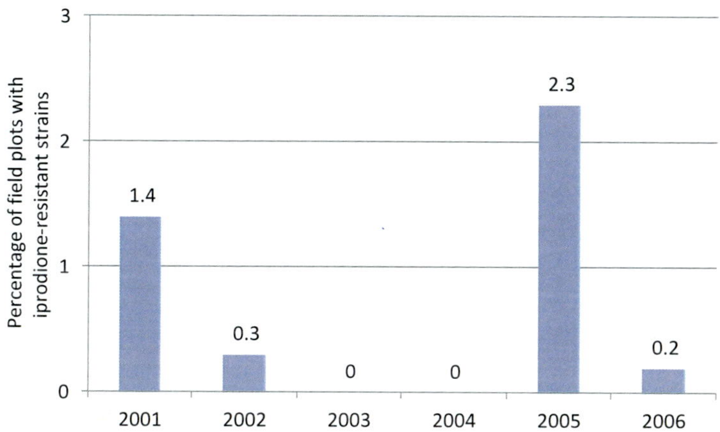

Fig. 24.3 Frequency of oilseed rape fields with at least one iprodione-resistant strain of *S. sclerotiorum*

In each case, the rare detections were unrelated to the use of this chemical group in the field, suggesting that they could have randomly emerged in local populations of *S. sclerotiorum*. Since 2007, procymidone, vinclozolin, and iprodione were withdrawn for toxicological reasons, and nowadays, no further fungicides from this group are registered for controlling SSR on oilseed rape. Therefore, the monitoring was also given up.

The first field-resistant isolate was introduced as a resistance check in monitoring tests. This strain had gradually lost its ability to produce sclerotia after successive subcultures, suggesting that resistance would have a negative impact on the fitness of this resistant strain which therefore could be less competitive compared to the population of susceptible strains as also suggested for some *B. cinerea*-resistant phenotypes. Otherwise, resistant mutants of *S. sclerotiorum* can easily be induced in the laboratory (Gindrat 1993; Liu et al. 2009; Duan et al. 2014). In a recent study, the laboratory-induced iprodione-resistant mutants lost the ability of producing sclerotia after ten generations on PDA medium. These mutants also showed an increase in sensitivity to osmotic stress. The putative target site of dicarboximides is supposed to be a two-component histidine kinase, encoded by the group III HisK gene (*Shk1*, syn. *os-1*). It contains an osmotic sensing domain, six 90-amino acid repeat motifs (also named HAMP domains), a kinase core, and response regulator domains. Sequence analysis of this histidine kinase gene has shown either a deletion or single point mutations suggesting that these modifications may be associated with dicarboximide resistance and also involved in sclerotial development and virulence (Liu et al. 2009). In two field mutants of *S. sclerotiorum*, various mutations in the amino acid repeat region of the histidine kinase gene led to different degrees of sensitivity to dicarboximides and phenylpyrroles (Alberoni et al. 2010). Further disruption of *Shk1* in laboratory mutants resulted in resistance to phenylpyrrole and dicarboximide fungicides and increased sensitivity to osmotic and oxidative stresses. The *Shk1* mutant also showed a significant reduction in hyphal growth and was unable to produce sclerotia (Duan et al. 2013). In addition, the *S. sclerotiorum Shk1* gene appears to be the ortholog of *B. cinerea Bos1* (Duan et al. 2014).

24.4 Demethylation Inhibitor Fungicides (DMIs)

Among the inhibitors of sterol synthesis, only the class of inhibitors in 14α-demethylase (DMIs) is used in oilseed rape. Most are triazole compounds (e.g., flutriafol, flusilazole, difenoconazole, metconazole, tebuconazole), but prochloraz (imidazole) and prothioconazole (triazolinethione) are also included in this class of fungicides. Their target site is the cytochrome P450 14α-demethylase (CYP51). Multiple mechanisms of resistance to DMI fungicides have been described involving (1) mutations in the *Cyp51* gene, (2) *Cyp51* overexpression, and (3) an upregulation of membrane transporters leading to an increase in the efflux of DMIs. A combination of these mechanisms has been reported in field isolates of *Mycosphaerella graminicola* in winter wheat displaying low to high resistance to DMIs (Leroux and Walker 2010), but not in *S. sclerotiorum*, according to current literature.

Some DMI fungicides are also very effective against secondary diseases of oilseed rape, for example, prochloraz for controlling light leaf spot (*Pyrenopeziza brassicae)* or flusilazole towards powdery mildew caused by *Erysiphe cruciferarum*. For over 20 years, DMI compounds were associated with carbendazim. Higher was the rate of carbendazim and more effective were the mixtures for controlling SSR (Kaczmar et al. 2000). Unfortunately, owing to a restricted use of carbendazim before its withdrawal, all mixtures of carbendazim + DMI were no longer reregistered. However, tebuconazole and metconazole have remained on the market as solo products, and prothioconazole has recently been introduced in the panel of tools for disease control in oilseed rape. Since 2007, DMI tests were also added in the monitoring conducted every year. Up to now, no resistant strains of *S. sclerotiorum* have been detected. Sometimes, few isolates are detected as less susceptible at the discriminatory concentration used in the test, suggesting that a shift in sensitivity of *S. sclerotiorum* to DMI fungicides could occur. But under field conditions, no significant decline has ever been reported in SSR control.

Therefore, these fungicides are involved in resistance management strategies. In practice, tebuconazole and metconazole which have more irregular efficacy than boscalid or prothioconazole are recommended when *Sclerotinia* risk is low or moderate. This *Sclerotinia* risk can be reduced by using biological control and a program combining biological control with *Coniothyrium minitans* in autumn, and a complementary application of tebuconazole at the flowering stage has proved to be as effective as a preventive application of boscalid (Penaud and Michi 2009). DMIs are also good partners to associate with QoIs and SDHIs for managing fungicide resistance.

24.5 Fungicides Affecting Specifically Respiration

24.5.1 Inhibitors of Mitochondrial Complex III (QoIs)

QoI fungicides are able to inhibit mitochondrial electron transfer by binding to the Qo site (an outer ubiquinol oxidizing pocket) of cytochrome *bc1* complex or complex III. In most pathogens, resistance to QoIs (strobilurins) is conferred by a single point mutation in the mitochondrial cytochrome b gene *cyt b* resulting in peptide sequence changes that prevent fungicide binding (Gisi et al. 2002).

In French oilseed rape crops, the first QoI fungicide (azoxystrobin) was introduced in the early 2000s. Until now, no QoI resistance has yet been detected in field isolates of *S. sclerotiorum*, maybe due to a weak selection pressure or a negative fitness of resistant isolates. Azoxystrobin used to be applied in mixture with DMI fungicide, rather than alone, which could have prevented resistance selection.

In most pathogens, the resistance to QoI fungicides is mainly due to the amino acid substitution from glycine to alanine at position 143 (G143A) in the *cyt b* which is also associated with a failure in disease control (Ishii 2012). In some fungi, QoI resistance has never been recorded due to the presence of an intron inserted directly after codon 143. In those cases, G143A change prevents the splicing of the intron, leading to a deficient cytochrome b, which is lethal (Grasso et al. 2006). In a few cases of our monitoring, some sclerotia of *S. sclerotiorum* have failed to germinate, and some have given colonies which have poorly grown on SHAM-amended media with or without 1 mg/L of azoxystrobin. These field isolates could be resistant, and they need to be investigated for mutations and the intron structure of the *cyt b* gene that could explain why QoI-resistant mutants of *S. sclerotiorum* are rare or unsuccessfully induced in vitro (Wang et al. 2014b). In some other fungi, other substitutions in the *cyt b* such as F129L and G137R have also been reported to confer a low to medium level of resistance, so that the efficacy of QoI fungicides is not affected in practice, but such substitutions have not yet been recorded in *S. sclerotiorum.*

24.5.2 Inhibitors of Mitochondrial Complex II (SDHIs)

The SDHI fungicides are able to inhibit the mitochondrial electron transfer from succinate to coenzyme Q by binding to the complex II or succinate dehydrogenase (SDH) complex. The enzyme is a component of the inner mitochondrial membrane and consists of four nucleus-encoded proteins SDHA, B, C, and D. The binding site of ubiquinone is formed by residues of the subunits B, C, and D and accounts for the target site of carboxamide SDHI compounds. Therefore, resistance to SDHIs is conferred by single point mutations in the *Sdh* genes encoding subunits B, C, or D. Many mutations have been reported in field populations of different pathogens (Sierotzki and Scalliet 2013). In France, the first strains of *S. sclerotiorum* resistant to boscalid were detected within two seasons after commercial use of boscalid in 2008.

Although poor epidemics of SSR occurred, the number of detected strains resistant to boscalid has gradually increased since 2011 (Fig. 24.4).

Moreover a reduced disease control was first observed at the field level in 2014. Most field-resistant strains of *S. sclerotiorum* exhibit moderate resistance to boscalid, and sequence analysis has revealed a point mutation in the *SdhD* gene, involving the replacement of histidine by arginine at position 132 (H132R) (Glättli et al. 2009). For fungi as *Alternaria alternata* or some strains of *B. cinerea*, the SdhD H132R mutants have shown positive cross-resistance between all novel SDHIs at low frequency which could be explained by a negative fitness cost (Leroux et al. 2010; Avenot and Michailides 2010). For other field-resistant strains not bearing the SdhD H132R change, a single point mutation has also been detected in the *SdhC* and *SdhB* genes respectively (Walker unpublished). Boscalid-resistant mutants of *S. sclerotiorum* have also been induced in vitro, carrying a mutation in the *SdhB* gene (Wang et al. 2014c).

In the mids-2000s, boscalid was the first SDHI registered for the control of SSR on oilseed rape in France. Due to carbendazim resistance of *S. sclerotiorum* and the withdrawal of benzimidazole and dicarboximide fungicides, boscalid has quickly entered into widespread use. As it came on the market in 2007, the monitoring was then focused on this newly registered compound (Moinard et al. 2009). The method used for detecting boscalid-resistant field isolates is based on classical mycelial growth on minimal agar medium with succinate as carbon source and amended with one single discriminatory dose of boscalid. This method is proved as reliable as Stammler's microtiter method (Stammler et al. 2007; Leroux unpublished; Wang et al. 2009).

Since its commercial use on French oilseed rape crops, boscalid was firstly applied alone once per year because only one application is effective to prevent oilseed rape infection if it is sprayed when the first petals are falling on leaves.

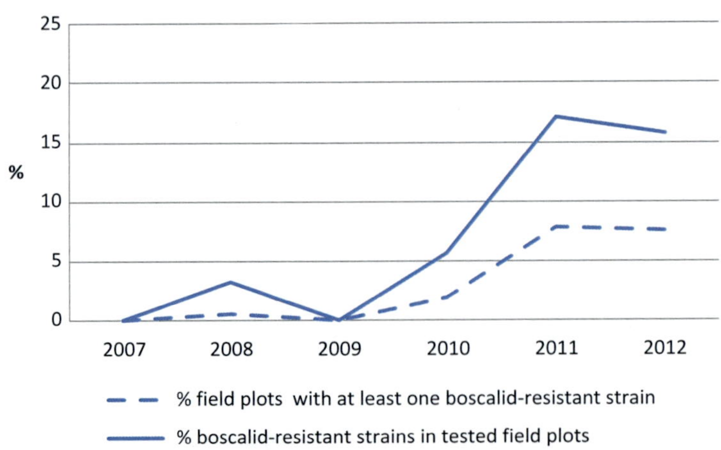

Fig. 24.4 Occurrence of boscalid-resistant strains of *S. sclerotiorum*. Results of monitoring conducted by ONPV, CETIOM, ANSES, INRA, and chemical companies

Due to its efficacy, boscalid has become the most used active ingredient towards SSR which is applied alone and more and more in mixture with other fungicides. The most popular mixture is boscalid + metconazole and provides not only a high efficacy on SSR but also a better yield increase. When another application is sometimes required, alternation with other no cross-resistant fungicides to limit the selection pressure is recommended. In order to develop anti-resistance strategies, new SDHIs are formulated with products carrying other modes of action. However mixtures with QoIs could present a risk to select double-resistant strains as it has already been reported on other crops. Nevertheless, this risk could be low if *S. sclerotiorum* is demonstrated as a pathogen with rare QoI resistance. At last, the recent introduction of the mixture of fluopyram + prothioconazole provides another tool in resistance management. Though cross-resistance between boscalid and fluopyram is suspected, prothioconazole offers a different mode of action with a high intrinsic activity against *S. sclerotiorum* strain populations so that SDHI-resistant strains could be controlled.

24.6 Conclusions

Oilseed rape diseases are controlled as far as possible by host resistance. Due to a lack of cultivars resistant to *S. sclerotiorum*, chemical control remains the main way to control SSR in rapeseed. Usually, only one foliar spray is necessary for controlling the disease. As all the available fungicides are preventive with little or no curative effects on *S. sclerotiorum*, they need to be applied just before petals fall on leaves, between 20 and 30 % bloom. Sometimes the treatment can be repeated for controlling other diseases such as powdery mildew or black spot. Over 15 years, SSR was well controlled by benzimidazole fungicides used solo or in mixture with sterol demethylation inhibitors or dicarboximide fungicides. But because of intensive use, resistance to carbendazim emerged and generalized. Therefore, strategies for managing resistance were developed. They were firstly based on the use of the two remaining fungicide groups at full doses. It is also advised to choose the fungicide according to *Sclerotinia* risk levels. When *Sclerotinia* risk was high, the most effective dicarboximide fungicides were preferred to DMIs. Farming practices are also recommended such as to enhance crop rotation including less susceptible crops and to eliminate susceptible weeds. In addition, biological control using *Coniothyrium minitans* to reduce soil infestation can also help to limit the risk of *Sclerotinia* and thus to avoid spraying or at least to allow the alternation of fungicides with different modes of action. After the withdrawal of dicarboximide fungicides, two new groups SDHIs and QoIs have fortunately been registered. But because they are single-target site fungicides, a risk of developing resistance is to be feared. Therefore, strategies must be considered for preventing resistance. With one treatment per season, the alternation strategy of different modes of action can only be promoted at the scale of the crop rotation. But when two treatments are required for controlling not only SSR but also other diseases, alternation of active ingredients with different modes of action can easily be implemented with the available three different modes of action. The strategy of fungicide mixtures with different modes of action is expanding in chemical companies in order to prevent or delay the emergence of resistance to their active ingredients prone to resistance such as single-target site fungicides. But this strategy of mixtures requires suitable partners at an appropriate rate to provide effective disease control without selecting for resistance. At last, the best strategy of resistance management of *S. sclerotiorum* should be to limit chemical control by improving *Sclerotinia* prediction and to spray only when it is needed.

References

Alberoni G, Collina M, Lanen C, Leroux P, Brunelli A (2010) Field strains of *Stemphylium vesicarium* with a resistance to dicarboximide fungicides correlated with changes in two-component histidine kinase. Eur J Plant Pathol 128:171–184

Avenot HF, Michailides TJ (2010) Progress in understanding molecular mechanisms and evolution of resistance to succinate dehydrogenase inhibiting (SDHI) fungicides in phytopathogenic fungi. Crop Prot 29:643–651

Davidse L, Ishii H (1995) Biochemical and molecular aspects of benzimidazoles, *N*-phenylcarbamates and *N*-phenylformamidoximes and the mechanisms of resistance to these compounds. Modern selective fungicides, pp 305–322

Duan YB, Ge CY, Liu SM, Wang JX, Zhou MG (2013) A two-component histidine kinase ShK1 controls stress response, sclerotial formation and fungicide resistance in *Sclerotinia sclerotiorum*. Mol Plant Pathol 14:708–718

Duan YB, Ge CY, Zhou MG (2014) Molecular and biochemical characterization of *Sclerotinia sclerotiorum* laboratory mutants resistant to dicarboximide and phenylpyrrole fungicides. J Pest Sci 87:221–230

Gindrat D (1993) La sclérotiniose: sensibilité du *Sclerotinia sclerotiorum* au carbendazime et à la vinclozoline. Rev Suisse Agric 25:115–119

Gisi U, Sierotzki H, Cook A, McCaffery A (2002) Mechanisms influencing the evolution of resistance to Qo inhibitor fungicides. Pest Manag Sci 58:859–867. doi:10.1002/ps.565

Gladders P, Evans N, Marcroft S, Pinochet X (2006) Dissemination of information about strategies and changes in farming practices of the exploitation of resistance to *Leptosphaeria maculans* (Phoma stem canker) in oilseed rape cultivars. Eur J Plant Pathol 114:117–126

Glättli A, Stammler G, Schlehuber S (2009) Mutations in the target proteins of succinate-deshydrogenase inhibitors (SDHI) and 14α-demethylase inhibitors (DMI) conferring changes in the sensitivity – structural insights from molecular modelling. In: AFPP (ed) 9ème conférence internationale sur les maladies des plantes, Tours, pp 670–681

Gossen B, Rimmer S (2001) First report of resistance to benomyl fungicide in *Sclerotinia sclerotiorum*. Plant Dis 85:1206

Grasso V, Palermo S, Sierotzki H, Garibaldi A, Gisi U (2006) Cytochrome b gene structure and consequences for resistance to Qo inhibitor fungicides in plant pathogens. Pest Manag Sci 62:465–472

Ishii H (2012) Resistance to QoI and SDHI fungicides in Japan. In: Thind TS (ed) Fungicide resistance in crop protection, risk and management. CABI, Oxfordshire, pp 223–234

Kaczmar MJ, Wilson V, Leroux P (2000) Sclérotiniose du colza, le carbendazime en sursis. Phytoma-LDV 529:31–33

Leroux P, Walker A-S (2010) Multiple mechanisms account for resistance to sterol 14α-demethylation inhibitors in field isolates of *Mycosphaerella graminicola*. Pest Manag Sci 67:44–59. doi:10.1002/ps.2028

Leroux P, Fritz R, Debieu D, Albertini C, Lanen C, Bach J, Gredt M, Chapeland F (2002) Mechanisms of resistance to fungicides in field strains of *Botrytis cinerea*. Pest Manag Sci 58:876–888

Leroux P, Gredt M, Leroch M, Walker A-S (2010) Exploring mechanisms of resistance to respiratory inhibitors in field strains of Botrytis cinerea the causal agent of gray mold. Appl Environ Microbiol 76:6615–6630

Li HX, Zhou MG, Lu YJ (2002) Using polymerase chain reaction for detection of carbendazim resistance in *Sclerotinia sclerotiorum*. Mycosystema 3:12

Liu X, Yin Y, Yan L, Michailides TJ, Ma Z (2009) Sensitivity to iprodione and boscalid of *Sclerotinia sclerotiorum* isolates collected from rapeseed in China. Pestic Biochem Physiol 95:106–112

Moinard J, Leroux P, Penaud A, Remuson F, Walker A-S, Micoud A, Huguet B (2009) Résistance de *Sclerotinia sclerotiorum* aux fongicides utilisés sur colza: bilan de la surveillance du territoire de 2000 à 2008. In: AFPP (ed) 9ème conférence internationale sur les maladies des plantes, Tours, France, pp 771–780

Pan Y, Wang Z, Wu HZ (1997) Resistance to carbendazim and stability of the resistance in *Sclerotinia sclerotiorum*. Jiangsu J Agric Sci 13:32–35

Penaud A, Michi H (2009) *Coniothyrium minitans*, un agent de lutte biologique au service de la protection intégrée. OCL 16:158–163

Penaud A, Huguet B, Wilson V, Leroux P (2001) Benzimidazole resistance of *Sclerotinia sclerotiorum* in French oilseed rape crops. Presented at Sclerotinia Workshop, York

Penaud A, Huguet B, Wilson V, Leroux P (2003) Fungicide resistance of *Sclerotinia sclerotiorum* in French oilseed rape crops. In: Proceedings of 11th International Rapeseed Congress, 6–10 July, Copenhagen, vol 4, pp 1097–1098

Penaud A, Wagner D, Makowsky D, Ruck L (2013a) Evaluation of a decision making tool for controlling Sclerotinia stem rot in WOSR. IOBC-WPRS Bull 92:199–204

Penaud A, Micoud A, Moinard J, Remuson F, Walker A-S, Leroux P (2013b) Monitoring the sensitivity of *Sclerotinia sclerotiorum* to fungicides. IOBC-WPRS Bull 92:205–209

Sierotzki H, Scalliet G (2013) A review of current knowledge of resistance aspects for the next-generation succinate dehydrogenase inhibitor fungicides. Phytopathology 103:880–887. doi:10.1094/phyto-01-13-0009-rvw

Souliac L, Leroux P (1995) Sclérotiniose du colza: Faut-il revoir la stratégie de lutte? Phytoma-LDV 474:29–31

Stammler G, Benzinger G, Peakman J (2007) A rapid and reliable method for monitoring the sensitivity of *Sclerotinia sclerotiorum* to boscalid. J Phytopathol 155:746–748

Wang JX, Ma HX, Chen Y, Zhu XF, Yu WY, Tang ZH, Chen CJ, Zhou MG (2009) Sensitivity of *Sclerotinia sclerotiorum* from oilseed rape crops to boscalid in Jiangsu province of China. Crop Prot 28:882–886

Wang Y, Hou YP, Chen CJ, Zhou MG (2014a) Detection of resistance in *Sclerotinia sclerotiorum* to carbendazim and demethachlon in Jiangsu province of China. Austr Plant Pathol 43:307–312

Wang Y, Duan YB, Zhou MG (2014b) Control of *Sclerotinia sclerotiorum* infection in oilseed rape with strobilurin fungicide SYP-7017. Can J Plant Pathol 36:354–359

Wang Y, Duan YB, Zhou MG (2014c) Molecular and biochemical characterization of boscalid resistance in laboratory mutants of *Sclerotinia sclerotiorum*. Plant Pathol. doi:10.1111/ppa.12246

Yin Y, Liu X, Shi Z, Ma Z (2010) A multiplex allele-specific PCR method for detection of carbendazim-resistant *Sclerotinia sclerotiorum*. Pestic Biochem Physiol 97:36–42

Chapter 25
Cucurbit Powdery Mildew in the USA

Margaret Tuttle McGrath

Abstract Fungicide resistance has been an issue with managing powdery mildew of cucurbit crops since resistance was detected to MBC fungicides. This was the first chemical class with resistance risk registered for this use and one of the first cases of resistance in the USA. The pathogen has developed resistance or reduced sensitivity to fungicide chemical classes registered subsequently following a few years of commercial use, first with DMIs, then QoIs, and next SDHIs. Resistance has been associated with failure to control powdery mildew with FRAC codes 1, 3, 7, and 11 fungicides. FRAC 1 and 11 fungicides are still labeled for this use, but are not recommended due to suspected widespread occurrence of resistance. FRAC 3 fungicides usually are effective when applied at high label dose. FRAC 7 fungicides usually are effective as well. Cross-resistance within chemical classes is the norm except that fluopyram is dissimilar from other SDHIs tested. Quinoxyfen and cyflufenamid, which were registered for this use in 2007 and 2012, respectively, are currently considered the most effective fungicides for managing cucurbit powdery mildew. They are recommended for use in alternation also including FRAC 3 and 7 fungicides. With four targeted chemistries to alternate among, farmers can implement a good program for managing powdery mildew and fungicide resistance. Pathogen sensitivity to fungicides is being determined with bioassays using seedlings for field populations and leaf disks for isolates. Challenges that resistance can impose on effectively managing a disease have been illustrated with cucurbit powdery mildew, arguably the most important pathosystem of a vegetable crop plagued by resistance.

Keywords *Podosphaera xanthii* • *Sphaerotheca fuliginea* • Powdery mildew • Cucurbit crop • Squash • Cucumber • Melon • Pumpkin • Fungicide resistance

---

M.T. McGrath (✉)
Plant Pathology & Plant-Microbe Biology Section, Cornell University LIHREC,
3059 Sound Avenue, Riverhead, NY, 11901 USA
e-mail: mtm3@cornell.edu

H. Ishii, D.W. Hollomon (eds.), *Fungicide Resistance in Plant Pathogens*,
DOI 10.1007/978-4-431-55642-8_25

25.1 Introduction

Powdery mildew of cucurbit crops is arguably the most important pathosystem of a vegetable crop that is afflicted by fungicide resistance. The crop group is very important. Crops in this group produce a diversity of fruit types consumed as fruit (melons) as well as vegetables (cucumber and squashes) and used for other purposes (Halloween pumpkins and gourds). They are grown throughout the USA mostly outdoors. Cucumbers are also grown in greenhouses. Powdery mildew is arguably the most important disease of cucurbits. It occurs routinely wherever cucurbits are grown. This is partly because the pathogen produces an abundance of spores easily dispersed by wind potentially for long distances. Also it does not need high humidity and/or leaf wetness in order to be able to infect, in contrast with other fungal pathogens. The pathogen rarely attacks fruit directly, but left unmanaged leaves typically die prematurely resulting in reduced yield or fruit quality (including poor flavor and storability).

Fungicides with potential for resistance development are the main management tool for powdery mildew. Resistant cultivars are available for some cucurbit crop types, but the degree of suppression is not always sufficient to achieve a commercially acceptable level of control when employed as the sole management practice, and the pathogen has potential to evolve new strains able to overcome genetic resistance; thus an integrated management program is recommended even with cultivars displaying a high level of suppression. These are the only management practices for powdery mildew because the pathogen cannot be avoided and cultural practices to make the environment less favorable for disease development is not a viable strategy with this disease. To effectively manage powdery mildew, fungicides are needed on the abaxial surface of leaves where the pathogen develops best. It is difficult to deliver a contact fungicide directly to this surface even with specialized sprayers such as those equipped with electrostatic or air-assist technology (McGrath 2004). Mobile fungicides able to move through leaves (via systemic or translaminar activity) or around them (vapor activity) are key to successful management of cucurbit powdery mildew. These fungicides tend to have single-site mode of action and thus are at risk for resistance development.

The cucurbit powdery mildew pathogen has demonstrated it is prone to developing resistance. The primary pathogen in the USA is *Podosphaera xanthii*. It was one of the first pathogens demonstrated to have developed resistance in the USA. Strains of the pathogen resistant to MBC (FRAC code 1) fungicides were detected in 1967 before the first product in this chemical group was registered for this use. Since then, resistance or reduced sensitivity has been detected to most chemistry subsequently registered following a few years of commercial use. Biological features that render *P. xanthii* prone to resistance development include its short cycle (about 1 week) and abundant production of spores able to be dispersed over long distances. It also reproduces sexually, which, while not considered important in the disease cycle due to the abundance of conidia and low survival of ascospores, could be important if sexual recombination yielded a strain with a new combination of resistance genes or resistance plus fitness genes.

25.2 History of Fungicide Use and Resistance Development

25.2.1 MBC (Methyl Benzimidazole Carbamate) Fungicides (FRAC Code 1)

Benomyl was the first mobile fungicide developed with activity for powdery mildew and one of the first fungicides with consequent single-site mode of action (McGrath 2001). Ability of a fungicide to move to the abaxial surface of leaves where powdery mildew develops best, especially with large-leaved cucurbit crops like pumpkin, revolutionized control of this disease. Resistance to benomyl was detected in 1967 (Schroeder and Provvidenti 1969), 5 years before the first product (Benlate) was registered for commercial use for cucurbit powdery mildew. At that time, resistance was a new phenomenon and consequently its impact was unknown; thus the product was registered anyway. Control failures began occurring almost immediately. The first documented case was in a university fungicide efficacy experiment in 1973, the seventh year of such experiments and second year of commercial use (McGrath 2001). Pathogen sensitivity to benomyl was not examined; thus it can only be assumed that changes in efficacy are due to resistance. Resistance to MBC fungicides is qualitative. Although the wind-dispersed spores of the pathogen enable resistant strains to be widely dispersed, impact on efficacy was not widespread for several years. Benomyl was effective in the Midwestern USA (Indiana and Michigan) until the mid-1980s (McGrath 2001). Benomyl resistance was found in several areas of the USA when examined in the early 1990s (McGrath et al. 1996).

At this time, benomyl is no longer registered for this use while another MBC fungicide, thiophanate-methyl, formulated as Topsin M, is still labeled for cucurbit powdery mildew and thus available for use in production fields. Cross-resistance occurs among MBC fungicides. These fungicides are no longer recommended because resistance is thought to be widespread. Resistant strains have been found commonly where examined in recent years, despite limited use of thiophanate-methyl for any disease in cucurbit crops (McGrath et al. 2009).

25.2.2 DMI (Sterol Demethylation Inhibitor) Fungicides (FRAC Code 3)

Triadimefon, the first active ingredient in this group for cucurbit powdery mildew, was registered in the USA for this use in April 1984. It was the second fungicide for this disease with single-site mode of action and thus risk of resistance developing. Just 2 years after the start of commercial use, the first reported control failure documented through university fungicide efficacy experiments occurred (McGrath 2001). During the early 1990s, there were more reports of poor to ineffective control in several states and resistant strains of the pathogen were detected in several areas of the USA (McGrath et al. 1996). Research conducted then documented how quickly selection of resistant strains can occur following treatment; for example, within 2 weeks, the proportion of isolates resistant to triadimefon in a commercial production field shifted from undetectable to 96 % while resistance was not detected in another planting where this fungicide was not used (McGrath 1996).

While cross-resistance exists among DMI fungicides, resistance is quantitative, and there are inherent differences in activity among chemistry; thus DMI fungicides developed after triadimefon effectively controlled cucurbit powdery mildew. Myclobutanil was the second DMI fungicide developed. During its development, pathogen sensitivity to DMIs was shifting. As a result, the lowest rate in the range put on the label, which was effective early in the development of DMI resistance, was no longer adequately effective after myclobutanil was registered. However, when applied at a high rate, myclobutanil was effective in university experiments where triadimefon was ineffective and triadimefon-resistant pathogen strains were common. Myclobutanil was granted US federal (Section 3) registration in 2000. It was allowed to be used for controlling cucurbit powdery mildew in some states beginning in 1998 under an emergency exemption from registration (FIFRA Section 18) granted because both benomyl and triadimefon, the only mobile fungicides registered for this use at the time, were inadequately effective due to resistance. Triflumizole was the third DMI fungicide registered for controlling cucurbit powdery mildew. It was registered in 2002. The amount of active ingredient applied when used at highest label rate is twice that of myclobutanil; therefore, triflumizole is often the recommended fungicide. Myclobutanil and triflumizole have provided effective control of powdery mildew in most field efficacy evaluations conducted over those years until 2009 (McGrath and Hunsberger 2011). None of the DMI fungicides developed recently, which are difenoconazole, tebuconazole, and metconazole, have exhibited greater inherent activity than the DMIs currently registered, unlike the situation with myclobutanil being substantially more active than triadimefon.

Triflumizole has been highly effective for controlling cucurbit powdery mildew where tested recently. Typically it has been tested at its highest label rate, which provided a level of control similar to the best fungicide tested including in experiments conducted in 2013 (Gugino and Grove 2014; Rideout et al. 2014). Myclobutanil in these experiments was less effective (Gugino and Grove 2014; Matheron and Porchas 2014); however, at highest labeled rate, myclobutanil is applied at about half the dose of active ingredient as triflumizole. Difenoconazole and tetraconazole, which is not labeled yet for this use, were less effective than myclobutanil (Matheron and Porchas 2014). The dose in the spray tank when these fungicides are applied at the highest labeled rate at 50 gal per acre is 263 ppm difenoconazole (formulated as Inspire Super), 300 ppm myclobutanil (Rally), 363 ppm tetraconazole (Tebuzol), and 527 ppm triflumizole (Procure). Triflumizole has not been as consistently effective as newer chemistry (quinoxyfen) in efficacy experiments conducted in New York (Table 25.1).

DMI fungicides continue to be a recommended component of chemical control programs for cucurbit powdery mildew. Triadimefon is no longer available for this use; the other DMI fungicides are still registered. DMI fungicides are recommended for use in alternation with other effective chemistry.

Table 25.1 Efficacy for cucurbit powdery mildew of fungicides at risk for resistance development determined in replicated experiments conducted under field conditions with pumpkin in New York from 2005 to 2013[a]

		Control of powdery mildew on abaxial leaf surfaces (%)[b]									
Fungicide	Product	2005	2006	2007	2008	2009	2010	2011	2012	2013	2014
Pyraclostrobin	Cabrio		20 ab[c]								
Pyraclostrobin + boscalid	Pristine		89 efg	58 e	94 d-g	80 bcd	50 abc		40 abc	93 bcd	54 b
Boscalid	Endura	56 bc	90 fg	60 ef							
Penthiopyrad	Fontelis				82 b-e		50 abc	60 ab	33 abc		
Triflumizole	Procure M[d]	76 def	35 b	75 fg		40 ab	50 abc				
Triflumizole	Procure H[d]			78 g	93 d-g	72 cd	69 bcd	22 ab	57 bc	95 bcd	70 b
Quinoxyfen	Quintec		89 efg	81 g	99 g	86 d	95 d	41 ab	96 e	99 d	96 cd
Cyflufenamid	Torino							62 ab			
Metrafenone	Vivando						82 cd			97 bcd	98 bc
Alternation[e]	*Alternation*								89 de	98 cd	95 cd

[a]Reports on these experiments have been published (McGrath and Davey 2006, 2007; McGrath and Fox 2008, 2009, 2010; McGrath and Hunsberger 2011, 2012; McGrath and LaMarsh 2013, 2014b, unpublished). Fungicides were applied on 7-day intervals in all experiments

[b]Control of powdery mildew on lower leaf surfaces based on comparing area under disease progress curve (AUDPC) values for the fungicide to the non-treated control. AUDPC is a summation measure of severity ratings taken on several dates over the assessment period

[c]Values with an "a" indicate that the AUDPC value was not significantly different from the non-treated control for all years except 2011, when mean separation for the control was "ab." Efficacy of all fungicides tested in 2011 was likely affected by applications being started after powdery mildew was well established

[d]M = middle label rate. H = high label rate. Other products were applied at the highest label rate when labeled for use over a rate range

[e]Quinoxyfen was applied on odd application numbers alternated with triflumizole and pyraclostrobin + boscalid applied on the even times

25.2.3 QoI (Quinone Outside Inhibitor) Fungicides (FRAC Code 11)

QoI fungicides were the third chemical class of mobile fungicides developed with activity for cucurbit powdery mildew in the USA (McGrath 2001). Azoxystrobin was registered in the USA in the spring of 1999. It could be used in some states in 1998 where an emergency exemption was granted due to fungicide resistance. With myclobutanil also available, this was the first time in the USA that there was more than one effective class of fungicide with single-site mode of action that could be used together for cucurbit powdery mildew. At the time, resistance risk for QoIs was thought to be low; thus a strong resistance management program was not thought necessary to implement. In particular, it was not thought necessary to tank-mix azoxystrobin with a contact fungicide. However, the label did have use restrictions for resistance management, compelling farmers to implement a strict alternation (consecutive applications were not allowed; a fungicide in a different chemical group had to be used for the subsequent application). Trifloxystrobin was registered in the fall of 1999 and pyraclostrobin in 2002. Control failures with QoIs were reported from several states throughout the USA in 2002. This was associated with resistance to QoIs (McGrath and Shishkoff 2003). Resistant strains were confirmed to be present in Georgia, North Carolina, Virginia, and New York. Impact on control was dramatic, with failure occurring where QoIs were highly effective the previous year. Resistance to QoIs is qualitative. Cross-resistance occurs among the QoIs.

QoI-resistant strains of *P. xanthii* have been detected commonly where investigated in the USA. Monitoring work has been ongoing in New York since 2002. Occurrence of resistance has also been examined in New Jersey, Pennsylvania, Indiana, and Ohio (McGrath et al. 2009). While QoI fungicides generally are no longer recommended for cucurbit powdery mildew in the USA, they are still labeled for this use and there continued to be selection pressure to maintain QoI resistance in the *P. xanthii* pathogen population until recently because the only fungicide available with a SDHI active ingredient labeled for cucurbit powdery mildew from 2003 until 2012 also contains a QoI fungicide (Pristine containing boscalid and pyraclostrobin).

Surprisingly, pyraclostrobin and trifloxystrobin exhibited some efficacy when tested alone in a recent university fungicide evaluation experiment in Arizona (Matheron and Porchas 2014). These fungicides have not been tested recently in other locations. Pyraclostrobin exhibited no control, not even at the assessment after the second application, in an experiment conducted in New York in 2006 (Table 25.1) (McGrath and Davey 2007).

25.2.4 SDHI (Succinate Dehydrogenase Inhibitor) Fungicides (FRAC Code 7)

In 2003 the fourth chemical class of mobile fungicides at risk for resistance development became available for managing cucurbit powdery mildew in the USA. The first product, Pristine, contained boscalid plus pyraclostrobin. A product with just boscalid was not labeled for this disease for resistance management. Unfortunately, by the time Pristine was registered, resistance had already developed to the QoI premix partner in this product. The Pristine label dictated restrictions on use as additional steps for managing resistance: consecutive applications were not allowed and alternation had to be with a non-QoI fungicide, which was changed to "fungicide with a different mode of action" when other SDHI fungicides were developed.

Strains fully resistant to boscalid were first detected in 2008 (McGrath, *unpublished*). These strains were able to tolerate label rates (500 ppm) in a leaf-disk bioassay (McGrath and Fox 2010). Control failure in a fungicide evaluation conducted in 2009 in New Jersey was associated with their presence (Wyenandt, *personal communication*) and in 2010 in New York (McGrath and Hunsberger 2011). Pathogen strains have exhibited a range in sensitivity to boscalid.

Other SDHI fungicides have subsequently been registered in the USA. Two products with fluopyram (combined with tebuconazole or trifloxystrobin) were registered in February 2012, but only labeled at the time for use on watermelon due to government restrictions on tolerances in the cucurbit crop group until additional studies were conducted. Penthiopyrad was registered in March 2012. In 2014, the expansion of the fluopyram product labels to the full cucurbit crop group was in review by the US Environmental Protection Agency (EPA), which makes regulatory decisions about pesticides in the USA; also, additional SDHI fungicides are on track to be registered in the future. One product contains fluxapyroxad formulated with pyraclostrobin.

Cross-resistance was documented to occur among all SDHI fungicides tested except fluopyram. Fluopyram is a pyramide while the others are in the carboxamide group. Boscalid-resistant isolates were able to grow on leaf tissue treated with 500 ppm of the four carboxamide fungicides tested but only 10 ppm (not 50 ppm) of fluopyram (McGrath and LaMarsh 2014a). Isolates able to tolerate 500 ppm of these fungicides would not be controlled by them applied at label rates in production fields.

Efficacy of SDHI fungicides for cucurbit powdery mildew has varied from year-to-year and among locations. For example, penthiopyrad was moderately effective in an evaluation conducted in Florida in 2013, with several fungicides exhibiting better control, while it was very effective in an evaluation conducted in California in 2013 (Fan 2014; Nguyen et al. 2014). In fungicide evaluations conducted in New York, boscalid (formulated as Pristine) was ineffective in 2010 and 2012 whereas it provided more than 90 % control in 2008 and 2013 (Table 25.1). Quinoxyfen was consistently highly effective in these evaluations providing 95–99 % control (Table 25.1).

Occurrence of resistant strains of the pathogen has varied from year-to-year and among locations.

25.2.5 Aza-Naphthalenes (FRAC Code 13)

Registration of quinoxyfen (Quintec) in 2007 brought a fifth chemical class of mobile fungicides at risk for resistance development to the USA for managing cucurbit powdery mildew. It was only labeled for use on melon until 2009 when pumpkin, winter squash, and gourd were added to the label. This fungicide had not been labeled for use on edible-peel cucurbit crops, which include cucumber, summer squash, and zucchini. Use restrictions on the first label to manage resistance permitted a farmer to use quinoxyfen almost exclusively: "Do not make more than four applications of Quintec per crop nor more than two consecutive applications of Quintec before alternating to a fungicide that has a different mode of action and that is effective on local populations of the target powdery mildew pathogen." For some crops, a farmer might make five applications of a fungicide with targeted activity for powdery mildew; four of those could be quinoxyfen. However, the recommended spray interval of 10–14 days is longer than the 7-day interval used by many farmers, which could compel use of a strict alternation. The current label has a section on "Disease Resistance Management" with more restrictive language than what is in the separate "Specific Use Restrictions" section: "Do not make more than 50 % of the total number of powdery mildew sprays with Quintec." "When more than one application of Quintec is made per crop, at least one of the applications of Quintec must be a tank mixture with a product that has a different mode of action and that is effective on local populations of the target powdery mildew pathogen."

Quinoxyfen has continued to be highly effective in university fungicide evaluations (Table 25.1). However, isolates of *Podosphaera xanthii* able to detect 80 ppm quinoxyfen have been detected (McGrath 2014). Quinoxyfen was registered for use on grapes 4 years before cucurbits. Decline in efficacy of quinoxyfen for grape powdery mildew has been detected recently in two states and associated with presence of resistant strain (tolerating 100 ppm quinoxyfen) (Baudoin 2014). If this related pathogen with a similar history of developing resistance can serve as an indicator, then practical resistance can be expected to occur with cucurbit powdery mildew.

25.2.6 Phenylacetamide (FRAC Code U6)

The most recent fungicide with risk for resistance development to become available in the USA for managing cucurbit powdery mildew is cyflufenamid. It was registered for this use in July 2012. With only two applications permitted to a crop, this fungicide has the most restrictive label for compelling resistance management of all fungicides available for this disease in the USA. Resistance in *Podosphaera xanthii* has been detected outside of the USA in research plantings where cyflufenamid was applied four times in each of three consecutive years (Pirondi et al. 2014).

25.2.7 Aryl Phenyl Ketone (FRAC Code U8)

There are two fungicides in the aryl phenyl ketone group that are expected to be available in the near future for managing cucurbit powdery mildew in the USA. Metrafenone is in the benzophenone chemical group. A product with this ingredient (Vivando) was registered in 2011 with initial approval for use on grapes. Pyriofenone is in the benzoylpyridine chemical group. An application has been submitted for registration of a product with this ingredient to be labeled for use on cucurbits; a decision is expected in 2016. Both of these products have exhibited good efficacy in fungicide evaluations conducted by pathologists at universities. These fungicides are considered to have medium risk for resistance development. Isolates of the wheat powdery mildew pathogen that are resistant or less sensitive to this chemical group have been detected outside the USA (Felsenstein et al. 2010).

25.3 Procedures for Monitoring Resistance

Three approaches are being used in the USA to monitor resistance to fungicides in *Podosphaera xanthii*. Isolates are being tested using a leaf-disk bioassay. Pathogen populations are being examined using an in-field seedling fungicide sensitivity bioassay. These approaches have not been used extensively. The most commonly used approach is indirect and entails assessing whether the efficacy of fungicides under field use is at the level expected for the chemistry based on past performance; however, monitoring for resistance is usually not the reason each fungicide is being tested, but rather it is being included in the fungicide evaluation for comparison with new fungicides being tested. Consequently, these evaluations often do not explicitly include a fungicide from each FRAC group tested alone. Detecting resistance to one chemistry based on control failure in commercial production fields is difficult because farmers use several fungicides in combination. In this situation, lack of control with one fungicide due to resistance likely will be undetectable due to the control provided by the other products. Even when there is some loss of control, detection is difficult without other fungicide treatments for comparison as in a replicated experiment.

25.3.1 Leaf-Disk Bioassay for Assessing Fungicide Sensitivity in Podosphaera xanthii

A bioassay with treated leaf disks has been used to determine sensitivity to fungicides of isolates of *Podosphaera xanthii* in order to examine baseline sensitivity and shifts in sensitivity, detect resistance, and see cross-resistance. For this bioassay, *Cucurbita pepo* seedlings at the cotyledon stage are sprayed to coverage with fungicide suspensions using atomizer bottles connected to an air compressor operated at 30 psi (McGrath et al. 1996). Treated plants are left to dry over night in a fume hood, and then leaf disks are cut with a #9 cork borer and placed with adaxial surface up on water agar in segmented Petri dishes. Six disks treated with the same fungicide concentration are placed in each section. Non-treated disks are placed in one of the four sections. Each disk is inoculated in its center by transferring spores of the isolate to be tested from a leaf culture (cotyledon on water agar in a Petri

Fig. 25.1 Leaf-disk bioassay conducted in segmented Petri dish with sensitive reaction in section of dish on the *right* (no growth of tested isolate on disks treated with 100 ppm boscalid) and resistant reactions in sections on the *left* (80 ppm myclobutanil) and the *top* (10 ppm quinoxyfen). Non-treated disks are in the *bottom* plate section

dish). The tool used to transfer spores is a disposable pipette whose tip had been melted in a Bunsen burner flame to form a fine, sealed tip suitable for selecting small clumps of spores and for sterilizing in alcohol between isolates. Assay plates are incubated under constant light in the laboratory. About 9 and 14 days after inoculation, disks are examined for pathogen growth (Fig. 25.1). A dissecting microscope is used when growth is not evident with the unaided eye. An isolate is rated sensitive to a fungicide concentration when it is unable to grow and produce conidia on any of the six disks receiving the treatment, tolerant if there is growth on fewer than half of the disks, and resistant if there is growth on most of the disks. Percent of each disk with visible powdery mildew is estimated in order to be able to determine whether growth was suppressed by the fungicide. Average percent growth on the fungicide-treated disks is compared to the non-treated disks. The ability of an isolate to produce conidia when growing on fungicide-treated leaf tissue, and thereby multiple, is considered an important measure of tolerance/resistance; therefore, the focus has been on the presence of growth rather than quantity in assessments. This bioassay is being used exclusively in New York by the author.

25.3.2 In-Field Seedling Bioassay for Assessing Fungicide Sensitivity in Podosphaera xanthii

A bioassay with treated seedlings was developed to assess fungicide sensitivity of a pathogen population (McGrath and Shishkoff 2001). For the protocol currently in use, seedlings are sprayed with various fungicide concentrations; after drying overnight, they are placed for at least 4 h in fields where powdery mildew is developing (Fig. 25.2) and then kept in a greenhouse until symptoms develop about 10 days later

Fig. 25.2 In-field seedling bioassay used to assess fungicide sensitivity in a pathogen population entails leaving fungicide-treated and non-treated seedlings for at least 4 h in a planting affected by powdery mildew

(Lebeda et al. 2010). Severity is visually estimated on each leaf. Severity on treated seedlings is compared to non-treated ones to estimate frequency of the pathogen population able to tolerate each fungicide concentration tested. This seedling bioassay has been used for populations of *Podosphaera xanthii* in commercial production fields and in research plots. This bioassay has been used primarily in New York. In 2008 this bioassay was also conducted in Indiana, Ohio, and Pennsylvania. In the populations examined in the four states, 49–94 % of the individuals were estimated to be resistant to FRAC code 1 fungicides, 64–100 % were estimated to be resistant to FRAC code 11 fungicides, 4–25 % tolerated 120 ppm myclobutanil (FRAC code 3 fungicide), 8–62 % tolerated 50 ppm boscalid (FRAC code 7), and 0–1 % tolerated 10 ppm quinoxyfen (FRAC code 13) (McGrath et al. 2009).

25.3.3 Fungicide Evaluations to Detect Decline in Efficacy

Efficacy of new fungicides is being assessed in replicated experiments conducted by applied plant pathologists typically located at land-grant universities in the USA. While the experiment objectives usually do not include examining whether occurrence of resistance in the pathogen is affecting fungicide efficacy, registered fungicides being used routinely by farmers are often included in the experiment as standards for comparison with the new fungicide(s). Monitoring resistance development through monitoring fungicide efficacy has been a goal of experiments conducted in New York by the author. Each fungicide tested is applied five to seven times on a weekly schedule with a tractor sprayer to replicated fungicide plots. Powdery mildew severity is also assessed weekly by rating severity on both surfaces of leaves. Some results from some of the experiments conducted in New York are presented in Table 25.1.

25.4 Challenges of Fungicide Resistance in the Cucurbit Powdery Mildew Pathogen

The challenges that fungicide resistance poses with cucurbit powdery mildew extend from prediction through management to detection. Some are also applicable to other pathosystems. While it is generally understood that fungicides with single-site mode of action are at risk for resistance development, it can be hard to predict how quickly resistance might develop and whether the chemistry should be considered low, medium, or high risk. The QoI fungicides exemplify these challenges. Risk was not considered high when the first product was registered, type of resistance was expected to be quantitative, and the cucurbit powdery mildew pathogen was expected to be the first to develop resistance.

Managing resistance has been challenged primarily by pathogen ability to move long distances and, until recently, lack of adequate tools. Unless a regional resistance management program is implemented, farmers do not have much incentive to manage resistance when the actions of other farmers can affect their success due to pathogen movement between farms during a season. Incentive is much greater for a pathogen that resides on their farm with limited movement among farms. In addition to there being limited perceived benefit for a farmer to manage resistance in *Podosphaera xanthii*, doing so can increase cost of managing powdery mildew and potentially reduce control if other fungicides used in the program are more expensive or not as effective. Therefore, another benefit of implementing a resistance management program is emphasized to farmers, which is minimizing loss due to poor control if pathogen strains are present with resistance to the main fungicide being applied. This benefit directly impacts them.

Farmers have had to learn that implementing resistance management programs is done preventively and thus begins before resistance develops, which is in contrast with the chemical control programs they have been encouraged to adopt as a component of Integrated Pest Management (IPM). These programs utilize insect or disease occurrence thresholds to determine when to start applications; therefore, treatment starts after the pest problem has begun to develop. The IPM approach may itself challenge resistance management because the disease is initially allowed to develop unhindered; thus selection pressure for fungicide resistance begins on an established pathogen population (Beckerman et al. 2013).

Until recently, managing resistance has been challenged by lack of tools, in particular other chemistry to use in alternation with an at-risk fungicide. From 1972 to 1997, farmers had at most only one effective fungicide with mobile activity needed to achieve effective control of cucurbit powdery mildew. Thus it was not possible to effectively implement a standard element of a resistance management program, which is to use more than one chemical class. Contact fungicides with their very low risk of resistance developing (e.g., chlorothalonil) have been recommended tank-mixed with fungicides with higher risk; however, they are not as suitable as other higher-risk fungicides because activity of contact fungicides is limited to where spray is deposited on foliage during the application, which is almost exclusively the adaxial surface of leaves, especially the large leaves of many cucurbit crops. Thus at-risk fungicides typically are the only ones acting on the pathogen (including exerting selective pressure for resistance) where the pathogen develops best. Chlorothalonil was applied with triadimefon in an experiment conducted in 1993 in which the frequency of triadimefon-resistant isolates shifted from 3 % after the first application to 71 % 21 days later, which provides no evidence that the contact fungicide was slowing selection of resistant isolates (McGrath and Shishkoff 2001).

In a few states in 1998, farmers were able to use two new at-risk fungicides (azoxystrobin and myclobutanil) before they were granted federal registration as a result of US EPA issuing an emergency exemption from registration (Section 18). This was a landmark event that has not occurred again. While most emergency exemptions granted since in the USA have been because of the impact of resistance on ability to manage a disease, exemptions have not again been issued for more than one product at the same time because of resistance. In most states, azoxystrobin was the only fungicide available in 1999 not affected by resistance for managing cucurbit powdery mildew. 2000 was the first year US farmers had two effective mobile fungicides to use in a resistance management program. Resistance to QoI fungicides was detected in 2002, which seems to be quick occurrence; however, resistance developed sooner than the fourth year of commercial use elsewhere in the world, suggesting that maybe resistance was delayed in the USA because there were two products to use in alternation plus use restrictions that compelled an alternation program be used with azoxystrobin. Presently (2014), farmers have fungicides in three or four chemical classes (depending on the crop type) to use for managing powdery mildew and resistance.

Isolates of *Podosphaera xanthii* have been detected with resistance or reduced sensitivity to multiple chemistries. This is not a new phenomenon: it was documented with the first two fungicides the pathogen developed resistance to. Most isolates (78 %) collected in September 1991 and September 1992 from fungicide-treated research plots and commercial fields in New York were resistant to both benomyl and triadimefon (McGrath 1996). More recently, one of the isolates collected near the end of the 2010 growing season was resistant to FRAC code 1 and 11 fungicides and to boscalid (FRAC 7), and it was able to grow on leaf disks treated with 80 ppm myclobutanil (FRAC 3) and 40 ppm quinoxyfen (FRAC 13), which were the highest concentrations of these fungicides tolerated by any of the isolates tested that year (McGrath 2013).

Detecting resistance has become more difficult. Resistance has been detected in the USA when efficacy of the fungicide declines. When the pathogen population includes strain(s) fully resistant to a fungicide (unsuppressed by the fungicide; field resistance) and it is the only effective mobile fungicide available, its use will select for resistance, and with consecutive applications, the frequency of resistant individuals likely will increase to a level that control failure occurs, thereby facilitating detection. Currently there are multiple mobile fungicides available. Applying these in alternation will reduce selection pressure, potentially not increasing the frequency of resistant strains to the point of control failure. And even if there is a high frequency of resistant strains, control achieved with the other fungicides could mask lack of control with the fungicide affected by resistance. Thus the best opportunity to detect resistance is in a fungicide evaluation experiment where fungicides are applied alone. Limited research is being conducted to monitor sensitivity to fungicides in the pathogen.

25.5 Current Situation

Powdery mildew is being managed effectively in cucurbit crops in the USA by applying at-risk fungicides in alternation to a resistant cultivar when applicable. Resistance to powdery mildew has become a common trait in many cucurbit crop types. Resistance provides a high level of suppression in some types (notably cucumber), variable in others (melon), and a low level in most (squash and pumpkin). Genetic resistance is the only viable cultural practice for cucurbit powdery mildew.

Currently farmers have unprecedented three or four (depending on crop type) groups of effective fungicide chemistry with targeted activity for powdery mildew to choose from, thereby enabling them to implement a theoretically good resistance management program with up to four different modes of action. Quinoxyfen (FRAC code 13) is not labeled for use on edible-peel cucurbit crop types, which includes zucchini and summer squashes. Quinoxyfen has continued to consistently be highly effective in fungicide evaluations conducted by university pathologists (through 2014) (Table 25.1). However, pathogen strains with reduced sensitivity have been detected, in particular in research plots where quinoxyfen has been used exclusively, documenting the pathogen's ability to adapt to this chemistry. The primary fungicide being recommended now in the USA is cyflufenamid (FRAC code U6), which is the newest chemistry. Label restriction on use to two applications compels farmers to implement a resistance management program. Farmers' increased knowledge about and experience with resistance in this pathogen help to ensure they will implement a good program. It is recommended that fungicides in FRAC groups 3 and 7 be integrated into an alternation program with cyflufenamid and quinoxyfen. Although strains of *Podosphaera xanthii* with reduced sensitivity or resistance to fungicides in FRAC groups 3 and 7 have been detected in the USA, these fungicides usually have exhibited good, albeit variable, efficacy when tested alone in field experiments conducted at universities (Table 25.1). In addition to temporal variation in efficacy, there has been spatial variation, most notably between experiments conducted in eastern and western states, which is not surprising as these pathogen populations are geographically separated. Triflumizole and fluopyram currently are considered the best choices based on efficacy relative to other fungicides in these groups and lack of cross-resistance with fluopyram (Ishii et al. 2011). Alternation programs with multiple fungicide chemistry have often been at least as effective as the best single fungicide in university experiments (Gugino and Grove 2014; Rideout et al. 2014). It is also recommended to farmers, for managing resistance and minimizing the impact on control of resistant strains, that they apply a contact protectant fungicide such as sulfur (when powdery mildew is the only disease occurring) or chlorothalonil (when other fungal diseases are also a concern) or copper (when bacterial diseases are also a concern). Additional fungicides are in development, including in a new chemical group; thus the prospect is good that farmers will continue to have a diversity of chemistry to use in alternation to manage fungicide resistance.

25.6 Conclusions

Fungicide resistance has been and is expected to continue to be an important issue pertaining to effectively managing powdery mildew in cucurbit crops. Fungicides are expected to continue to be the most important management tool for this important disease. Fungicides at risk for resistance development are essential for effective management due to their mobility enabling them to move to the abaxial surface of leaves where the pathogen develops best. This pathogen has proven well its ability to develop resistance to fungicides. Currently farmers in the USA have effective fungicides in four chemical groups to use to manage powdery mildew and fungicide resistance in cucurbit crops. Fungicides in a fifth group are in the process of being registered. This unprecedented diversity in chemistry to use together in a fungicide management program provides an opportunity at the production level to examine how well an integrated program can delay resistance development.

References

Beckerman JL, Sundin GW, Rosenberger DA (2013) Do some IPM concepts contribute to the development of fungicide resistance? Lessons learned from the apple scab pathosystem in the United States. Pest Manag Sci 12/2013. doi: 10.1002/ps.3715

Baudoin A (2014) First confirmation of resistance to quinoxyfen in grape powdery mildew in North America. Phytopathology 104(Suppl 3):S3.160 (Abstr)

Gugino B, Grove TL (2014) Efficacy of select fungicides and fungicide programs for the management of powdery mildew on pumpkin, 2013. Plant Dis Manag Rep 8:V206

Fan X (2014) Field evaluation of fungicides to control squash powdery mildew, 2013. Plant Dis Manag Rep 8:V236

Felsenstein F, Semar M, Stammler G (2010) Sensitivity of wheat powdery mildew (*Blumeria graminis* f.sp. *tritici*) towards metrafenone. Gesunde Pflanzen 62:29–33

Ishii H, Miyamoto T, Ushio S, Kakishima M (2011) Lack of cross-resistance to a novel succinate dehydrogenase inhibitor, fluopyram, in highly boscalid-resistant isolates of *Corynespora cassiicola* and *Podosphaera xanthii*. Pest Manag Sci 67:474–482

Lebeda A, McGrath MT, Sedlákova B (2010) Fungicide resistance in cucurbit powdery mildew fungi. Chapter 11. In: Carisse O (ed) Fungicides. InTech Publishers, Vienna, pp 221–246. ISBN 978-953-307-266-1

Matheron M, Porchas M (2014) Evaluation of fungicides for management of powdery mildew on muskmelon, 2013. Plant Dis Manag Rep 8:V197

McGrath MT (1996) Increased resistance to triadimefon and to benomyl in *Sphaerotheca fuliginea* populations following fungicide usage over one season. Plant Dis 80:633–639

McGrath MT (2001) Fungicide resistance in cucurbit powdery mildew: experiences and challenges. Plant Dis 85:236–245

McGrath MT (2004) Efficacy of Bravo for powdery mildew control in muskmelon and pumpkin as affected by nozzle type and sprayer, 2003. Fungic Nematic Tests 59:V057

McGrath MT (2013) Efficacy of fungicides with resistance risk for cucurbit powdery mildew and fungicide sensitivity of *Podosphaera xanthii* in New York. Phytopathology 103(Suppl. 2):S2.93 (Abstr)

McGrath MT (2014) Status of fungicide sensitivity in *Podosphaera xanthii* and impact on managing cucurbit powdery mildew in New York. Phytopathology 104(Suppl 3) (in press) (Abstr)

McGrath MT, Davey JF (2006) Evaluation of fungicide programs for managing powdery mildew of pumpkin, 2005. Fungic Nematic Tests 61:V038

McGrath MT, Davey JF (2007) Evaluation of fungicides for management of powdery mildew on pumpkin, 2006. Plant Dis Manag Rep 1:V144

McGrath MT, Egel DS, Jasinski J et al (2009) Fungicide sensitivity and resistance of the cucurbit powdery mildew pathogen in New York, Pennsylvania, Ohio, and Indiana in 2008. Phytopathology 99:S82 (Abstr)

McGrath MT, Fox GM (2008) Evaluation of fungicides for managing powdery mildew in pumpkin, 2007. Plant Dis Manag Rep 2:V142

McGrath MT, Fox GM (2009) Efficacy of fungicides for managing cucurbit powdery mildew and treatment impact on pathogen sensitivity to fungicides, 2008. Plant Dis Manag Rep 3:V125

McGrath MT, Fox GM (2010) Efficacy of fungicides for managing cucurbit powdery mildew and treatment impact on pathogen sensitivity to fungicides, 2009. Plant Dis Manag Rep 4:V147

McGrath MT, Hunsberger LK (2011) Efficacy of fungicides for managing cucurbit powdery mildew and pathogen sensitivity to fungicides, 2010. Plant Dis Manag Rep 5:V104

McGrath MT, Hunsberger LK (2012) Efficacy of fungicides for managing cucurbit powdery mildew and pathogen sensitivity to fungicides, 2011. Plant Dis Manag Rep 6:V080

McGrath MT, LaMarsh KA (2013) Efficacy of fungicides for managing powdery mildew in pumpkin, 2012. Plant Dis Manag Rep 7:V013

McGrath MT, LaMarsh, KA (2014a) Cross resistance among some SDHI fungicides documented in *Podosphaera xanthii*, the cucurbit powdery mildew pathogen. Phytopathology 104(Suppl. 1):S1.4. http://apsjournals.apsnet.org/doi/pdf/10.1094/PHYTO-104-3-S1.1 (Abstr)

McGrath MT, LaMarsh KA (2014b) Efficacy of fungicides for managing powdery mildew in pumpkin, 2013. Plant Dis Manag Rep 8:V204

McGrath MT, Shishkoff N (2001) Resistance to triadimefon and benomyl: dynamics and impact on managing cucurbit powdery mildew. Plant Dis 85:147–154

McGrath MT, Shishkoff N (2003) First report of the cucurbit powdery mildew fungus (*Podosphaera xanthii*) resistant to strobilurin fungicides in the United States. Plant Dis 87:1007

McGrath MT, Staniszewska H, Shishkoff N et al (1996) Fungicide sensitivity of *Sphaerotheca fuliginea* populations in the United States. Plant Dis 80:697–703

Nguyen TT, Bay AA, Abramian WD et al (2014) Management of cucurbit powdery mildew with organic and conventional fungicides, 2013. Plant Dis Manag Rep 8:V233

Pirondi A, Nanni IM, Brunelli A et al (2014) First report of resistance to cyflufenamid in *Podosphaera xanthii*, causal agent of powdery mildew, from melon and zucchini fields in Italy. Plant Dis 98:1581–1581

Rideout S, Waldenmaier CM, Gu G et al (2014) Evaluation of selected fungicides for the management of powdery mildew in acorn squash, 2013. Plant Dis Manag Rep 8:V251

Schroeder WT, Provvidenti R (1969) Resistance to benomyl in powdery mildew of cucurbits. Plant Dis Rep 53:271–275

Chapter 26
Grapevine Gray Mold in France

Anne-Sophie Walker and Pierre Leroux

Abstract Chemical control remains the main approach for decreasing the incidence of gray mold, a disease of many crops, including grapevine, caused principally by *Botrytis cinerea*. Fungicides with seven different modes of action are currently authorized in French vineyards, but specific resistance has developed against five of these modes of action. Multidrug resistance caused by fungicide efflux has been detected and affects all the classes of fungicide used. Here, we present the history and current status of resistance to the various botryticides in French vineyards. We also discuss resistance management options, based on decreases in the number of botryticide applications per season and the alternation of single products from different classes of molecules with different biochemical modes of action.

Keywords *Botrytis cinerea* • Gray mold • Botryticide • Fungicide resistance • Resistance mechanism • Resistance monitoring • Resistance management

26.1 Introduction

The three main fungal diseases attacking the aerial organs of grapevine are downy mildew, powdery mildew, and gray mold. Gray mold is also observed on many other dicotyledonous crops and is caused by *Botrytis cinerea* and *Botrytis pseudocinerea*. In France, these two cryptic species can be found in sympatry on grapevine, but most of the damage to bunches is caused by *B. cinerea* (Walker et al. 2011). Attacks close to harvest time, resulting in a large number of rotten berries, can greatly decrease yield and modify wine taste and color.

Chemical control is still the chief method for decreasing the incidence of gray mold in grapevine. In the 1960s and 1970s, up to four botryticide applications per year were recommended: at stages A (end of flowering), B (bunch closure), C (veraison), and D (3 weeks before harvest). Following the introduction of highly effective botryticides, programs involving only three treatments (at stages A, B, and

A.-S. Walker (✉) • P. Leroux
INRA AgroParisTech, UMR 1290 BIOGER, Avenue Lucien Brétignières, F-78850 Thiverval-Grignon, France
e-mail: walker@versailles.inra.fr

H. Ishii, D.W. Hollomon (eds.), *Fungicide Resistance in Plant Pathogens*, DOI 10.1007/978-4-431-55642-8_26

C) were considered satisfactory in the 1990s. Nowadays, appropriate prophylactic measures are favored, and only two applications, at stages A and B or A and C, are permitted to achieve the control of gray mold in situations at risk, particularly in Champagne vineyards (Leroux 1995; Walker et al. 2013). Synthetic botryticides include multisite toxicants, anti-microtubule agents, compounds affecting osmoregulation, methionine biosynthesis inhibitors, sterol biosynthesis inhibitors, and substances interfering with mitochondrial respiration (Leroux 2004). The use of these fungicides has been restricted by the emergence of *B. cinerea* strains resistant to most of them and by toxicological considerations. We describe the history and current status of resistance to the various botryticides used in French vineyards in this chapter. Most of the data were obtained in the Champagne region, where fungicides have always been used intensively because of the high disease pressure. We describe the characteristics of *B. cinerea* strains displaying either specific resistance to the different classes of botryticides or multidrug resistance. Resistance management options, based on a decrease in the number of botryticide applications per season and the alternation of single products from different classes of fungicides with different biochemical modes of action, are discussed.

26.2 Multisite Toxicants

The first botryticides to be introduced were multisite toxicants. These compounds are toxic to conidia, as they block several thiol-containing enzymes involved in the breakdown of reserve molecules and cellular respiration. They include thiram (dithiocarbamate), captan, dichlofluanid, folpet (chloroalkyl mercaptan derivatives), and chlorothalonil (phthalonitrile). Despite their multisite effect in *B. cinerea*, only a few cases of practical resistance have been reported on ornamental plants and vegetables. It has been suggested that their reaction with nonessential thiol components, such as glutathione in particular, leads to the detoxification of several multisite toxicants (Leroux et al. 2002). Strains of *B. cinerea* with reduced sensitivity to this class of fungicides have never been detected in French vineyards. Nevertheless, the practical use of these fungicides was restricted for toxicological reasons and because these compounds have only weak efficacy in the field and an inhibitory effect on must fermentation. Thiram was the last multisite botryticide used in French vineyards, and its application was restricted to stage A. However, multisite toxicants remain essential for the management of fungicide resistance in *Plasmopara viticola*, the causal agent of grapevine downy mildew (Corio-Costet 2012).

26.3 Anti-Microtubule Agents

Carbendazim and its precursors, benomyl and thiophanate-methyl, were introduced into European vineyards in the late 1960s, for the control of gray mold and powdery mildew. These benzimidazole fungicides prevent microtubule assembly by binding

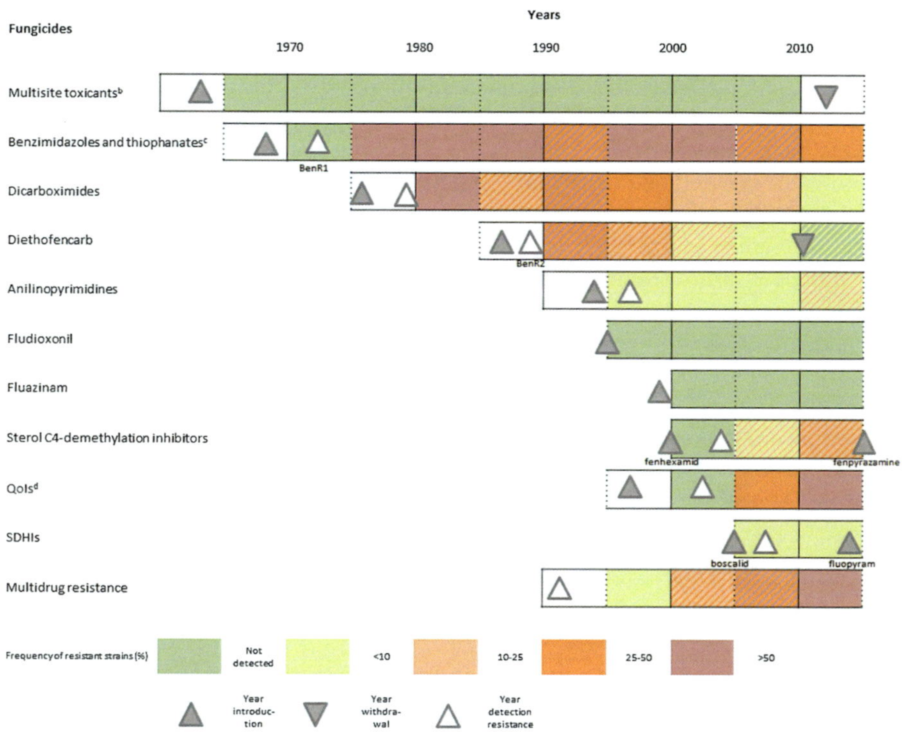

Fig. 26.1 Evolution of resistance in *B. cinerea* populations from Champagne vineyards (France)[a]. [a]Results from the monitoring conducted by INRA and CIVC; see Walker et al. (2013). [b]Several multisite toxicants have been introduced before 1965 and the last one, withdrawn in 2012, was thiram. [c]BenR1, strains resistant to benzimidazoles and hypersusceptible to phenylcarbamates; BenR2, strains resistant to benzimidazoles and phenylcarbamates. [d]QoIs were introduced in the mid-1990s against downy and powdery mildew only

to β-tubulin (Davidse and Ishii 1995). In *B. cinerea*, highly resistant strains (phenotype BenR1; Leroux et al. 2002) were selected soon after the commercial release of these compounds. For instance, BenR1 strains had become generalized in Champagne vineyards after only 3 years of use, with four treatments per season. This problem led to the cessation of benzimidazole treatments in the mid-1970s. Benzimidazole resistance did not decline over the next decade, despite the absence of selection pressure, suggesting that resistance has little or no cost in BenR1 strains (Fig. 26.1; Leroux 1995). The discovery of negative cross-resistance to anti-microtubule phenylcarbamate derivatives in BenR1 led to the renewed use of carbendazim in combination with diethofencarb in 1987. This mixture, initially recommended for use twice per season, rapidly selected *B. cinerea* strains resistant to both carbendazim and diethofencarb (phenotype BenR2; Leroux et al. 2002). Despite subsequent limitation to a single application per season, the frequency of BenR2 strains increased strongly in the early 1990s (Fig. 26.1; Leroux 1995). The use of carbendazim + diethofencarb mixtures has thus been abandoned since 1993. This led to a rapid decrease in the frequency of BenR2 strains to low levels, suggesting that resistance was associated with high fitness penalties. As for BenR1 strains, the observed decline in these strains since the mid-2000s (Fig. 26.1; Walker et al. 2013) coincided with the introduction of zoxamide for the control of downy mildew (Corio-Costet 2012). This benzamide, the first commercial anti-microtubule agent to be used against oomycetes, is also active against *B. cinerea.* Its fungitoxicity is highest in BenR1 strains (Young and Slawecki 2005; Malandrakis et al. 2011). Benzimidazole resistance is conferred by allelic mutations in the gene encoding β-tubulin, leading to a G198A substitution in BenR1 strains and a F200Y substitution in BenR2 strains (Leroux et al. 2002). Field strains harboring G198K/V substitutions have also been detected on other crops (Banno et al. 2008).

Finally, there is a lower risk of resistance to benzimidazole botryticides in practice if they are used in combination with diethofencarb, favoring a limited competitiveness phenotype (i.e., BenR2). These anti-microtubule agents should not be used after veraison, to prevent the excessive accumulation of residues. In the past, a single application of the carbendazim + diethofencarb mixture was recommended, at stage A. However, this mixture is no longer authorized for use in France. Thiophanate-methyl is still available but is of limited practice value due to the residual presence of BenR1 strains.

26.4 Compounds Affecting Osmoregulation

Dicarboximides (i.e., iprodione, procymidone, vinclozolin) replaced benzimidazoles for the control of gray mold in the late 1970s. The phenylpyrrole fludioxonil has a structure very similar to that of the antibiotic pyrrolnitrin (Ajouz et al. 2011). It was introduced into French vineyards in the mid-1990s. These botryticides induce an accumulation of glycerol in fungal cells, suggesting a probable effect on osmotic signal transduction (Fujimura et al. 2000). This hypothesis is supported by the demethylation that *B. cinerea* laboratory mutants highly resistant to dicarboximides and phenylpyrroles are also hypersusceptible to osmotic stress. This phenotype is associated with qualitative modifications to the class III histidine kinase Bos1 (Leroux 2004; Cui et al. 2004). The *N*-terminus of this cytoplasmic osmo-sensing protein contains a coiled-coil region formed by six tandem repeats of about 90 amino acids (also known as HAMP domains). Many of the substitutions in laboratory mutants concern amino acids located in the helical regions of HAMP domains (Fillinger et al. 2012). The general characteristics of these mutants are similar to those of null mutants, which display not only fungicide resistance and osmosensitivity but also enhanced sensitivity to oxidative stress, low levels of sporulation, and restricted pathogenicity. These pleiotropic effects suggest that Bos1 is probably involved in several signal transduction systems leading to glycerol accumulation and several other metabolic disorders (Liu et al. 2008). It remains to be determined whether dicarboximides and phenylpyrroles inhibit histidine kinase activity by binding to the HAMP domains of Bos1.

B. cinerea field strains resistant to dicarboximides have low to moderate levels of resistance, and unlike laboratory mutants, they remain susceptible to phenylpyrroles and their osmotic sensitivity is not abnormal. However, this resistance phenomenon is also determined by mutations in the gene encoding Bos1, resulting in substitutions in the loop between the second and third HAMP domains. In most cases, a hydrophilic residue (i.e., arginine, asparagine, or serine) replaces the hydrophobic isoleucine at position 365. I365V+Q369H, E369P+N373S, and V368F+Q369H+T447S combinations and the insertion of a serine residue between codons 271 and 272 have also been observed (Banno et al. 2008; Cui et al. 2004; Leroux et al. 2002). Dicarboximides were initially registered for use in France without restrictions. In Champagne, between 1977 and 1982, most plots were treated four times per season, leading to the development of practical resistance to these botryticides. In other French vineyards, the situation was less critical, because dicarboximides were applied only twice per season, on average (Leroux and Clerjeau 1985). It was also shown that a single dicarboximide fungicide application at stage B or C had a limited effect on fungal populations. A brief period in which dicarboximide sprays were not used led to a large decrease in the frequency of resistant strains, suggesting the association of resistance with a high fitness penalty. The limitation of dicarboximide use was recommended, together with the combination of these botryticides with a multisite toxicant, such as thiram (Leroux 1995). However, since the introduction of new botryticides in the mid-1990s, the use of dicarboximides (alone or in mixtures) has gradually decreased in all French vineyards, and this decrease has been accompanied by a decrease in the frequency of resistant strains (Fig. 26.1; Walker et al. 2013). Procymidone and vinclozolin have recently been withdrawn for toxicological reasons. Iprodione (the weakest dicarboximide against gray mold on grapevine) is the only compound from this group still authorized, and its use is now restricted to a few vineyards in the south of France.

Fludioxonil has been available in France since 1995, for use either alone or in combination with cyprodinil. Monitoring in grapevine and other crops has not resulted in the detection of *B. cinerea* strains moderately to highly resistant to this phenylpyrrole fungicide. Such mutants, which also display cross-resistance to dicarboximides, are easily selected in the laboratory. Their absence in natural conditions probably reflects their weak capacity to act as saprophytes and parasites (Liu et al. 2008). Despite the low risk of resistance in practice, this phenylpyrrole fungicide was registered in France with the restriction that it should be used only once per season on grapevine. Due to its high persistence, it can be applied only at stage A or B. It is currently one of the most widely used botryticides in French vineyards (Walker et al. 2013).

26.5 Methionine Biosynthesis Inhibitors

Anilinopyrimidines were introduced as botryticides in the mid-1990s, for use alone (mepanipyrim, pyrimethanil) or in mixtures with fludioxonil (cyprodinil). In laboratory conditions, their inhibitory effects are greatest when *B. cinerea* is grown on minimal media. Moreover, several amino acids, including methionine in particular, can antagonize their fungitoxicity. Biochemical studies have suggested that cystathionine β-lyase may be the target of anilinopyrimidines. These compounds also have the common property of preventing the secretion of enzymes involved in the infection process. It has been suggested that anilinopyrimidines affect protein secretory pathways at a step involving the Golgi complex (Leroux et al. 2002). Field strains of *B. cinerea* moderately to highly resistant to anilinopyrimidines have been detected on several crops. The molecular characterization of genes encoding key enzymes in methionine biosynthesis (i.e., cystathionine β-lyase, cystathionine γ-lyase, cystathionine β-synthase, cystathionine γ-synthase) in resistant strains demonstrated the presence of several single-nucleotide polymorphisms, some leading to amino-acid changes, but none associated with anilinopyrimidine resistance (Leroux 2004; De Miccolis Angelini et al. 2012). Additional studies are required to identify the primary target of anilinopyrimidine fungicides and the product of the gene involved in specific resistance to this class of botryticides.

The first case of practical resistance to anilinopyrimidines in *B. cinerea* was recorded in a long-term trial conducted in a Swiss vineyard (Forster and Staub 1996). Such resistance was subsequently reported in various commercial crops treated several times during a single season (Leroux 2004; Bardas et al. 2008). However, no decrease in performance has ever been detected in French vineyards, because of the poor establishment of anilinopyrimidine resistance. This favorable situation results from the restrictions imposed on the use of anilinopyrimidine-containing botryticides since their introduction, with no more than one application per year permitted on grapevine. Programs including cyprodinil + fludioxonil mixtures and the use of mepanipyrim, pyrimethanil, and fludioxonil can no longer be implemented. Moreover, only mepanipyrim and pyrimethanil can be applied at stage C, a stage frequently used for the application of fungicide treatments in Champagne.

26.6 Sterol Biosynthesis Inhibitors (SBIs)

The only SBIs used against gray mold in grapevine are the hydroxyanilide fenhexamid and the recently introduced pyrazolinone fenpyrazamine. Both target 3-ketoreductase, one of the enzymes involved in sterol C4-demethylation, encoded by the *Erg27* gene (Debieu et al. 2001). Even before fenhexamid was released onto the market, *Botrytis* strains resistant to this hydroxyanilide were detected at low frequency in French vineyards. These strains correspond to the minor species *B. pseudocinerea* (HydR1 phenotype). This natural resistance, which is strongly expressed at the mycelial growth stage, results mostly from fenhexamid detoxification, due to overproduction of the cytochrome P450 monooxygenase Cyp684 (Billard et al. 2011). Under field conditions, preventive applications of fenhexamid can control *B. pseudocinerea*, probably because this species is susceptible at the conidial stage (Suty et al. 1999). According to a survey of Champagne vineyards, *B. pseudocinerea* is not subject to fenhexamid selection. In *B. cinerea*, strains resistant to fenhexamid were first detected in Champagne four years after the release of this fungicide onto the market. This acquired resistance was shown to be mediated by target modification, due to the mutation of *Erg27*. The most frequent mutants, with the highest resistance levels, harbor a substitution in which phenylalanine is replaced by an isoleucine, serine, or leucine residue (F412I/S/L) at the *C*-terminus of the putative transmembrane domain of the 3-ketoreductase (Fillinger et al. 2008). Since its introduction into French vineyards in 2000, fenhexamid has been authorized for use only once per year. As positive cross-resistance occurs with fenpyrazamine (Tanaka et al. 2012); the restriction to one application per year also applies to C4-demethylation inhibitors. Despite these restrictions, resistance has tended to progress in most French vineyards since the mid-2000s. For instance, in Champagne, where fenhexamid is usually applied at stage A, resistant strains are now present in more than 80 % of plots, but their mean frequency within plots rarely exceeds 20 % (Fig. 26.1; Walker et al. 2013). A resistance cost was recently demonstrated, mostly expressed at the conservation stage, for strains with F412I/S/L substitutions (Billard et al. 2012), possibly counterbalancing fenhexamid selection. No significant decrease in the efficacy of fenhexamid against gray mold has yet been reported. The discontinuous use of sterol C4-demethylation inhibitors, through alternation with fungicides with other modes of action over a cycle of several years, is recommended.

26.7 Fungicides Specifically Affecting Respiration

26.7.1 Uncouplers of Oxidative Phosphorylation

The dinitroaniline derivative fluazinam uncouples oxidative phosphorylation in mammalian mitochondria, through a protonophoric cycle involving its secondary amine group (Guo et al. 1991). However, it has not yet been demonstrated that this mechanism is responsible for its fungitoxicity. In Japan, fluazinam was introduced for use against gray mold in 1990, and practical resistance was reported 6 years later, in bean crops (Tamura 2000). The genetic and biochemical bases of fluazinam resistance in *B. cinerea* have yet to be demonstrated, but it is thought that this resistance involves a detoxification-based mechanism (Leroux et al. 2002). Fluazinam was released onto the market for use in French vineyards in 1999, and as a precaution, its use was restricted to one spray per season. Field monitoring has yet to detect any significant shift in sensitivity to fluazinam in *B. cinerea*, suggesting that this strategy of resistance management is reliable in grapevine.

26.7.2 Inhibitors of Mitochondrial Complex III (QoIs)

Complex III, or cytochrome *bc*1, which transfers electrons from coenzyme Q to cytochrome *c*, is the target site of synthetic strobilurins. These fungicides bind to the Qo site (an outer ubiquinol oxidizing pocket) of cytochrome *b*, the only subunit of complex III encoded by a mitochondrial gene (Gisi et al. 2002). In French vineyards, QoIs are used exclusively for the control of downy and powdery mildews and are not authorized for use against gray mold, because they are not reliably effective against this fungus (probably because there is an effective alternative oxidase in the mitochondria of *B. cinerea* (Leroux 2004)). However, some strobilurins, including azoxystrobin or pyraclostrobin, are used against gray mold in several other crops. As reported for many phytopathogenic fungi, strobilurin resistance in *B. cinerea* field strains results from a G143A substitution in cytochrome *b*. However, the *Cytb* gene of *B. cinerea* may contain 0–4 introns. Strobilurin resistance is never recorded in the presence of a 1205-bp intron just downstream from codon 143, which prevents resistance selection (Leroux et al. 2010; Yin et al. 2012). In Champagne vineyards, strobilurin resistance was first detected in 2006 and is now considered to be generalized in populations of *B. cinerea*, *P. viticola*, and, to a lesser extent, *E. necator*. No specific management policy has been implemented to control strobilurin resistance in *B. cinerea*, because no QoI is currently registered in France for use against gray mold on grapevine. However, the novel QoI pyribencarb may be introduced in the future, because this benzyl carbamate seems to be effective against strobilurin-resistant strains of *B. cinerea* (Kataoka et al. 2010; Ishii 2012).

26.7.3 Inhibitors of Mitochondrial Complex II (SDHIs)

Complex II, or succinate dehydrogenase, transfers electrons from succinate to coenzyme Q and is the only enzyme of the TCA cycle integrated into the mitochondrial membrane. It consists of four subunit proteins (SDHA, B, C, and D), all encoded by nuclear genes. SDHI fungicides have diverse structures, but all include a carboxamide essential for their binding to the ubiquinone reducing pocket formed by the B, C, and D subunits. Early SDHIs, such as the oxathiin-carboxamides carboxin and oxycarboxin, were introduced in the late 1960s, mostly for use against basidiomycetes. In the last decade, novel SDHIs with activity extending to ascomycetes and deuteromycetes have been released onto the market. These new compounds include boscalid (pyridine-carboxamide), fluopyram (benzamide), and several pyrazole carboxamides (e.g., bixafen, fluxapyroxad, isopyrazam, penthiopyrad; Avenot and Michailides 2010; Glättli et al. 2011). Boscalid and fluopyram are available for use in French vineyards either alone against gray mold or in mixtures with a strobilurin against powdery mildew.

In German and French vineyards, the first strains of *B. cinerea* resistant to SDHIs were detected 2 years after the introduction of boscalid. Most were also simultaneously resistant to strobilurins, as observed in other crops treated with boscalid + pyraclostrobin mixture (Leroux et al. 2010; Bardas et al. 2010). Most field strains of *B. cinerea* displaying moderate to strong resistance to boscalid harbor a point mutation in the *SdhB* gene, leading, in particular, to H272Y/R substitutions, as in many other fungi (Leroux et al. 2010; Sierotzki and Scalliet 2013). The SdhB H272R mutants remain susceptible to fluopyram and some pyrazole carboxamides (i.e., bixafen, fluxapyroxad), whereas the SdhB H272Y mutants are resistant to all pyrazole carboxamides but hypersusceptible to fluopyram and other benzamides. Other substitutions conferring positive cross-resistance to all novel SDHIs have been observed in field strains, but at lower frequencies: SdhB H272L, P225L/F/T, N230I, and SdhD H132R. The highest resistance levels were recorded for the first three substitutions. Some resistant strains have also been found to be devoid of mutations of the *Sdh* genes (Lalève et al. 2014b; Leroux et al. 2010; Veloukas et al. 2013).

Since its release on the market in 2005 for use in French vineyards, the use of boscalid has been regulated. This compound can be applied no more than once per year against gray mold. In Champagne, its average rate of use was initially about 0.5 sprays per season, but it was applied on only about 10 % of the plots in the early 2010s, because of the availability of other effective botryticides. Despite this weak selection pressure, SDHI resistance is tending to spread and can now be found on about 30 % of plots, but at a low frequency (about 5 %; Fig. 26.1; Walker et al. 2013). Higher rates of SDHI resistance have been reported on other crops, such as strawberries, apples, and vegetables, decreasing the efficacy of boscalid + pyraclostrobin mixtures (Bardas et al. 2010). Thus, the combination of SDHIs with strobilurins does not constitute a reliable strategy for the management of resistance to these two types of respiratory inhibitors in *B. cinerea*. Moreover, as resistance may evolve quickly, restricting use to a single SDHI application per season against gray mold in French vineyards is a justifiable precaution. As boscalid and fluopyram have different cross-resistance patterns, the alternation of these compounds in consecutive years should be considered. Despite these precautions, and because SDHIs are also used against powdery mildew, monitoring is required to follow the evolution of resistance in populations and to identify the resistant genotypes most likely to be selected. Finally, a fitness penalty has been observed in some resistant genotypes, and this may help to delay the evolution of resistance, particularly if anti-resistance strategies are implemented (Lalève et al. 2014a).

26.8 Multidrug Resistance (MDR)

Multidrug resistance can be defined as simultaneous resistance to various unrelated toxicants. It often involves an increase in drug efflux, due to the constitutive overexpression of plasma membrane transporters with low substrate specificities (Kretschmer 2012). Three MDR phenotypes have been characterized in *B. cinerea*, differing in terms of their pattern of cross-resistance (Kretschmer et al. 2009; Leroux and Walker 2013). MDR1 strains, displaying resistance to carbendazim, diethofencarb, dicarboximides, phenylpyrroles, anilinopyrimidines, fluazinam, pyrazole carboxamides, and some strobilurins, harbor activating mutations or deletions in the gene encoding the transcription factor BcMrr1. This transcription factor controls expression of the gene encoding the ABC transporter BcatrB. Moreover, an additional 3-bp deletion in *Bcmrr1* has been shown to confer a stronger MDR1h phenotype in adapted gray mold populations growing on strawberry (*Botrytis* group S; Leroch et al. 2013). In MDR2 strains, simultaneous resistance to diethofencarb, dicarboximides, anilinopyrimidines, inhibitors of sterol C4- or C14-demethylation, novel SDHIs, and some strobilurins results from overexpression of the MFS transporter BcmfsM2. This overexpression results from the insertion of a retroelement-derived sequence, together with a deletion in the *BcmfsM2* promoter. Finally, MDR3 strains display the broadest cross-resistance pattern. They are natural recombinants between MDR1 and MDR2 strains (Kretschmer et al. 2009; Leroux and Walker 2013).

The first MDR1 and MDR2 strains were detected in Champagne vineyards in 1993, and they gradually spread until the late 1990s. They were found in most plots in the early 2000s and became generalized simultaneously with the emergence of MDR3 strains, in the middle of the first decade of this century, with an overall frequency of more than 50 % (Fig. 26.1; Walker et al. 2013). During this period, the most common spray program consisted of fenhexamid (stage A), fludioxonil (stage B), and pyrimethanil (stage C). This program probably selected MDR strains, but it remained effective, possibly because resistance levels were low to moderate (Leroux and Walker 2013). The management of this type of resistance may involve the use of full doses of individual products and their rotation between seasons. It has been shown in *B. cinerea* that transporter modulators (i.e., inhibitors of efflux pumps) such as chlorpromazine or verapamil and the uncouplers CCCP or fluazinam prevent the cellular efflux of several fungicides (Kretschmer 2012; Kretschmer et al. 2009; Leroux and Walker 2013). Consequently, the combination of such substances with fludioxonil, anilinopyrimidines, and inhibitors of C4-demethylation or SDHIs may be of practical value for the control of gray mold (Leroux and Walker 2013).

26.9 Conclusions

The use of synthetic fungicides remains the most effective way to control gray mold in grapevine, but other methods can help to limit disease development. Several prophylactic measures have been recommended, such as decreasing vine vigor (e.g., adapted rootstock and grape varieties; controlled fertilization, mostly nitrogen; permanent grass cover between the rows); favoring grape aeration (e.g., pruning, desuckering, lifting, leaf removal in the area of the grape cluster or green harvest); limiting berry wounding due to insects (especially berry moths), powdery mildew, or mechanical tools; and harvesting earlier than scheduled (Elmer and Michailides 2004). In cases of low-level infestation, other control methods, such as the use of biological agents, mineral salts, or plant extracts (among others, extracts from giant knotweed or seaweed), may be implemented (Jacometti et al. 2010). In France, two microbial agents containing either *Aureobasidium pullulans* or *Bacillus subtilis* and potassium bicarbonate are registered for use against gray mold in organic and conventional farming systems (Walker et al. 2013). Chemical control of gray mold in the French vineyards at greatest risk can currently be achieved with one or two botryticide sprays per year. The available active ingredients can be grouped into seven classes on the basis of their patterns of cross-resistance: anti-microtubule agents (thiophanate-methyl), dicarboximides (iprodione), phenylpyrroles (fludioxonil), anilinopyrimidines (cyprodinil, mepanipyrim, pyrimethanil), sterol C4-demethylation inhibitors (fenhexamid, fenpyrazamine), uncouplers of oxidative phosphorylation (fluazinam), and succinate dehydrogenase inhibitors (boscalid, fluopyram). The two oldest classes (anti-microtubule agents and dicarboximides), which triggered practical resistance following their intensive use, are now very seldom used. By contrast, the remaining botryticides, introduced since 1990, are authorized for use only once per year, this restriction being extended to all members of a given class. Given the small number of treatments per season, limitations on the use of each botryticide class at the multiseasonal scale, in addition to the alternation of active ingredients with different modes of action, would appear to be a suitable approach. The mixture strategy, commonly recommended for the management of fungicide resistance in *P. viticola* (Corio-Costet 2012), does not seem to be appropriate for the management of resistance in gray mold on grapevine, particularly because fungicides often have lower intrinsic activities against *B. cinerea* than against other fungi, rendering strategies involving the use of reduced doses in mixtures ineffective. For botryticides to be effective, applications must be optimized by targeting the grape clusters and applying full doses. As a result of the implementation of such strategies, no failure of control linked to botryticide resistance has been recorded in France since the mid-1990s. However, resistance monitoring remains essential, for detecting the emergence of new resistances, observing the spatial and temporal dynamics of known resistances and estimating resistance frequencies regionally. All this information is important for the local adaptation of resistance management in *B. cinerea*. As an example, *Botrytis* monitoring is organized yearly by representatives of the French Ministry of Agriculture. Resistance tests are carried out by INRA (Institut National de la Recherche Agronomique) and ANSES (Agence Nationale de Sécurité Sanitaire de l'alimentation, de l'environnement et du travail). Additionally, some tests can also be ordered by the technical institutes from local vineyards (e.g., Champagne). Results are provided to winegrowers and practitioners by these authorities and integrated yearly to the official recommendations of disease control (Walker et al. 2013).

References

Ajouz S, Walker AS, Fabre F, Leroux P, Nicot PC, Bardin M (2011) Variability of *Botrytis cinerea* sensitivity to pyrrolnitrin, an antibiotic produced by biological control agents. Biocontrol 56(3):353–363. doi:10.1007/s10526-010-9333-7

Avenot HF, Michailides TJ (2010) Progress in understanding molecular mechanisms and evolution of resistance to succinate dehydrogenase inhibiting (SDHI) fungicides in phytopathogenic fungi. Crop Prot 29(7):643–651. doi:10.1016/j.cropro.2010.02.019

Banno S, Fukumori F, Ichiishi K, Okada H, Uekusa M, Fujimura M (2008) Genotyping of benzimidazole resistance, dicarboximide resistant mutations in *Botrytis cinerea* using real-time polymerase chain reaction assays. Phytopathology 98:397–404

Bardas GA, Myresiotis CK, Karaoglanidis GS (2008) Stability and fitness of anilinopyrimidine-resistant strains of *Botrytis cinerea*. Phytopathology 98(4):443–450. doi:10.1094/phyto-98-4-0443

Bardas GA, Veloukas T, Koutita O, Karaoglanidis GS (2010) Multiple resistance of *Botrytis cinerea* from kiwifruit to SDHIs, QoIs and fungicides of other chemical groups. Pest Manag Sci 66(9):967–973. doi:10.1002/ps.1968

Billard A, Walker AS, Fillinger S, Leroux P, Lachaise H, Beffa R, Debieu D (2011) Le fenhexamid et la vigne, mécanismes de résistance décryptés chez l'agent de la pourriture grise. Phytoma – La Défense des Végétaux 648:40–45 (in French)

Billard A, Fillinger S, Leroux P, Lachaise H, Beffa R, Debieu D (2012) Strong resistance to the fungicide fenhexamid entails a fitness cost in *Botrytis cinerea*, as shown by comparisons of isogenic strains. Pest Manag Sci 68(5):684–691. doi:10.1002/ps.2312

Corio-Costet MF (2012) Fungicide resistance in *Plasmopara viticola* in France and anti-resistance measures. In: Thind TS (ed) Fungicide resistance in crop protection. Risk and management. CABI, Oxfordshire, pp 157–171

Cui W, Beever RE, Parkes SL, Templeton MD (2004) Evolution of an osmosensing histidine kinase in field strains of *Botryotinia fuckeliana* (*Botrytis cinerea*) in response to dicarboximide fungicide usage. Phytopathology 94(10):1129–1135. doi:10.1094/phyto.2004.94.10.1129

Davidse L, Ishii H (1995) Biochemical and molecular aspects of benzimidazoles, *N*-phenylcarbamates and *N*-phenylformamidoximes and the mechanisms of resistance to the compounds. In: Lyr HGF (ed) Modern selective fungicides. Gustav Fischer, Iena, pp 309–322

De Miccolis AR, Pollastro S, Faretra F (2012) Genetics of fungicide resistance in *Botryotinia fuckeliana* (*Botrytis cinerea*). In: Thind TS (ed) Fungicide resistance in crop protection. Risk and management. CABI, Oxfordshire, pp 237–250

Debieu D, Bach J, Hugon M, Malosse C, Leroux P (2001) The hydroxyanilide fenhexamid, a new sterol biosynthesis inhibitor fungicide efficient against the plant pathogenic fungus *Botryotinia fuckeliana* (*Botrytis cinerea*). Pest Manag Sci 57(11):1060–1067

Elmer PAG, Michailides TJ (2004) Epidemiology in *Botrytis cinerea* in orchard and vine crops. In: Elad Y, Williamson B, Tudzynski P, Delen N (eds) *Botrytis*: biology, pathology and control. Kluwer Academic Publishers, Dordrecht, pp 243–262

Fillinger S, Leroux P, Auclair C, Barreau C, Al Hajj C, Debieu D (2008) Genetic analysis of fenhexamid-resistant field isolates of the phytopathogenic fungus *Botrytis cinerea*. Antimicrob Agents Chemother 52(11):3933–3940. doi:10.1128/aac.00615-08

Fillinger S, Ajouz S, Nicot PC, Leroux P, Bardin M (2012) Functional and structural comparison of pyrrolnitrin- and iprodione-induced modifications in the class III histidine-kinase Bos1 of *Botrytis cinerea*. PLoS ONE 7(8):e42520. doi: 10.1371/journal.pone.0042520

Forster B, Staub T (1996) Basis for use strategies of anilinopyrimidine and phenylpyrrole fungicides against *Botrytis cinerea*. Crop Prot 15(6):529–537. doi:10.1016/0261-2194(96)00021-x

Fujimura M, Ochiai N, Ichiishi A, Usami R, Horikoshi K, Yamaguchi I (2000) Sensitivity to phenylpyrrole fungicides and abnormal glycerol accumulation in *Os* and *Cut* mutant strains of *Neurospora crassa*. J Pestic Sci 25(1):31–36

Gisi U, Sierotzki H, Cook A, McCaffery A (2002) Mechanisms influencing the evolution of resistance to Qo inhibitor fungicides. Pest Manag Sci 58(9):859–867. doi:10.1002/ps.565

Glättli A, Grote T, Stammler G (2011) SDH-inhibitors: history, biological performance and molecular mode of action. In: Dehne D, Deising H, Fraaije B et al (eds) Modern fungicides and antifungal compounds VI. DPG, Braunschweig, pp 159–170

Guo Z, Miyoshi H, Komyoji T, Haga T, Fujita T (1991) Uncoupling activity of a newly developed fungicide, fluazinam [3-chloro-*N*-(3-chloro-2,6-dinitro-4-trifluoromethylphenyl)-5-

trifluoromethyl-2-pyridinamine]. Biochim Biophys Acta (BBA) – Bioenergetics 1056(1):89–92

Ishii H (2012) Resistance to QoI and SDHI fungicides in Japan. In: Thind TS (ed) Fungicide resistance in crop protection, risk and management. CABI, Oxfordshire, pp 223–234

Jacometti M, Wratten S, Walter M (2010) Alternative to synthetic fungicides for *Botrytis cinerea* management in vineyards. Aust J Grape Wine Res 16:154–172

Kataoka S, Takagaki M, Kaku K, Shimizu T (2010) Mechanism of action and selectivity of a novel fungicide, pyribencarb. J Pestic Sci 35(2):99–106. doi:10.1584/jpestics.G09-40

Kretschmer M (2012) Emergence of multi-drug resistance in fungal pathogens: a potential threat to fungicide performance in agriculture. In: Thind TS (ed) Fungicide resistance in crop protection: risk and management. CABI, Oxfordshire, pp 251–267. doi:10.1079/9781845939052.0251

Kretschmer M, Leroch M, Mosbach A, Walker A, Fillinger S, Mernke D, Schoonbeek H, Pradier J, Leroux P, De Waard MA, Hahn M (2009) Fungicide-driven evolution and molecular basis of multidrug resistance in field populations of the grey mould fungus *Botrytis cinerea*. PLoS Pathog 5(12), e1000696

Lalève A, Fillinger S, Walker AS (2014a) Fitness measurement reveals contrasting costs in homologous recombinant mutants of Botrytis cinerea resistant to succinate dehydrogenase inhibitors. Fungal Genet Biol (in press). doi:http://dx.doi.org/10.1016/j.fgb.2014.03.006

Lalève A, Gamet S, Walker A-S, Debieu D, Toquin V, Fillinger S (2014b) Site-directed mutagenesis of the P225, N230 and H272 residues of succinate dehydrogenase subunit B from Botrytis cinerea highlights different roles in enzyme activity and inhibitor binding. Environ Microbiol 16(7):2253–2266. doi:10.1111/1462-2920.12282

Leroch M, Plesken C, Weber RWS, Kauff F, Scalliet G, Hahn M (2013) Gray mold populations in German strawberry fields are resistant to multiple fungicides and dominated by a novel clade closely related to *Botrytis cinerea*. Appl Environ Microbiol 79(1):159–167. doi:10.1128/aem.02655-12

Leroux P (1995) Progress and problems in the control of *Botrytis cinerea* in grapevine. Pestic Outlook 6(5):13–19

Leroux P (2004) Chemical control of *Botrytis* and its resistance to chemical fungicides. In: *Botrytis*: biology, pathology and control. Kluwer Academic Publishers, Dordrecht, pp 195–222

Leroux P, Clerjeau M (1985) Resistance of *Botrytis cinerea* Pers. and *Plasmopara viticola* (Berl. and de Toni) to fungicides in French vineyards. Crop Prot 4(2):137–160

Leroux P, Walker A (2013) Activity of fungicides and modulators of membrane drug transporters in field strains of *Botrytis cinerea* displaying multidrug resistance. Eur J Plant Pathol 135(4):683–693. doi:10.1007/s10658-012-0105-3

Leroux P, Fritz R, Debieu D, Albertini C, Lanen C, Bach J, Gredt M, Chapeland F (2002) Mechanisms of resistance to fungicides in field strains of *Botrytis cinerea*. Pest Manag Sci 58(9):876–888

Leroux P, Gredt M, Leroch M, Walker A-S (2010) Exploring mechanisms of resistance to respiratory inhibitors in field strains of *Botrytis cinerea*, the causal agent of gray mold. Appl Environ Microbiol 76(19):6615–6630. doi:10.1128/aem.00931-10

Liu WW, Leroux P, Fillinger S (2008) The HOG1-like MAP kinase Sak1 of *Botrytis cinerea* is negatively regulated by the upstream histidine kinase Bos1 and is not involved in dicarboximide- and phenylpyrrole-resistance. Fungal Genet Biol 45(7):1062–1074. doi:10.1016/j.fgb.2008.04.003

Malandrakis A, Markoglou A, Ziogas B (2011) Molecular characterization of benzimidazole-resistant *B. cinerea* field isolates with reduced or enhanced sensitivity to zoxamide and diethofencarb. Pestic Biochem Physiol 99(1):118–124. doi:10.1016/j.pestbp.2010.11.008

Sierotzki H, Scalliet G (2013) A review of current knowledge of resistance aspects for the next-generation succinate dehydrogenase inhibitor fungicides. Phytopathology 103(9):880–887. doi:10.1094/phyto-01-13-0009-rvw

Suty A, Pontzen R, Stenzel K (1999) Fenhexamid – sensitivity of *Botrytis cinerea*: determination of baseline sensitivity and assessment of the resistance risk. Pflanzenschutz-Nachrichten Bayer 52(2):149–161 (in German)

Tamura O (2000) Resistance development of grey mould on beans towards fluazinam and relevant countermeasures. In: Abstract of the 10th symposium of research committee of fungicide resistance, Okayama, Japan, pp 7–16

Tanaka S, Ishikawa R, Armengaud P, Senechal Y (2012) La fenpyrazamine, une nouvelle génération de fongicide pour lutter contre *Botrytis cinerea* en vigne. Paper presented at the 10ème conférence internationale sur les maladies des plantes, Tours, France, 3–5 décembre (in French)

Veloukas T, Markoglou AN, Karaoglanidis GS (2013) Differential effect of SdhB gene mutations on the sensitivity to SDHI fungicides in *Botrytis cinerea*. Plant Dis 97(1):118–122. doi:10.1094/pdis-03-12-0322-re

Walker A-S, Gautier A, Confais J, Martinho D, Viaud M, Le Pecheur P, Dupont J, Fournier E (2011) *Botrytis pseudocinerea*, a new cryptic species causing gray mold in French vineyards in sympatry with *Botrytis cinerea*. Phytopathology 101(12):1433–1445. doi:10.1094/phyto-04-11-0104

Walker A-S, Micoud A, Rémuson F, Grosman J, Gredt M, Leroux P (2013) French vineyards provide information that opens ways for effective resistance management of *Botrytis cinerea* (grey mould). Pest Manag Sci 69(6):667–678. doi:10.1002/ps.3506

Yin YN, Kim YK, Xiao CL (2012) Molecular characterization of pyraclostrobin resistance and structural diversity of the cytochrome b gene in *Botrytis cinerea* from apple. Phytopathology 102(3):315–322. doi:10.1094/phyto-08-11-0234

Young DH, Slawecki RA (2005) Cross resistance relationship between zoxamide, carbendazim and diethofencarb infield isolates of *Botrytis cinerea* and other fungi. In: Dehne D, Deising H, Fraaije B et al (eds) Modern fungicides and antifungal compounds. BCPC, Alton, pp 125–131

Chapter 27
Fungicide Resistance in *Venturia inaequalis*, the Causal Agent of Apple Scab, in the United States

Kerik D. Cox

Abstract Apple scab, caused by the fungal pathogen *Venturia inaequalis*, is one of the most economically important diseases of apple in the United States and other temperate climate production regions worldwide. In the absence of durable host resistance, applications of fungicides made throughout the growing season at weekly intervals have been the foundation of apple scab management for the past 70 years. In this regard, apple producers continually seek to reduce the number of fungicide applications and improve the timing of applications by capitalizing on postinfection activity. As older broad-spectrum fungicides with postinfection activity were taken off from the market due to health concerns in the 1960s, only the "modern" single-site fungicides introduced in the later part of the twentieth and beginning of the twenty-first centuries could offer postinfection activity. Unfortunately, the introduction of new single-site fungicide classes in the United States occurred in a slow successive manner and led to the unfortunate practice of overreliance and development of fungicide resistance in many of the fungicide classes effective for the management of apple scab.

The current status of resistance to several classes of fungicides in *V. inaequalis* in the United States is presented and discussed in this chapter. Due to the fact that apple scab is more prevalent in the temperate climate apple production regions of the northeastern United States compared to the semiarid production region of the western United States, research efforts on fungicide resistance have primarily focused on the eastern United States. To date, resistance to dodine, methyl benzimidazole carbamates (MBCs), sterol biosynthesis inhibitors (SBIs), and quinone-outside inhibitors (QoIs) has been reported in orchard populations of *V. inaequalis*. Unfortunately, the mechanisms for resistance to these fungicides in *V. inaequalis* have only been elucidated for the MBC and QoI fungicides. Hence, detection of resistance still relies on traditional microbiological methods for many fungicide classes. Although resistance has not been discovered for anilinopyrimidine (AP)

K.D. Cox (✉)
Department of Plant Pathology and Plant-Microbe Biology, Cornell University, Geneva, NY 14456, USA
e-mail: kdc33@cornell.edu

H. Ishii, D.W. Hollomon (eds.), *Fungicide Resistance in Plant Pathogens*,
DOI 10.1007/978-4-431-55642-8_27

and succinate dehydrogenase inhibitor (SDHI) fungicides, the effectiveness of these classes against apple scab is considerably less than that of SBIs, QoIs, dodine, and MBC in the absence of resistance. Hence, producers will continue to rely heavily on premixes of different fungicide classes and broad-spectrum protectants applied on close intervals to avoid management failure and crop loss.

Keywords *Venturia inaequalis* • Apple scab • Practical resistance • Dodine • Methyl benzimidazole carbamate • Sterol biosynthesis inhibitor • Quinone-outside inhibitor • Anilinopyrimidine • Succinate dehydrogenase inhibitor

27.1 Introduction

Apple scab, caused by the fungus *Venturia inaequalis* (Cke.) Wint., is one of the most economically important diseases of apples (*Malus × domestica*) in the eastern United States where cool rainy weather conditions in the spring are conducive to infection (MacHardy 1996). Under severe infection conditions, diseased apple trees can defoliate prematurely and fail to produce flowers in the following spring. The most economically important symptoms of the disease are scabby lesions that develop on fruit at harvest, which render fruit unmarketable for fresh consumption (MacHardy et al. 2001). Downgrading from fresh market to "utility"-grade apples, which are often used in processing for juice concentrate, can result in losses on the order of 0.11 USD per kilogram, which translates to a loss of nearly 12,500 USD per hectare (Anonymous 2012). Indeed, the profits producers receive for "utility"-grade apples in the United States are currently lower than the costs of production. This situation has rapidly declined in the last decade due to the influx of inexpensive juice concentrates imported from abroad. In addition to quality concerns, the costs associated with the management of apple scab further limit the profitability of apple production. Although there are no published cost estimates for apple scab management, it is the general consensus that eastern US apple producers spend on average 550 USD per hectare on fungicides alone each year. With approximately 34,000 ha planted to apples in the eastern United States, the cost of apple scab management alone could be as high as 18.6 million USD per year.

The development of durable apple scab-resistant cultivars to reduce reliance on fungicides has been a major goal of apple management programs for decades (MacHardy 1996). Cultivars fully resistant to apple scab have been commercially available for more than half a century, but grower and consumer acceptance has been limited due to horticultural quality concerns (MacHardy 1996; Merwin et al. 1994). In the absence of commercially acceptable cultivars with resistance, fungicides have served as the foundation for the management of apple scab. Fungicide applications occur at weekly intervals from bud break to early summer, which coincides with the presence of inoculum in the form of ascospores and conidia (MacHardy 1996; MacHardy et al. 2001). The number of fungicide applications needed to manage apple scab may be a low as 10 or as many as 20 depending on the fungicides being used, the level of cultivar-specific minor gene resistance, and the amount of rainfall during the growing season (Jones 1991; MacHardy et al. 2001). Since their introduction in the 1940s, the multisite, broad-spectrum protectant fungicides captan and mancozeb have served the basis of apple scab fungicide programs in the United States. However, the introduction of single-site fungicides with curative or "postinfection" properties in the 1950s (e.g., dodine and the methyl benzimidazole carbamates (MBCs) and later sterol biosynthesis inhibitors (SBIs), anilinopyrimidines (APs), and quinone-outside inhibitors (QoIs)) allowed growers to continue to pursue postinfection application paradigms for apple scab management when organomercury and other environmentally unsafe fungicides were no longer allowed. Indeed, up to the turn of the twenty-first century, fungicide applications were often made following infection periods to reduce the number of fungicide applications by capitalizing on the postinfection properties of single-site fungicides. During this time, the frequent reliance on both protective and postinfection modes of systemic fungicides led to a succession of resistance to each newly introduced fungicide class in the United States from the 1950s to date (Chapman et al. 2011; Jones 1991; Lesniak et al. 2011; Szkolnik and Gilpatrick 1969; Wicks 1974). The AP fungicides released in the late 1990s, the DMI fungicide difenoconazole released in 2010, and the succinate dehydrogenase inhibitor (SDHI) fungicides released after 2012 are the only fungicide classes to which resistance has not been reported at the time of this manuscript.

27.2 Development of Resistance in the Fungicide Classes Used in the Management of Apple Scab

27.2.1 Methyl Benzimidazole Carbamate Fungicides

Methyl benzimidazole carbamate (MBC) fungicides were first registered for use on apples in the 1970s, and within 5 years of registration, MBC resistance in orchard populations of *V. inaequalis* was reported in Michigan (Jones and Walker 1976). A few years later, resistance to MBC fungicides was reported in New York (Katan et al. 1983), and reports of widespread resistance to MBC fungicides in populations of *V. inaequalis* increased over the years to the point where they no longer recommended for the use of apple scab. Indeed, the MBC fungicide benomyl is not registered for the management of apple scab in commercial orchards. Interestingly, the MBC fungicide thiophanate-methyl is still used and effective for managing summer diseases of apple. These diseases include the preharvest diseases such as flyspeck, sooty blotch, bitter rot, and white and black fruit rots. Although it has not been investigated, it is believed that the use of MBC fungicides for apple summer diseases exerts a sufficient selective pressure to maintain MBC resistance in orchard populations of *V. inaequalis*. Moreover, resistance to MBC fungicides has been shown to be stable in *Botrytis cinerea* on grape and in other pathosystems despite whether or not MBC fungicides are used for management (Leroux and Clerjeau 1985).

Resistance to MBC fungicides is typically attributed of point mutations in the coding regions of the β-tubulin target gene at the site of benzimidazole binding. In other fungal pathosystems, several point mutations have been reported (Ma and Michailides 2005), but only three mutations at codons 198, 200, 240 have been associated with a moderate, high, and low to moderate level of resistance, respectively, in *V. inaequalis* (Koenraadt et al. 1992; Quello et al. 2010). In surveys of apple orchards in Indiana, Quello et al. (2010) observed that *V. inaequalis* populations from ornamental crab apples were fairly sensitive to MBC fungicides, while isolates from commercial apple orchards were often still resistant to MBC fungicides. Hence, it's still likely that resistance to MBC fungicides is still fairly widespread in commercial apple orchards. Such discoveries provide impetus for future investigations of MBC sensitivity for populations of *V. inaequalis* in newly planted orchards isolated from previous orchard plantings. It is possible that with resistance screening, MBC fungicides could be reintroduced into commercial apple production in certain scenarios. Such was the case with dodine as discussed below.

27.2.2 Dodine

Dodine has been used to manage apple scab in the eastern United States since the late 1950s. By the late 1960s dodine resistance had begun to develop in the region, and shortly after, widespread use of the dodine for the management of apple was discontinued (Olaya et al. 1998). Investigations at the turn of the century in Michigan and New York indicated that the frequency of dodine-resistant isolates was reduced in commercial orchards where the dodine use had been discontinued (Köller and Wilcox 2000). Unfortunately, the frequency of dodine-resistant isolates rapidly increased in these orchards when dodine applications were resumed for apple scab management (Köller and Wilcox 2000). Along these lines, Carisse and Jobin (2010) found dodine-resistant isolates in apple orchards in Quebec orchards 30 years after dodine use was discontinued. Such studies suggest that dodine resistance is fairly stable in apple scab populations in eastern North America.

While the aforementioned studies have established that dodine resistance is stable in apple orchards in which dodine resistance originated (Carisse and Jobin 2010; Köller and Wilcox 2000), dodine resistance may not be as prevalent in the eastern United States as believed. New apple operations and new orchards are continually being established on land that has not been exposed to dodine for more than 25 years. A study examining multiple resistance in *V. inaequalis* from five apple orchards in Michigan and Indiana from 2007 to 2009 indicated that the frequency of dodine resistance was only 5.2 % of the isolates collected (Chapman et al. 2011). Examinations of dodine sensitivity in 233 orchard populations of *V. inaequalis* from New York and New England by the programs of Köller from 2004 to 2006 and Cox from 2007 to 2011 using the methodology developed by Köller and Wilcox (2000) have revealed an apparent decline in the identification of orchard populations of *V. inaequalis* with practical resistance to dodine in the region (Fig. 27.1). Practical resistance is defined as greatly diminished levels of disease control (i.e., a control failure) following proper fungicide application (Köller 1991; Köller et al. 2004; Frederick et al. 2014). Given that dodine resistance is stable in isolates of *V. inaequalis* (Carisse and Jobin 2010), there may be several explanations for a lack of practical resistance to dodine in the eastern United States. The first is that the observed decline in dodine resistance in populations of *V. inaequalis* over the two decades is not due to isolates losing acquired resistance to dodine, but is due to the dilution of dodine-resistant isolates within populations. Such a phenomenon could result from continual selection for isolates with resistance to chemistries that were introduced after dodine use was discontinued. For example, selective pressures resulting from excessive applications of DMI and QoI fungicides applied over the last two decades may have contributed to a dilution of dodine-resistant isolates in orchard populations in the region. A second hypothesis is that older orchards with dodine resistance have been abandoned and new orchards have been established in areas where dodine-resistant *V. inaequalis* wasn't present. Regardless, apple extension specialists in New York and New England are beginning to recommend moderate dodine use (1–2 applications per season) in orchards planted without a history of dodine resistance.

To date, the mechanism of dodine resistance in *V. inaequalis* is largely unknown, but development appears to occur in a graduate or quantitative manner such that population members become increasingly resistant to dodine as exposure increases. Such resistance responses are typically governed by the contributions of several genes or multi-gene families. Interestingly, a dodine-resistant phenotype in *Nectria haematococca* was mapped to four genes using forward genetics (i.e., mating) (Kappas and Georgopoulos 1970). However, the sequence of these four genes and the number and sequence of genes involved in dodine resistance in *V. inaequalis* are likely to remain unknown until investigations using next-generation sequencing approaches (e.g., sequencing by synthesis, or Illumina sequencing) are conducted at the genome and transcriptome levels.

27.2.3 DMI Fungicides

In the 1980s, the demethylation inhibitor (DMI) fungicides, a type of sterol biosynthesis inhibitor fungicide, were registered for use on apple scab. Compared to the fungicide chemistries previously discussed, it took more than 20 years before resistance to the DMI fungicides was well established in *V. inaequalis* populations in the eastern United States (Köller et al. 1997; Köller and Wilcox 2000). Reduced sensitivity to DMIs in isolates of *V. inaequalis* was first observed in Michigan in 1985 (Stanis and Jones 1985), but the first confirmed resistance to myclobutanil in orchard populations of *V. inaequalis* was reported in Nova Scotia and Michigan in 1988 and

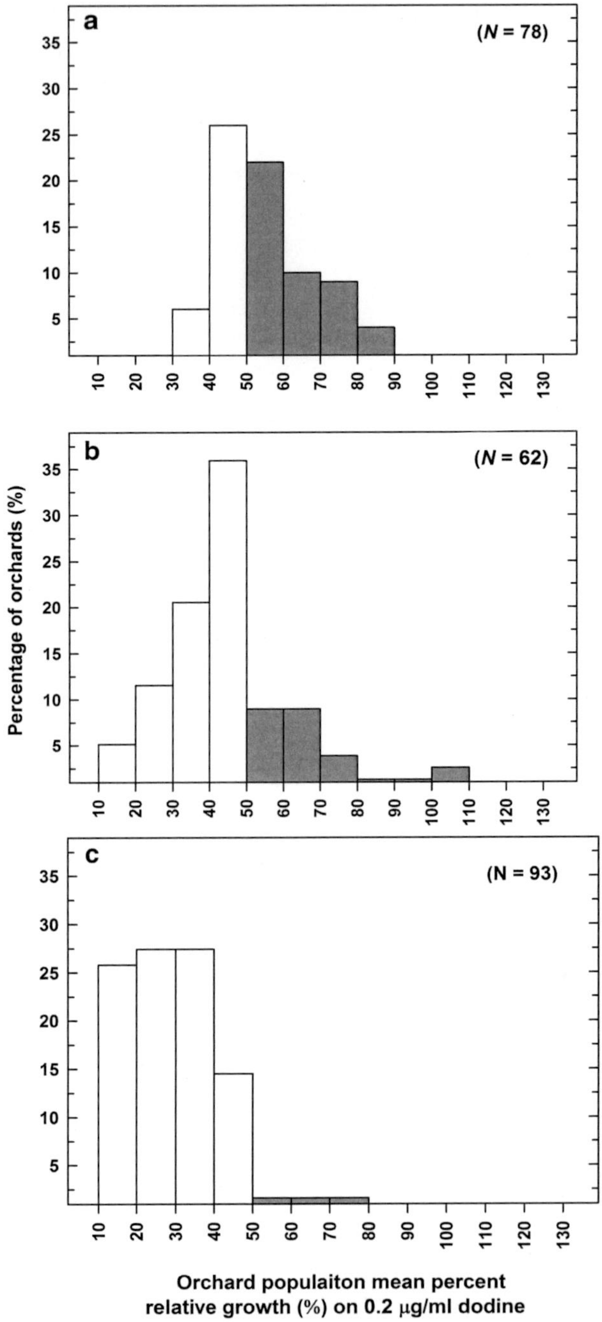

Fig. 27.1 Frequency distributions of population mean sensitivity to the fungicide dodine for *V. inaequalis* populations from orchards in the eastern United States from 2004 to 2006 (**a**), 2007 to 2009 (**b**), and 2010 to 2012 (**c**). Values are population means of relative growth values calculated from more than 25 *V. inaequalis* single-lesion isolates. Relative growth values for the isolates comprising the population mean were expressed as percent microcolony growth on medium amended with the analytical-grade dodine (0.2 μg ml^{-1}) relative to that on non-fungicide-amended medium. On all graphs, the bars "with" and "without" shading represent orchard populations that would be considered having and lacking practical resistance to dodine, respectively, on the basis of statistical comparisons to orchards with confirmed practical resistance

1991, respectively (Köller et al. 1997). By the turn of the twenty-first century (Köller and Wilcox 2000), DMI resistance in *V. inaequalis* was fairly widespread in Michigan orchards, but orchard populations in New York and New England remained sensitive to DMI fungicides. Soon after the turn of the twenty-first century, surveys of *V. inaequalis* populations from orchards in New York and New England conducted from 2004 to 2012 confirmed DMI resistance in New York, Massachusetts, Vermont, New Hampshire, Maine, Rhode Island, and Connecticut (Cox et al. 2009b). At the same time, DMI resistance was reported in orchard populations of *V. inaequalis* in Indiana, Pennsylvania, and Virginia (Marine et al. 2007; Pfeufer et al. 2010; Chapman et al. 2011).

Although DMI resistance has been reported in many orchard populations of *V. inaequalis* in the eastern United States, few studies have investigated the phenomenon of practical resistance to DMI fungicides. Given the existence of research orchards with practical resistance to the DMI fungicide myclobutanil (Cox et al. 2008, 2009a, 2010, 2011), it has been possible to identify a threshold by which practical resistance can be represented. In this regard, a statistical comparison of myclobutanil sensitivity distributions of 93 *V. inaequalis* populations (1779 isolates total) to the aforementioned population standard using a previously published methodology (Frederick et al. 2014; Köller et al. 1997) revealed that 73 of the 93 orchards would have practical resistance to myclobutanil (Fig 27.2A).

Interestingly, initial reports of DMI resistance in *V. inaequalis* were based on responses of field isolates to myclobutanil and fenarimol, the most commonly used DMI fungicides for apple scab management. When myclobutanil and fenarimol resistance was considered fairly widespread in the eastern United States in 2009, several DMI fungicides previously only available in Europe were registered for the management of apple scab. The newly registered DMI chemistries included flutriafol, fenbuconazole, and difenoconazole. At the time of registration, *V. inaequalis* populations in North America had not been exposed to these fungicides. In field trials conducted by the author in orchards with practical resistance to myclobutanil, products containing difenoconazole and fenbuconazole were able to manage apple scab as effectively or more effectively than the protectant fungicide standards (Cox et al. 2008, 2009a, 2010, 2011). By comparison, products containing flutriafol performed similarly to myclobutanil and displayed practical resistance (Cox et al. 2008, 2009a, 2010, 2011). The fact that resistance response to DMI fungicides is considered incremental, quantitative, or dose-dependent suggests that difenoconazole and fenbuconazole have a higher intrinsic activity against *V. inaequalis* than flutriafol and myclobutanil. Interestingly, these differences in activity are often observed when *V. inaequalis* isolates are grown in vitro amended with identical doses of different DMI fungicides (Table 27.1). Moreover, an examination of 93 *V. inaequalis* orchard populations ($n > 25$) for sensitivity to myclobutanil and difenoconazole at an identical dose of 1.0 μg/ml (1779 isolates) revealed that although the majority of populations would appear to have practical resistance to myclobutanil, few of the populations had practical resistance to difenoconazole (Fig. 27.2b). Given the higher intrinsic activity of difenoconazole, it is tempting to speculate as to whether difenoconazole will be overcome by continual selection for

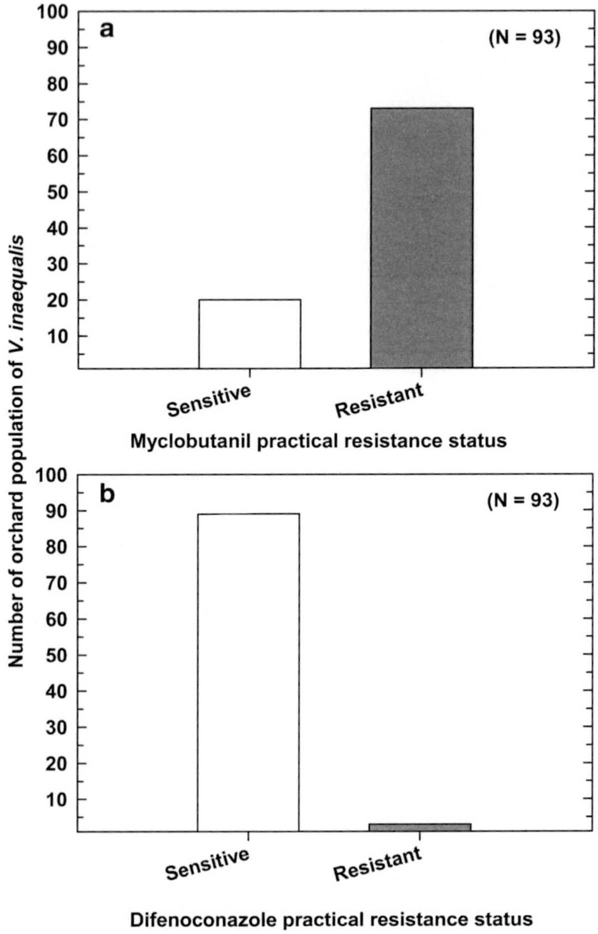

Fig. 27.2 Classification of DMI practical resistance status for 93 *V. inaequalis* populations from orchards in the eastern United States from 2010 to 2012. Values represent the number of orchards considered having or lacking practical resistance to myclobutanil (**a**) or difenoconazole (**b**), respectively, on the basis of statistical comparisons to orchards with confirmed practical resistance. Comparisons are based on relative growth values for more than 25 isolates per population tested at an identical discriminatory dose of 0.1 μg ml^{-1} for myclobutanil and difenoconazole. On all graphs, the bars "with" and "without" shading represent orchard populations that would be considered having and lacking practical resistance, respectively

isolates with increasingly higher resistance or whether there is a metabolic ceiling in *V. inaequalis* where the fungal metabolism is unable to dispense the fungicide. Unfortunately, there is nothing other than anecdotal evidence to support the claim of a metabolic ceiling. In a research orchard at the New York Agricultural Experiment Station with stable practical resistance to myclobutanil (Cox et al. 2008, 2009a, 2010, 2011), difenoconazole has provided a high level of control of apple scab (>90 %) after receiving a minimum of four applications of difenocon-

Table 27.1 In vitro sensitivity to the DMI fungicides myclobutanil and difenoconazole for isolates of *V. inaequalis* collected from a commercial apple orchard in Vermont in 2012

Isolate	Mean % relative growth on 0.1 μg ml^{-1} myclobutanil[a]	Mean % relative growth on 0.1 μg ml^{-1} difenoconazole[a]
5-3-12	57.2	13.8
5-4-12	54.6	11.3
5-5-12	59.0	29.1
5-8-12	114.8	7.4
5-9-12	80.2	36.1
5-10-12	77.8	16.7
5-11-12	61.0	22.8
5-14-12	61.4	27.1
5-15-12	86.9	20.5
5-17-12	66.1	9.5
5-18-12	53.2	31.7
5-19-12	78.4	28.4
5-20-12	98.8	48.5
5-21-12	105.3	20.0
5-24-12	78.6	12.9
5-27-12	57.6	23.8
5-29-12	69.0	21.2
5-31-12	110.8	13.8
5-34-12	72.0	37.4
5-38-12	109.2	45.2
5-39-12	66.4	19.7
5-40-12	89.4	31.0
5-41-12	96.9	45.4
5-42-12	105.9	55.4
5-43-12	78.6	4.3
5-44-12	116.5	29.2
5-46-12	83.4	16.0
5-47-12	101.4	25.8
5-48-12	95.2	7.5
5-49-12	84.1	23.8
5-50-12	71.2	40.5

[a]In vitro sensitivity was expressed as percent microcolony growth on medium amended with the same dose of with analytical-grade myclobutanil or difenoconazole relative to that of the non-fungicide-amended medium. Values are the means of five randomly selected single-conidium microcolonies for each isolate

azole each year for the last seven years. In addition, the distribution and magnitude of myclobutanil- and difenoconazole-resistant phenotypes have remained unchanged despite application practice. It should be noted that it is difficult to examine the temporal development within an apple production season as apple scab lesions with viable conidia only persist for 1–2 months before they are killed by consecutive days where the temperatures are in excess of 29 °C. Hence, the best indication of population sensitivity at the end of season is the sensitivity of isolate-derived lesions found on the cluster leaves, which best represent the overwintering population. In surveys of populations from the research orchards at the New York Agricultural Experiment Station, the fungicide sensitivity profiles of isolates derived from lesions on cluster leaves have always been similar to those derived from terminal leaves later in the season (Cox *unpublished*).

Although resistance to DMI fungicide has been documented in *V. inaequalis* in the United States for more than 25 years, the mechanism of resistance is still not fully understood. Certainly, the target site for DMI fungicides, the 14α-demethylase (*CYP51A1*) gene, was previously elucidated (Schnabel and Jones 2001). The authors did not find coding-frame point mutations associated with DMI-resistant isolates of *V. inaequalis*, but did note that the *CYP51A1* gene was overexpressed in DMI-resistant isolates from three commercial orchards in Michigan. In one orchard, *CYP51A1* overexpression coincided with the presence of a 533 bp insertion upstream of the *CYP51A1* gene (Schnabel and Jones 2001). Such discoveries aren't surprising as CYP51 overexpression is reported to be a mechanism of DMI fungicide resistance in other fungi (Hamamato et al. 2000; Ma et al. 2006; Nikou et al. 2009; Cools et al. 2012).

Later investigations (*in preparation for publication*) involving the characterization of 133 *V. inaequalis* isolates representing a wide range of DMI resistance phenotypes from orchard populations from the midwestern and eastern United States revealed the presence of numerous novel upstream insertion events. The indels were positioned upstream of the *CYP51A1* start codon and ranged 0.5–2 kb in size. The indels contained several predicted promoters and some were nested within each other (Beckerman and Cox unpublished). Unfortunately, such indels seem to exert little on *CYP51A1* gene expression (Villani et al. 2012). Interestingly, the most highly resistant isolates had more indels or indels with more predicted promoter sequences than those with low to moderate DMI resistance or baseline isolates, which often had no indels. Despite these discoveries, the lack of conclusive information on *CYP51A1* gene expression and information of multidrug efflux pumps, which may also play a role in DMI resistance, will keep the molecular basis for DMI resistance unknown until transcriptome information is readily available.

27.2.4 QoI Fungicides

The QoI fungicides trifloxystrobin (Flint) and kresoxim-methyl (Sovran) were registered for use on apple scab around the beginning of the twenty-first century (Ypema and Gold 1999). *V. inaequalis* seems to have both a qualitative (complete or vertical) and a quantitative (incremental or horizontal) resistance response to QoI fungicides whereby isolates may display a variety of sensitivity phenotypes (quantitative response) when exposed to a QoI fungicide in the absence of target-site (*cyt b* gene) mutations (Frederick et al. 2014; Köller et al. 2004; Färber et al. 2002: Olaya et al.

1998). This quantitative response only occurs at doses below the point at which the alternative respiration pathway becomes activated. This quantitative resistance was first discovered in *V. inaequalis* isolates from a commercial orchard in Michigan that had been using QoI fungicides for only 2 years (Köller et al. 2004). By 2008, qualitative resistance to QoI fungicides conferred by the cytochrome b gene G143A mutation was confirmed in *V. inaequalis* isolates in numerous commercial apple orchards in Michigan (Lesniak et al. 2011). Over the next 2 years, Lesniak et al. (2011) surveyed 81 commercial orchards and found that nearly 67 % *V. inaequalis* isolates collected had qualitative resistance to QoI fungicides. In New York and New England orchards, quantitative resistance to QoI fungicides at levels that could cause concern for practical resistance was reported as early as 2004 (Frederick et al. 2014). However, it wasn't until 2007 that qualitative resistance conferred by the G143A *cyt b gene* mutation was detected in five isolates from a commercial orchard in New York (Frederick et al. 2014). A later examination of 120 orchard populations of *V. inaequalis* from New York and New England revealed that only 34 of 120 populations were found to have practical resistance to QoI fungicide trifloxystrobin, and of these populations, only 13 had isolates with the *cyt b* G143A mutation (Frederick et al. 2014). While qualitative QoI resistance in New York and New England isn't as prevalent as DMI resistance, careful monitoring and strategic use of QoI fungicides are needed to prevent the further development of QoI-resistant populations of *V. inaequalis*.

27.2.5 Anilinopyrimidine and Succinate Dehydrogenase Inhibitor Fungicides

Anilinopyrimidine (AP) fungicides were first labeled on apple for the management of apple scab in the United States near the turn of the twenty-first century (Köller 1999). Compared to their efficacy on *Botrytis* spp., the AP fungicides aren't as effective in inhibiting ascospore germination and cuticle penetration in *V. inaequalis* (Köller et al. 2005; Kunz et al. 1998; Daniels et al. 1994). The use of AP fungicides has not been popular for apple scab management in New York and New England because their field performance did not substantially improve over protectant fungicides (e.g., captan and EBDCs) (Köller et al. 2005). However, AP fungicides are popular in northern apple production regions of New York and New England where cold rainy springs mandate the use of local systemic fungicides with good activity in cold weather. To date, resistance to AP fungicides has not been reported in the United States (Köller et al. 2005). Interestingly, Kunz et al. (1998) reported that populations of *V. inaequalis* remained sensitive after a total of 43 AP applications made over 4 years. Köller et al. (2005) reported a loss of sensitivity to AP fungicides for *V. inaequalis* in several commercial New York orchards. Since these eight *V. inaequalis* populations had never received applications of AP fungicides, it was unlikely that they represent selection for novel mechanisms of resistance. Because of limited use for apple scab in the eastern United States, there have been no efforts to further document the development of fungicide resistance to AP fungicides.

The succinate dehydrogenase inhibitor (SDHI) fungicides were first registered in 2005 for use on apples in the United States as Pristine, a premixed fungicide product from BASF. Pristine is a mixture of the SDHI fungicide boscalid (pyridinecarboxamide) with the QoI fungicide pyraclostrobin. Since the introduction of Pristine, few growers have used it for the management of apple scab as it has a limited number of allowed uses and is fairly effective against flyspeck, sooty blotch, and other late-season disease of apple. A few years after the introduction of Pristine, several other SDHI fungicides were registered for apples and others are still being developed for use in apples in the United States. The newly released SDHI fungicides include two pyrazole carboxamide fungicides penthiopyrad and fluxapyroxad and one pyridinyl-ethyl-benzamide fungicide fluopyram. Similar to the registration of boscalid, the SHDI fungicides fluxapyroxad and fluopyram are being formulated and sold as premixed fungicides containing a QoI or AP fungicide. Given the rapid development of fungicide resistance in recent years for many of the fungicide chemistries (citations from above), the agrochemical industry in the United States is releasing more products as premixes with fungicides as a means of providing built-in fungicide resistance management.

Resistance to SDHI fungicides and target-site mutations in the succinate dehydrogenase gene complex has been reported for several other pathogens of fruit crops such as *Alternaria alternata* (Avenot et al. 2009) and *Botrytis cinerea* (Yin et al. 2011; Veloukas et al. 2011), but not pathogens of apples. Many of the SHDI fungicides are still being registered for the management of apple scab in the United States and only a few have been used for 2–3 seasons depending on the state. Thus, there have been no confirmed reports of SDHI resistance in isolates of *V. inaequalis* to date. Aside from baseline sensitivity testing of isolated populations of *V. inaequalis* and cloning of the *SDHB* gene coding the iron sulfur subunit of the succinate dehydrogenase complex, little has been accomplished.

27.3 Conclusions

Since the early twentieth century, apple producers have sought to reduce the number of applications and improve the timing of fungicides used to manage apple scab. As older broad-spectrum fungicides with postinfection activity were taken off from the market due to health concerns, only the modern single-site fungicides introduced in the later of part of twentieth and beginning of the twenty-first centuries could offer postinfection activity. The introduction of new single-site fungicide chemistries in the United States occurred in a slow successive manner and led to the unfortunate practice of overreliance on the newest fungicide class and the development of fungicide resistance to each class in succession. Presently, resistance to dodine, MBCs, SBIs, and QoIs has been reported in field populations of *V. inaequalis* in nearly all of the apple production regions in the eastern United States where temperate climate favors the disease. Although resistance to the anilinopyrimidine (AP) and succinate dehydrogenase inhibitor (SDHI) fungicides has not been observed, the protective and postinfection activities of these fungicide classes against apple scab are considerably less than those of the SBIs, QoIs, dodine, and MBCs. In the absence of fungicides with strong postinfection activity, growers will need to increase the frequency of fungicide applications as they endeavor to control the highly endemic apple scab pressure. Growers opt to use fungicides against populations of *V. inaequalis* with resistance to several fungicide classes, risk management failure, and potentially, complete crop loss.

To date, only multisite, broad-spectrum protectant fungicides like ethylenebisdithiocarbamates (EBDCs) and phthalimides (captan) are not at risk for fungicide resistance. These contact fungicides only work in a protective mode and must be applied prior to apple scab infections. Additionally, many of the broad-spectrum protectant fungicides have long preharvest intervals that occur before growers finish managing preharvest diseases. Given the concerns of fungicide resistance in single-site fungicides, growers rely almost exclusively on broad-spectrum protectant fungicides for the management of apple scab during the beginning of the season. The consequence of this practice is that additional fungicide inputs will be needed during rainy springs and new disease problems may emerge as a consequence of curtailing single-site fungicide use. Indeed, broad-spectrum protectant fungicides are less effective against rusts and other anthracnose leaf spots. If the trend of resistance development to each newly introduced class of fungicides with postinfection activity continues, apple producers in the United States will continue to face the considerable challenge of managing apple scab with limited options.

References

Anonymous (2012) Noncitrus fruits and nuts, 2011 preliminary summary. U. S. D. o. Agriculture, ed. National Agricultural Statistic Service

Avenot HF, Sellam A, Michailides TJ (2009) Characterization of mutations in the membrane-anchored AaSDHC and AaSDHD of succinate dehydrogenase from *Alternaria alternata* isolates conferring field resistance to the fungicide boscalid. Plant Pathol 58:1134–1143

Carisse O, Jobin T (2010) Resistance in dodine populations of *Venturia inaequalis* in Quebec, Canada. Online. Plant Health Prog. doi: 10.1094/PHP-2010-0614-01-RS

Chapman KS, Sundin GW, Beckerman JL (2011) Identification of resistance to multiple fungicides in field populations of *Venturia inaequalis*. Plant Dis 95:921–926

Cools HJ, Bayon C, Atkins S, Lucas JA, Fraaije BA (2012) Overexpression of the sterol 14 alpha-demethylase gene (MgCYP51) in *Mycosphaerella graminicola* isolates confers a novel azole fungicide sensitivity phenotype. Pest Manage Sci 68:1034–1040

Cox KD, Russo NL, Villani SM, Parker DM, Köller W (2008) QoI qualitative resistance and CYP51A1 upstream anomalies in NY populations of the apple scab pathogen *Venturia inaequalis*. Phytopathology 98:S42

Cox KD, Villani SM, Ellis MA, Poon CK, Craig-Kuhn MC (2009a) Evaluation of fungicides in an orchard with a DMI-resistant apple scab population, 2008. Plant Dis Manage Rep PF039

Cox KD, Villani SM, Köller W (2009b) Practical implications of fungicide resistance in Northeastern US populations of the apple scab pathogen *Venturia inaequalis*. Phytopathology 99:S25

Cox KD, Villani SM, Raes JJ (2010) Evaluation of fungicides for control of apple scab in a 'McIntosh' 'Cortland' orchard with a DMI-resistant apple scab population, 2009. Plant Dis Manage Rep PF019

Cox KD, Villani SM, Raes JJ (2011) Evaluation of fungicides for control of apple scab in a 'McIntosh' 'Cortland' orchard with a DMI-resistant apple scab population, 2010. Plant Dis Manage Rep PF032

Daniels A, Birchmore RJ, Winter EH (1994) Activity of pyrimethanil on *Venturia inaequalis*. In: The BCPC conference pests & diseases 1994, conference proceedings. British Crop Protection Council, Farnham, Surrey, UK, pp 525–532

Färber RB, Chin KM, Leadbitter N (2002) Sensitivity of *Venturia inaequalis* to trifloxystrobin. Pest Manag Sci 58:261–267

Frederick ZA, Villani SM, Cooley DR, Biggs AR, Raes JJ, Cox KD (2014) Prevalence of quinone outside inhibitor resistance and the stability of qualitative QoI resistance in *Venturia inaequalis*. Plant Dis 98:1122–1130

Hamamato H, Hasegawa K, Nakaune R, Lee YJ, Makizumi Y, Akutsu K, Hibi T (2000) Tandem repeat of a transcriptional enhancer upstream of the sterol 14 alpha-demethylase gene (CYP51) in *Penicillium digitatum*. Appl Environ Microbiol 66:3421–3426

Jones A (1991) Fungicide resistance: past experience with benomyl and dodine and future concerns with sterol biosynthesis inhibitors. Plant Dis 65:990–992

Jones A, Walker RJ (1976) Tolerance of *Venturia inaequalis* to dodine and benzimidazole fungicides in Michigan. Plant Dis 60:40–44

Kappas A, Georgopoulos SG (1970) Genetic analysis of dodine resistance in *Nectria haematococca*. Genetics 66:6-7-622

Katan T, Shabi E, Gilpatrick JD (1983) Genetics of resistance to benomyl in *Venturia inaequalis* isolates from Israel and New York. Phytopathology 73:600–603

Koenraadt H, Somervile SC, Jones AL (1992) Characterization of mutations in the beta-tubulin gene of benomyl resistant field strains of *Venturia inaequalis* and other plant pathogenic fungi. Phytopathology 82:1348–1354

Köller W (1991) Fungicide resistance in plant pathogens. In: CRC handbook of pest management in agriculture (2nd edn). CRC Press, Boca Raton, pp 525–532

Köller W (1999) Chemical approaches to managing plant pathogens. In: Ruberson JR (ed) Handbook of pest management. Marcel Dekker, New York, pp p337–p376

Köller W, Wilcox WF (2000) Interactive effects of dodine and the DMI fungicide fenarimol in the control of apple scab. Plant Dis 84:863–870

Köller W, Wilcox WF, Barnard J, Jones AL, Braun PG (1997) Detection and quantification of resistance of *Venturia inaequalis* populations to sterol demethylation inhibitors. Phytopathology 87:184–190

Köller W, Parker DM, Turechek WW, Avila-Adame C (2004) A two-phase resistance response of *Venturia inaequalis* populations to the QoI fungicides kresoxim-methyl and trifloxystrobin. Plant Dis 88:537–544

Köller W, Wilcox WF, Parker DM (2005) Sensitivity of *Venturia inaequalis* populations to anilinopyrimidine fungicides and their contribution to scab management in New York. Plant Dis 89:357–365

Kunz S, Lutz B, Deising H, Mendgen K (1998) Assessment of sensitivities to anilinopyrimidine- and strobilurin-fungicides in populations of the apple scab fungus *Venturia inaequalis*. J Phytopathol 146:231–238

Leroux P, Clerjeau M (1985) Resistance of *Botrytis cinerea* Pers. and *Plasmopara viticola* (Berk & Curt.) Berl. and de Toni to fungicides in French vineyards. Crop Prot 4:137–160

Lesniak KE, Proffer TJ, Beckerman JL, Sundin GW (2011) Occurrence of QoI resistance and detection of the G143A mutation in Michigan populations of *Venturia inaequalis*. Plant Dis 95:927–934

Ma Z, Michailides TJ (2005) Advances in understanding molecular mechanisms of fungicide resistance and molecular detection of resistant genotypes in phytopathogenic fungi. Crop Prot 24:853–863

Ma ZH, Proffer TJ, Jacobs JL, Sundin GW (2006) Overexpression of the 14 alpha-demethylase target gene (CYP51) mediates fungicide resistance in *Blumeriella jaapii*. Appl Environ Microbiol 72:2581–2585

MacHardy WE (1996) Apple scab: biology, epidemiology, and management. The American Phytopathological Society, St. Paul

MacHardy WE, Gadoury DM, Gessler C (2001) Parasitic and biological fitness of *Venturia inaequalis*: relationship to disease management strategies. Plant Dis 85:1036–1051

Marine SC, Schmale III DG, Yoder KS (2007) Resistance to myclobutanil in populations of *Venturia inaequalis* in Winchester, Virginia. Online. Plant Health Prog. doi: 10.1094/PHP-2007-1113-01-RS

Merwin IA, Brown SK, Rosenberger DA, Cooley DR, Berkett LP (1994) Scab-resistant apples for the Northeastern United States: new prospects and old problems. Plant Dis 78:4–10

Nikou D, Malandrakis A, Konstantakaki M, Vontas J, Markoglou A, Ziogas B (2009) Molecular characterization and detection of overexpressed C-14 alpha-demethylase-based DMI resistance in *Cercospora beticola* field isolates. Pestic Biochem Physiol 95:18–27

Olaya G, Zheng D, Köller W (1998) Differential responses of germinating *Venturia inaequalis* conidia to kresoxim-methyl. Pestic Sci 54:230–236

Pfeufer EE, Travis JW, Nugugi HK (2010) Resistance to DMI fungicides in *Venturia inaequalis* from Pennsylvania. Phytopathology 100:S100

Quello KL, Chapman KS, Beckerman JL (2010) In situ detection of benzimidazole resistance in field isolates of *Venturia inaequalis* in Indiana. Plant Dis 94:744–750

Schnabel G, Jones A (2001) The 14α – demethylase (CYP51A1) gene is overexpressed in *Venturia inaequalis* strains resistant to myclobutanil. Phytopathology 91:102–110

Stanis VF, Jones AL (1985) Reduced sensitivity to sterol-inhibiting fungicides in field isolates of *Venturia inaequalis*. Phytopathology 75:1098–1101

Szkolnik M, Gilpatrick JD (1969) Apparent resistance of *Venturia inaequalis* to dodine in New York apple orchards. Plant Dis Rep 53:861–864

Veloukas T, Leroch M, Hahn M, Karaoglanidis GS (2011) Detection and molecular characterization of boscalid-resistant *Botrytis cinerea* isolates from strawberry. Plant Dis 95:1302–1307

Villani SM, Freier J, Cox KD (2012) Characterizing the role of CYP51A1 overexpression in myclobutanil- and difenoconazole-resistant *Venturia inaequalis*. Phytopathology 102:S42

Wicks T (1974) Tolerance of the apple scab fungus to benzimidazole fungicides. Plant Dis Rep 58:886–889

Yin YN, Kim YK, Xiao CL (2011) Molecular characterization of boscalid resistance in field isolates of *Botrytis cinerea* from apple. Phytopathology 101:986–995

Ypema HL, Gold RE (1999) Kresoxim-methyl: modification of a naturally occurring compound to produce a new fungicide. Plant Dis 83:4–19

Chapter 28
Resistance in Postharvest Pathogens of Citrus in the United States

James E. Adaskaveg and Helga Förster

Abstract Among citrus postharvest pathogens, fungicide resistance in the United States to date has only been reported for species of *Penicillium*. Except for the inorganic salts, widespread resistance has developed to all of the older fungicides such as the still registered phenylphenols *o*-phenylphenol and sodium *o*-phenylphenate, the methyl benzimidazole carbamate, thiabendazole, and the demethylation inhibitor imazalil. The almost simultaneous introduction of several new compounds in the early 2000s that include the anilinopyrimidine pyrimethanil, the phenylpyrrole fludioxonil, and the quinone outside inhibitor azoxystrobin offered a unique opportunity in keeping the development of resistance to a minimum. Fungicide modes of action could be mixed and rotated from the first introduction before resistance had occurred. Resistance to pyrimethanil, however, has developed in some packinghouse populations of *P. digitatum* because end users did not follow these guidelines and used the fungicide exclusively. For azoxystrobin, resistance in *P. digitatum* has only been described for laboratory mutants and for fludioxonil only in mass platings of conidia on selective media in the laboratory or in packinghouse air samplings. Thus, practical resistance to azoxystrobin and fludioxonil has not occurred. Natural resistance frequencies and molecular mechanisms for thiabendazole, imazalil, pyrimethanil, fludioxonil, and azoxystrobin have been studied, and resistant pathogen isolates have been evaluated for their fitness. Anti-resistance strategies focus on sanitation of fruit, equipment, and storage rooms; limitation of pathogen sporulation and spore dispersal; use of fungicide mixtures, pre-mixtures, and rotations; as well as the early detection of resistance.

Keywords Postharvest decays • *Penicillium* spp. • Green mold • Postharvest fungicides • Resistance development • Resistance management

---

J.E. Adaskaveg (✉) • H. Förster
Department of Plant Pathology and Microbiology, University of California, Riverside, CA, USA
e-mail: jim.adaskaveg@ucr.edu

H. Ishii, D.W. Hollomon (eds.), *Fungicide Resistance in Plant Pathogens*,
DOI 10.1007/978-4-431-55642-8_28

28.1 Introduction

The main citrus-growing areas in the United States are Florida, California, Texas, and Arizona. In 2012, Florida produced 63 % of the total citrus crop with oranges and grapefruit accounting for about 70 % and 65 % of the country's production, respectively (http://www.agmrc.org/commodities). California produced 34 % of the citrus, providing 92 % of the lemon and 80 % of the tangerine crop. With current major disease threats to the Florida industry (i.e., citrus greening – Huanglongbing, bacterial canker), production in this state is expected to decline dramatically over the next years. Although California ranks second in production, postharvest fungicide use on citrus surpasses that of Florida. This is because Florida oranges are mainly used for juicing, while most of the California crop is used for fresh fruit consumption and has to arrive decay-free and with high quality to the final consumer. Additionally, California lemon fruit may be stored for up to 3 or 4 months in the packinghouse before marketing based on economic needs and incentives. With extensive handling of fruit during harvest, transport to packinghouse, grading, packing, storage, and final marketing that often includes long-distance transport to worldwide destinations, fruit may become infected with decay organisms at any stage of the processing chain and thus have to be protected from decay development. This is most effectively done in an integrated postharvest decay management program that includes the application of postharvest fungicides (Adaskaveg et al. 2002).

Postharvest decay is one of the most important factors diminishing value and limiting marketing of fresh citrus. Losses of fresh citrus by decay after harvest are more costly than those occurring before harvest because of the added expenses for harvesting, postharvest handling, treatments, shipping, and storage. Occurrence and severity of postharvest decays depend on many factors including growing region and environmental conditions, fruit variety, tree condition, cultural practices, preharvest treatments, harvest methods, and postharvest handling practices.

28.2 Major Postharvest Decays of Citrus

Green mold, caused by the fungus *Penicillium digitatum*, is the most important postharvest decay of citrus fruit. It occurs year-round on all citrus varieties worldwide, but especially in locations with arid and subtropical climates (Eckert and Eaks 1989). Blue mold, caused by *P. italicum*, is generally less prevalent, but may become a major problem under conditions that suppress development of green mold such as storage temperatures below 10 °C (Brown and Eckert 2000). In California, green and blue molds, as well as Alternaria decay, may also be major problems when orange, mandarin, and lemon fruits are de-greened with ethylene immediately after harvest in non-refrigerated storage rooms (e.g., 20–25 °C) without preharvest or postharvest fungicide and sanitation treatments (Eckert and Eaks 1989). In other regions such as Florida, de-greening results in higher incidence of other decays such as anthracnose caused by *Colletotrichum* species and stem-end rots (see below).

Sour rot, caused by *Galactomyces citri-aurantii* (anamorph, *Geotrichum citri-aurantii*), ranks second in postharvest decay losses after Penicillium decays. In California, sour rot is particularly problematic on lemons grown in the coastal regions and on mandarins that are often stored for extended periods at relatively high temperatures of 12–14 °C and high relative humidity (92–98 %) to obtain marketable rind color (Eckert 1959; Suprapta et al. 1996). Overall, outbreaks of the disease are sporadic with highest incidences on those fruits that are harvested during prolonged wet conditions (Baudoin and Eckert 1982; Eckert 1959). In storage, the disease may result in complete collapse and liquefaction of infected fruit. Juices dripping from infected fruit can readily spread the pathogen to healthy fruit (Eckert and Eaks 1989).

Other postharvest decays of citrus fruits include brown rot (caused by several species of *Phytophthora*, mostly *P. citrophthora*, *P. syringae*, *P. parasitica*, *P. hibernalis*, and *P. palmivora*), gray mold (mostly on lemons; caused by *Botrytis cinerea*), anthracnose (caused by *Colletotrichum gloeosporioides*), Septoria spot (caused by *Septoria citri*), Alternaria fruit rot (caused by *Alternaria alternata*), whisker mold (caused by *P. ulaiense*), and Trichoderma rot (caused by *Trichoderma viride*) (Timmer et al. 2000). Stem- and blossom-end rots are more prevalent in the humid climate of Florida and may be caused by *Alternaria citri, Botryosphaeria rhodina*, *Phomopsis citri*, and other species (Timmer et al. 2000).

28.3 Fungicides Registered for Postharvest Use on Citrus in the United States and Development of Resistance in Postharvest Pathogen Populations

Due to their high incidence and severity, Penicillium decays have been the primary targets in the development of citrus postharvest fungicide treatments. All compounds registered to date have at least some activity against these decays (Table 28.1). The recently registered propiconazole is the only fungicide that also has high activity against sour rot. In other parts of the world, however, guazatine is available and is effective against sour rot and Penicillium decays (Rippon and Morris 1981).

Some of the first treatments used to control postharvest decays of citrus fruits were alkaline solutions of borax, sodium carbonate (soda ash), and sodium bicarbonate. In California, their beneficial action to reduce the incidence of *Penicillium* molds of citrus was first realized over 80 years ago (Barger 1928). In addition to these inorganic salts and the phenylphenols, citrus postharvest fungicides represent six FRAC (Fungicide Resistance Action Committee; http://www.frac.info/publication/anhang/FRAC) groups that all target a single-site mode of action: the methyl benzimidazole carbamates (MBCs; FRAC 1), the sterol demethylation inhibitors (DMIs; FRAC 3), the anilinopyrimidines (FRAC 9), the quinone outside inhibitors

Table 28.1 Fungicides registered in the United States as postharvest treatments on citrus to prevent decays caused by fungi

Fungicide class	Chemicals and trade name	FRAC group	Main targets of activity
Inorganic salt	Sodium borate (borax – sodium tetraborate), sodium carbonate (soda ash), sodium bicarbonate (baking soda)	–	*Penicillium* spp.
Phenol derivative	Phenylphenols: *o*-phenylphenol (OPP), sodium *o*-phenylphenate (tetrahydrate) (SOPP)	–	*Geotrichum* and *Penicillium* spp., stem-end rots, *Trichoderma* spp.
Methyl benzimidazole carbamate (MBC)	2-(4-Thiazolyl) benzimidazole (thiabendazole – TBZ, Mertect 340, Alumni, Decco Salt No. 19)	1	*Penicillium* spp., stem-end rots
Sterol demethylation inhibitor (DMI) – triazole	Propiconazole (Mentor)	3	*Penicillium* spp., *Geotrichum citri-aurantii*
Sterol demethylation inhibitor (DMI) – imidazole	Imazalil (Deccocil, Fungaflor, Freshgard, and others)	3	*Penicillium* spp.
Anilinopyrimidine	Pyrimethanil (Penbotec)	9	*Penicillium* spp.
Quinone outside inhibitor (QoI, strobilurin)	Azoxystrobin (Diploma)[a]	11	*Penicillium* spp.
Phenylpyrrole	Fludioxonil (Graduate)	12	*Penicillium* and *Botrytis* spp., stem-end rots
Phosphonate	Phosphorous acid (potassium and calcium phosphite – Fungi-Phite, ProPhyt, and others)	33	Mostly *Phytophthora* spp., some activity against *Penicillium* spp.
MBC + phenylpyrrole	TBZ + fludioxonil (Graduate Max)	1 + 12	*Penicillium* and *Botrytis* spp., stem-end rots
DMI + anilinopyrimidine	Imazalil + pyrimethanil (Philabuster)	3 + 9	*Penicillium* and *Botrytis* spp., stem-end rots, anthracnose
QoI + phenylpyrrole	Azoxystrobin + fludioxonil (Graduate A+)	11 + 12	*Penicillium* and *Botrytis* spp., stem-end rots

[a]Only sold in a pre-mixture at this time (i.e., Graduate A+)

(QoIs, strobilurins; FRAC 11), the phenylpyrroles (FRAC 12), and the phosphonates (FRAC 33) (Adaskaveg et al. 2004; Adaskaveg and Förster 2010).

Resistance in *Penicillium* populations has developed in selected populations to all of the older fungicides, except for the inorganic salts (Table 28.2). The phenylphenols, thiabendazole, and imazalil each were introduced years apart and were then often used exclusively after resistance to the previously registered compounds had already occurred: *o*-phenylphenol in the 1930s, thiabendazole in the 1970s, and

Table 28.2 Resistance development in decay organisms of citrus against postharvest fungicides in the United States

Fungicide class	FRAC group	Fungicide	Year introduced on citrus	Currently registered	First occurrence of postharvest resistance		
					Year	Decay organism	Reference
Phenol derivative	–	OPP, SOPP	1936	Yes	1959	*Penicillium digitatum*	Harding (1962)
Aromatic hydrocarbon	14	Biphenyl	1944	No	1962	*Penicillium* spp.	Harding (1962)
Aliphatic amine	–	2-Aminobutane	1962	No	1976	*Penicillium* spp.	Harding (1976)
Methyl benzimidazole carbamate (MBC)	1	Benomyl	1967	No	1973	*P. digitatum*	Wild (1983)
		Thiabendazole	1967	Yes	1970	*Penicillium* spp.	Harding (1972) and Muirhead (1974)
Sterol demethylation inhibitor (DMI) – imidazole	3	Imazalil	1980	Yes	1982 (lab, subsequently in packinghouses)	*P. italicum*	De Waard et al. (1982)
					1987 (lab, subsequently in packinghouses)	*P. digitatum*	Eckert (1987)
DMI – triazole	3	Propiconazole	2008	Yes	Present at introduction[a]	*P. digitatum*	McKay et al. (2012)
Anilinopyrimidine	9	Pyrimethanil	2005	Yes	2009	*P. digitatum*	Kanetis et al. (2010)
Quinone outside inhibitor (QoI, strobilurin)	11	Azoxystrobin	2008	Yes	2011 (lab)[b]	*P. digitatum*	Zhang et al. (2009)
Phenylpyrrole	12	Fludioxonil	2006	Yes	2010 (packinghouse air samplings)[b]	*P. digitatum*	Kanetis et al. (2010)
					2012[c]	*P. italicum*	Adaskaveg *unpublished*

[a]Cross-resistance to imazalil
[b]Resistance only reported from laboratory mutants or in packinghouse air samplings, no practical resistance to date
[c]Resistance found at a single location, no practical resistance to date

Fig. 28.1 Lemon fruit inoculated with isolates of *Penicillium digitatum* sensitive (**a**, **b**) or resistant (**c**, **d**) to pyrimethanil and not treated (**a**, **c**) or treated with 1000 mg/L pyrimethanil (**b**, **d**)

imazalil in the 1980s. The phenylphenols are no longer widely used because of costly disposal of re-collected, spent material. Currently, resistance to thiabendazole and imazalil is very common. The molecular mechanism for MBC resistance in *P. digitatum* was determined to be a point mutation at codon 198 of the β-tubulin gene where glutamine was replaced with lysine (Koenraadt et al. 1992) or at codon 200 where thymine was replaced by adenine (Schmidt et al. 2006). For the DMI imazalil, several molecular mechanisms have been characterized in *P. digitatum*. These include a tandem repeat of a transcriptional enhancer in the promoter region of *CYP51* that leads to overexpression of the gene (Hamamoto et al. 2001) and a unique 199-bp insert within the transcriptional enhancer unit (Ghosoph et al. 2007).

In contrast to the older compounds, the newer fungicides pyrimethanil, fludioxonil, and azoxystrobin were introduced almost simultaneously. This offered a unique opportunity in keeping the development of resistant pathogen populations to a minimum because fungicide modes of action (FRAC groups) could be mixed and rotated. Still, some packers chose to exclusively use a single fungicide based solely on economic reasons or, simply, use the least costly treatment. This resulted in the development of pyrimethanil resistance in populations of *P. digitatum*, rendering this treatment ineffective in some packinghouses in California (Adaskaveg, *unpublished*) (Fig. 28.1). Pyrimethanil-resistant isolates of the pathogen were also readily obtained in mass platings of conidia on selective media in the laboratory or in packinghouse air samplings, and EC_{50} values for mycelial growth were >8 μg/ml (Kanetis et al. 2010). Anilinopyrimidines are classified as amino-acid and protein synthesis inhibitors by FRAC (http://www.frac.info/publication/anhang/FRAC), and biosynthesis of methionine and other amino acids was originally described as the mode of action for pyrimethanil in *B. cinerea* (Masner et al. 1994). This was questioned in subsequent investigations (Leroux et al. 2002; Fritz et al. 2003; Kanetis et al. 2008b). In studies with *P. digitatum*, growth of sensitive isolates was inhibited by pyrimethanil even in the presence of methionine, indicating a mode of action different from methionine biosynthesis also in this pathogen (Kanetis et al. 2008b).

Isolates of *P. digitatum* resistant to fludioxonil, which targets the mitogen-activated protein kinase pathway in this organism (Kanetis et al. 2008b), have only been obtained in mass selections of conidia in the laboratory or from packinghouse air samplings (Kanetis et al. 2010). Isolates obtained from selection plates could be placed into two categories: moderately resistant isolates with EC_{50} values ≤1 μg/ml and highly resistant isolates with EC_{50} values >1 μg/ml (Fig. 28.2). All were pathogenic in fruit inoculation studies although less virulent than the wild type. Treatments with fludioxonil were not effective or were reduced in their efficacy when fruits were inoculated with resistant isolates as compared to the wild-type sensitive isolate. To date, however, no fludioxonil-resistant isolates of *P. digitatum* have been obtained from decayed fruit in packinghouses. A first detection of fludioxonil resistance in *P. italicum* from treated lemon fruit, however, occurred in 2012 in California with measured EC_{50} values greater than 10 μg/ml (Adaskaveg *unpublished*).

Resistance in *P. digitatum* to azoxystrobin currently has only been reported from UV-induced laboratory mutants and the molecular mechanism was identified as a G143A mutation in the cytochrome b gene (Zhang et al. 2009). This pathogen was rated as likely to develop high levels of azoxystrobin resistance based on the genetic stability of resistant mutants and absence of a type I intron in the cytochrome b gene directly after codon 143 which has been correlated with a high potential for QoI resistance (Sierotzki et al. 2007; Zhang et al. 2009). Still, despite extensive surveys, no resistance to azoxystrobin was found in *Penicillium* spp. in citrus packinghouse air samplings or in isolates from treated, decayed fruit (Kanetis et al. 2010). More recently, however, some isolates of *P. digitatum* from decayed fruit in packinghouse monitoring surveys showed a five- to tenfold reduction in sensitivity against this fungicide (Adaskaveg *unpublished*).

Thus, among citrus postharvest pathogens, fungicide resistance has only been reported for species of *Penicillium*. No resistance to the DMI propiconazole has been observed in *G. citri-aurantii* although this pathogen has been exposed to the DMI imazalil for many years (McKay et al. 2012). Additionally, in mass platings of conidia and in soil population enrichment assays with *G. citri-aurantii*, no isolates with reduced sensitivity could be recovered. This is in contrast to experiments with propiconazole-sensitive isolates of *P. digitatum* where resistant isolates were readily obtained in platings of conidia from propiconazole-sensitive isolates (McKay et al. 2012). Propiconazole-resistant isolates of *Penicillium* spp., however, commonly occur due to cross-resistance with imazalil. Table 28.3 summarizes the current

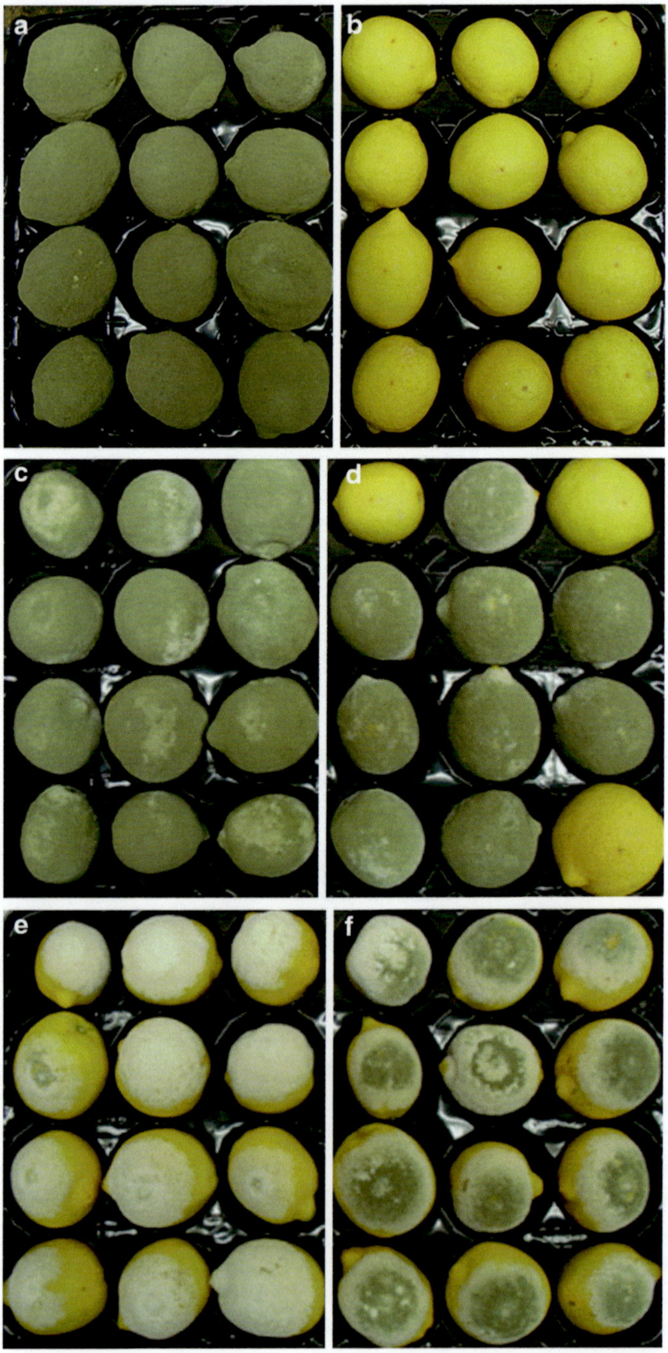

Fig. 28.2 Lemon fruit inoculated with isolates of *Penicillium digitatum* sensitive (**a**, **b**), moderately resistant (**c**, **d**), or highly resistant (**e**, **f**) to fludioxonil and not treated (**a**, **c**, **e**) or treated with 1000 mg/L fludioxonil (**b**, **d**, **f**)

Table 28.3 Resistance potential to four citrus postharvest fungicides in populations of *P. digitatum* and characteristics of resistant isolates[a]

Characteristic	Azoxystrobin	Fludioxonil	Pyrimethanil	Propiconazole
Laboratory selection of resistant isolates	No	Yes	Yes	Yes
Packinghouse selection of resistant isolates	No	Yes	Yes	Yes
Practical resistance (packinghouse)	No	No	Yes (locally)	Yes
Natural resistance frequency	–	$9.5 \times 10^{-7} - 1.5 \times 10^{-6}$	$7.3 \times 10^{-6} - 6.2 \times 10^{-5}$	$8.0 \times 10^{-8} - 1.2 \times 10^{-7}$
Resistance factor	–	MR: 3–26; HR: >1,500[b]	>250	14.3 – 81.6
Pathogenicity of resistant isolates	–	Similar to wild type, sometimes reduced	Similar to wild type	Similar to wild type
Fitness of resistant isolates	–	HR: not fit; MR: reduced fitness as compared to wild type	Similar to wild type	Not determined

[a]Data from Kanetis et al. (2010), Förster and Adaskaveg (2012), and McKay et al. (2012)

[b]*MR* moderately resistant isolates, *HR* highly resistant isolates

occurrence of resistance in *P. digitatum* to the newer fungicides azoxystrobin, fludioxonil, pyrimethanil, and propiconazole and characteristics of resistant isolates that are further discussed below.

Multiple resistance (resistance to different classes of fungicides that have unique modes of action) has also been found in *P. digitatum* populations. For example, resistance was present to biphenyl and SOPP (Harding 1962) or to 2-aminobutane, SOPP, TBZ, and benomyl (Davé et al. 1980). In another report, multiple resistance developed in populations of *Penicillium* spp. from citrus where thiabendazole and imazalil were introduced after resistance had already developed against the previously registered biphenyl and the phenylphenols (Holmes and Eckert 1999).

28.4 Factors Contributing to Resistance Development in Postharvest Decay Pathogens of Citrus

Resistant pathogen populations develop gradually by selection and subsequent proliferation of rare naturally occurring individuals in the population that are less sensitive to a specific chemical (Brent and Hollomon 1998). The frequency of the less sensitive individuals in the population (i.e., the natural resistance frequency) and the risk for resistance development are determined by the pathogen and the fungicide's mode of action. With their enormous asexual reproduction potential, *Penicillium* species are ranked at high risk for resistance development (Brent and Hollomon 1998). Resistance frequencies have been quantified for some pathogen-fungicide combinations. Thus, the resistance frequency for the MBC benomyl in unselected orchard populations of *P. digitatum* was determined at approximately 10^{-7} to 10^{-8} (Wild 1980; Eckert 1988). In mass platings of conidia of this pathogen, ranges of resistance frequencies in laboratory and packinghouse populations for the newer postharvest fungicides fludioxonil, pyrimethanil, and propiconazole were 9.5×10^{-7} to 1.5×10^{-6}, 7.3×10^{-6} to 6.2×10^{-5}, and 8.0×10^{-8} to 1.2×10^{-7}, respectively (Kanetis et al. 2010; McKay et al. 2012; Table 28.3). In contrast, similar tests conducted with *G. citri-aurantii* did not yield any resistant individuals, indicating that perhaps this fungus is less prone to resistance development (McKay et al. 2012). All of the newer postharvest fungicides are single-site mode of action compounds, and these intrinsically have a much higher risk for resistance development than multisite mode of action compounds (Hewitt 1998; Kendall and Hollomon 1998). The speed of resistance development after the initial selection depends largely on the presence of continued selection pressure (e.g., stability of fungicide residue on treated fruit during storage, repeated applications of the same fungicide class to the same fruit lot) and the competitive fitness of the less sensitive isolates.

Resistant pathogen isolates have been evaluated for their fitness based on pathogenicity (ability to cause decay), decay rate (growth rate of pathogen), virulence (severity of decay), and sporulation potential (reproductive ability). For the qualitative, single-step type of resistance that is the result of a mutation in a single or, at most, a small number of major genes (Kendall and Hollomon 1998) and that is typical for the MBC fungicides, resistant populations generally remain stable in the absence of selection pressure (Brent and Hollomon 2007). Thus, there are limited fitness penalties in isolates with qualitative resistance, but exceptions are known such as with phenylamide resistance in *Phytophthora infestans* (Cooke et al. 2006). Isolates of *P. digitatum* resistant to MBC compounds, however, showed a somewhat reduced virulence, and the resistant biotype decreased in frequency in the absence of selection pressure in laboratory competition studies as well as in packinghouses (Wild 1980; Eckert 1988). In contrast, current packinghouse populations of *P. digitatum* with thiabendazole resistance (EC_{50} values >7.8 μg/ml) have been found to be stable and persistent (Adaskaveg 2004).

Fitness penalties commonly occur in pathogens with quantitative, multistep resistance that is the result of mutation of several genes and that is typical for DMI fungicides. This type of resistance rapidly reverts to a more sensitive condition in the absence of selection pressure (Brent 1995). In agar and fruit assays with *P. digitatum*, the frequency of resistant isolates declined over two disease cycles when agar plates or lemon fruits were inoculated with mixtures of imazalil-sensitive and imazalil-resistant strains (Holmes and Eckert 1995). This reduced fitness of resistant isolates results in an annual fluctuation of imazalil resistance in populations of *P. digitatum* in citrus packinghouses. Thus, there is a gradual increase in imazalil resistance over the year, with the lowest incidence of resistance during the off-peak season in late fall and early winter and the highest incidence during the main harvest and storage periods in late winter and spring (Adaskaveg 2004).

Fitness and competitiveness of resistant isolates of *P. digitatum* have also been evaluated for the newer fungicides fludioxonil and pyrimethanil. Pyrimethanil-resistant isolates with EC_{50} values of 2 to >10 mg/L (baseline, <0.7 mg/L) and fludioxonil-resistant isolates with EC_{50} values of 0.1 to >10 mg/L (baseline, <0.02 mg/L) were pathogenic on citrus fruit. Isolates resistant to pyrimethanil were mostly similar in growth, virulence, and sporulation to wild-type sensitive isolates (Fig. 28.1), whereas isolates resistant to fludioxonil displayed a range of growth, virulence levels, and sporulation capacities (Fig. 28.2). Resistant isolates were sometimes less virulent and most isolates showed reduced sporulation as compared to wild-type sensitive isolates (Kanetis et al. 2010). Isolates highly resistant to fludioxonil were not recovered from decaying fruit after co-inoculation with sensitive isolates; recovery rates for isolates moderately resistant to this fungicide were 0–2.1 % (Förster and Adaskaveg 2012). Isolates highly resistant to pyrimethanil, in contrast, were recovered at frequencies between 14.8 and 84.1 %. Thus, these studies demonstrated fitness differences for different modes of action among resistant isolates of *P. digitatum*. Fitness penalties exist for fludioxonil resistance but are less apparent for pyrimethanil resistance indicating that without proper anti-resistance strategies, resistant isolates may become predominant after repeated fungicide applications. Isolates from fludioxonil and pyrimethanil selection plates were also stable in their fungicide sensitivity characteristics after several passages on non-amended agar and disease could not be sufficiently controlled after treatment with the respective fungicides (Kanetis et al. 2010).

Postharvest handling practices can also affect the risk of resistance development. As indicated above, some California citrus fruits, especially lemons, may be stored for long time periods in the packinghouse before marketing. Decay may develop in storage, leading to an initial selection process on the fungicide-treated fruit. When fruits are taken out of storage before shipment to market, they are repacked, and airborne inoculum from decayed fruit is readily dispersed to healthy fruit or to fruit coming from the field if strict sanitation practices are not followed which include the spatial separation of these different packinghouse activities. This inoculation potential is exacerbated when fruits were treated with a postharvest fungicide that has no anti-sporulation activity such as pyrimethanil. Fruits may then receive a second fungicide application, sometimes with the same fungicide that was used before storage. In California lemon production, the risk of fungicide resistance development in *Penicillium* populations is further increased by the year-round availability of susceptible fruit in the packinghouse because fruits are harvested, processed, and stored almost continuously throughout the year without a break (Adaskaveg et al. 2002).

28.5 Resistance Management

The overall goal of fungicide resistance management is to achieve sustainable, highly effective disease management with the use of fungicides now and in the future. Specific objectives are to minimize the development of new resistance in pathogen populations (i.e., to fungicides against which resistance was never reported previously and in new fungal species – fungicide combinations), to detect the development of resistance at an early stage, and to mitigate the impact of resistance in managing diseases so that crops can be economically produced and distributed to the consumer. To achieve these goals, decay management programs generally have to be modified at several levels. Resistance has not developed to date in commercial situations in the United States against several of the newer postharvest fungicides (i.e., azoxystrobin and fludioxonil in *Penicillium* species, propiconazole in *G. citri-aurantii*) and the challenge is to protect their efficacy. Table 28.4 summarizes postharvest practices that help mitigating the development of resistance. Below we discuss some of those practices that are directly related to fungicide use and reduction of pathogen populations (e.g., sanitation, anti-sporulation methods). Information on other practices that are beyond the scope of this chapter can be found in several detailed publications (see Kader 2002).

Early detection of resistance and of the composition of the pathogen population allows the targeted adjustment of fungicide usage. This is done by implementing routine monitoring programs where fungicide sensitivities of the current pathogen population are compared to those of a population that was sampled before the use of a new fungicide (baseline sensitivity). Baseline sensitivities for isolates of *P. digitatum* and *P. italicum* in California have been established for azoxystrobin, fludioxonil, and pyrimethanil (Kanetis et al. 2008a) and for *G. citri-aurantii* for propiconazole (McKay et al. 2012). Additionally, new high-throughput monitoring methods have

Table 28.4 Resistance management strategies used in a packinghouse

Overall strategy	Specific practice	Benefit
Optimize health of the fruit	Minimize injuries during harvest and handling	Most postharvest pathogens are wound pathogens and minimizing injuries reduces decay
	Keep fruit at optimal temperature during storage, transportation, and marketing	Senescence is delayed and fruits are less susceptible to decay
	Limit storage times	Senescence is reduced and fruits are less susceptible to decay. The chance of decay development is reduced
Use sanitizing methods (chemical and physical)	Sanitize healthy and decayed fruit, handling equipment, and storage rooms	Pathogen populations are reduced and contamination of healthy fruit is minimized
	Filter air to reduce spore load and regulate air flow	Inoculum dispersal among different sections of the packinghouse is reduced
	Remove cull fruit from packinghouse	Resistance that was potentially selected on treated fruit is moved away from packinghouse
	Spatially separate incoming fruit from the field from the repacking area in the packinghouse	Prevent introduction of new pathogens or pathogens that have been selected for resistance
Optimize fungicide use	Use rotations and mixtures or pre-mixtures whenever possible before resistance selection occurs	Resistance risk is reduced when pathogens are exposed to more than one fungicide mode of action
	Limit the total number of fungicide applications of any one class ideally to one per fruit lot	Resistance risk is reduced in single exposures
	Use labeled rates	Sublethal rates may select for less sensitive pathogen individuals
	Use fungicides that inhibit sporulation on fruit that will be stored	Selection for resistance is reduced when low numbers of pathogen propagules present
Optimize fungicide efficacy	Use optimal application methods (e.g., high volume drenches are better than low-volume sprays)	Fungicide coverage determines efficacy and minimizes pathogen survivors
	Use heated (50–55 °C) fungicide solutions	Heated solutions often are more effective
	Integrate use of alkaline salts with fungicide programs	Pathogen population is reduced. Resistance development to these salts is unlikely
	Use compatible sanitizers in recirculating fungicide solutions	Accumulation of viable inoculum that may infect healthy fruit is prevented
	Use fungicide mixtures, pre-mixtures, and rotations	Exposure of pathogen to multiple fungicide classes reduces resistance frequency and potential
Early detection of resistance	Routinely monitor pathogen populations on decayed fruit for fungicide sensitivity	Early detection of resistance increases the chance that its proliferation can be stopped

Fig. 28.3 Plate technique for spore air sampling as fruits with high decay incidence are conveyed in bins from storage rooms for packing and shipping to markets. Fungicide-amended plates are exposed for approximately 1–2 min. Note white spray bar (*arrow*) for applying a chlorine sanitation solution to fruit as they are emptied from bins to the conveyer line to reduce airborne spores

been developed that allow the rapid determination of EC_{50} values of *Penicillium* species populations in packinghouses (Adaskaveg et al. 2004; Kanetis et al. 2010) or of individual isolates (Förster et al. 2004). Spiral plating procedures are used in these methods where fungicide concentration gradients are created in agar plates using a spiral plater, and spore populations are either exposed to the plates in air samplings (Figs. 28.3 and 28.4a) or individual isolates are streaked along the concentration gradients (Fig. 28.4b).

Sanitation of fruit, equipment, and fruit storage rooms with multisite, broad-spectrum toxicants is one of the first strategies in resistance management that can be employed upon arrival of fruit at the packinghouse. Washing fruit with sanitizing oxidation treatments (e.g., sodium hypochlorite, peroxyacetic acid, ozonated water) and cleaning of equipment with oxidizers or quaternary ammonium compounds after or between handling fruit loads are ways broad-spectrum toxicants are utilized (Adaskaveg et al. 2002, 2004; Adaskaveg and Förster 2010). Proper sanitation reduces the amount of pathogen propagules exposed to the fungicide and thus reduces the chance for selection of resistance. The resistance frequency to any one fungicide in a pathogen population does not change, but reducing the total number of spores results in a proportional reduction in the level of resistant isolates present.

Sanitation also includes the regulation of air flow and spore dispersal in the packinghouse by filtration to reduce the spore load and by directing the direction of air flow. Additionally, culled fruit should be removed from the packinghouse before sporulation occurs and should not be disposed of in a citrus orchard. Stored or packed fruits that were previously fungicide treated and that developed decay and pathogen sporulation should be sprayed with sodium hypochlorite (Fig. 28.3) and isolated in an area where the contaminated air can be exhausted away, and fruits should never be repacked at the same packinghouse location where newly

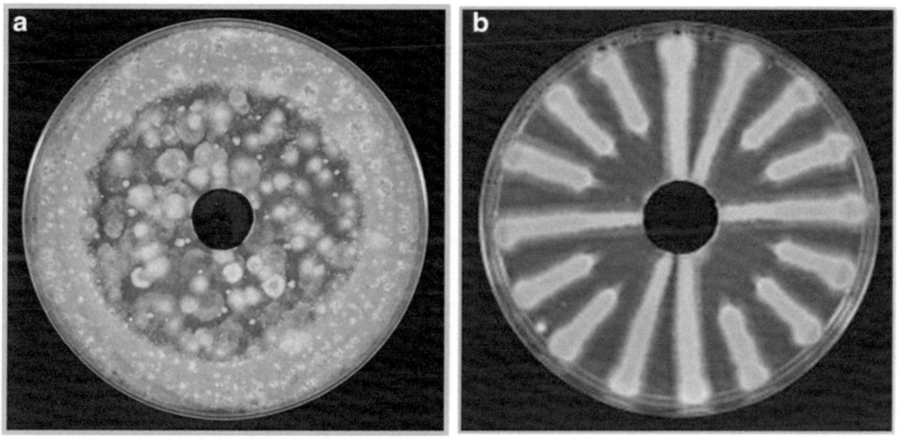

Fig. 28.4 Monitoring methods for fungicide sensitivity in populations of *Penicillium digitatum*. Plates with radial fungicide concentration gradients (pyrimethanil was used in the plates shown) were exposed to spore masses during air samplings in a packinghouse (**a**) or were inoculated with spore suspensions of individual isolates along the concentration gradient (**b**). Plates were prepared using the spiral gradient dilution method and highest fungicide concentrations are toward the center and lowest concentrations toward the edge of the plate (Förster et al. 2004). In (**a**), a high density of sensitive and resistant isolates is growing in a ring at lower concentrations, whereas numerous resistant colonies are growing at higher concentrations in the inner ring. In (**b**) three resistant isolates (replicated in opposite streaks) still grow at the highest concentrations, whereas five sensitive isolates are inhibited at lower concentrations

harvested fruits are being processed. Any decay on treated fruit may be the result of selection of resistance to the fungicide that was used, and thus, the dissemination of surviving subpopulations should be minimized by physically separating the repacking process.

Postharvest fungicides that inhibit sporulation of the pathogen on decaying escape fruit can further reduce pathogen population sizes in the packinghouse and minimize the chance of selection. Fungicides with anti-sporulation activity are imazalil and propiconazole (when decay is caused by strains sensitive to DMI compounds), as well as fludioxonil and azoxystrobin (Kanetis et al. 2007). These treatments preferentially should be used for fruit being stored, whereas those without anti-sporulation activity should be used on packed fruit that are leaving the facility for marketing.

Fungicides with single-site mode of action should not be used alone on a continuous basis. The use of rotations, mixtures, or pre-mixtures of compounds with different modes of action is an excellent strategy to reduce the risk of resistance development (Table 28.4). Ideally, rotations of mixtures or pre-mixtures should be used for fruit crops that are treated more than once, such as some citrus fruits (especially lemons) in California. Pre-mixtures are increasingly being registered for postharvest use in the United States, not only because of resistance management but also to increase the spectrum of activity of a treatment. Currently available pre-mixtures for citrus include imazalil/pyrimethanil (trade name, Philabuster), fludioxonil/thiabendazole (Graduate Max), and azoxystrobin/fludioxonil (trade name, Graduate A+) (Table 28.1), but additional ones are in development such as fludioxonil/azoxystrobin/propiconazole (Adaskaveg and Förster 2010). Unfortunately, in addition to the amounts of individual pesticides, recent regulatory restrictions in some export markets are limiting the total number of pesticides on a crop and thus impede the use of these anti-resistance strategies.

Mixtures and pre-mixtures ideally should be used starting with the introduction of the new fungicide or before resistance is detected. This is because a pathogen population that is already resistant to one of the mixture components will be more easily selected for resistance to the second fungicide component. Mixtures should also be used in each application and at effective rates for each component in the mixture. Furthermore, each mixture component should have a similarly high efficacy and performance against target populations.

References

Adaskaveg JE (2004) Evaluation of new postharvest treatments to reduce postharvest decays and improve fruit quality in citrus packinghouse operations. Citrus Research Board, 2004 Annual Report, pp 64–65

Adaskaveg JE, Förster H (2010) New developments in postharvest fungicide registrations for edible horticultural crops and use strategies in the United States. In: Prusky D, Gullino ML (eds) Post-harvest pathology: plant pathology in the 21st century: contributions to the 9th international congress. Springer, Dordrecht, pp 107–117

Adaskaveg JE, Förster H, Sommer NF (2002) Principles of postharvest pathology and management of decays of edible horticultural crops. In: Kader A (ed) Postharvest technology of horticultural crops, 4th edn, Publication 3311. University of California Agriculture and Natural Resources, Oakland, pp 163–195

Adaskaveg JE, Kanetis L, Soto-Estrada A, Förster H (2004) A new era of postharvest decay control in citrus with the simultaneous introduction of three new ‘reduced-risk’ fungicides. Proc Int Soc Citric III:999–1004

Barger WR (1928) Sodium bicarbonate as citrus fruit disinfectant. Calif Citrogr 13:164,172–174

Baudoin ABAM, Eckert JW (1982) Factors influencing the susceptibility of lemons to infection by *Geotrichum candidum*. Phytopathology 72:1592–1597

Brent KJ (1995) Fungicide resistance in crop pathogens: How can it be managed? FRAC monogr No. 1. GIFAP, Brussels

Brent KJ, Hollomon DW (1998) Fungicide resistance: the assessment of risk, FRAC monogr. No. 2. GCPF, Brussels

Brent KJ, Hollomon DW (2007) Fungicide resistance in crop pathogens: how can it be managed? 2nd edn, FRAC monogr No. 1. GCPF, Brussels

Brown GE, Eckert JW (2000) Penicillium decays – blue mold. In: Timmer LW, Garnsey SM, Graham JH (eds) Compendium of citrus diseases, 2nd edn. American Phytopathological Society Press, St. Paul, p 41

Cooke LR, Carlisle DJ, Donaghy C, Quinn M, Perez FM, Deahl KL (2006) The Northern Ireland *Phytophthora infestans* population 1998–2002 characterized by genotypic and phenotypic markers. Plant Pathol 55:320–330

Davé BA, Kaplan HJ, Petrie JF (1980) The isolation of *Penicillium digitatum* Sacc. strains tolerant to 2-AB, SOPP, TBZ, and benomyl. Proc Fla State Hortic Soc 93:344–347

De Waard MA, Groeneweg H, Van Nistelrooy JGM (1982) Laboratory resistance to fungicides which inhibit ergosterol biosynthesis in *Penicillium italicum*. Neth J Plant Pathol 88:99–112
Eckert JW (1959) Lemon sour rot. Calif Citrogr 45:30–36
Eckert JW (1987) Biotypes of *Penicillium digitatum* with reduced sensitivity to imazalil. Phytopathology 77:1728 (Abstr)
Eckert JW (1988) Dynamics of benzimidazole-resistant *Penicillia* in the development of postharvest decays of citrus and pome fruits. In: Delp CJ (ed) Fungicide resistance in North America. APS Press, St. Paul, pp 31–35
Eckert JW, Eaks IL (1989) Postharvest disorders and diseases of citrus fruit. In: Reuther W, Calavan EC, Carman GE (eds) The citrus industry, volume V – crop protection, postharvest technology, and early history of citrus research in California. Publication no. 3326. University of California, Division of Agricultural and Natural Resources, Oakland, pp 179–260
Förster H, Adaskaveg JE (2012) Fitness and competitiveness of *Penicillium* spp. isolates with reduced sensitivity to fludioxonil and pyrimethanil. Phytopathology 102(Suppl 6):S6.9
Förster H, Kanetis L, Adaskaveg JE (2004) Spiral gradient dilution, a rapid method for determining growth responses and EC_{50} values in fungus-fungicide interactions. Phytopathology 94:163–170
Fritz R, Lanen C, Chapeland-Leclerc F, Leroux P (2003) Effect of the anilinopyrimidine fungicide pyrimethanil on the cystathionine β-lyase of *Botrytis cinerea*. Pestic Biochem Physiol 77:54–65
Ghosoph JM, Schmidt LS, Margosan DA, Smilanick JL (2007) Imazalil resistance linked to a unique insertion sequence in the PdCYP51 promoter region of *Penicillium digitatum*. Postharv Biol Technol 44:9–18
Hamamoto H, Hasegawa K, Nakaune R, Lee YJ, Akutsu K, Hibi T (2001) PCR-based detection of sterol demethylation inhibitor-resistant strains of *Penicillium digitatum*. Pest Manag Sci 57:839–843
Harding PR Jr (1962) Differential sensitivity to sodium orthophenylphenate by biphenyl-sensitive and biphenyl-resistant strains of *Penicillium digitatum*. Plant Dis Rep 46:100–104
Harding PR Jr (1972) Differential sensitivity to thiabendazole by strains of *Penicillium italicum* and *P. digitatum*. Plant Dis Rep 56:256–260
Harding PR Jr (1976) R23979, a new imidazole derivative effective against postharvest decay of citrus by molds resistant to thiabendazole, benomyl, and 2-aminobutane. Plant Dis Rep 60:643–646
Hewitt HG (1998) Fungicides in crop protection. CAB International, Cambridge, UK, 221 pp
Holmes GJ, Eckert JW (1995) Relative fitness of imazalil-resistant and -sensitive biotypes of *Penicillium digitatum*. Plant Dis 79:1068–1073
Holmes GJ, Eckert JW (1999) Sensitivity of *Penicillium digitatum* and *P. italicum* to postharvest citrus fungicides in California. Phytopathology 89:716–721
Kader AA (2002) Postharvest technology of horticultural crops, 3rd edn, DANR Publication 3311. University of California, Oakland
Kanetis L, Förster H, Adaskaveg JE (2007) Comparative efficacy of the new postharvest fungicides azoxystrobin, fludioxonil, and pyrimethanil for managing citrus green mold. Plant Dis 91:1502–1511
Kanetis L, Förster H, Adaskaveg JE (2008a) Baseline sensitivities for new postharvest fungicides against *Penicillium* spp. on citrus and multiple resistance evaluations in *P. digitatum*. Plant Dis 92:301–310
Kanetis L, Förster H, Jones CA, Borkovich KA, Adaskaveg JE (2008b) Characterization of genetic and biochemical mechanisms of fludioxonil and pyrimethanil resistance in field isolates of *Penicillium digitatum*. Phytopathology 98:205–214
Kanetis L, Förster H, Adaskaveg JE (2010) Determination of natural resistance frequencies in *Penicillium digitatum* using a new air-sampling method and characterization of fludioxonil- and pyrimethanil-resistant isolates. Phytopathology 100:738–746
Kendall SJ, Hollomon DW (1998) Fungicide resistance. In: Hutson D, Miyamoto J (eds) Fungicidal activity: chemical and biological approaches to plant protection. Wiley, New York, pp 87–108

Koenraadt H, Somerville SC, Jones AL (1992) Characterization of mutations in the beta-tubulin gene of benomyl-resistant field strains of *Venturia inaequalis* and other plant pathogenic fungi. Phytopathology 82:1348–1354

Leroux P, Fritz R, Debieu D, Albertini C, Lanen C, Bach J, Gredt M, Chapeland F (2002) Mechanisms of resistance to fungicides in field strains of *Botrytis cinerea*. Pest Manag Sci 58:876–888

Masner P, Muster P, Schmid J (1994) Possible methionine biosynthesis inhibition by pyrimidinamine fungicides. Pestic Sci 42:163–166

McKay AH, Förster H, Adaskaveg JE (2012) Toxicity of selected fungicides to the postharvest pathogens *Galactomyces citri-aurantii, G. geotrichum,* and *Penicillium digitatum* and resistance potential to propiconazole. Plant Dis 96:87–96

Muirhead TF (1974) Resistance to benzimidazole fungicides in blue mold of citrus in Queensland. Aust J Exp Agric Anim Husb 14:698–701

Rippon LE, Morris SC (1981) Guazatine control of sour rot in lemons, oranges and tangors under various storage conditions. Sci Hortic 14:245–251

Schmidt LS, Ghosoph JM, Margosan DA, Smilanick JL (2006) Mutation at β-tubulin codon 200 indicated thiabendazole resistance in *Penicillium digitatum* collected from California citrus packinghouses. Plant Dis 90:765–770

Sierotzki H, Frey R, Wullschleger J, Palermo S, Karlin S, Godwin J, Gisi U (2007) Cytochrome b gene sequence and structure of *Pyrenophora teres* and *P. tritici-repentis* and implications for QoI resistance. Pest Manag Sci 63:225–233

Suprapta DN, Arai K, Iwai H, Matsuo T (1996) Change in susceptibility of satsuma mandarin fruit to sour rot pathogen (*Geotrichum candidum* citrus race) with relation to biochemical changes during maturation and storage. Mycoscience 37:209–216

Timmer LW, Garnsey SM, Graham JH (2000) Compendium of citrus diseases, 2nd edn. The American Phytopathological Society Press, St. Paul

Wild BL (1980) Resistance to citrus green mold *Penicillium digitatum* Sacc. Doctoral thesis, University of California, Riverside, 89 pp

Wild BL (1983) Double resistance by citrus green mould Penicillium digitatum to the fungicides guazatine and benomyl. Ann. Appl. Biol. 103:237-241

Zhang Z, Zhu Z, Ma Z, Li H (2009) A molecular mechanism of azoxystrobin resistance in *Penicillium digitatum* UV mutants and a PCR-based assay for detection of azoxystrobin-resistant strains in packing- or store-house isolates. Int J Food Microbiol 131:157–161

Chapter 29
Banana Diseases

Andreas Mehl and Friederike Manger-Jacob

Abstract Bananas and plantains are highly susceptible for diseases due to the lack of sexual recombination and production in monocultures. Most important fungal pathogens are *Fusarium oxysporum* f. sp. *cubense* (FOC), the cause of Panama disease, and *Mycosphaerella musicola* and *M. fijiensis*, causing yellow and black sigatoka, respectively. Control of FOC by fungicides is generally quite difficult and, consequently, fungicide resistance management measures are not suitable. In particular, *M. fijiensis* bears a high resistance risk and its control requires sound strategies to ensure sufficient activity of most important fungicidal mode of actions. Important resistance cases toward specific fungicides such as strobilurins or azoles, major resistance mechanisms, and resistance management guidelines according to the FRAC Banana Working Group are discussed, including recommendations for the new class of SDHI fungicides. For proper sigatoka control and resistance management, restrictions for the use of single-site inhibitors, mixtures, or alternation with the maximum of available and effective non-cross-resistant fungicides are recommended, including the use of multisite compounds and biological products.

Keywords *Mycosphaerella fijiensis* • Black sigatoka • Panama disease • Fungicide resistance

29.1 Introduction

Bananas (*Musa* spp.) are treelike herbaceous plants with fruits belonging botanically to the berries (Morton 1987). The modern banana we know today originates from ancestors in Southeast Asia and Western Pacific. The fruits of these banana plants had large seeds which made them uneatable for humans. Over the years, triploid hybrids evolved, proliferating no longer via seeds but via suckers and developing seedless fruits through parthenocarpy. With these hybrids, the breeding

A. Mehl (✉) • F. Manger-Jacob
Bayer CropScience AG, Research Disease Control, Alfred-Nobel-Str. 50, Building 6240, 40789 Monheim, Germany
e-mail: andreas.mehl@bayer.com

H. Ishii, D.W. Hollomon (eds.), *Fungicide Resistance in Plant Pathogens*, DOI 10.1007/978-4-431-55642-8_29

Table 29.1 Banana production and export of the main banana-producing countries in 2011 (www.faostat.org)

Production (million t)	Country	Country	Export (million t)
28.5	India	Ecuador	5.8
10.7	China	The Philippines	2.5
9.1	The Philippines	Costa Rica	1.9
7.4	Ecuador	Columbia	1.8
7.3	Brazil	Guatemala	1.4
2.7	Guatemala	Honduras	0.4
2.1	Costa Rica	Dominican Republic	0.3
2	Columbia	China	0.04
0.8	Dominican Republic	India	0.04
0.7	Honduras	Brazil	0.01

of banana cultivars started in Southeast Asia. Bananas can be subdivided into the dessert banana, the banana we know, which has a sweet taste and can be consumed raw, and into the plantains, a starchy fruit which has to be cooked before eating.

In the tropical and subtropical regions, bananas and especially plantains represent, after rice, wheat, and maize, the fourth most important food crop. For over 400 million people, they represent a steadily staple food and are cultivated in more than 120 countries predominantly in new industrialized countries (Thangavelu and Mustaffa 2012). The main banana- and plantain-producing countries are India, China, the Philippines, Ecuador, and Brazil providing 60 % of the worldwide production (www.faostat.fao.org). In 2011, approximately 102 million metric tons of bananas and plantains were harvested, of which the majority was not produced for export but for local trades, especially the plantains which play almost no role for export trade. Table 29.1 compares the production values of bananas with the export quantities, visualizing that, for example, India and China are the main banana producers but deliver the world market only with a relatively modest amount of bananas.

The export trade of dessert bananas had its inception in the middle of the nineteenth century starting from Central America and the Caribbean. The favored banana cultivar at that time was ‘Gros Michel’ due to the good taste of the banana and the thick peel, which made this cultivar suitable for long transports by ship. In the 1950s, an epidemic spread of *Fusarium* wilt almost eradicated the cultivar ‘Gros Michel’ in South America. As a consequence, this banana cultivar was replaced in the worldwide export trade at the beginning of the 1970s by the banana cultivar ‘Cavendish’, which was resistant toward *Fusarium* wilt and which dominates the export market since that time. The formerly market leader ‘Gros Michel’ is today only of little importance for the world trade of bananas.

In 2011, 17.9 million tons of bananas have been exported, mainly from South America (www.fao.org). Most important importers of bananas are the USA, the EU, and Japan.

29.2 Biology

The treelike banana plant belongs botanically to the herbs. A pseudostem consisting of tightly packed leaves forms a trunk. Within this pseudostem, the floral stem, also called true stem, develops from the terminal point of the rhizome and forms the photosynthetically active leaves and the inflorescence consisting of two to three types of flowers. The female flower appears first, forming the fruits; then, depending on the cultivar, a neutral flower arises, followed by the male flower, which produces pollen (Morton 1987).

Modern banana cultivars have three sets of chromosomes and are hybrids originating from the two wild-type bananas *Musa acuminate* (genome A) and *Musa balbisiana* (genome B). All hybrids obtained from these crosses are called *Musa x paradisiaca* (www.kew.org). The advantage of these asexually proliferating bananas is mainly based on the fact that their fruits contain no seeds: the fruits develop parthenocarpically. After the growing season, the complete aerial parts of the plants die, but offshoots developing from the rhizome of the plants generate suckers.

Cultivation of bananas was, mainly in the past, accomplished by replanting of suckers. Modern banana plantations use tissue culture plantlets originating from single-cell tissue cultures produced in specialized laboratories. Both methods comprise pros and cons. The use of suckers is much cheaper compared to the very expensive method of cultivation plantlets under sterile conditions. On the other hand, single-cell tissues are well controlled and free of pathogens (www.promusa.org). However, both cultivation methods have the same problem: that they use genetical clones of banana plants and due to the absence of sexual recombination, the uniform plants are very vulnerable to pests and diseases.

The majority of banana plantations are monocultures, and thanks to low seasonal temperature fluctuation in the tropical regions, bananas can be harvested during the whole year. Although this characteristic is an economical benefit for the farmers, it has some disadvantages, because plants are continually exposed to the influence of pests and pathogens (Marín et al. 2003).

29.3 Fungal Banana Diseases

As mentioned above, banana plantations are particularly susceptible for diseases due to the lack of sexual recombination and the cultivation in monocultures. In this chapter, we want to highlight the most important fungal diseases jeopardizing the modern banana production: *Fusarium* wilt, also called Panama disease, caused by *Fusarium oxysporum* f. sp. *cubense* (FOC), yellow sigatoka, and particularly black sigatoka, caused by the ascomycete *Mycosphaerella fijiensis* Morelet. Besides these fungal diseases, bacterial and viral pathogens can infect banana plants causing diseases such as *Xanthomonas* wilt, a lethal disease caused by the bacterium *Xanthomonas campestris* f.sp. *musacearum* or the banana bunchy top virus (BBTV) disease (Marín et al. 2003). However, none of these viral or bacterial diseases has such a devastating impact as *Fusarium* wilt and black sigatoka.

29.3.1 **Fusarium *Wilt***

Fusarium wilt is the most important disease of bananas and was first described in Australia in 1876 (Ploetz 2006). The disease is caused by the hyphomycete *F. oxysporum* f. sp. *cubense* (FOC) (Stover 1962) which is part of the large family of *F. oxysporum,* a soilborne fungus of which the majority lives in endophytical or saprophytical alliances with plants. However, some species are very virulent, mainly FOC, causing *Fusarium* wilt in bananas.

The infection has its origin at the root of the plant where the fungus enters the xylem and destroys the vascular system leading to a dehydration of the plant. Symptoms are wilting leaves which turn yellow mostly after 5–6 months after planting (Thangavelu and Mustaffa 2012). The chlamydospores of the pathogen can survive several decades in the field being propagated via running water and soil (Stover 1962). Today, the species can be subdivided into four races which show very different pathogenicities toward different plant species and cultivars. Races 1, 2, and 4 represent banana pathogens, and race 3 is a pathogen of *Heliconia.*

Race 1 is pathogenic to 'Gros Michel' and 'Silk' and race 2 to 'Bluggoe' and race 4 is virulent for 'Cavendish' and for all bananas for which races 1 and 2 are pathogenic as well as for plantains. It has recently been subdivided into subtropical race 4 (ST4) and tropical race 4 (TR4) (Zhang et al. 2013).

In the 1950s, a widespread epidemic of *F. oxysporum* f. sp. *cubense* wiped out almost all plantations in South America cultivating the banana 'Gros Michel', which was the most exported banana at that time. Since Panama was one of the most affected countries, the disease was named Panama disease (Jeger et al. 1996), and as we know today, the responsible pathogen was most likely FOC race 1. The destructive dimension was amplified by the fact that the farmers used phenotypically healthy suckers from infected plants, not knowing that these suckers already were infected with the pathogen. Planting these suckers on uncontaminated fields tremendously supported the epidemic spread of *Fusarium* wilt and led to economically great damage for the farmers. After this experience, farmers worldwide switched to another banana sort called 'Cavendish', which was not susceptible to *Fusarium* wilt. However, in the 1990s, a second epidemic spread of Panama disease caused by *F. oxysporum* f. sp. *cubense* TR4 infested also 'Cavendish' bananas, having its origin in Taiwan but spreading quickly through Southeast Asia such as the Philippines, China, South Africa, and Australia (Lu et al. 2013). South America is currently free of TR4.

This new type of *Fusarium* wilt also affects the most important export cultivar of bananas on which the export trades depend and plantains representing an important staple food in many regions, leading to even greater problems for the banana-producing countries.

Chemical control of *Fusarium* wilt is quite difficult. Fungicides assuring reliable protection are not available yet. In the past decades, several protocols for fungicide application have been published, describing methods for dip treatments of the banana sucker in fungicide mixtures before planting (Deacon 1984) or stem injections with carbendazim (Lakshmanan et al. 1987). Although some promising results were achieved, none of these fungicide applications could be implemented elsewhere. More promising is the approach started by Huang et al. (2012) using Chinese leek in a rotation system with banana plants. The antifungal effect of leek is well known and could be transferred by Huang in a plant rotation system leading to stable banana yields. This method is now established in some Chinese banana-growing regions by now (Huang et al. 2012).

Chemical treatment of soil can lead to some environmental damages and may therefore not always the preferred treatment of choice, whereas biological control of FOC via *Trichoderma, Pseudomonas, Streptomyces*, and nonpathogenic *Fusarium* strains has been investigated intensively, resulting in promising banana yields in both greenhouse and field trials (Marois et al. 1981; Sivan and Chet 1986; Larkin and Fravel 1998; Thangavelu and Mustaffa 2012). However, none of these methods has reached marketability yet.

Besides these approaches to combat the pathogen directly, there are also many universities and organizations working on the molecular detection of FOC in soil before banana plants are planted (Zhang et al. 2013) and on the genetic modification of bananas in order to enhance resistance toward FOC (e.g., Hwang and Ko 2004; Paul et al. 2011; Yip et al. 2011). But due to the poor acceptance of genetically modified food within the general public, the chances of success for these developments are at the moment relatively small.

In fighting against further proliferation of *Fusarium* wilt, it is advantageous that *Fusarium* is a soilborne fungus which is not propagated through the air. Consequently, good hygienic measures in banana plantations can prevent infection with FOC, and practical guidelines for farmers have been recommended by the Food and Agriculture Organization of the United Nations (FAO) to protect noninfected fields from contamination with FOC (www.FAO.org):

- Raising awareness at all levels and adoption of appropriate risk assessment, surveillance, and early warning systems
- Implementation of phytosanitary measures to prevent the spread of the disease through agricultural practices, irrigation and drainage systems, transportation, vehicles, containers, tools, or visitors
- Preventive measures, including quarantines, the use of disease-free planting materials, prevention of movement of infected soil and planting materials into and out of farms, and disinfection of vehicles
- Capacity building in national plant protection organizations (NPPOs) in planning, extension, and research, including the use of rapid and accurate diagnostic tools

- Training of technical officers, producers, and farm workers in disease identification, prevention, and management under field conditions and appropriate instructions to visitors

These recommendations are helpful for those regions yet not affected by this disease like South America, but if soil is contaminated with *F. oxysporum* f. sp. *cubense* TR4, these fields cannot be used for banana cultivation for decades.

29.3.2 Yellow and Black Sigatoka

Yellow sigatoka was described first in 1902 in Java (Zimmermann 1902) and is caused by the ascomycete *Mycosphaerella musicola*. Yellow sigatoka (sigatoka leaf spot) was the most important foliar disease of bananas till the broadening of black sigatoka and was named after Sigatoka Valley where the disease propagated epidemically between 1912 and 1923 (Marín et al. 2003).

Black sigatoka was also first described in Fiji in 1963 and is caused by *M. fijiensis* Morelet, which is also known as black leaf streak. Today black sigatoka has almost completely supplanted yellow sigatoka in the main banana-producing regions (Balint-Kurti et al. 2001). Black sigatoka is the most damaging foliar disease in banana production and has clearly a higher economic importance compared to yellow sigatoka.

M. fijiensis infects the leaves of the plant both with conidia and ascospores. In conditions of high humidity, both can develop on the leaf until they penetrate the leaves through the stomata. The pathogen grows intercellular with biotrophic nutrition leading to black lesions on the leaf after 3–4 weeks (Marín et al. 2003). The hyphae can outgrow from the stomata again and either penetrate other stomata nearby or form conidia which infect surrounding stomata. Early symptoms of the disease are streaks on the underside of the leaf. Later in the course of the disease, they disperse to black necrosis in which the epithelium dies and becomes dry. Due to necrotic areas, the photosynthetic effort is reduced, leading to significantly decreased yield and lower fruit weights of bananas harvested (Saraiva et al. 2013). Additionally, the green life stage of the bananas is shortened leading to postharvest yield loss of 30–50 % (Stover 1983). The main proliferation is generated by ascospores, developing in perithecia on the upper surface of the leaf, being dispersed by wind (Craenen and Ortiz 1996). Due to the fact that sigatoka occurs mainly on leaves and ascospores are produced on leaves as well, partial or complete deleafing of contaminated leaves minimizes ascospore production and reduces premature fruit ripening (Chillet et al. 2013). In contrast to *Fusarium* wilt, black sigatoka can be controlled with contact and systemic fungicides, allowing the farmers to prevent black sigatoka infestation effectively.

29.4 Systemic Fungicides, Sensitivity Status, and Resistance Management

There are several systemic fungicides available comprising many classes of active agents with different modes of action. The following systemic single-target fungicides are currently used to control black sigatoka in bananas (www.frac.info):

DMIs	Bitertanol, difenoconazole, epoxiconazole, fenbuconazole, myclobutanil, propiconazole, tebuconazole, tetraconazole, triadimenol
Qols	Azoxystrobin, pyraclostrobin, trifloxystrobin
Amines	Spiroxamine, fenpropimorph, fenpropidin, tridemorph
Anilinopyrimidines	Pyrimethanil
Benzimidazoles	Benomyl, carbendazim, thiophanate, thiophanate-methyl
SDHIs	Boscalid, fluopyram, fluxapyroxad, isopyrazam
Guanidines	Dodine
N-Phenylcarbamates	Diethofencarb

Due to the fact that all of these fungicides are single-target fungicides, they bear a considerable risk for the development of resistance. Consequently, the application of these components is regularized by resistance management guidelines.

29.4.1 DMI Fungicides

Decreased sensitivity of *M. fijiensis* toward DMI fungicides has been described to be based on several point mutations in the *cyp51* gene of *M. fijiensis*, close to the putative substrate-binding site such as Y136F, A313G, Y461D, Y463D, Y463H, or Y463N (Canas-Gutiérrez et al. 2009), and on overexpression of the *cyp51* gene (Diaz-Trujillo et al. 2011).

Overall, the shifting of DMI sensitivity over the baseline sensitivity has been observed up to 2009. Sensitivity has since stabilized at a higher level compared to pre-2009 in Belize, Ecuador, Colombia, Guatemala, Costa Rica, Honduras, Panama, and the Philippines (www.frac.info). In order to further stabilize the DMI sensitivity, the following FRAC guidelines are recommended for the use of DMI fungicides against black sigatoka (www.frac.info):

- DMIs should be used only in mixtures with other, non-cross-resistant modes of action, all partners at manufacturers' recommended effective rates.
- DMI fungicides are recommended to be used in full alternation with other, non-cross-resistant modes of action.

- Apply a maximum of 8 applications containing DMI fungicides, but not more than 50 % of the total number of sprays.
- Applications containing DMI fungicides should preferably start at the onset of the annual disease progress curve.

29.4.2 QoI Fungicides

Already few years after the market introduction of QoI fungicides, decreased control of black sigatoka was observed mainly in Central American banana-growing areas and in the Philippines, and characterization of field isolates of *M. fijiensis* showed very high resistance factors of QoI-resistant isolates to different inhibitors of the cytochrome bc1 enzyme complex, both in mycelial growth inhibition and NADH consumption assays (Sierotzki et al. 2000). Cross-resistance was observed among all QoIs, including strobilurin B and myxothiazol. In sensitive, but not in resistant isolates, mixtures of QoIs and SHAM, an inhibitor of the alternative oxidase (AOX), were more active than the components alone, indicating that the alternative pathway is essential in metabolism but not causal for resistance. In the cell-free NADH-consumption assay, the Qo inhibitors affected the sensitive but not the resistant isolates, suggesting that AOX was not active in submitochondrial particles. The sequencing of the cytochrome *b* gene of sensitive and resistant *M. fijiensis* isolates revealed a difference in the nucleotide sequence leading to a single amino acid change from glycine to alanine at position 143 in the resistant isolates. Resistant isolates do not seem to contain a mixture of mutated and nonmutated DNA, indicating a complete selection of resistant mitochondria and a maternally donated mode of resistance (Sierotzki et al. 2000).

In 2014, the frequency of resistance to QoIs was reported to be stable with no further spread observed in all countries, where sensitivity monitoring programs have been carried out. In areas where reduced sensitivity was observed in the past and QoIs have not been used since 2003, the sensitivity improved again, e.g., in Guatemala and the Philippines (www.frac.info).

Due to the high risk of resistance of QoI fungicides, their use is strongly regulated by the FRAC Banana Working Group, and only a maximum of 3 applications of QoIs in mixture with a non-cross-resistant mixture partner is recommended. More details for the use recommendations are given at the FRAC website (www.frac.info).

29.4.3 Amine and Anilinopyrimidine Fungicides

According to the sensitivity monitoring results presented during the 2014 FRAC Banana Working Group meeting, the sensitivity of *M. fijiensis* to amines is at high levels and did not change significantly during 2011 and 2014 in all regions studied.

Pyrimethanil is the only active ingredient from the group of anilinopyrimidines, which is currently used in banana cropping. For anilinopyrimidine sensitivity, a stable situation was further reported.

Use recommendations are as well given at the FRAC website (www.frac.info).

29.4.4 Benzimidazole Fungicides

Resistance cases toward benzimidazole fungicides (methyl benzimidazole carbamates) have been already described more than 30 years ago (Stover 1979). Analysis of resistant isolates showed a change of adenine to cytosine in codon 198 of the target site encoding β-tubulin gene (Canas-Gutiérrez et al. 2006).

Today, resistance to benzimidazoles is widespread at high levels and field performance is affected in all banana-growing regions with reported MBC resistance. Similar to QoIs, MBC fungicides should be applied only in mixtures and in full alternation with other, non-cross-resistant modes of action and not more than 3 times a year (www.frac.info).

29.4.5 SDHI Fungicides

From publications available that describe the known resistance mechanisms to SDHI fungicides, it has been shown that resistance is mostly based on single target-site mutations. In contrast to the situation with QoI or MBC fungicide resistance, a multitude of mutations have been detected up to now (e.g., SDHB-H253L; SDHB-H272Y, R; SDHB-P225L, T, F; SDHC-N80K; SDHC-H134R; or SDHD-S89P) which often occur at different positions or subunits of the target enzyme succinate dehydrogenase, dependent on the pathogen. It has been reported that strains of *Alternaria alternata*, *Corynespora cassiicola*, *Didymella bryoniae*, and *Podosphaera xanthii*, carrying specific amino acid substitutions in their respective SDH b subunit proteins, do not show complete cross-resistance between all members of the SDHI fungicide class (Amiri et al. 2014; Avenot and Michailides 2007, 2009, 2010; Avenot et al. 2008, 2012; Fraajie et al. 2012; Glaettli et al. 2009; Ishii et al. 2008, 2011; Lalève et al. 2014; Miyamoto et al. 2010; Veloukas et al. 2011, 2014; Yin et al. 2011). Thus, the consequences of each of the different mutations for the biological properties of the individual pathogens as well as for the performance of the entire group of SDHI fungicides are not yet fully understood.

With *M. fijiensis*, after establishing the baseline sensitivity, follow-up monitoring showed for the first time isolates with a reduced in vitro sensitivity, originating from Costa Rica and Ecuador (www.frac.info). Further studies are ongoing to clarify the variability of sensitivity and presence of target-site mutations and the relevance for field use.

The following FRAC guidelines are currently recommended for the foliar use of SDHI fungicides against black sigatoka (www.frac.info):

- Apply SDHI fungicides only in mixtures with other, non-cross-resistant modes of action, all partners at manufacturers' recommended effective rates.
- SDHI fungicides should be used in full alternation with other, non-cross-resistant modes of action. No consecutive SDHI applications (blocks) should be applied.
- Apply a maximum of 3 applications containing SDHI fungicides but not more than 33 % of the total number of sprays.
- Applications containing SDHI fungicides should preferably start at the onset of the annual disease progress curve and be applied at times of lower disease pressure.
- Applications should be separated by at least 3 months of an SDHI-free period.

29.4.6 Guanidine and N-*Phenylcarbamate Fungicides*

During the FRAC Banana Working Group meeting in 2014, baseline sensitivity data for dodine was presented for several countries in Latin America and the Philippines. Data show sensitivity with a broad variation irrespective of the origin (no difference between wild areas and farms) (www.frac.info).

For diethofencarb, first sensitivity baseline data were as well presented from the Philippines.

Use recommendations for both fungicide classes are given at the FRAC website (www.frac.info).

29.5 Multisite Fungicides, Sensitivity Status, and Resistance Management

Additionally to the above-listed systemic fungicides, the following multisite fungicides are frequently used as protectants in banana plantations: mancozeb, chlorothalonil, propineb, thiram, captan, and metiram (www.frac.info).

In contrast to systemic fungicides, multisite fungicides remain on the surface of the plant, preventing the ascospore germination. This kind of fungicides can be removed by rain quite easily and does not exhibit any protection against pathogens on new leaves. Thus, treatment has to be repeated very often to ensure protection. On the other hand, in return for its unspecific mode of action, these fungicides do not cause resistance problems in pathogens, and application of these agents is not restricted by resistance management guidelines.

Consequently, multisite fungicides (and other fungicides of low resistance risk such as microbial disrupters of pathogen cell membranes) can be applied for control of black sigatoka solo or in mixtures with partners at manufacturers' recommended effective rates. There are no limitations or restrictions concerning the number of application, the timing, or the sequence as long as it is within the limits of the manufacturer's labels (www.frac.info).

Generally, fungicides for application on banana plants are mostly prepared in an oil or oil–water emulsion, since oil delays the development of the early stages of the pathogen and enhances the penetration of the systemic fungicide into the leaves.

29.6 Outlook

The breeding of banana species resistant toward black sigatoka is an urgent purpose, and in 1987, the International Institute of Tropical Agriculture (IITA) initiated a research program aiming at the development of resistant banana and plantain strains generated by the crossing of resistant diploid wild-type strains with susceptible triploid strains, and it has produced its first positive results in Uganda (Vuylsteke et al. 1993). Besides this classical breeding strategy, genetic engineering might be a tool for the creation of disease-resistant bananas. First field trials with pathogen-resistant strains have been accomplished in Australia, Africa, and Israel, showing resistance to both *Fusarium* wilt and black sigatoka. But due to the lack of confidence within the general public toward genetically modified plants, the establishment of fruits originating from genetically modified plants seems today somewhat problematic.

In summary, there is further need for fungicides with new biochemical mode of action to control sigatoka diseases and *Fusarium* wilt and to ensure the implementation of sound resistance management strategies.

References

Amiri A, Heath SM, Peres NA (2014) Resistance to fluopyram, fluxapyroxad and penthiopyrad in *Botrytis cinerea* from strawberry. Plant Dis 98:532–539

Avenot HF, Michailides TJ (2007) Resistance to boscalid fungicide in *Alternaria alternata* isolates from pistachio in California. Plant Dis 91:1345–1350

Avenot HF, Michailides TJ (2009) Monitoring the sensitivity to boscalid of *Alternaria alternata* populations from California pistachio orchards. Phytopathology 99:S6 (Abstr)

Avenot HF, Michailides TJ (2010) Progress in understanding molecular mechanisms and evolution of resistance to succinate dehydrogenase inhibiting (SDHI) fungicides in phytopathogenic fungi. Crop Prot 29:643–651

Avenot HF, Sellam A, Karaoglanidis G, Michailides TJ (2008) Characterization of mutations in the iron-sulphur subunit of succinate dehydrogenase correlating with boscalid resistance in *Alternaria alternata* from California pistachio. Phytopathology 98:437–742

Avenot HF, Thomas A, Gitaitis RD, Langston DB, Stevenson KL (2012) Molecular characterization of boscalid- and penthiopyrad-resistant isolates of *Didymella bryoniae* and assessment of their sensitivity to fluopyram. Pest Manag Sci 68:645–651

Balint-Kurti PJ, May GD, Churchill ACL (2001) Development of a transformation system for Mycosphaerella pathogens of banana: a tool for the study of host/pathogen interactions. Microbiol Lett 195:9–15

Canas-Gutiérrez GP, Patino LF, Rodríguez-Arango E, Arango R (2006) Molecular characterization of benomyl-resistant isolates of *Mycosphaerella fijiensis*, collected in Colombia. J Phytopathol 154:403–409

Canas-Gutiérrez GP, Angarita-Velásquez MJ, Restrepo-Flórez JM, Rodríguez P, Moreno CX, Arango R (2009) Analysis of the CYP51 gene and encoded protein in propiconazole-resistant isolates of *Mycosphaerella fijiensis*. Pest Manag Sci 65:892–899

Chillet M, Castelan FC, Abadie C, Hubert O, de Lapeyre de Bellaire L (2013) Necrotic leaf removal, a key component of integrated management of *Mycosphaerella* leaf spot diseases to improve the quality of banana: the case of Sigatoka disease. Fruits 68:271–277

Craenen K, Ortiz R (1996) Effect of the black Sigatoka resistance locus bs_1 and ploidy level on fruit and bunch traits of plantain-banana hybrids. Euphytica 87:97–101

Deacon JW (1984) Panama disease of banana in South Africa. Hortic Sci 1:29–31

Diaz-Trujillo C, Cordovez da Cunha V, Chong P, Kema GHJ, Arango-Isaza R, Guzman M, Stergiopoulos I, de Wit PJGM (2011) Azole resistance is related to overexpression of the CYP51 gene in *Mycosphaerella fijiensis*. Book of abstracts EPS PhD Autumn School 'Host-Microbe Interactomics', Wageningen, The Netherlands, 01–03 November 2011, pp 26–27

Fraajie BA, Bayon C, Atkins S, Cools HJ, Lucas JA, Fraajie MW (2012) Risk assessment studies on succinate dehydrogenase inhibitors, the new weapons in the battle to control Septoria leaf blotch in wheat. Mol Plant Pathol 13:263–275

Glaettli A, Stammler G, Schlehuber S (2009) Mutations in the target proteins of succinate-dehydrogenase inhibitors (SDHI) and 14α-demethylase inhibitors (DMI) conferring changes in the sensitivity-structural insights from molecular modelling. In: 9th international conference on plant diseases, Tours, France, pp 670–681

Huang YH, Wang RC, Li CH, Zuo CW, Wei YR, Zhang L, Yi GJ (2012) Control of Fusarium wilt in banana with Chinese leek. Eur J Plant Pathol 134:87–95

Hwang SC, Ko WH (2004) Cavendish banana cultivars resistant to Fusarium wilt acquired through somaclonal variation in Taiwan. Plant Dis 88:580–588

Ishii H, Fountaine J, Miyamoto T, Nishimura K, Tomita Y (2008) Occurrence of a mutation in the succinate dehydrogenase gene found in some isolates of cucumber Corynespora leaf spot fungus resistant to boscalid. Jpn J Phytopathol 74:38–39 (Japanese abstr)

Ishii H, Miyamoto T, Ushio S, Kakishima K (2011) Lack of cross-resistance to a novel succinate dehydrogenase inhibitor, fluopyram, in highly boscalid-resistant isolates of *Corynespora cassiicola* and *Podosphaera xanthii*. Pest Manag Sci 67:474–482

Jeger MJ, Waller JM, Johanson A, Gowen SR (1996) Monitoring in banana pest management. Crop Prot 4:391–397

Lakshmanan P, Selvaraj P, Mohan S (1987) Efficiency of different methods for the control of Panama disease. Trop Pest Manag 33:373–376

Lalève A, Gamet S, Walker AS, Debieu D, Toquin V, Fillinger S (2014) Site-direct mutagenesis of the P225, N230 and H272 residues of the succinate dehydrogenase subunit B from *Botrytis cinerea* highlights different roles in enzyme activity and inhibitor binding. Environ Microbiol 16:2253–2266

Larkin R, Fravel D (1998) Efficacy of various fungal and bacterial biocontrol organisms for the control of Fusarium wilt of tomato. Plant Dis 82:1022–1028

Lu Y, Liao D, Pu J, Qi Y, Xie Y (2013) Proteome analysis of resistant and susceptible Cavendish banana roots following inoculation with *Fusarium oxysporum* f. sp. *cubense*. Physiol Mol Plant Pathol 84:163–171

Marín DH, Romero RA, Guzmàn M, Sutton TB (2003) Black Sigatoka: an increasing threat to banana cultivation. Plant Dis 87:208–222

Marois JJ, Mitchel DJ, Somada RM (1981) Biological control of Fusarium crown and root rot of tomato under field condition. Phytopathology 12:1257–1260

Miyamoto T, Ishii H, Stammler G, Koch A, Ogawara T, Tomita Y, Fountaine JM, Ushio S, Seko T, Kobori S (2010) Distribution and molecular characterization of *Corynespora cassiicola* isolates resistant to boscalid. Plant Pathol 59:873–881

Morton J (1987) Banana. In: Fruits of warm climates. JF Morton, Miami, pp 29–46. ISBN: 0-9610184-1-0

Paul JY, Becker DK, Dickman MB, Harding RM, Kanna HK (2011) Apoptosis-related genes confer resistance to Fusarium wilt in transgenic 'Lady Finger' bananas. Plant Biotechnol J 9:1141–1148

Ploetz RC (2006) Fusarium wilt of banana is caused by several pathogens referred to as *Fusarium oxysporum* f. sp. *cubense*. Phytopathology 96:653–656

Saraiva LA, Castelan FP, Shitakubo R, Hassimotto NM, Purgatto E, Chillet M, Cordenunsi BR (2013) Black leaf streak disease affects starch metabolism in banana fruit. J Agric Food Chem 61:5582–5589

Sierotzki H, Parisi S, Steinfeld U, Tenzer I, Poirey S, Gisi U (2000) Mode of resistance to respiration inhibitors at the cytochrome bc1 complex of *Mycosphaerella fijiensis*. Pest Manag Sci 56:833–841

Sivan A, Chet I (1986) Biological control of Fusarium spp. in cotton, wheat and muskmelon by *Trichoderma harzianum*. J Phytopathol 116:39–47

Stover RH (1962) Fusarium Wilt (Panama Disease) of bananas and other *Musa* species. Commonwealth Mycological Institute, Kew

Stover RH (1979) Field observations on benomyl tolerance in ascospores of *Mycosphaerella fijiensis* var. difformis. Trans Br Mycol Soc 72:518–519

Stover RH (1983) Effect du Cercospora noir sur le plantains an Amérique central. Fruits 38:326–329 (in French)

Thangavelu R, Mustaffa MM (2012) Current advances in the Fusarium wilt disease management in banana with emphasis on biological control. In: Cumagun CJR (ed) Plant pathology, pp 273–298. doi: 10.5772/33775

Veloukas T, Leroch M, Hahn M, Karaoglanidis GS (2011) Detection and molecular characterization of boscalid-resistant *B. cinerea* isolates from strawberry. Plant Dis 95:1302–1307

Veloukas T, Kalogeropoulou P, Markoglou AN, Karaoglanidis GS (2014) Fitness and competitive ability of *Botrytis cinerea* field isolates with dual resistance to SDHI and QoI fungicides, associated with several sdhB and cytb G143A mutations. Phytopathology 104:347–356

Vuylsteke D, Ortiz R, Pasberg-Gauhl C, Gauhl F, Gold C, Ferris S, Speijer P (1993) Plantain and banana research at the International Institute of Tropical Agriculture. Hort Sci 28:874 and 970–971

Yin YN, Kim YK, Xiao CL (2011) Molecular characterization of boscalid resistance in field isolates of *Botrytis cinerea* from apple. Phytopathology 101:986–995

Yip MK, Lee SW, Su KC, Lin YH, Chen TY (2011) An easy and efficient protocol in the production of *pflp* transgenic banana against Fusarium wilt. Plant Biotechnol Rep 5:245–254

Zhang X, Zhang H, Pu J, Qi Y, Yu Q, Xie Y, Peng J (2013) Development of a real-time fluorescence loop-mediated isothermal amplification assay for rapid and quantitative detection of *Fusarium oxysporum* f. sp. *cubense* Tropical Race 4 in soil. PLos One 8:1–10. doi:10.1371/journal.pone.0082841

Zimmermann A (1902) Über einigen tropischen Kulturpflanzen beobachtete Pilze. Zentralbl Bakteriol Parasitenk Infektionskh Hyg 8:219 (in German)

Concluding Remarks

For growers, fungicide resistance may seem a problem without a solution. But as the chapters in this book reveal, 50 years of research has expanded the understanding of the problem, and despite the development of resistance to various fungicides, chemical control remains a key component in many disease control strategies. A combination of chemical, biochemical and genetic approaches, including identifying how natural anti-fungal compounds work, has resulted in the timely introduction of a succession of products with novel modes of action, so essential for anti-resistance strategies. Environmental and toxicological demands have undoubtedly slowed down the introduction of novel products, but the new modes of action continue to be exploited. Although the triazolo-pyrimidylamine ametoctradin (Initium™, see Chap. 9) and the isoxazoline oxathiapiprolin (Ji and Csinos 2015) belong to two new fungicide groups, which apparently are not cross-resistant to existing QoI fungicides, detailed information on their modes of action are still to be published.

Knowledge of the biochemical mechanisms of resistance, coupled with the identification of the DNA changes, has allowed resistance monitoring to embrace the rapid developments in molecular diagnostic techniques. Alongside bioassays, these techniques not only have had a substantial impact on the detection of resistance, but they have been useful in following the evolution of resistance in pathogen populations and evaluating the performance of anti-resistance strategies. But a challenge remains to link these accurate laboratory techniques with "in field" monitoring, if they are to provide useful guidance to growers for "in season" spray decisions.

Modelling the evolution of resistance and prediction of the impact of different anti-resistance treatment strategies has a long history. But recently, emphasis has shifted from rather theoretical approaches to a more realistic modelling of data collected from field experiments. Behaviour of resistance within natural pathogen populations depend on the relative fitness of resistant individuals in the absence of the at-risk fungicide. Obtaining meaningful values for fitness under the range of environmental conditions experienced in field populations remains an obstacle to the use of modelling to suggest how best to minimise resistance risk.

H. Ishii, D.W. Hollomon (eds.), *Fungicide Resistance in Plant Pathogens*,
DOI 10.1007/978-4-431-55642-8

To prevent the loss of key modes of action, growers not only need to use fungicides responsibly, but manufacturers, researchers and advisors must not give mixed messages. Legislators keen to ban fungicide groups on social rather than scientific backgrounds must be made aware that this will increase the resistance risk for the fewer remaining modes of action. Integrated strategies involving both chemical and biological control agents, and using agronomy and resistant crop varieties to reduce disease pressure and the need for fungicide treatments, are seen as a way forward to maintain sustainable and effective disease control. But success is in the hands of growers who will need to translate these strategies into profitable production systems.

Derek William Hollomon
Hideo Ishii

Reference

Ji P, Csinos AS (2015) Effect of oxathiapiprolin on asexual life stages of *Phytophthora capsici* and disease development on vegetables. Ann Appl Biol 166:229–235. doi:10.1111/1276

Index

A

A143 allele, 162
ABC transporter *atrB*, 290
Acquired resistance, 14, 224
Affinity decrease, hydroxyanilide, 226
A313G, 473
Agarose gel electrophoresis, 301
Agroecosystems, 78
Allele G143A, 267
Allele marker monitoring, 263
Allele-specific PCR (ASPCR), 26–27, 301–302, 347
Alternaria alternata, 43, 109, 189, 301, 358
Alternaria brassicicola, 186
Alternaria solani, 109, 124–125, 301, 359
Alternation, fungicides, 70, 390
Alternative oxidase (AOX), 124, 345
A2 mating types, 153
Ametoctradin, 121
Amino-acid changes, 217
Amisulbrom, 121
Amplification, 295
Amplification refractory mutation system-PCR (ARMS-PCR), 321
Anastomosis, 349
Anilinopyrimidines (APs), 239, 282, 423
Anti-microtubule agents, 420–422
Anti-resistance
  management, 82
  strategies, 307, 325–326
Anti-tubulin fungicides, 94
Apple scab fungus, 300
Aromatic hydrocarbons, 181
Ascomycetes, 119
AS quantitative PCR, 300
ATP-binding cassette (ABC) transporters, 192, 234
AtrB, 192
Azoles, 200, 235
Azoxystrobin, 42, 121, 155, 160, 162, 302, 318, 344, 395

B

Bakanae disease, 350
Barley industry, 330
Baseline, fungicide resistance, 261–262
Basidiomycetes, 119
BcOS1, 186
Benomyl, 297, 345
Benzamides, 97–99
Benzimidazoles, 282, 382–383, 475
  fungicides, 297
  resistance, 314
Benzovindiflupyr, 106
Binding site, 110
Bioassays, 295
Biofungicides, 351
Biological control, 397
Biological tests, 257–261
Bipolaris leaf spot disease, 349
Bixafen, 106
Blackberries, 282
Black dot, 368–369
Black leaf streak, 472
Black scurf, 369–371
Black sigatoka, 469
Blasticides, 343
Blumeria, 313
Blumeria graminis, 120

H. Ishii, D.W. Hollomon (eds.), *Fungicide Resistance in Plant Pathogens*,
DOI 10.1007/978-4-431-55642-8

Blumeria graminis f. sp. *hordei (Bgh)*, 329
Blumeria graminis f. sp. *tritici*, 39
Bordeaux mixture, 4
Boscalid, 43, 106, 284, 395–396
Boscalid resistance, 301
Botryotinia fuckeliana, 303
Botrytis cinerea, 36, 109, 127, 220, 235, 282, 297, 317
Botrytis pseudocinerea, 222
Brown spot disease, 349
Bst polymerase, 295

C
Candida albicans, 237
Candida spp, 245
Captan, fungicides, 282
Carbendazim, 9, 297, 315, 390
Carboxamides, 106
Carboxin, 107
Carboxylic acid amide (CAA)
  fungicide mandipropamid, 299
  fungicides, 164–168, 358
  resistance, 271–272
Carpropamid, 300, 342
Cavendish, 468
C14-demethylation *(Cyp51)* gene, 39, 200, 238, 265, 334, 384, 473
  overexpression, 208, 305
  paralogues, 209
  point mutations, 205–206
Cellulose synthase gene *(CesA3)*, 164, 165, 304
Cercospora beticola, 39
Cercospora kikuchii, 37
Characteristics of oomycetes, 147
Chemical diversity, 79
Chlorothalonil, 44
Chromosomal abnormalities, 168
Citrus bacterial canker, 304
Cochliobolus heterostrophus, 186
Cochliobolus miyabeanus, 349
Co-formulated products, 159
Colchicine site, 99
Collection isolates, 256–257
Colletotrichum coccodes, 368
Colletotrichum gloeosporioides, 41
Combinations, mechanisms, 211
Commercial kits, 306
Competition assays, 43
Competitive ability, 343
Competitiveness, 114
Complex II, 106
Complex III, 395
Corynespora cassiicola, 42, 109, 300
Co-segregation, 165
Cross-resistance, 24, 44, 234, 265–266, 303, 317, 351, 403
Cryptic species, 217
Cultural methods, 295
Curative, 72–73
Cyazofamid, 121
Cyp684, 223
Cyprodinil, 239, 284
Cytochrome *b (cyt b)*gene, 120, 125, 160, 299, 345
Cytochrome P450 monooxygenase, 223
Cytoplasmic genes, 15, 20

D
Daf1, 186
Decline of resistance, 37, 167
Degradation, 351
Demethylase inhibitors (DMI), 394
  fungicides, 200, 473–474
  resistance, 238, 264, 305
Deoxynivalenol (DON), 314
Deuteromycetes, 119
Dicarboximide, 36
Dicarboximides, 282, 392, 422
Diclocymet, 342
Didymella bryoniae, 109
Diethofencarb, 38, 96
Difenoconazole, 324
Dikaryon, 23
Dimethomorph, 164
Diploid/polyploid populations, 22
Discriminant dosees, 261–262
Discriminatory concentration, 286
Disease resistance inducers, 343
Disruptive resistance, 266–270
Dodine, 437
Dominance, 23
DON biosynthesis, 318
Dose, 65–66
Dot-blotting, 297
Double-stranded DNA, 298
Downy mildews, 121, 259–261
Drug efflux transporters, 234
Drug transporters, 203–204
Dry rot, 365–366
Dynamic turnover of fungicide mixtures, 87

E
Early blight, 358
EC_{50} values, 127
EDDP, 342

Electron transfer pathway, 113, 345
Electrophoresis, 298
Emergence, 132
Endonuclease activity, 297
Epidemics, 385
Erg27 gene mutations, 225
Ergosterol biosynthesis inhibitors (EBIs), 200
Erg27 polymorphism, 223
Erwinia amylovora, 304
Erysiphe cruciferarum, 394
Erysiphe necator, 43, 252, 301
Esterase, 126
Ethaboxam, 97
Ethidium-bromide, 298, 300
Ethylenebisdithiocarbamates (EBDC), 383

F
Famoxadone, 121
Fenamidone, 121
Fenhexamid, 219, 284, 424
Fenoxanil, 342
Fenpiclonil, 182
Fenpyrazamine, 219
Field resistance, 343, 414
Fimbrin gene *FgFim*, 323
Fitness, 24, 114, 137, 306
  costs, 37, 80, 227
  defects, 244
  parameters, 37
  penalties, 36, 459
F129L mutation, 160, 345
Fluazinam, 358
Fludioxonil, 182, 235, 287
Fluopyram, 44, 106, 303, 397
Fluoxastrobin, 121
Flusilazole, 394
Fluxapyroxad, 106
Formulation, 9
F412 substitutions, 226
Fuberidazole, 315
Fungal species, 217
Fungicide Resistance Action Committee (FRAC), 3, 106, 120, 135, 402
Fungicides
  alternation, 86
  efficacy, 244, 307
  metabolization, 224
  mixtures, 86–87
  resistance, 324
Fusarium, 313, 365
Fusarium asiaticum, 315
Fusarium graminearum, 95, 235–236, 301, 315
Fusarium head blight (FHB), 297, 313
Fusarium oxysporum, 304, 469
Fusarium oxysporum f. sp. *cubense* (FOC), 304

G
Gain-of-function mutations, 237
Gangrene, 365–366
G143A substitution, 160
Gene replacement, 27
Genetic diversity, 35
Genetic homeostasis, 36
Genetics of fungicide resistance, 14
Genome sequencing, 303
Genotype, 322
Germination bioassay, 270
Gibberella fujikuroi, 350
Gibberella zeae, 305
Gibberellic acid, 350
Glycine by alanine at codon 143 (G143A), 299, 345, 391
Governing principles, 52–53, 64
Grapes, 282
Grapevine powdery mildew, 301
Gray leaf spot, 345
Gray mold, 282
Green mold, 449
Grey mould, 281
Griseofulvin, 4
Group III HisK, 393
Growth, fungicide sensitivity assays, 258
G1105S mutation, 165
Guidelines, fungicide use, 346

H
HAMP domain, 189
Haploid, 22
Helminthosporium solani, 366
Heme b, 112–113
Heterokaryosis, 23
Heteroplasmy, 300
  stages, 132, 161
  states, 21
  status, 347
Heterozygous isolates, 167
Hexaconazole, 351
High-throughput diagnosis, 297, 347
His-Asp phosphorelay, 183
Histidine kinase, 181
Homologous recombination, 45
Homoplasmic status, 300
Homozygous isolates, 167

Hot water treatment, 351
Huanglongbing (HLB), 306
Hybridization, 295
Hydroxyanilide fenhexamid, 239
Hydroxyanilides, 282
Hydroxynaphthol blue, 305–306
HydR3 strains, 225

I
Imazalil, 449
Iminoctadine albesilate, 44
Inheritable resistance, 167
Inheritance, 132
Inhibitor, AOX, 345
Inhibitors of melanin biosynthesis group (MBI-D), 175
Instability, 39
Integrated pest management (IPM), 412
Intrinsic resistance, 217
Intron-exon structure, 125, 126
Ipconazole, 350
Iprodione, 182, 239, 392
Isofetamid, 106
Isoprothiolane, 342
Isopyrazam, 59, 106
Isotianil, 343

J
Jinggangmycin, 324

K
Kasugamycin, 341
Kresoxim-methyl, 121

L
L1109, 166
Labelled dose, 65
Laboratory mutants, 26
Late blight, 121, 357
Leaf blast disease, 346
Less fit, resistant strains, 343
Loop-mediated isothermal amplification (LAMP), 321, 347
Low affinity, 220

M
Magnaporthe grisea, 42
Magnaporthe oryzae, 235, 297, 341
Major facilitator superfamily (MFS) transporters, 192–193, 227, 234
Major gene, 40
Mandipropamid, 155, 164
MAP kinase cascade, 183
Maternal inheritance, 161
Mating types, 132, 148
Maximum permitted dose, 66
MBI-D fungicides, 341–342
MBI-R fungicides, 343
MBI targeting scytalone dehydratase (MBI-Ds), 297
MDR1, 290
MDR1, 239
MDR2, 239
MDR3, 239
MDR1h, 241, 290
MDR modulators, 245
Measure of fitness, 80
Mefenoxam, 150, 155
Melanin, 175
Melting temperature, 304
Mendelian analysis, 26
Mendelian segregation, 151, 155
Metabolic instability, 121
Metalaxyl, 36, 150, 357
Metconazole, 394
Methionine biosynthesis inhibitors, 342, 423–424
Methyl benzimidazole carbamate (MBC), 435
  resistance development dynamics, 316
  resistance management, 316
  resistance monitoring, 316
Metominostrobin, 344
mfsM2, 193
Microbial disrupters, 476
Micro-titre plates, 297
Microtubules, 94
Migration, 134, 152
Mildews, 261–274
Minor genes, 40
Mitochondria, 300, 345
Mitochondrial haplotypes, 268
Mitosis, 132
Mixing partner, 68
Mixtures, 67–70, 323, 390
Mixture *versus* alternation, 71–72
Mlo alleles, 330
Mode of action (MOA), 390
Molecular beacons, 297
Molecular diagnosis, 320–321, 347–348
Molecular genetic monitoring, 114
Monilinia fructicola, 39
Monitoring resistance, 261–274, 345
Monogenic and oligogenic resistance, 15
Mrr1, 192
Multi-allelic resistance, 20

Multidrug resistance (MDR), 5, 15, 191, 203–204, 226, 427–428
Multi-drug transporter, 96
Multifungicide resistance, 281
Multiple *cyp51s*, 209
Multiple mutations, 297
Multiple resistance, 15
Multiplex PCR, 302
Multiplication, 256–257
Multiresistant (MR) strains, 244
Multisite inhibitors, 234
Multisite toxicants, 420
Mutagenesis, 26
Mutation frequency, 151
Mutations, 120
Mycosphaerella fijiensis, 469
Mycosphaerella graminicola, 236, 300, 302
Mycotoxin, 24, 314
Mycovellosiella nattrassii, 42
Myxothiazol, 124

N
Narrow spectrum, 221
Natural/innate resistance, 14, 222
Negative cross-resistance, 38, 97, 350
Neurospora crassa, 183
Nivalenol (NIV), 314
Nuclear genes, 15
Number of applications, 67
Nursery box treatment, 342

O
Obligate pathogens, 296
Oculimacula yallundae, 238
Oilseed rape, 389
Oligonucleotide probes, 296
On-site detection, 306
Oomycete control, 149
Oomycete fungicides, 149–150
Oomycetes, 146
Oospore, 260–261
Organophosphorus fungicide, 342
Orysastrobin, 42, 344
Osmoregulation, 422–423
Osmotic stress, 181
Overexpression, 39
Overexpression, fungicides, 290

P
Panama disease, 469
Pathogenic fitness, 154
Pathosystem, 402
PCR-Luminex system, 297, 347
PCR-restriction fragment length polymorphism (PCR-RFLP), 41, 298–300, 345
Pear scab, 301
Pefurazoate, 351
Penflufen, 106, 349
Penicillium digitatum, 238
Penicillium spp., 449
Penthiopyrad, 106
Perennial ryegrass, 345
Perithecium, 132
Peronosporales, 121, 146, 164
Phenamacril, 318, 323
Phenylamide (PA) fungicides, 36, 150–160, 357
Phenylcarbamate diethofencarb, 317
N-Phenylformamidoxime, 350
Phenylpyrroles, 282, 422
Phenylpyrrol fludioxonil, 239
Phthalide, 343
Phythophthora erythroseptica, 156
Phytopathological Society of Japan, 346
Phytophthora capsici, 157
Phytophthora cinnamomi, 156
Phytophthora erythroseptica, 369
Phytophthora infestans, 36, 146, 153, 357–358
Phytophthora ramorum, 304
Picoxystrobin, 121
Pink rot, 369
Plant physiology, 120
Plasmopara viticola, 41, 128, 134, 146, 155, 252, 299
Pleiotropic effects, 24
Ploidy, 21
Podosphaera xanthii, 41, 109
Point mutations, 345
Polygenic resistance, 20
Polymerase chain reaction (PCR), 6
Polyscytalum pustulans, 367
Population changes, 152
Population genetics, 25
Portable devices, 306
Position 412, amino-acids, 217
Postharvest decay management, 450
Postharvest fungicides, 449
Potato tuber diseases, 361–371
Powdery mildew, 329
Practical resistance, 300, 437
Pre-mixtures, 449
Primer-introduced restriction analysis PCR (PIRA-PCR), 300–301, 320
Principal highly resistant, 217
Probenazole, 343
Prochloraz, 59, 351

Procymidone, 182
Progeny tests, 40
Progressive resistance, 262–266
Promoter insertion, 305
Prophylactic measures, 428
Propiconazole, 460
n-Propyl gallate, 346
Protective, 72–73
Prothioconazole, 5, 394
Pseudoperonospora cubensis, 146
Puccinia, 313
PvCesA3 gene, 299
Pyraclostrobin, 43, 283, 284
Pyraoxystrobin, 121
Pyrenopeziza brassicae, 304
Pyrenophora teres, 109, 126
Pyrenophora tritici-repentis, 133, 239
Pyribencarb, 121
Pyricularia grisea, 345
Pyrimethanil, 239, 284
Pyroquilon, 343
Pyrosequencing, 112, 303
Pyrrolnitrin, 182
Pythiales, 146
Pythium aphanidermatum, 304

Q
Qo heme, 119
Qualitative, 403
Quantitative, 404
Quantitative modelling, 54–58
Quantitative PCR technologies, 127
Quantitative polymerase chain reaction (qPCR), 43
Quencher, 297
Quinone inside inhibitor (QiI), 119
Quinone outside inhibitors, 282
Quinone outside inhibitors (QoIs), 160–163, 384–385, 395
  fungicides, 299, 358, 474
  resistance, 269–270, 299

R
Ramularia collo-cygni, 128, 134
Real-time polymerase chain reaction (RT-PCR), 302
Recessive nuclear gene, 164, 168
Recessivity, 23
Recommendations, 139
Relatedness among oomycetes, 148
Research Committee on Fungicide Resistance, 346
Resistance, 109–110, 475
  acquisition, 252
  detection, 254
  factors, 42, 262
  fitness, 317
  levels, 322
  management, 49–61, 254, 296, 390, 473
  management for PAs, 158–159
  management for QoI fungicides, 163
  management tactics, 53
  mechanisms, 296
  risk, 22, 139
Resistance alleles, 295
Resistance development, 449
Resistance epidemiology, 324
Resistance in postharvest pathogens of citrus, 449–464
Resistance to mefenoxam, 157
Respiration, 425–427
Respiration inhibitor, 119
Restriction enzymes, 295
Restriction sites, 298
Reversion of resistance, 41
Reversion to sensitivity, 39
Rhizoctonia, 313
Rhizoctonia cerealis, 324
Rhizoctonia solani, 134, 344–345, 369
Rhizoxin, 99
Rice, 341
Rice blast fungus, 297, 342
Risk assessment, 273
RNA biosynthesis, 125–126, 150
RNApolI gene, 151
Rotation/mixture, resistance management, 284
Rotations, 449
RT ASPCR, 303
Rust fungi, 125

S
Saccharomyces cerevisiae, 236
Salicylhydroxamic acid (SHAM), 287, 345
Sampling, fungicide resistance, 255
Sanitation, resistance management, 462
Scab, 314
Sclerotinia monitoring, 390
Sclerotinia risk, 397
Sclerotinia sclerotiorum, 109, 317, 389
Sclerotinia stem rot (SSR) genotypes, 153, 389
Scytalone dehydratase, 342
SDH-A, 107
SDH activity, 45
SdhB gene, 44, 107, 301, 303

SDH-C, 107
SDH complex, 395
SDH-D, 107
SdhD H132R, 396
Sedaxane, 106
Seed-borne oomycete pathogens, 158
Seed disinfection, 351
Segregation of sensitivity/resistance, 152
Segregation pattern, 132, 164
Selection pressure, 36, 300, 343
Selectors, multifungicide resistance, 291
Sensitivity distribution, 127
Sequencing, 295
Sexual populations, 163
Sexual recombination, 22, 128, 153, 154, 157, 161, 163
Sharp eyespot, 313
Sheath blight, 344–345
Silver scurf, 366–368
Single nucleotide polymorphisms (SNPs), 295
Single-point mutation, 175
Single resistances, 290
Site-direct mutagenesis, 27, 45
Site-specific fungicides, 234
Skin spot, 366–368
Soilborne oomycetes, 156
Sour rot, 451
Southern hybridisation, 297
Splicing, 125, 160–161
Spontaneous mutations, 26
Sporangia, 259–260
Sporulation, 258, 317
Spread of resistant, 127
Stability of resistance, 347
Staining, 298
Stakeholder engagement, 54–55
Stem canker, 369–371
Stemphylium vesicarium, 190
Sterol 14α-demethylase, 334
Sterol biosynthesis inhibitors (SBIs), 200, 218, 424–425, 437
Sterol *Cyp51* deficiency, 203
Sterol DMI, 238
Sterol 3-ketoreductase (Erg27), 219
Stewardship programmes, 58–59
Stigmatellin, 137
Strategy, fungicides, 429
Strawberries, 240–241, 282
Streptavidin, 297
Stress tolerance, 122
Stringency, 297
Stripe rust, 313
Strobilurin fungicides, 267–269
Strobilurins, 121, 426
Succinate dehydrogenase, 106, 301
Succinate dehydrogenase-inhibiting (SDHI), 59, 106, 282, 426–427
  boscalid, 239
  fungicides, 349–350, 395, 476
  resistant strains, 397
Succinate-ubiquinone reductase, 106
Sugar beet, 379
Systemic acquired resistance (SAR), 343

T
Taq DNA polymerase, 296
Target modifications, 217
Target overexpression, 234
Target site mutations, 105
Target site resistance, 234
Tebuconazole, 318, 336, 394
Thiabendazole, 315, 365
Thifluzamide, 350
Thiophanatemethyl, 36, 283, 297, 315, 350
Thiram, 282
Tiadinil, 343
Toxophore, 121, 123
Transcription factor, 192
Transporter inhibitors, 245
Transposable element, 39
Triadimefon, 324
Triazoles, 330, 384
Trichothecenes, 318
Tricyclazole, 343
Trifloxystrobin, 121, 300, 345
Triflumizole, 350
1,3,8-Trihydroxynaphthalene reductase, 343
Triphenyltin hydroxide (TPTH), 383
Tropical race 4 (TR4), 470
β-Tubulin gene, 36, 95, 297, 305, 317, 319

U
Ubiquinone, 112
Ubiquitin pocket, 124
Ustilago maydis, 107

V
Validamycin, 324
Venturia inaequalis, 126, 300
Venturia nashicola, 36, 301

Vineyards, 242
Virulence, 323
V1009L, 166

W
Wheat blast, 349
Wheat powdery mildew, 313
Wild-type, resistances, 296

Y
YBA agar medium, 44
Y461D, 473
Y463D, 473
Yellow sigatoka, 472
Y136F, 473
Y463H, 473
Yield loss, 321
Y463N, 473

Z
Zarilamide, 98
Zoxamide, 97
Zymoseptoria tritici, 109, 124, 236, 245, 305
Zymoseptoria tritici (Mycosphaerella graminicola), 56

FSC
www.fsc.org
MIX
Papier | Fördert gute Waldnutzung
FSC® C083411

Kolibri 360 GmbH
Ferdinand-Jühlke-Straße 7
99095 Erfurt, Deutschland
produktsicherheit@kolibri360.de